▲1961年8月，在颐和园。基础数学学科座谈会函数论组合影。

▶1956年7月，在伏尔加河上游览。左为周毓麟，右为袁兆鼎。

▼1953年8月，师专师训班数学组合影。前排站立者左起严士健、孙永生、张禾瑞、郝钢新，以及董延闿(右三)。

▲1981年8月，北京师范大学数学系领导与首届硕士毕业生合影。

◀1983年1月，在美国College Station A&M大学访问时合影。

▶1987年11月，前苏联Steklov研究所Stechkin教授在北京师范大学数学系做学术报告，孙永生教授现场翻译。

◀1992年5月参加北京大学数学力学系博士论文答辩会后，孙永生教授与程民德教授（前右）等合影。

▶1993年9月，与以色列数学家Pinkns教授（右二）等合影。

▼1989年9月，前苏联数学家Korneichuk教授(中)、Tikhomirov教授访问北京师范大学数学系，孙永生教授与之交谈。

◀1996年6月，陪同三位俄罗斯数学家登长城。

▶1993年6月，北京师范大学数学系博士论文答辩会后合影。

▼1999年5月，北京师范大学数学系《五教授执教五十周年庆祝会》上，孙永生教授与夫人曾美美合影。

孙永生文集

逼近与恢复的优化

李仲来／主编

北京师范大学出版社

2005·北京

图书在版编目（CIP）数据

逼近与恢复的优化　孙永生文集/孙永生著，李仲来主编．—北京：北京师范大学出版社，2005

ISBN 7-303-07493-7

Ⅰ．逼…　Ⅱ．孙…　Ⅲ．最佳逼近　Ⅳ．O174.41

中国版本图书馆 CIP 数据核字（2005）第 031563 号

北京师范大学出版社出版发行

（北京新街口外大街 19 号　邮政编码：100875）

http：//www．bnup．com．cn

出版人：赖德胜

北京新丰印刷厂印刷　全国新华书店经销

开本：155 mm×235 mm　印张：31.5　字数：520 千字

2005 年 8 月第 1 版　2006 年 6 月第 2 次印刷

印数：1001～2000　定价：58.00 元

自 序

这本论文选集是从我在 1958 年至 2002 年间发表的论文中筛选出来的，其中有些论文是我和我的学生们合作的。它基本上展现了我学习函数逼近论的历史轨迹。

20 世纪 50 年代中期，我在莫斯科大学力学数学系做研究生，师从 S. Stechkin 教授学习函数逼近论。在他的热情指导下，我完成了学位论文，这是我在逼近论方向入门的工作，其主要成果以及嗣后得到的一些有关结果包括在本文集的前面几篇论文中。回国后，我的科研工作经历了一段曲折的道路。1958 年至 1976 年间政治运动连续不断，教学科研工作被迫陷于停顿竟达 13 年之久，“文革”后，科研工作从 1978 年开始恢复。我当时已年近 50 岁了，做科研的黄金时期已经逝去，50 年代学到的一点点东西几乎完全丢光了，我不得不从零开始。我的能力平庸，但对学数学还有一股追求新知的热情和干劲。当务之急是尽快地摸索出一条前进的方向。函数逼近论好比一株百年老树，旧干新枝纷繁密茂，我十分困惑，真有“绕树三匝，何枝可依”之感。除了学习 Stechkin 和 Nikolskii 的新近著作，我先后选择了国际上三个著名学派来学习：前苏联学派的 N. Korneichuk 和 V. Tikhomirov 教授的逼近论的极值理论，主要是宽度论；美国学派 Karlin，Micchelli，Pinkus 教授的样条，宽度和最优恢复论；Traub，Wozniakowskii 教授的信息复杂性理论。学习过程持续了若干年，大致在 20 世纪 80 年代，我已经历了一个不断“脱毛”的过程，到 80 年代末，接近了经典逼近论、计算数学、计算机科学的数学理论这三个方向的汇合点，这是一个多方向交叉渗透的新的研究领域。在这个领域的研究中我是一个晚到者，然而我决心要在这一广阔领域中为中国的数学工作者争得一席之地。本选集的其他各组论文不过是我（和我的合作者）在长途跋涉过程中迈出的第一步。其核心内容是实现宽度和最优恢复研究从 **T** 到 **R** 上的过渡。步伐虽小，然而我感到它是坚实的。它见证着我和我的合作者追求的执着，凝聚着我们的心血，渗透着我们的汗水。语云：敝帚自珍。用这四个字来譬喻这本选集再恰当不过了。

深切怀念扶持、诱掖和提携我的老前辈傅种孙教授、张禾瑞教授、程民德教授和 Stechkin 教授。老前辈徐利治教授多年来一直给予我热情的鼓励和提携，老朋友陆善镇教授多年来给予我一贯的支持，我对他们表示衷心的感谢。

衷心感谢俄罗斯的朋友们，特别是 V. M. Tikhomirov 教授。我们有过多次交流，我从他们那里学到了很多。

我特别感谢王昆扬、黄达人、房艮孙、李淳、陈迪荣、李落清、刘永平、汪成咏、汪和平等同志，他们是我的朋友、同行和战友。在多年的合作中，我们共同学习，研讨问题，结成了忘年交，他们给了我很多帮助，我从他们那里学到了很多东西。

北京师范大学领导、数学系领导多年来给了我许多支持和鼓励。王昆扬、李仲来教授和北师大出版社王松浦女士为了本书的出版做了很多工作，我谨向他们致以衷心感谢。

孙永生

2003 - 12 - 27

目 录

四　全实轴上光滑函数类的逼近及其最优恢复

五　带有 Gauss 测度的 *B* 空间内点集的平均逼近问题及多元问题

附录

Contents

Ⅲ. Generalization of Landau's Inequality and its Applications

Ⅳ. Approximation and Best Recovery of the Smooth Function Classes on the Real Line

Ⅴ. Average Approximation of a Subset in a *B*- space with Gaussian Measure and Multivariate Problems/

Appendix/

一、可微函数类的逼近常数精确计算问题

I. Exact Evaluation of the Constants of the Best Approximation for the Differen-tiable Function Class

苏联科学院院刊
1961,数学卷,25:143 ~ 152

周期可微函数用三角多项式的最佳逼近*

Best Approximation of Periodic Differentiable Functions by Trigonometric Polynomials

本文建立了对于 $r>1$ 和一切 α 的周期可微函数类 $W^{(r)}(\alpha)$ 用 $n-1$ 阶三角多项式的最佳逼近的精确上界.

§1. 引言

在文[6]中我们研究了寻求对于 $r \geqslant 6$ 和一切 α 的周期可微函数类 $W^{(r)}(\alpha)$ 用 $n-1$ 阶三角多项式的最佳逼近的精确上界的问题,而在文[7]中考虑了对于 $r>2$ 和一切 α 的同样的问题。此外,在文[7]中还解决了当 $r>2$ 时对于类 $W^{(r)}$ 的类似的问题.①

本文中我们将对于 $r>1$ 和一切 α 解决我们对于函数类 $W^{(r)}(\alpha)$ 所提出的问题.也就是说,我们将证明下述定理.

定理 1 (基本的).设 $f \in W^{(r)}(\alpha)$, $r>1$ 且 α 是实数,即

$$f(x)=\frac{a_0}{2}+\frac{1}{\pi}\int_0^{2\pi} K(t-x)\varphi(t)\mathrm{d}t,$$

其中

$$K(t)=\sum_{k=1}^{\infty} k^{-r}\cos\left(kt-\frac{\alpha\pi}{2}\right),$$

而 φ 是满足下述条件的可测函数:

---

* 苏联科学院院刊于 1958 年 8 月 12 日收到本文.

① 本文投出之后,V.K.Dziadyk 的一篇论文问世,其中用另外的方法解决了对于函数类 $W^{(r)}$ 所考查的问题.

$$\operatorname*{ess\,sup}_{0\leqslant t\leqslant 2\pi}|\varphi(t)|\leqslant 1,\int_0^{2\pi}\varphi(t)\mathrm{d}t=0.$$

那么

$$\begin{aligned}E_n[W^{(r)}(\alpha)]&=\sup_{f\in W^{(r)}(\alpha)}E_n(f)_c\\&=\sup_{f\in W^{(r)}(\alpha),f\perp T_{n-1}}\|f\|_c\\&=\frac{1}{\pi}\min_{T_{n-1}}\int_0^{2\pi}|K(t)-T_{n-1}(t)|\mathrm{d}t\\&=\frac{4}{\pi}\frac{K_{r,\alpha}}{n^r}(n=1,2,\cdots),\end{aligned}$$

其中

$$K_{r,\alpha}=\left|\sum_{\nu=0}^{\infty}\frac{\sin\left[(2\nu+1)\beta\pi-\frac{\alpha\pi}{2}\right]}{(2\nu+1)^{r+1}}\right|,$$

$0\leqslant\beta<1$ 且 $\beta\pi$ 是下述方程的根

$$H(\beta\pi)=\sum_{\nu=0}^{\infty}\frac{\cos\left[(2\nu+1)\beta\pi-\frac{\alpha\pi}{2}\right]}{(2\nu+1)^{r+1}}=0.$$

这个定理的证明的核心部分是确定这样的事实，即特征函数

$$W(k)=\int_0^{\infty}\sum_{\nu=0}^{\infty}\frac{\sin\left[(2\nu+1)\beta\pi-\frac{\alpha\pi}{2}+ku\right]}{(2\nu+1+u)^r}\mathrm{d}u,$$

其中 $0\leqslant\beta<1$ 且 $\beta\pi$ 是方程 $H(\beta\pi)=0$ 的根，对于 $r>1$ 和一切 α 在实轴上没有零点.

我们将沿用文[6] 中引入的全部记号和定义. 此外，我们指出，在文[6] 的 §3 和 §4 中在 $r>2$ 的限制之下所证明的全部引理，在条件 $r>1$ 之下仍成立. 由于当从 $r>2$ 的情形过渡到 $r>1$ 的情形时，证明几乎不变，我们就不再叙述这些证明，并且在引用这些引理时，认为它们对于 $r>1$ 已经证明.

本文中所用的证明方法适用于 $\beta\neq 0$ 的情形. 由于 $\beta=0$ 的情形对应于 $\alpha=1$ 的情形，而这种情况下所考察的问题已为 B. Nagy[3] 所解决，所以我们的研究不再考虑这种情况.

§2. 特征函数 $W(k)$ 之研究

我们引入文[6] 的引理 1 对于情形 $r>1$ 的类比. 令(见[1],[6])

$$W_n(t) = \frac{1}{2}\sum_{\nu=0}^{\infty}\frac{\sin\left[(2\nu+1)\beta\pi - \frac{\alpha\pi}{2}\right]}{(2\nu+1)^r n^r} + \sum_{j=1}^{\infty}\sum_{\nu=0}^{\infty}\frac{\sin\left[(2\nu+1)\beta\pi - \frac{\alpha\pi}{2}\right]}{[(2\nu+1)n+j]^r}\cos jt + \sum_{j=1}^{\infty}\sum_{\nu=0}^{\infty}\frac{\cos\left[(2\nu+1)\beta\pi - \frac{\alpha\pi}{2}\right]}{[(2\nu+1)n+j]^r}\sin jt, \tag{2.1}$$

其中 $r > 1, n = 1,2,3,\cdots$.

引理 1 设 $r > 1, 0 \leqslant \alpha < 2$,

$$W(k) = \int_0^{\infty}\sum_{\nu=0}^{\infty}\frac{\sin\left[(2\nu+1)\beta\pi - \frac{\alpha\pi}{2} + ku\right]}{(2\nu+1+u)^r}\mathrm{d}u, \tag{2.2}$$

其中 $0 < \beta < 1$, 而 $\beta\pi$ 是方程 $H(\beta\pi) = 0$ 的根. 那么

$$\lim_{n\to\infty} n^{r-1} W_n\left(\frac{k}{n}\right) = W(k) \tag{2.3}$$

关于 $|k| \leqslant k_0$ 一致成立, 其中 k_0 是任意的正数.

证明 首先指出, 函数

$$\left|\sum_{\nu=0}^{\infty}\frac{\sin(2\nu+1)\beta\pi}{(2\nu+1+u)^r}\right|, \left|\sum_{\nu=0}^{\infty}\frac{\cos(2\nu+1)\beta\pi}{(2\nu+1+u)^r}\right|$$

都是 u 的单调减函数. 只要把周知的关于具有单调减系数函数的符号的定理(见[8]) 应用于它们的导函数, 就可以确定这个结论.

设 $|k| \leqslant k_0$. 那么

$$n^{r-1} W_n\left(\frac{k}{n}\right) - W(k) = n^{r-1}\sum_{j=1}^{Nn}\sum_{\nu=0}^{\infty}\frac{\sin\left[(2\nu+1)\beta\pi - \frac{\alpha\pi}{2} + k\frac{j}{n}\right]}{[(2\nu+1)n+j]^r} - W(k) + n^{r-1}\sum_{j=Nn+1}^{\infty}\sum_{\nu=0}^{\infty}\frac{\sin\left[(2\nu+1)\beta\pi - \frac{\alpha\pi}{2} + k\frac{j}{n}\right]}{[(2\nu+1)n+j]^r} + \frac{n^{r-1}}{2}\mathrm{d}_0^{(n)}, \tag{2.4}$$

其中 N 是任意的自然数, 而

$$\mathrm{d}_0^{(n)} = \frac{1}{n^r}\sum_{\nu=0}^{\infty}\frac{\sin\left[(2\nu+1)\beta\pi - \frac{\alpha\pi}{2}\right]}{(2\nu+1)^r}. \tag{2.5}$$

我们来证明, 对于任给的 $\delta > 0$ 可以选取这样的数 n_0 和 N_0, 使得当

$n \geqslant n_0, N \geqslant N_0$ 时,

$$\left| n^{r-1} \sum_{j=Nn+1}^{\infty} \sum_{\nu=0}^{\infty} \frac{\sin\left[(2\nu+1)\beta\pi - \dfrac{\alpha\pi}{2} + k\dfrac{j}{n}\right]}{[(2\nu+1)n+j]^r} + \frac{n^{r-1}}{2} d_0^{(n)} \right| < \frac{\delta}{2} \tag{2.6}$$

关于 $|k| \leqslant k_0$ 一致成立.

我们指出(见[8]),对于一切自然数 n 和 j,

$$\sum_{\nu=0}^{\infty} \frac{\sin(2\nu+1)\beta\pi}{[(2\nu+1)n+j]^r} > 0, \text{当} 0 < \beta < 1,$$

且

$$\sum_{\nu=0}^{\infty} \frac{\cos(2\nu+1)\beta\pi}{[(2\nu+1)n+j]^r} \begin{cases} \geqslant 0, \text{当} 0 < \beta \leqslant \dfrac{1}{2}, \\ < 0, \text{当} \dfrac{1}{2} < \beta < 1. \end{cases}$$

因此

$$\left| \sum_{j=Nn+1}^{\infty} \cos k\frac{j}{n} \sum_{\nu=0}^{\infty} \frac{\sin(2\nu+1)\beta\pi}{[(2\nu+1)n+j]^r} \right| \leqslant \sum_{j=Nn+1}^{\infty} \sum_{\nu=0}^{\infty} \frac{\sin(2\nu+1)\beta\pi}{[(2\nu+1)n+j]^r}.$$

使用 Abel 变换,我们得到:

$$\sum_{\nu=0}^{\infty} \frac{\sin(2\nu+1)\beta\pi}{[(2\nu+1)n+j]^r} = \sum_{\nu=0}^{\infty} \frac{\sin^2(\nu+1)\beta\pi}{\sin\beta\pi} \left\{ \frac{1}{[(2\nu+1)n+j]^r} - \frac{1}{[(2\nu+3)n+j]^r} \right\}$$

$$\leqslant \frac{2nr}{\sin\beta\pi} \sum_{\nu=0}^{\infty} \frac{\sin^2(\nu+1)\beta\pi}{[(2\nu+1)n+j]^{r+1}}.$$

因此,

$$\sum_{j=Nn+1}^{\infty} \sum_{\nu=0}^{\infty} \frac{\sin(2\nu+1)\beta\pi}{[(2\nu+1)n+j]^r} \leqslant \frac{2nr}{\sin\beta\pi} \sum_{j=Nn+1}^{\infty} \sum_{\nu=0}^{\infty} \frac{\sin^2(\nu+1)\beta\pi}{[(2\nu+1)n+j]^{r+1}}$$

$$= O\left(\frac{1}{n^{r-1}N^{r-1}}\right).$$

完全一样地求得

$$\left| \sum_{j=Nn+1}^{\infty} \sum_{\nu=0}^{\infty} \frac{\cos(2\nu+1)\beta\pi}{[(2\nu+1)n+j]^r} \right| = O\left(\frac{1}{n^{r-1}N^{r-1}}\right).$$

把两个估计式合起来,我们有:

$$n^{r-1} \sum_{j=Nn+1}^{\infty} \cos k\frac{j}{n} \sum_{\nu=0}^{\infty} \frac{\sin\left[(2\nu+1)\beta\pi - \dfrac{\alpha\pi}{2}\right]}{[(2\nu+1)n+j]^r} = O\left(\frac{1}{N^{r-1}}\right) \tag{2.7}$$

和

$$n^{r-1}\sum_{j=Nn+1}^{\infty}\sin k\,\frac{j}{n}\sum_{\nu=0}^{\infty}\frac{\cos\left[(2\nu+1)\beta\pi-\dfrac{\alpha\pi}{2}\right]}{[(2\nu+1)n+j]^{r}}=O\left(\frac{1}{N^{r-1}}\right).\quad(2.8)$$

从(2.5),(2.7) 和(2.8) 我们得到(2.6).

我们来估计差

$$\left|n^{r-1}\sum_{j=1}^{Nn}\sum_{\nu=0}^{\infty}\frac{\sin\left[(2\nu+1)\beta\pi-\dfrac{\alpha\pi}{2}+k\,\dfrac{j}{n}\right]}{[(2\nu+1)n+j]^{r}}-W(k)\right|$$

$$\leqslant\left|n^{r-1}\sum_{j=0}^{Nn}\sum_{\nu=0}^{\infty}\frac{\sin\left[(2\nu+1)\beta\pi-\dfrac{\alpha\pi}{2}+k\,\dfrac{j}{n}\right]}{[(2\nu+1)n+j]^{r}}-\int_{0}^{N}\sum_{\nu=0}^{\infty}\frac{\sin\left[(2\nu+1)\beta\pi-\dfrac{\alpha\pi}{2}+ku\right]}{(2\nu+1+u)^{r}}\mathrm{d}u\right|+$$

$$\left|\int_{N}^{\infty}\sum_{\nu=0}^{\infty}\frac{\sin\left[(2\nu+1)\beta\pi-\dfrac{\alpha\pi}{2}+ku\right]}{(2\nu+1+u)^{r}}\mathrm{d}u\right|.$$

由于积分

$$\int_{0}^{\infty}\sum_{\nu=0}^{\infty}\frac{\sin(2\nu+1)\beta\pi}{(2\nu+1+u)^{r}}\,\mathrm{d}u,\int_{0}^{\infty}\sum_{\nu=0}^{\infty}\frac{\cos(2\nu+1)\beta\pi}{(2\nu+1+u)^{r}}\mathrm{d}u$$

都是收敛的,所以可以选取 $N\geqslant N_0$,使得

$$\left|\int_{N}^{\infty}\sum_{\nu=0}^{\infty}\frac{\sin\left[(2\nu+1)\beta\pi-\dfrac{\alpha\pi}{2}+ku\right]}{(2\nu+1+u)^{r}}\mathrm{d}u\right|<\frac{\delta}{4}$$

关于一切 k 一致成立.固定这个 N,由于 $|k|\leqslant k_0$,则函数

$$\sum_{\nu=0}^{\infty}\frac{\sin\left[(2\nu+1)\beta\pi-\dfrac{\alpha\pi}{2}+ku\right]}{(2\nu+1+u)^{r}}$$

在区间 $0\leqslant u\leqslant N$ 上的全变差将一致囿于某数 $c(k_0,N)$,从而根据 Polya 和 Szegö 的定理[5],

$$\left|n^{r-1}\sum_{j=1}^{Nn}\sum_{\nu=0}^{\infty}\frac{\sin\left[(2\nu+1)\beta\pi-\dfrac{\alpha\pi}{2}+k\,\dfrac{j}{n}\right]}{[(2\nu+1)n+j]^{r}}-\int_{0}^{N}\sum_{\nu=0}^{\infty}\frac{\sin\left[(2\nu+1)\beta\pi-\dfrac{\alpha\pi}{2}+ku\right]}{(2\nu+1+u)^{r}}\,\mathrm{d}u\right|\leqslant\frac{c(k_0,N)}{n}.$$

因此,如果我们选取 $n_1 \geqslant n_0$ 使得

$$\frac{c(k_0, N)}{n_1} < \frac{\delta}{4},$$

则当 $n \geqslant n_1$ 时,有

$$\left| n^{r-1} W_n\left(\frac{k}{n}\right) - W(k) \right| \leqslant \delta$$

关于 $|k| \leqslant k_0$ 一致成立. 引理证毕.

引理 2 设 $r > 1$ 且 $0 \leqslant \alpha < 2$. 那么

1. 若 $k > 0$, 则

$$W(k) = -k^{r-1} r \int_0^{\pi} H(\beta\pi + u)\Phi(u)\mathrm{d}u; \tag{2.9}$$

2. 若 $k < 0$, 则

$$W(k) = |k|^{r-1} r \int_0^{\pi} H(\beta\pi - u)\Phi(u)\mathrm{d}u, \tag{2.10}$$

其中

$$H(u) = \sum_{\nu=0}^{\infty} \frac{\cos\left[(2\nu+1)u - \frac{\alpha\pi}{2}\right]}{(2\nu+1)^r},$$

$$\Phi(u) = \sum_{s=0}^{\infty} \frac{(-1)^s}{(u + s\pi + |k|)^{r+1}}.$$

证明 我们有:

$$W(k) = \int_0^{\infty} \sum_{\nu=0}^{\infty} \frac{\left[(2\nu+1)\beta\pi - \frac{\alpha\pi}{2} + ku\right]}{(2\nu+1+u)^r}\mathrm{d}u$$

$$= \int_0^{\infty} \sum_{\nu=0}^{N} \frac{\sin\left[(2\nu+1)\beta\pi - \frac{\alpha\pi}{2} + ku\right]}{(2\nu+1+u)^r}\mathrm{d}u +$$

$$\int_0^{\infty} \sum_{\nu=N+1}^{\infty} \frac{\sin\left[(2\nu+1)\beta\pi - \frac{\alpha\pi}{2} + ku\right]}{(2\nu+1+u)^r}\mathrm{d}u.$$

使用 Abel 变换, 易见

$$\lim_{N\to\infty} \int_0^{\infty} \sum_{\nu=N+1}^{\infty} \frac{\sin\left[(2\nu+1)\beta\pi - \frac{\alpha\pi}{2} + ku\right]}{(2\nu+1+u)^r}\mathrm{d}u = 0$$

关于 k 一致成立. 因此

$$W(k)=\sum_{\nu=0}^{\infty}\int_0^{\infty}\frac{\sin\left[(2\nu+1)\beta\pi-\frac{\alpha\pi}{2}+ku\right]}{(2\nu+1+u)^r}\mathrm{d}u$$

$$=\sum_{\nu=0}^{\infty}\frac{k^{r-1}}{(2\nu+1)^{r-1}}\int_0^{\infty}\frac{\sin\left[(2\nu+1)(\beta\pi+u)-\frac{\alpha\pi}{2}\right]}{(u+k)^r}\mathrm{d}u.$$

逐项分部积分，我们得到对于 $k>0$，

$$W(k)=k^{r-1}\sum_{\nu=0}^{\infty}\frac{1}{(2\nu+1)^{r-1}}\left\{\frac{\cos\left[(2\nu+1)\beta\pi-\frac{\alpha\pi}{2}\right]}{(2\nu+1)k^r}-\right.$$

$$\left.\frac{r}{2\nu+1}\int_0^{\infty}\frac{\cos\left[(2\nu+1)(\beta\pi+u)-\frac{\alpha\pi}{2}\right]}{(u+k)^{r+1}}\mathrm{d}u\right\}$$

$$=\frac{1}{k}H(\beta\pi)-k^{r-1}r\int_0^{\infty}\frac{H(\beta\pi+u)}{(u+k)^{r+1}}\mathrm{d}u$$

$$=-k^{r-1}r\sum_{s=0}^{\infty}\int_{s\pi}^{(s+1)\pi}\frac{H(\beta\pi+u)}{(u+k)^{r+1}}\mathrm{d}u$$

$$=-k^{r-1}r\sum_{s=0}^{\infty}\int_0^{\pi}H(\beta\pi+u)\frac{(-1)^s}{(u+s\pi+k)^{r+1}}\mathrm{d}u$$

$$=-k^{r-1}r\int_0^{\pi}H(\beta\pi+u)\Phi(u)\mathrm{d}u.$$

公式(2.10) 对于 $k<0$ 的情形类似地证明.

引理 3 当 $r>1, 0\leqslant\alpha<2$ 时函数 $W(k)$ 在实轴上无零点.

证明 由于当 $u\geqslant 0$ 时 $\Phi(u)>0$，所以引理的结论可借助下述事实从公式(2.9) 和(2.10) 得出：

$$H(\beta\pi+u)<0,\text{当}0<\alpha<1,$$

且

$$H(\beta\pi+u)>0,\text{当}1<\alpha<2.$$

(见[6]，72 页)

如果 $k=0$，那么

$$W(0)=\int_0^{\infty}\sum_{\nu=0}^{\infty}\frac{\sin\left[(2\nu+1)\beta\pi-\frac{\alpha\pi}{2}\right]}{(2\nu+1+u)^r}\mathrm{d}u.$$

在这种情况下，对于任何 $u\geqslant 0$(见[8])，

$$\sum_{\nu=0}^{\infty}\frac{\sin\left[(2\nu+1)\beta\pi-\frac{\alpha\pi}{2}\right]}{(2\nu+1+u)^r}\begin{cases}>0,\text{当}0<\alpha<1,\\<0,\text{当}1<\alpha<2.\end{cases}$$

引理 4　设 $r > 1$ 且 $0 \leqslant \alpha < 2$. 存在这样的数 $k = k(r) > 0$ 和 $n_0(r,\alpha) > 0$, 使得当 $n > n_0(r,\alpha)$ 时,

$$\varepsilon W_n(t) > 0 \quad \left(\frac{k}{n} \leqslant t \leqslant 2\pi - \frac{k}{n}\right),$$

其中

$$\varepsilon = \operatorname{sign} \sum_{\nu=0}^{\infty} \frac{\sin\left[(2\nu+1)\beta\pi - \frac{\alpha\pi}{2}\right]}{(2\nu+1)^{r+1}},$$

也就是说, 当 $0 < \alpha < 1$ 时, $\varepsilon = 1$, 而当 $1 < \alpha < 2$ 时, $\varepsilon = -1$.

此引理的证明与文[6] 的引理 5 的证明无异.

引理 5　设 $r > 0, 0 \leqslant \alpha < 2$, 那么对于一切 n,

$$\varepsilon W_n(t) > 0, 0 \leqslant t \leqslant 2\pi, \varepsilon = \pm 1.$$

证明　从引理 1, 3 和 4 推出, 引理 5 的结论对于足够大的 n 成立. 进而, 证明借助于文[6] 的引理 4 而完成.

§ 3. 基本定理的证明

我们来构造对于函数

$$K(t) = \sum_{k=1}^{\infty} k^{-r} \cos\left(kt - \frac{\alpha\pi}{2}\right)$$

与其在 $2n$ 个点

$$\frac{\beta}{n}\pi, \left(\frac{\beta}{n} + \frac{1}{n}\right)\pi, \cdots, \left(\frac{\beta}{n} + \frac{2n-1}{n}\right)\pi \tag{3.1}$$

处重合的 $n-1$ 阶插值多项式.

周知(见[2]) 这是可以做到的. 类似于此事在文[6] 中所做的那样, 我们得到

$$K(t) - U_{n-1}^*(t) = -2\sin(nt - \beta\pi) W_n(t), \tag{3.2}$$

其中 $U_{n-1}^*(t)$ 是所需的多项式. 从(3.2) 和引理 5 看到, 差 $K(t) - U_{n-1}^*(t)$ 当 $0 \leqslant t \leqslant 2\pi$ 时在(3.1) 诸点处改变符号, 并且仅在这些点处改变符号. 因此

$$\begin{aligned}\sup_{f \in W^{(r)}(\alpha)} E_n(f)_c &\leqslant \frac{1}{\pi} \min_{T_{n-1}} \int_0^{2\pi} |K(t) - T_{n-1}(t)| \,\mathrm{d}t \\ &\leqslant \frac{1}{\pi} \int_0^{2\pi} |K(t) - U_{n-1}^*(t)| \,\mathrm{d}t \\ &= \frac{1}{\pi} \left| \int_0^{2\pi} [K(t) - U_{n-1}^*(t)] \operatorname{sign} \sin(nt - \beta\pi) \,\mathrm{d}t \right|\end{aligned}$$

$$= \frac{4}{\pi}\cdot\frac{1}{n^r}\left|\sum_{\nu=0}^{\infty}\frac{\sin\left[(2\nu+1)\beta\pi-\frac{\alpha\pi}{2}\right]}{(2\nu+1)^{r+1}}\right|, n=1,2,\cdots$$

另一方面,在文[6]中已证,当 $r>1$ 时,

$$\sup_{f\in W^{(r)}(\alpha)} E_n(f)_c \geqslant \frac{4}{\pi}\cdot\frac{1}{n^r}\left|\sum_{\nu=0}^{\infty}\frac{\sin\left[(2\nu+1)\beta\pi-\frac{\alpha\pi}{2}\right]}{(2\nu+1)^{r+1}}\right|, n=1,2\cdots$$

定理证毕.

我们从基本定理做出一些推论.

定理 2 设 $f\in W^{(r)}, r>1$,即

$$f(x)=\frac{a_0}{2}+\frac{1}{\pi}\int_0^{2\pi}K_r(t-x)\varphi(t)\mathrm{d}t,$$

其中
$$K_r(t)=\sum_{k=1}^{\infty}k^{-r}\cos\left(kt-\frac{r\pi}{2}\right),$$

而 $\varphi(t)$ 是满足下述条件的可测函数:

$$\operatorname*{ess\,sup}_{0\leqslant t\leqslant 2\pi}|\varphi(t)|\leqslant 1,\int_0^{2\pi}\varphi(t)\mathrm{d}t=0.$$

那么

$$\sup_{f\in W^{(r)}}E_n(f)_c=\sup_{f\in W^{(r)},f\perp T_{n-1}}\|f\|_c=\frac{1}{\pi}\min_{T_{n-1}}\int_0^{2\pi}|K_r(t)-T_{n-1}(t)|\mathrm{d}t$$

$$=\frac{4}{\pi}\cdot\frac{1}{n^r}\left|\sum_{\nu=0}^{\infty}\frac{\sin\left[(2\nu+1)\beta\pi-\frac{r\pi}{2}\right]}{(2\nu+1)^{r+1}}\right|, n=1,2,\cdots$$

其中 $0\leqslant\beta<1$, $\beta\pi$ 是下面的方程的根:

$$H_r(\beta\pi)=\sum_{\nu=0}^{\infty}\frac{\cos\left[(2\nu+1)\beta\pi-\frac{r\pi}{2}\right]}{(2\nu+1)^r}=0.$$

定理 3 设 $f(x)\in\widetilde{W}^{(r)}, r>1$,即

$$f(x)=\frac{a_0}{2}+\frac{1}{\pi}\int_0^{2\pi}\widetilde{K}_r(t-x)\varphi(t)\mathrm{d}t,$$

其中
$$\widetilde{K}_r(t)=\sum_{K=1}^{\infty}K^{-r}\sin\left(kt-\frac{r\pi}{2}\right),$$

$\varphi(t)$ 是满足下面条件的可测函数:

$$\operatorname*{ess\,sup}_{0\leqslant t\leqslant 2\pi}|\varphi(t)|\leqslant 1,\int_0^{2\pi}\varphi(t)\mathrm{d}t=0.$$

那么

$$\sup_{f\in\widetilde{W}^{(r)}} E_n(f)_c = \sup_{f\in\widetilde{W}^{(r)},\, f\perp T_{n-1}} \| f \|_c = \frac{1}{\pi}\min_{T_{n-1}}\int_0^{2\pi}\left|\widetilde{K}_r(t) - T_{n-1}(t)\right| dt$$

$$= \frac{4}{\pi}\cdot\frac{1}{n^r}\left|\sum_{\nu=0}^{\infty}\frac{\cos\left[(2\nu+1)\,\tilde{\beta}\,\pi - \frac{r\pi}{2}\right]}{(2\nu+1)^{r+1}}\right|, n = 1, 2, \cdots$$

其中 $0 \leqslant \tilde{\beta} < 1$, $\tilde{\beta}\,\pi$ 是下面方程的根:

$$\widetilde{H}_r(\beta\pi) = \sum_{\nu=0}^{\infty}\frac{\sin\left[(2\nu+1)\,\tilde{\beta}\,\pi - \frac{r\pi}{2}\right]}{(2\nu+1)^r} = 0.$$

定理 4　如果周期为 2π 的函数 f 具有 Weyl 意义下的连续的 r 阶导数 $f^{(r)}(r > 1)$, 那么

$$E_n(f)_c \leqslant \frac{4}{\pi}\cdot\frac{1}{n^r}\left|\sum_{\nu=0}^{\infty}\frac{\sin\left[(2\nu+1)\,\beta\pi - \frac{r\pi}{2}\right]}{(2\nu+1)^{r+1}}\right| E_n(f^{(r)})_c, n = 1, 2, \cdots \tag{3.3}$$

如果共轭函数 $\tilde{f}$ 具有 Weyl 意义下的连续的 r 阶导数 $\tilde{f}^{(r)}(r > 1)$, 那么

$$E_n(f)_c \leqslant \frac{4}{\pi}\cdot\frac{1}{n^r}\left|\sum_{\nu=0}^{\infty}\frac{\cos\left[(2\nu+1)\,\tilde{\beta}\,\pi - \frac{r\pi}{2}\right]}{(2\nu+1)^{r+1}}\right| E_n(\tilde{f}^{(r)})_c, n = 1, 2, \cdots \tag{3.4}$$

在不等式(3.3) 和(3.4) 中的常数不可改进.

最后, 我们指出, 根据 S.M.Nikolskii[4] 的一个定理, 定理 1 ~ 3 转化为依 L 度量的逼近的情形.

参考文献

[1] V.K.Dziadyk.具有有界的 s 阶导数($0 < s < 1$) 的周期函数类上的最佳逼近.苏联科学院院刊, 1953, 数学卷, 17:135 ~ 162

[2] M.G.Krein.关于周期函数的最佳逼近的定理.苏联科学院报告, 1938, 18: No.4 ~ 5, 245 ~ 249

[3] B.Nagy.Über gewisse Extremal fragen bei transformierten trigonometrischen Entwicklungen.I.Periodischen Fall, Berichte der math.-phys.Kl.Acad, der Wiss.zu Leipzig, 1938, 90:103 ~ 134

[4] S.M.Nikolskii.函数借助三角多项式平均逼近.苏联科学院院刊, 1946, 数学卷(10):207 ~ 256

[5] G. Polya, G. Szegö. 分析中的问题和定理. 卷 1. 莫斯科：科技出版社, 1956, 80
[6] 孙永生. 周期可微函数借助三角多项式的最佳逼近. 苏联科学院院刊, 1959, 数学卷(23)：67 ~ 92
[7] 孙永生. 周期可微函数借助三角多项式的最佳逼近. 北京师范大学学报, 1959, (3)：21 ~ 25
[8] L. Fejér. Trigonometrische Reihen und Potenzreihen mit mehrfach monotoner Koeffizientenfolge. Trans. AMS, 1936, (39)：18 ~ 59
[9] V. K. Dziadyk. 由作为绝对单调函数的积分的核所定义的周期函数类上的最佳逼近. 苏联科学院院刊, 1959, 数学卷(23)：933 ~ 950

(王昆扬译)

数学学报
1962, 6, 12(2)

用三角多项式近迫周期可微函数*

Best Approximation of Periodic Differentiable Functions by Trigonometric Polynomials

§1. 引言

本文系作者以前的工作[7] [8] 的继续. S. B. Stechkin 在[6] 中引入了函数类 $W^{(r)}(\alpha)$ 并且提出了计算该函数类的三角多项式最佳一致逼近问题. 他解决了 $0 < r < 1, r \leqslant \alpha \leqslant 2 - r$ 的情形. 作者在[8] 中, 在 $r > 1, \alpha$ 是任何实数的条件下解决了这一问题. 在本文中我们给出这一问题在条件 $0 < \alpha < r \leqslant 1$ 或 $0 < r \leqslant 1, 2 - r < \alpha < 2$ 下的解①.

首先给出 $W^{(r)}(\alpha)$ 的定义. 我们把能够如下地表示的周期 2π 的连续函数类叫做 $W^{(r)}(\alpha)$:

$$f(x) = \frac{a_0}{2} + \frac{1}{\pi}\int_0^{2\pi} \Psi_{r,\alpha}(t - x)\varphi(t)\mathrm{d}t, \tag{1.1}$$

此处

$$\Psi_{r,\alpha}(t) = \sum_{k=1}^{\infty} k^{-r}\cos\left(kt - \frac{\pi\alpha}{2}\right), \tag{1.2}$$

$r > 0, \alpha$ 是任意实数, 而 $\varphi(t)$ 满足以下条件:

$$\operatorname*{ess\,sup}_{0\leqslant t\leqslant 2\pi} |\varphi(t)| \leqslant 1, \int_0^{2\pi} \varphi(t)\mathrm{d}t = 0.$$

此外, $T_{n-1}(x)$ 表次数不高于 $n-1$ 的三角多项式, 符号 $E_n(f)$ 的意义是:

---

* 数学学报于 1961 年 12 月 21 日收到本文.

① 由于周期性, 条件 $0 \leqslant \alpha < 2$ 并不妨碍一般性.

$$E_n(f) = \min_{T_{n-1}} \max_x | f(x) - T_{n-1}(x) |.$$

同样我们规定

$$E_n(f)_L = \min_{T_{n-1}} \int_0^{2\pi} | f(x) - T_{n-1}(x) | \mathrm{d}x.$$

在[7][8] 中用过的一切其他符号和定义在本文中仍然采用，为节省篇幅起见不再说明.

我们所要考虑的情形的特点在于：当 $0 < \alpha < r < 1, x = 0$ 时，$\Psi_{r,\alpha}(x) = +\infty$，因此比较其他情形复杂一些. 要建立的主要结果是：

定理 1 设 $0 < \alpha < r \leqslant 1, f(x) \in W^{(r)}(\alpha)$，则

$$\sup_{f \in W^{(r)}(\alpha)} E_n(f) = \sup_{\substack{f \in W^{(r)}(\alpha) \\ f \perp T_{n-1}}} \| f \|_c = \frac{1}{\pi} E_n(\Psi_{r,\alpha})_L = \frac{4}{\pi} \cdot \frac{K_{r,\alpha}}{n^r}, n = 1,2,3,\cdots \tag{1.3}$$

此处

$$K_{r,\alpha} = \left| \sum_{\nu=0}^{\infty} \frac{\sin\left[(2\nu+1)\beta\pi - \frac{\pi\alpha}{2}\right]}{(2\nu+1)^{r+1}} \right|, \tag{1.4}$$

$\frac{1}{2} < \beta < 1$，而 $\beta\pi$ 是下列方程的根：

$$H(\beta\pi) = \sum_{\nu=0}^{\infty} \frac{\cos\left[(2\nu+1)\beta\pi - \frac{\pi\alpha}{2}\right]}{(2\nu+1)^r} = 0. \tag{1.5}$$

至于情形 $0 < r \leqslant 1, 2 - r < \alpha < 2$，那么它可以归结为前者. 对此有

定理 1′ 设 $0 < r \leqslant 1, 2 - r < \alpha < 2, f(x) \in W^{(r)}(\alpha)$，则

$$\sup_{f \in W^{(r)}(\alpha)} E_n(f) = \sup_{\substack{f \in W^{(r)}(\alpha) \\ f \perp T_{n-1}}} \| f \|_c = \frac{1}{\pi} E_n(\Psi_{r,\alpha})_L = \frac{4}{\pi} \cdot \frac{K_{r,\alpha}}{n^r}, n = 1,2,3,\cdots \tag{1.6}$$

此处

$$K_{r,\alpha} = \left| \sum_{\nu=0}^{\infty} \frac{\sin\left[(2\nu+1)\beta\pi - \frac{\pi\alpha}{2}\right]}{(2\nu+1)^{r+1}} \right|, \tag{1.7}$$

$0 < \beta < \frac{1}{2}$，而 $\beta\pi$ 是方程 $H(\beta\pi) = 0$ 的根. $H(\beta\pi)$ 与(1.5) 同.

最后，对于函数类 $\widetilde{W}^{(r)}$[6]，当 $\frac{1}{2} < r < 1$ 时，有

$\left(0 < r \leqslant \frac{1}{2}\right.$ 的情形，见[6]$\left.\right)$

定理 2 设 $f(x) \in \widetilde{W}^{(r)}, \frac{1}{2} < r < 1$，亦即

$$f(x) = \frac{1}{\pi}\int_0^{2\pi}\widetilde{K}_r(t-x)\varphi(t)\mathrm{d}t, \tag{1.8}$$

此处

$$\widetilde{K}_r(t) = \sum_{k=1}^{\infty} k^{-r}\sin\left(kt + \frac{\pi r}{2}\right), \tag{1.9}$$

$$\operatorname*{ess\,sup}_{0\leqslant t\leqslant 2\pi} |\varphi(t)| \leqslant 1, \int_0^{2\pi}\varphi(t)\mathrm{d}t = 0, \text{则}$$

$$\sup_{f\in\widetilde{W}^{(r)}} E_n(f) = \sup_{\substack{f\in\widetilde{W}^{(r)}\\ f\perp T_{n-1}}} \|f\|_c = \frac{1}{\pi}E_n(\widetilde{K}_r)_L = \frac{4}{\pi}\cdot\frac{K_{r,r+1}}{n^r}, n = 1,2,3,\cdots \tag{1.10}$$

$$K_{r,r+1} = \left|\sum_{\nu=0}^{\infty}\frac{\cos\left[(2\nu+1)\beta\pi - \frac{\pi r}{2}\right]}{(2\nu+1)^{r+1}}\right|, \tag{1.11}$$

$0 < \beta < \frac{1}{2}$，而 $\beta\pi$ 是下列方程的根：

$$\widetilde{H}(\beta\pi) = \sum_{\nu=0}^{\infty}\frac{\sin\left[(2\nu+1)\beta\pi - \frac{\pi r}{2}\right]}{(2\nu+1)^r} = 0. \tag{1.12}$$

以下主要的是证明定理 1.

§2. 几个引理

为了后面的需要，首先给出 L. Fejér 的一个关于具有单调下降系数的三角级数的定理.

引理 1 (**L. Fejér**).

1° 设 $\{\alpha_n\}$ 是四次单调下降的零序列，亦即

$\alpha_n \to 0, \alpha_n > 0, \Delta\alpha_n = \alpha_n - \alpha_{n+1} \geqslant 0, \Delta^2\alpha_n \geqslant 0, \Delta^3\alpha_n \geqslant 0, \Delta^4\alpha_n \geqslant 0$，则函数

$$F(\theta) = \frac{\alpha_0}{2} + \sum_{n=1}^{\infty}\alpha_n\cos n\theta$$

在区间 $0 < \theta < \pi$ 内单调下降.

2° 设$\{\beta_n\}$是二次单调下降的零序列，则函数

$$G(\theta)=\sum_{n=1}^{\infty}\beta_n\sin n\theta$$

在区间$0<\theta<\pi$上是非负的.

3° 设$\{\alpha_n\}$是三次单调下降的零序列，亦即$\alpha_n>0,\alpha_n\to 0$，$\Delta\alpha_n\geqslant 0,\Delta^2\alpha_n\geqslant 0,\Delta^3\alpha_n\geqslant 0$，则

$$f(\theta)=\sum_{\nu=1}^{\infty}\alpha_\nu\cos(2\nu-1)\theta$$

在$0<\theta<\frac{\pi}{2}$为正，在$\frac{\pi}{2}<\theta<\pi$上为负，并且在$0<\theta<\pi$内单调下降.

4° 设$\{\alpha_\nu\}$是一次单调下降的零序列，即$\alpha_\nu>0,\Delta\alpha_\nu\geqslant 0,\alpha_\nu\to 0$，则

$$g(\theta)=\sum_{\nu=1}^{\infty}\alpha_\nu\sin(2\nu-1)\theta$$

在$0<\theta<\pi$上是非负的.

引理的证明见[9].

引理 2 设$0<\alpha<r\leqslant 1$，而

$$H(t)=\sum_{\nu=0}^{\infty}\frac{\cos\left[(2\nu+1)t-\frac{\pi\alpha}{2}\right]}{(2\nu+1)^r},$$

则在区间$\frac{\pi}{2}<t<\pi$内存在一个唯一的数$\beta\pi$使$H(\beta\pi)=0$.

证明 先考虑$0<\alpha<r<1$的情形. 令

$$\Phi(t)=\sum_{\nu=0}^{\infty}\frac{\cos(2\nu+1)t}{(2\nu+1)^r},\ \Psi(t)=\sum_{\nu=0}^{\infty}\frac{\sin(2\nu+1)t}{(2\nu+1)^r}.$$

由引理1的3°，4°，知$\Phi(t)\downarrow$在$0<t<\pi$内成立，并且$\Phi(t)\geqslant 0$，当$0<t\leqslant\frac{\pi}{2}$时，而$\Phi(t)\leqslant 0$当$\frac{\pi}{2}\leqslant t<\pi$时. 此外又知$\Psi(t)>0$在$0<t<\pi$内成立. 我们证明：$\Psi(t)\downarrow$在$0<t<\frac{\pi}{2}$上成立，而在$\frac{\pi}{2}<t<\pi$上则有$\Psi(t)\uparrow$. 注意，由于有$\Psi(\pi-t)=\Psi(t)$，故只需证明前者. 利用 Abel 变换，得

$$2\sin t\Psi(t)=\sum_{\nu=0}^{\infty}\left\{\frac{1}{(2\nu+1)^r}-\frac{1}{(2\nu+3)^r}\right\}[1-\cos(2\nu+2)t].$$

由此得

$$2\sin t\Psi'(t)+2\cos t\Psi(t)$$

$$= \sum_{\nu=0}^{\infty}(2\nu+2)\cdot\Delta\frac{1}{(2\nu+1)^r}\cdot\sin(2\nu+2)t$$

$$= \sum_{\nu=0}^{\infty}\{(2\nu+1)^{1-r}-(2\nu+3)^{1-r}\}\sin(2\nu+2)t+$$

$$\sum_{\nu=0}^{\infty}\left\{\frac{1}{(2\nu+1)^r}+\frac{1}{(2\nu+3)^r}\right\}\sin(2\nu+2)t.$$

但是

$$2\cos t\Psi(t)=\sum_{\nu=0}^{\infty}\left\{\frac{1}{(2\nu+1)^r}+\frac{1}{(2\nu+3)^r}\right\}\sin(2\nu+2)t,$$

故

$$2\sin t\Psi'(t)=\sum_{\nu=0}^{\infty}\{(2\nu+1)^{1-r}-(2\nu+3)^{1-r}\}\sin 2(\nu+1)t.$$

容易验证，序列$\{(2\nu+3)^{1-r}-(2\nu+1)^{1-r}\}$，$\nu=0,1,2,\cdots$是一个零序列，并且至少二次单调下降. 由此，根据引理 1 的 2°，得到 $\Psi'(t)<0$ 在 $0<t<\frac{\pi}{2}$ 内成立.

现在已经不难完成引理 2 的第一部分的证明. 事实上，$H(t)>0$ 在 $0<t\leqslant\frac{\pi}{2}$ 时成立. 但 $H\left(\frac{\pi}{2}\right)>0$. 又 $H(t)=\frac{1}{2}\{\Psi_{r,\alpha}(t)-\Psi_{r,\alpha}(\pi+t)\}$. 但是，当 $0<r<1$ 时，有[10]

$$\sum_{k=1}^{\infty}k^{-r}\cos kt\approx t^{r-1}\Gamma(1-r)\sin\frac{\pi r}{2},$$

$$\sum_{k=1}^{\infty}k^{-r}\sin kt\approx t^{r-1}\Gamma(1-r)\cos\frac{\pi r}{2}\quad(t\to 0+).$$

由此得，当 $t\to 2\pi-$ 时，成立着

$$\Psi_{r,\alpha}(t)\approx(2\pi-t)^{r-1}\Gamma(1-r)\sin\frac{\pi}{2}(r-\alpha).$$

故$\{\Psi_{r,\alpha}(\pi-)-\Psi_{r,\alpha}(2\pi-)\}=2H(\pi-)<0$. 因此，至少存在一个数 β：$\frac{1}{2}<\beta<1$，使 $H(\beta\pi)=0$. 再注意到在$\frac{\pi}{2}<t<\pi$上，$\Phi(t)\downarrow$ 而 $\Psi(t)\uparrow$，就知道这样的 β 是唯一的.

剩下的是 $0<\alpha<r=1$ 的情形. 此时，由于

$$\Phi(t)=\frac{1}{2}\ln\left(\cot\frac{t}{2}\right),\quad 0<t<\pi,$$

而

$$\Psi(t) = \frac{\pi}{4}\text{sign}\sin t, \quad 0 < t < 2\pi.$$

利用这两个函数的表示式立即可以推出我们所要求的论断. 引理 2 证毕.

推论 设 $0 < \alpha < r \leqslant 1$, 则

$$\text{sign}\, H(t) = \text{sign}\sin(\beta\pi - t),\ 0 \leqslant t \leqslant 2\pi.$$

此事只须注意 $H(t+\pi) = -H(t)$, 立即得到.

引理 3 设 $0 < \alpha < r \leqslant 1$, 而 $t_k = \frac{k-1}{n}\pi + \frac{\beta\pi}{n}$, n 是任一自然数, $k = 1,2,\cdots,2n$, 则存在一个次数不高于 $n-1$ 的三角多项式 $U_{n-1}^*(t)$, 满足条件

$$U_{n-1}^*(t_k) = \Psi_{r,\alpha}(t_k),\ k = 1,2,\cdots,2n,$$

并且

$$\Psi_{r,\alpha}(t) - U_{n-1}^*(t) = -2\sin(nt - \beta\pi)W_n(t), \tag{2.1}$$

此处

$$W_n(t) = \frac{1}{2}\sum_{\nu=0}^{\infty}\frac{\sin\left[(2\nu+1)\beta\pi - \frac{\pi\alpha}{2}\right]}{(2\nu+1)^r n^r} + \sum_{j=1}^{\infty}\sum_{\nu=0}^{\infty}\left\{\frac{\sin\left[(2\nu+1)\beta\pi - \frac{\pi\alpha}{2} + jt\right]}{[(2\nu+1)n + j]^r}\right\}. \tag{2.2}$$

关于引理 3, 我们不拟在此加以验证. 实际上, 三角多项式的存在根据 M. G. Krein 的一个定理(见[3] 或[7]) 立即可以证明, 而最后一个等式的成立, 可以用[2] 中的方法加以验证. (参考[7]) 在此处只想指出: 最后给出的三角级数在 $0 < t < 2\pi$ 上处处收敛. 这是由于, 这一级数的系数

$$\left\{\sum_{\nu=0}^{\infty}\frac{\sin(2\nu+1)\beta\pi}{[(2\nu+1)n + j]^r}\right\}, \left\{\left|\sum_{\nu=0}^{\infty}\frac{\cos(2\nu+1)\beta\pi}{[(2\nu+1)n + j]^r}\right|\right\}(n = 1,2,\cdots),$$

$$j = 0,1,2,\cdots$$

是单调下降的零序列.

问题的进一步考虑在于研究函数 $W_n(t)$ 的符号. 我们在下一节中证明:

$$W_n(t) > 0$$

对于一切 n 在 $0 \leqslant t \leqslant 2\pi$ 上成立.

§3. 对函数 $W_n(t)$ 的研究

引理4 设 $0<\alpha<r\leqslant 1$, $W_n(t)$ 由(2.2)所确定,则存在一适当大的正数 K_1 及一适当大的自然数 n_1,当 $n\geqslant n_1$ 时,有

$$W_n(t)>0 \quad 在\frac{K_1}{n}\leqslant t\leqslant 2\pi-\frac{K_1}{n}$$

上成立.

证明 在实质上与[7]中的引理5无异.为完整起见,给出一个扼要的证明.令

$$d_j^{(n)}=\sum_{\nu=0}^{\infty}\frac{\sin\left[(2\nu+1)\beta\pi-\frac{\pi\alpha}{2}\right]}{[(2\nu+1)n+j]^r},$$

$$\bar{d}_j^{(n)}=\sum_{\nu=0}^{\infty}\frac{\cos\left[(2\nu+1)\beta\pi-\frac{\pi\alpha}{2}\right]}{[(2\nu+1)n+j]^r},$$

$$n=1,2,3,\cdots,j=0,1,2,\cdots$$

而

$$W_n(t)=\frac{d_0^{(n)}}{2}+\sum_{j=1}^{\infty}d_j^{(n)}\cos jt+\sum_{j=1}^{\infty}\bar{d}_j^{(n)}\sin jt,$$

运用 Abel 变换两次,得

$$\frac{d_0^{(n)}}{2}+\sum_{j=1}^{\infty}d_j^{(n)}\cos jt=\frac{1}{4\sin^2\frac{t}{2}}\sum_{j=0}^{\infty}\Delta^2 d_j^{(n)}-\frac{1}{4\sin^2\frac{t}{2}}\sum_{j=0}^{\infty}\Delta^2 d_j^{(n)}\cos(j+1)t;$$

$$\sum_{j=1}^{\infty}\bar{d}_j^{(n)}\sin jt=\frac{\sin t}{4\sin^2\frac{t}{2}}\sum_{j=0}^{\infty}(j+1)\Delta^2\bar{d}_j^{(n)}-\frac{1}{4\sin^2\frac{t}{2}}\sum_{j=0}^{\infty}\Delta^2\bar{d}_j^{(n)}\sin(j+1)t.$$

但

$$\sum_{j=0}^{\infty}\Delta^2 d_j^{(n)}=d_0^{(n)}-d_1^{(n)}$$

$$= \frac{r}{n^{r+1}} \sum_{\nu=0}^{\infty} \frac{\sin\left[(2\nu+1)\beta\pi - \frac{\pi\alpha}{2}\right]}{(2\nu+1)^{r+1}} + O\left(\frac{1}{n^{r+2}}\right),$$

$$\sum_{j=0}^{\infty} (j+1)\Delta^2 \overline{d}_j^{(n)} = \overline{d}_0^{(n)} = 0,$$

故得

$$W_n(t) = \frac{r}{4n^{r+1}\sin^2\frac{t}{2}} \left\{ \sum_{\nu=0}^{\infty} \frac{\sin\left[(2\nu+1)\beta\pi - \frac{\pi\alpha}{2}\right]}{(2\nu+1)^{r+1}} + O\left(\frac{1}{n}\right) \right\} -$$

$$\frac{1}{4\sin^2\frac{t}{2}} \left\{ \sum_{j=0}^{\infty} \Delta^2 d_j^{(n)} \cos(j+1)t + \sum_{j=0}^{\infty} \Delta^2 \overline{d}_j^{(n)} \sin(j+1)t \right\}.$$

令
$$\lambda_j = \Delta^2 d_{j-1}^{(n)}, j = 0,1,2,\cdots \tag{3.1}$$

此处
$$d_{-1}^{(n)} = \sum_{\nu=0}^{\infty} \frac{\sin\left[(2\nu+1)\beta\pi - \frac{\pi\alpha}{2}\right]}{[(2\nu+1)n-1]^r}, n \geqslant 2.$$

仍运用 Abel 变换，得

$$\sum_{j=0}^{\infty} \Delta^2 d_j^{(n)} \cos(j+1)t = \sum_{j=0}^{\infty} \lambda_{j+1} \cos(j+1)t$$

$$= \left\{ \frac{\lambda_0}{2} + \sum_{\nu=1}^{\infty} \lambda_\nu \cos \nu t \right\} - \frac{\lambda_0}{2}$$

$$= \frac{1}{4\sin^2\frac{t}{2}} \sum_{\nu=0}^{\infty} \Delta^2 \lambda_\nu [1 - \cos(\nu+1)t] - \frac{\lambda_0}{2}.$$

注意

$$\lambda_0 = \Delta^2 d_{-1}^{(n)} = O\left(\frac{1}{n^{r+2}}\right),$$

又

$$|\Delta^2 \lambda_\nu| \leqslant \sum_{i=1}^{\infty} \left\{ \Delta_\nu^4 \frac{1}{[(\nu-1)+(2i-1)n]^r} \right\},$$

故

$$\sum_{\nu=0}^{\infty} |\Delta^2 \lambda_\nu| \leqslant \sum_{\nu=0}^{\infty} \sum_{i=1}^{\infty} \left\{ \Delta_\nu^4 \frac{1}{[(\nu-1)+(2i-1)n]^r} \right\}$$

$$= \sum_{i=1}^{\infty} \sum_{\nu=0}^{\infty} \left\{ \Delta_\nu^4 \frac{1}{[(\nu-1)+(2i-1)n]^r} \right\}$$

$$= O\left(\frac{1}{n^{r+3}}\right),$$

故得

$$\sum_{j=0}^{\infty}\Delta^2 d_j^{(n)}\cos(j+1)t = O\left(\frac{1}{4n^{r+3}\sin^2\frac{t}{2}}\right)+O\left(\frac{1}{n^{r+2}}\right). \quad (3.2)$$

用相同的方法又可得到

$$\sum_{j=0}^{\infty}\Delta^2 \bar{d}_j^{(n)}\sin(j+1)t = O\left(\frac{1}{4n^{r+2}\sin\frac{t}{2}}\right)+O\left(\frac{1}{4n^{r+3}\sin^2\frac{t}{2}}\right). \quad (3.3)$$

合并(3.1),(3.2),(3.3),得

$$W_n(t) = \frac{r}{4n^{r+1}\sin^2\frac{t}{2}}\left\{\sum_{\nu=0}^{\infty}\frac{\sin\left[(2\nu+1)\beta\pi-\frac{\pi\alpha}{2}\right]}{(2\nu+1)^{r+1}}+O\left(\frac{1}{n\sin\frac{t}{2}}\right)+O\left(\frac{1}{n^2\sin^2\frac{t}{2}}\right)+O\left(\frac{1}{n}\right)\right\}. \quad (3.4)$$

注意上面的估计对 t 是均匀的;又于 $0<\alpha<r\leqslant 1$ 时,有

$$\sum_{\nu=0}^{\infty}\frac{\sin\left[(2\nu+1)\beta\pi-\frac{\pi\alpha}{2}\right]}{(2\nu+1)^{r+1}}>0,$$

故由(3.4)立即看出引理4中所说的 $K_1>0$ 和 n_1 是存在的.引理4证毕.

引理4没有给出 $W_n(t)$ 在原点 $t=0$ 的邻近的变化情形.下面的引理解决这一问题.

引理5 设 $0<\alpha<r\leqslant 1$, 而 $K_2>0$ 是任意给定的正数, 并且 $|k|\leqslant K_2$, 则

$$\lim_{n\to\infty}4n^{r+1}\sin^2\left(\frac{k}{2n}\right)W_n\left(\frac{k}{n}\right)=W(k) \quad (3.5)$$

在 $|k|\leqslant K_2$ 上一致成立, 此处

$$W(k)=r\left\{\sum_{\nu=0}^{\infty}\frac{\sin\left[(2\nu+1)\beta\pi-\frac{\pi\alpha}{2}\right]}{(2\nu+1)^{r+1}}-(r+1)\int_0^{\infty}\sum_{\nu=0}^{\infty}\frac{\sin\left[(2\nu+1)\beta\pi-\frac{\pi\alpha}{2}+ku\right]}{(2\nu+1+u)^{r+2}}\mathrm{d}u\right\}. \quad (3.6)$$

证明　在引理 4 中证明了

$$4\sin^2\frac{t}{2}W_n(t)$$

$$=\sum_{j=0}^{\infty}\Delta^2 d_j^{(n)}-\left\{\sum_{j=0}^{\infty}\Delta^2 d_j^{(n)}\cos(j+1)t+\sum_{j=0}^{\infty}\Delta^2\overline{d}_j^{(n)}\sin(j+1)t\right\}.$$

注意

$$\sum_{j=0}^{\infty}\Delta^2 d_j^{(n)}=\frac{r}{n^{r+1}}\sum_{\nu=0}^{\infty}\frac{\sin\left[(2\nu+1)\beta\pi-\frac{\pi\alpha}{2}\right]}{(2\nu+1)^{r+1}}+O\left(\frac{1}{n^{r+2}}\right),$$

$$\Delta^2 d_j^{(n)}=\sum_{\nu=0}^{\infty}\Delta_j^2\frac{1}{[(2\nu+1)n+j]^r}\sin\left[(2\nu+1)\beta\pi-\frac{\pi\alpha}{2}\right],$$

$$\Delta^2\overline{d}_j^{(n)}=\sum_{\nu=0}^{\infty}\Delta_j^2\frac{1}{[(2\nu+1)n+j]^r}\cos\left[(2\nu+1)\beta\pi-\frac{\pi\alpha}{2}\right].$$

又由引理 1, 知对于一切 n 及 j, 成立着

$$\sum_{\nu=0}^{\infty}\Delta_j^2\frac{1}{[(2\nu+1)n+j]^r}\sin(2\nu+1)\beta\pi>0,$$

$$\sum_{\nu=0}^{\infty}\Delta_j^2\frac{1}{[(2\nu+1)n+j]^r}\cos(2\nu+1)\beta\pi<0,$$

故对于一切 t, 有

$$\left|\sum_{j=Nn+1}^{\infty}\cos(j+1)t\sum_{\nu=0}^{\infty}\Delta_j^2\frac{1}{[(2\nu+1)n+j]^r}\sin(2\nu+1)\beta\pi\right|$$

$$\leqslant\sum_{j=Nn+1}^{\infty}\sum_{\nu=0}^{\infty}\Delta_j^2\frac{1}{[(2\nu+1)n+j]^r}\sin(2\nu+1)\beta\pi.$$

但是

$$\sum_{\nu=0}^{\infty}\Delta_j^2\frac{1}{[(2\nu+1)n+j]^r}\sin(2\nu+1)\beta\pi$$

$$=\frac{1}{\sin\beta\pi}\sum_{\nu=0}^{\infty}\Delta_\nu\Delta_j^2\frac{1}{[(2\nu+1)n+j]^r}\sin^2(\nu+1)\beta\pi$$

$$\leqslant\frac{1}{\sin\beta\pi}\sum_{\nu=0}^{\infty}\Delta_\nu\Delta_j^2\frac{1}{[(2\nu+1)n+j]^r}=\frac{1}{\sin\beta\pi}\Delta_j^2\frac{1}{(n+j)^r},$$

故得

$$\left|\sum_{j=Nn+1}^{\infty}\cos(j+1)t\sum_{\nu=0}^{\infty}\Delta_j^2\frac{1}{[(2\nu+1)n+j]^r}\sin(2\nu+1)\beta\pi\right|$$

$$\leqslant\frac{1}{\sin\beta\pi}\sum_{j=Nn+1}^{\infty}\Delta_j^2\frac{1}{(n+j)^r}=\frac{r}{(nN)^{r+1}\sin\beta\pi}+O\left(\frac{1}{n^{r+2}N^{r+2}}\right).$$

同法可证

$$\begin{aligned}&\left|\sum_{j=Nn+1}^{\infty}\cos(j+1)t\sum_{\nu=0}^{\infty}\Delta_j^2\frac{1}{[(2\nu+1)n+j]^r}\cos(2\nu+1)\beta\pi\right|\\&\leqslant\left|\sum_{j=Nn+1}^{\infty}\sum_{\nu=0}^{\infty}\Delta_j^2\frac{1}{[(2\nu+1)n+j]^r}\cos(2\nu+1)\beta\pi\right|\\&\leqslant\frac{r}{(nN)^{r+1}\sin\beta\pi}+O\left(\frac{1}{n^{r+2}N^{r+2}}\right).\end{aligned}$$

由此得

$$n^{r+1}\sum_{j=Nn+1}^{\infty}\cos(j+1)t\sum_{\nu=0}^{\infty}\Delta_j^2\frac{1}{[(2\nu+1)n+j]^r}\sin\left[(2\nu+1)\beta\pi-\frac{\pi\alpha}{2}\right]$$
$$=O\left(\frac{1}{N^{r+1}}\right).\tag{3.7}$$

同理可证

$$n^{r+1}\sum_{j=Nn+1}^{\infty}\sin(j+1)t\sum_{\nu=0}^{\infty}\Delta_j^2\frac{1}{[(2\nu+1)n+j]^r}\cos\left[(2\nu+1)\beta\pi-\frac{\pi\alpha}{2}\right]$$
$$=O\left(\frac{1}{N^{r+1}}\right).\tag{3.8}$$

这里的N是任取的一个正数,上面所得估值$O(N^{-r-1})$与t,N及n无关.

然后,我们有

$$\begin{aligned}&n^{r+1}\sum_{j=0}^{Nn}\sum_{\nu=0}^{\infty}\Delta_j^2\frac{1}{[(2\nu+1)n+j]^r}\sin\left[(2\nu+1)\beta\pi-\frac{\pi\alpha}{2}+(j+1)t\right]\\&=n^{r+1}\sum_{j=0}^{Nn}\sum_{\nu=0}^{\infty}\frac{r(r+1)}{[(2\nu+1)n+j+1]^{r+2}}\sin\left[(2\nu+1)\beta\pi-\frac{\pi\alpha}{2}+(j+1)t\right]+\\&O\left(\frac{1}{n}\right),\end{aligned}\tag{3.9}$$

且此处的$O\left(\frac{1}{n}\right)$与$t$无关.

现在注意函数

$$\sum_{\nu=0}^{\infty}\frac{\sin\left[(2\nu+1)\beta\pi-\frac{\pi\alpha}{2}+ku\right]}{(2\nu+1+u)^{r+2}}\quad(0\leqslant u\leqslant N)$$

是u的有界变差函数,并且当$|k|\leqslant K_2$时,它在$[0,N]$上的总变量均匀有界:

$$\bigvee_0^N\left\{\sum_{\nu=0}^{\infty}\frac{\sin\left[(2\nu+1)\beta\pi-\frac{\pi\alpha}{2}+ku\right]}{(2\nu+1+u)^{r+2}}\right\}\leqslant C(K_2,N)<\infty.$$

由此，令 $t=\frac{k}{n}$，$|k|\leqslant K_2$，根据 Polya 和 Szegö 的一个定理[5, 第60页]，得到

$$n^{r+1}\sum_{j=0}^{Nn}\sum_{\nu=0}^{\infty}\frac{r(r+1)}{[(2\nu+1)n+j+1]^{r+2}}\sin\left[(2\nu+1)\beta\pi-\frac{\pi\alpha}{2}+k\left(\frac{j+1}{n}\right)\right]$$

$$=\frac{1}{n}\sum_{j=0}^{Nn}\sum_{\nu=0}^{\infty}\frac{r(r+1)}{\left[2\nu+1+\left(\frac{j+1}{n}\right)\right]^{r+2}}\sin\left[(2\nu+1)\beta\pi-\frac{\pi\alpha}{2}+k\left(\frac{j+1}{n}\right)\right]$$

$$=r(r+1)\int_0^N\sum_{\nu=0}^{\infty}\frac{\sin\left[(2\nu+1)\beta\pi-\frac{\pi\alpha}{2}+ku\right]}{(2\nu+1+u)^{r+2}}\mathrm{d}u+O\left(\frac{1}{n}\right),$$

此处

$$\left|O\left(\frac{1}{n}\right)\right|\leqslant\frac{C(K_2,N)}{n}. \tag{3.10}$$

由(3.7) ~ (3.10)，得

$$4n^{r+1}\sin^2\left(\frac{k}{2n}\right)W_n\left(\frac{k}{n}\right)-W(k)$$

$$=-r(r+1)\int_N^{\infty}\sum_{\nu=0}^{\infty}\frac{\sin\left[(2\nu+1)\beta\pi-\frac{\pi\alpha}{2}+ku\right]}{(2\nu+1+u)^{r+2}}\mathrm{d}u+$$

$$O\left(\frac{1}{n}\right)+O\left(\frac{1}{N^{r+1}}\right). \tag{3.11}$$

但是容易证明

$$\int_N^{\infty}\sum_{\nu=0}^{\infty}\frac{\sin\left[(2\nu+1)\beta\pi-\frac{\pi\alpha}{2}+ku\right]}{(2\nu+1+u)^{r+2}}\mathrm{d}u=O\left(\frac{1}{N^{r+1}}\right),$$

此估值与 k 无关. 由此得

$$4n^{r+1}\sin^2\left(\frac{k}{2n}\right)W_n\left(\frac{k}{n}\right)-W(k)=O\left(\frac{1}{n}\right)+O\left(\frac{1}{N^{r+1}}\right). \tag{3.12}$$

因此，对于任意小的数 $\varepsilon>0$，有适当大的正数 N，使(3.12) 中的 $|O(N^{-r-1})|<\varepsilon/2$，然后，固定此 N，可以取 $n_2>0$，使于 $n>n_2$ 时，在 $|k|\leqslant K_2$ 上有 $\left|O\left(\frac{1}{n}\right)\right|<\varepsilon/2$. 这样就得到

$$\left|4n^{r+1}\sin^2\left(\frac{k}{2n}\right)W_n\left(\frac{k}{n}\right)-W(k)\right|<\varepsilon$$

在 $|k|\leqslant K_2$ 上于 $n>n_2$ 时成立. 引理证毕.

引理 6 设 $0<\alpha<r\leqslant1$，则 $k\neq0$ 时，$W(k)>0$，$W(0)=0$.

证明 用分部积分得到

$$-(r+1)\int_0^{\infty}\sum_{\nu=0}^{\infty}\frac{\sin\left[(2\nu+1)\beta\pi-\frac{\pi\alpha}{2}+ku\right]}{(2\nu+1+u)^{r+2}}\mathrm{d}u$$

$$=\sum_{\nu=0}^{\infty}\int_0^{\infty}\sin\left[(2\nu+1)\beta\pi-\frac{\pi\alpha}{2}+ku\right]\mathrm{d}_u\frac{1}{(2\nu+1+u)^{r+1}}$$

$$=\sum_{\nu=0}^{\infty}\frac{\sin\left[(2\nu+1)\beta\pi-\frac{\pi\alpha}{2}+ku\right]}{(2\nu+1+u)^{r+1}}\Bigg|_0^{\infty}-$$

$$k\int_0^{\infty}\sum_{\nu=0}^{\infty}\frac{\cos\left[(2\nu+1)\beta\pi-\frac{\pi\alpha}{2}+ku\right]}{(2\nu+1+u)^{r+1}}\mathrm{d}u,$$

但因

$$\sum_{\nu=0}^{\infty}\frac{\sin\left[(2\nu+1)\beta\pi-\frac{\pi\alpha}{2}+ku\right]}{(2\nu+1+u)^{r+1}}\Bigg|_0^{\infty}=-\sum_{\nu=0}^{\infty}\frac{\sin\left[(2\nu+1)\beta\pi-\frac{\pi\alpha}{2}\right]}{(2\nu+1)^{r+1}},$$

故得

$$W(k)=-rk\int_0^{\infty}\sum_{\nu=0}^{\infty}\frac{\cos\left[(2\nu+1)\beta\pi-\frac{\pi\alpha}{2}+ku\right]}{(2\nu+1+u)^{r+1}}\mathrm{d}u. \tag{3.13}$$

现在对(3.13)右边的富里埃变换进行如[8]中引理2中所做的变形，可以得到

$$\begin{aligned}&1^\circ\quad k>0\text{ 时}, W(k)=-rk^{r+1}\int_0^{\pi}H(\beta\pi+u)S(u)\mathrm{d}u,\\&2^\circ\quad k<0\text{ 时}, W(k)=r\mid k\mid^{r+1}\int_0^{\pi}H(\beta\pi-u)S(u)\mathrm{d}u,\end{aligned}\tag{3.14}$$

此处

$$S(u)=\sum_{s=0}^{\infty}\frac{(-1)^s}{(s\pi+u+\mid k\mid)^{r+1}},\ \mid k\mid>0.$$

根据引理2的推论知，当 $0<\alpha<r\leqslant 1$ 时，

$$H(\beta\pi+u)<0,\qquad \text{当 } 0<u<\pi \text{ 时};$$

$$H(\beta\pi-u)>0,\qquad \text{当 } 0<u<\pi \text{ 时}.$$

由此得，当 $k\neq 0$ 时，$W(k)>0$. 最后，

$$W(0)=r\sum_{\nu=0}^{\infty}\frac{\sin\left[(2\nu+1)\beta\pi-\frac{\pi\alpha}{2}\right]}{(2\nu+1)^{r+1}}-$$

$$r(r+1)\int_0^\infty \sum_{\nu=0}^\infty \frac{\sin\left[(2\nu+1)\beta\pi - \frac{\pi\alpha}{2}\right]}{(2\nu+1+u)^{r+2}}\mathrm{d}u.$$

用部分积分法,有

$$-(r+1)\int_0^\infty \sum_{\nu=0}^\infty \frac{\sin\left[(2\nu+1)\beta\pi - \frac{\pi\alpha}{2}\right]}{(2\nu+1+u)^{r+2}}\mathrm{d}u$$

$$= -\sum_{\nu=0}^\infty \frac{\sin\left[(2\nu+1)\beta\pi - \frac{\pi\alpha}{2}\right]}{(2\nu+1)^{r+1}}.$$

由此即得 $W(0) = 0$.

推论 对于任意指定的正数 $0 < \delta < K_3$ 存在一个适当大的正数 $n_3(\delta, K_3)$,于 $n > n_3$ 时

$$W_n\left(\frac{k}{n}\right) > 0, \quad \delta \leqslant |k| \leqslant K_3.$$

这一推论重要的是:其中 δ 可以取得任意小,而 K_3 可以取得任意大.

引理 7 设 $0 < \alpha < r < 1, 0 < t < 2\pi$,则

$$\Psi_{r,\alpha}(t) = \frac{\Gamma(1-r)\sin\frac{\pi}{2}(r+\alpha)}{t^{1-r}} + A_1(t) + A_2(t) + A_3(t), \tag{3.15}$$

此处

$$1^\circ \quad A_1(t) = \begin{cases} 0 & \text{当 } 0 < t \leqslant \pi \text{ 时}, \\ \dfrac{\Gamma(1-r)\sin\frac{\pi}{2}(r-\alpha)}{(2\pi - t)^{1-r}} - \dfrac{\Gamma(1-r)\sin\frac{\pi}{2}(r+\alpha)}{t^{1-r}} + \\ \dfrac{2\Gamma(1-r)\cos\frac{\pi r}{2}\sin\frac{\pi\alpha}{2}}{\pi^{1-r}}, & \text{当 } \pi \leqslant t < 2\pi \text{ 时}, \end{cases}$$

是一个在 $0 < t < 2\pi$ 上单调上升的连续函数;

$$2^\circ \quad A_2(t) = \frac{2\Gamma(1-r)\cos\frac{\pi\alpha}{2}\sin\frac{\pi r}{2}}{\pi} g(t), 0 \leqslant t \leqslant 2\pi,$$

$g(t)$ 是 $0 \leqslant t \leqslant 2\pi$ 上的一个连续函数, $g(t) = g(2\pi - t)$,并且在 $0 \leqslant t \leqslant \pi$ 上单调上升;

$$3^\circ \quad A_3(t) = \frac{2\Gamma(1-r)\sin\frac{\pi\alpha}{2}\cos\frac{\pi r}{2}}{\pi}\left\{h(t) + \frac{\pi^r}{2}\operatorname{sign}\sin t\right\} + C_{r,\alpha},$$

$$C_{r,\alpha}=-\left\{\Gamma(1-r)\cos\frac{\pi r}{2}\sin\frac{\pi\alpha}{2}\cdot\frac{1}{\pi^{1-r}}+\frac{a_0}{2}\cos\frac{\pi\alpha}{2}\right\},$$

此处 $h(t)$ 是 $0\leqslant t\leqslant 2\pi$ 上有界的单调下降函数，a_0 是一个常数.

同样又成立着

$$\Psi_{r,\alpha}(t)=\frac{\Gamma(1-r)\sin\frac{\pi}{2}(r-\alpha)}{(2\pi-t)^{1-r}}+A_1^*(t)+A_2(t)+A_3^*(t),\tag{3.16}$$

此处

1° $$A_1^*(t)=\begin{cases}\dfrac{\Gamma(1-r)\sin\frac{\pi}{2}(r+\alpha)}{t^{1-r}}-\dfrac{\Gamma(1-r)\sin\frac{\pi}{2}(r-\alpha)}{(2\pi-t)^{1-r}}-\\ 2\Gamma(1-r)\cos\dfrac{\pi r}{2}\sin\dfrac{\pi\alpha}{2}\cdot\dfrac{1}{\pi^{1-r}},0<t\leqslant\pi\text{时},\\ 0,\qquad \pi\leqslant t<2\pi\text{时};\end{cases}$$

2° $$A_3^*(t)=\frac{2\Gamma(1-r)\sin\frac{\pi\alpha}{2}\cos\frac{\pi r}{2}}{\pi}\left\{h(t)+\frac{\pi^r}{2}\operatorname{sign}\sin t\right\}+C_{r,\alpha}^*,$$

$$C_{r,\alpha}^*=\Gamma(1-r)\cos\frac{\pi r}{2}\sin\frac{\pi\alpha}{2}-\frac{a_0}{2}\cos\frac{\pi\alpha}{2}.$$

证明 当 $t\to 0+$ 时有

$$\sum_{k=1}^{\infty}k^{-r}\cos kt\approx t^{r-1}\Gamma(1-r)\sin\frac{\pi r}{2},$$

$$\sum_{k=1}^{\infty}k^{-r}\sin kt\approx t^{r-1}\Gamma(1-r)\cos\frac{\pi r}{2}.$$

现在计算上列二渐近公式的余项.

令

$$\phi(t)=\begin{cases}\Gamma(1-r)\sin\dfrac{\pi r}{2}\cdot t^{r-1}, & 0<t\leqslant\pi\text{时},\\ \Gamma(1-r)\sin\dfrac{\pi r}{2}\cdot(2\pi-t)^{r-1}, & \pi\leqslant t<2\pi\text{时}.\end{cases}$$

把 $\phi(t)$ 展成富里埃级数. $k\geqslant 1$ 时有

$$a_k=\frac{2}{\pi}\int_0^{\pi}\phi(t)\cos kt\,\mathrm{d}t$$

$$=\frac{2\Gamma(1-r)\sin\frac{\pi r}{2}}{k^r\pi}\int_0^{k\pi}t^{r-1}\cos t\,\mathrm{d}t$$

$$= \frac{2\Gamma(1-r)\sin\frac{\pi r}{2}}{k^r\pi}\left\{\int_0^\infty t^{r-1}\cos t\,\mathrm{d}t - \int_{k\pi}^\infty t^{r-1}\cos t\,\mathrm{d}t\right\}.$$

但是

$$\int_0^\infty t^{r-1}\cos t\,\mathrm{d}t = \Gamma(r)\cos\frac{\pi r}{2},\ \Gamma(r)\Gamma(1-r) = \frac{\pi}{\sin r\pi},$$

故得

$$a_k = k^{-r} - \frac{2\Gamma(1-r)\sin\frac{\pi r}{2}}{k^r\pi}\int_{k\pi}^\infty t^{r-1}\cos t\,\mathrm{d}t,\ k\geqslant 1.$$

然后,利用

$$u^{r-1} = \frac{1}{\Gamma(1-r)}\int_0^\infty \mathrm{e}^{-zu}z^{-r}\mathrm{d}z,\ u>0,$$

得到

$$\begin{aligned} C_k &= \int_{k\pi}^\infty t^{r-1}\cos t\,\mathrm{d}t \\ &= \frac{1}{\Gamma(1-r)}\int_{k\pi}^\infty \cos u\int_0^\infty \mathrm{e}^{-zu}z^{-r}\mathrm{d}z\,\mathrm{d}u \\ &= \frac{1}{\Gamma(1-r)}\int_0^\infty z^{-r}\int_{k\pi}^\infty \mathrm{e}^{-zu}\cos u\,\mathrm{d}u\,\mathrm{d}z \\ &= \frac{(-1)^k}{\Gamma(1-r)}\int_0^\infty \frac{\mathrm{e}^{-zk\pi}z^{1-r}\mathrm{d}z}{1+z^2}. \end{aligned}$$

由此可见,序列$\{(-1)^kC_k\}$是一个完全单调下降的零序列.容易计算,有下列估计成立:

$$|C_k| = O\left(\frac{1}{k^{2-r}}\right).$$

由此,令

$$g(t) = \sum_{k=1}^\infty k^{-r}C_k\cos kt, \tag{3.17}$$

则由引理1之1°,知$g(t)$在$0\leqslant t\leqslant\pi$上单调上升,并且$g(t)$在$0\leqslant t\leqslant 2\pi$上连续,而

$$\sum_{k=1}^\infty k^{-r}\cos kt = \varphi(t) + \frac{2\Gamma(1-r)\sin\frac{\pi r}{2}}{\pi}g(t) - \frac{a_0}{2}, \tag{3.18}$$

$$a_0 = \frac{2}{\pi}\int_0^\pi \varphi(t)\mathrm{d}t.$$

然后令

$$\psi(t)=\begin{cases}\dfrac{\Gamma(1-r)\cos\dfrac{\pi r}{2}}{t^{1-r}}-\dfrac{\Gamma(1-r)\cos\dfrac{\pi r}{2}}{\pi^{1-r}}, & 0<t\leqslant\pi,\\[2ex] \dfrac{-\Gamma(1-r)\cos\dfrac{\pi r}{2}}{(2\pi-t)^{1-r}}+\dfrac{\Gamma(1-r)\cos\dfrac{\pi r}{2}}{\pi^{1-r}}, & \pi\leqslant t<2\pi,\end{cases}$$

$$\psi(0)=\psi(2\pi)=0.$$

把 $\psi(t)$ 展成富里埃级数. $k\geqslant 1$ 时, 有

$$b_k=\frac{2}{\pi}\int_0^{\pi}\psi(t)\sin kt\,\mathrm{d}t=\frac{1}{k^r}-\frac{2\Gamma(1-r)\cos\dfrac{\pi r}{2}}{\pi k^r}\int_{k\pi}^{\infty}t^{r-1}\sin t\,\mathrm{d}t,$$

$k=2s$ 时,

$$b_k=\frac{1}{k^r}-\frac{2\Gamma(1-r)\cos\dfrac{\pi r}{2}}{\pi k^r}\int_{k\pi}^{\infty}t^{r-1}\sin t\,\mathrm{d}t+\frac{4\Gamma(1-r)\cos\dfrac{\pi r}{2}}{\pi^{2-r}}\cdot\frac{1}{k},$$

$k=2s+1$ 时, 此处有

$$C'_k=\int_{k\pi}^{\infty}t^{r-1}\sin t\,\mathrm{d}t=\frac{(-1)^k}{\Gamma(1-r)}\int_0^{\infty}\frac{\mathrm{e}^{-kz\pi}z^{-r}\mathrm{d}z}{1+z^2},\ k\geqslant 1.$$

容易计算

$$|C'_k|=O\left(\frac{1}{k^{1-r}}\right),\text{而 } |k^{1-r}C'_k|\neq o(1).$$

令

$$h(t)=\sum_{k=1}^{\infty}C'_k k^{-r}\sin kt,$$

则 $h(t)$ 在 $0\leqslant t\leqslant 2\pi$ 上是有界函数, 而且它除了在 $t=\pi$ 上有第一类间断点外, 在其他点上都连续.

现在证明, $h(t)$ 是在 $0\leqslant t<\pi$ 上的单调下降函数. 此结论可以从 V. K. Dzjadyk 在[2] 中的一个结论推出. V. K. Dzjadyk 证明, 当 $0<\alpha=r<1$ 时,

$$\sum_{k=1}^{\infty}k^{-r}\cos\left(kt-\frac{\pi r}{2}\right)-\frac{\Gamma(1-r)\sin\pi r}{t^{1-r}}$$

$$=\Gamma(r)\{g(t)+h(t)\}-\frac{a_0}{2}\cos\frac{\pi r}{2}=A_r(t)\quad(0\leqslant t\leqslant\pi)$$

是负号有界并且单调下降函数. 由此, 从

$$\Gamma(r)h(t)=\left\{A_r(t)+\frac{a_0}{2}\cos\frac{\pi r}{2}\right\}+\{-\Gamma(r)g(t)\}\quad(0<t<\pi)$$

即得 $h(t)$ 在 $0\leqslant t<\pi$ 上的单调下降性. 总之, 我们得到

$$\sum_{k=1}^{\infty}k^{-r}\sin kt=\psi(t)+\frac{2\Gamma(1-r)\cos\frac{\pi r}{2}}{\pi}\left\{h(t)+\frac{\pi^r}{2}\operatorname{sign}\sin t\right\},$$

$$0\leqslant t\leqslant 2\pi, \tag{3.19}$$

并且括号中的函数已经是 $0\leqslant t\leqslant 2\pi$ 上的连续且单调下降函数. 由 (3.18), (3.19) 得

$$\Psi_{r,\alpha}(t)=\left\{\cos\frac{\pi\alpha}{2}\phi(t)+\sin\frac{\pi\alpha}{2}\psi(t)\right\}+\frac{2\Gamma(1-r)}{\pi}\left\{\cos\frac{\pi\alpha}{2}\sin\frac{\pi r}{2}g(t)+\sin\frac{\pi\alpha}{2}\cos\frac{\pi r}{2}\left[h(t)+\frac{\pi^r}{2}\operatorname{sign}\sin t\right]\right\}-\frac{a_0}{2}\cos\frac{\pi\alpha}{2}. \tag{3.20}$$

最后, 只须注意

$$\cos\frac{\pi\alpha}{2}\phi(t)+\sin\frac{\pi\alpha}{2}\psi(t)=$$

$$\begin{cases}\dfrac{\Gamma(1-r)\sin\frac{\pi}{2}(r+\alpha)}{t^{1-r}}-\dfrac{\Gamma(1-r)\cos\frac{\pi r}{2}\sin\frac{\pi\alpha}{2}}{\pi^{1-r}}, & 0<t\leqslant\pi,\\[2ex] \dfrac{\Gamma(1-r)\sin\frac{\pi}{2}(r-\alpha)}{(2\pi-t)^{1-r}}+\dfrac{\Gamma(1-r)\cos\frac{\pi r}{2}\sin\frac{\pi\alpha}{2}}{\pi^{1-r}}, & \pi\leqslant t<2\pi\end{cases}$$

即得到(3.15), (3.16).

引理 8 设 $0<\alpha<r<1$, 则下列渐近公式成立:

$$1^\circ\quad W_n(t)=\frac{\Gamma(1-r)\sin\frac{\pi}{2}(r+\alpha)}{2\sin\beta\pi}\{t^{r-1}[1+o(1)]\}, \tag{3.21}$$

当 $0<nt\leqslant\delta_1,\delta_1\to 0$ 时;

$$2^\circ\quad W_n(t)=\frac{\Gamma(1-r)\sin\frac{\pi}{2}(r-\alpha)}{2\sin\beta\pi}\{|t|^{r-1}[1+o(1)]\}, \tag{3.22}$$

当 $t<0, 0<|nt|\leqslant\delta_1', \delta_1'\to 0$ 时.

在(3.21), (3.22) 中 $o(1)$ 是对于一切适当大的自然数 n 均匀趋于零的量.

证明 我们来证明(3.21). 先假定 $0<nt\leqslant\delta_1, 0<\delta_1<\frac{1}{2}\beta\pi$. 利

用(3.15)

$$\Psi_{r,\alpha}(t)=\frac{\Gamma(1-r)\sin\frac{\pi}{2}(r+\alpha)}{t^{1-r}}+A_1(t)+A_2(t)+A_3(t).$$

取$\{t_k,k=2,\cdots,2n\}$作为结点,给(3.15)右边的每一函数在上述$2n-1$个结点上作次数不超过$n-1$的三角插值多项式,令$T_{n-1}(t)$是t^{r-1}的插值多项式,$T_{n-1}^{(i)}(t)$是$A_i(t)$的插值多项式,$i=1,2,3$,则$U_{n-1}^{*}(t)=\Gamma(1-r)\sin\frac{\pi}{2}(r+\alpha)T_{n-1}(t)+\sum_{i=1}^{3}T_{n-1}^{(i)}(t)$. 由(2.1)得

$$W_n(t)=\frac{\Gamma(1-r)\sin\frac{\pi}{2}(r+\alpha)}{2\sin(\beta\pi-nt)}\{t^{r-1}-T_{n-1}(t)\}+\sum_{i=1}^{3}\frac{A_i(t)-T_{n-1}^{(i)}(t)}{2\sin(\beta\pi-nt)}. \tag{3.23}$$

由于$0<nt<\frac{1}{2}\beta\pi$,有下面的估计成立:

$$\frac{1}{2\sin(\beta\pi-nt)}=O(1), \tag{3.24}$$

$$\frac{A_i(t)}{2\sin(\beta\pi-nt)}=O(1),i=1,2,3. \tag{3.25}$$

现在证明

$$\begin{cases}T_{n-1}(t)=O(n^{1-r}), & (3.26)\\ T_{n-1}^{(1)}(t)=O(n^{1-r}), & (3.27)\\ T_{n-1}^{(2)}(t)=O(1), & (3.28)\\ T_{n-1}^{(3)}(t)=O(1) & (3.29)\end{cases}$$

对于$0<nt\leqslant\delta_1$及一切适当大的n均匀成立. 为此目的,我们利用M. Reisz的一个插值公式:

$$T_{n-1}(t)=\frac{\sin(nt-\beta\pi)}{2n}\sum_{k=1}^{2n}(-1)^kT_{n-1}(t_k)\cot\frac{t-t_k}{2}.$$

这一公式在[2]中曾用来解决类似的问题. 注意

$$T_{n-1}(t_k)=t_k^{r-1},k=2,3,\cdots,2n.$$

并且,由于

$$\sum_{k=1}^{2n}(-1)^kT_{n-1}(t_k)=0,$$

得到

$$T_{n-1}(t_1)=\sum_{k=2}^{2n}(-1)^k t_k^{r-1}.$$

所以

$$T_{n-1}(t)=\frac{\sin(nt-\beta\pi)}{2n}\left\{T_{n-1}(t_1)\cot\frac{t-t_1}{2}+\sum_{k=2}^{2n}(-1)^k\cot\frac{t-t_k}{2}\cdot t_k^{r-1}\right\}.$$

注意,

$$|T_{n-1}(t_1)|<t_1^{r-1},$$

又因 $k=2,3,\cdots,n$ 时, $t_k^{r-1}\cot\dfrac{t_k-t}{2}\downarrow(0<nt\leqslant\delta_1)$, 故

$$\left|\sum_{k=2}^{n}(-1)^k t_k^{r-1}\cot\frac{t-t_k}{2}\right|\leqslant t_2^{r-1}\cot\frac{t_2-t}{2}<t_1^{r-1}\cot\frac{t_1-t}{2}.$$

又

$$\begin{aligned}&\left|\sum_{k=n+1}^{2n}(-1)^k t_k^{r-1}\cot\frac{t-t_k}{2}\right|\\&=\left|\sum_{s=1}^{n}(-1)^s t_{n+s}^{r-1}\tan\frac{t_s-t}{2}\right|\\&=\left|\frac{1}{\pi^{1-r}}\left\{\sum_{s=1}^{n}(-1)^s\tan\frac{t_s-t}{2}\right\}-\sum_{s=1}^{n}(-1)^s\{\pi^{r-1}-t_{s+n}^{r-1}\}\tan\frac{t_s-t}{2}\right|\\&\leqslant\pi^{r-1}\left|\sum_{s=1}^{n}(-1)^s\tan\frac{t_s-t}{2}\right|+\left|\sum_{s=1}^{n}(-1)^s\{\pi^{r-1}-t_{n+s}^{r-1}\}\tan\frac{t_s-t}{2}\right|\\&\leqslant\pi^{r-1}\tan\frac{t_n-t}{2}+(\pi^{r-1}-t_{2n}^{r-1})\tan\frac{t_n-t}{2}<\frac{2}{\pi^{1-r}}\tan\frac{t_n-t}{2}.\end{aligned}$$

但是容易验证

$$\tan\frac{t_n-t}{2}=O\left(\cot\frac{t_1-t}{2}\right).$$

故得

$$\begin{aligned}|T_{n-1}(t)|&\leqslant\frac{1}{2n}\left\{2t_1^{r-1}\cot\frac{t_1-t}{2}+O\left(\cot\frac{t_1-t}{2}\right)\right\}\\&=\frac{1}{2n}\cdot O\left(t_1^{r-1}\cot\frac{t_1-t}{2}\right)=O(t_1^{r-1})=O(n^{1-r}).\end{aligned}$$

此即(3.26).

现在证明(3.27). 注意此时 $T_{n-1}^{(1)}(t_k)=A_1(t_k)=0, k=2,3,\cdots,n$ 时. 故

$$T_{n-1}^{(1)}(t)=\frac{\sin(nt-\beta\pi)}{2n}\left\{T_{n-1}^{(1)}(t_1)\cot\frac{t-t_1}{2}+\sum_{k=n+1}^{2n}(-1)^k A_1(t_k)\cot\frac{t-t_k}{2}\right\},$$

此处

$$T_{n-1}^{(1)}(t)=\sum_{k=n+1}^{2n}(-1)^k A_1(t_k).$$

故

$$|T_{n-1}^{(1)}(t_1)|\leqslant A_1(t_{2n})=O(n^{1-r}),$$

$$\left|\sum_{k=n+1}^{2n}(-1)^k A_1(t_k)\cot\frac{t-t_k}{2}\right|\leqslant A_1(t_{2n})\tan\frac{t_n-t}{2}=O\left(n^{1-r}\cot\frac{t_1-t}{2}\right).$$

由此得

$$|T_{n-1}^{(1)}(t)|\leqslant\frac{1}{2n}\cdot O\left(n^{1-r}\cot\frac{t_1-t}{2}\right)=O(n^{1-r}).$$

(3.27) 证完.(3.28),(3.29) 用同样方法证明.例如,取(3.28),有

$$T_{n-1}^{(2)}(t)=\frac{\sin(nt-\beta\pi)}{2n}\left\{T_{n-1}^{(2)}(t_1)\cot\frac{t-t_1}{2}+\sum_{k=2}^{2n}(-1)^k A_2(t_k)\cot\frac{t-t_k}{2}\right\}.$$

但是

$$|T_{n-1}^{(2)}(t_1)|=\left|\sum_{k=2}^{2n}(-1)^k A_2(t_k)\right|=O(1),$$

$$\left|\sum_{k=2}^{n}(-1)^k A_2(t_k)\cot\frac{t-t_k}{2}\right|=O\left(\cot\frac{t_1-t}{2}\right),$$

$$\left|\sum_{k=n+1}^{2n}(-1)^k A_2(t_k)\cot\frac{t-t_k}{2}\right|=O\left(\tan\frac{t_n-t}{2}\right)=O\left(\cot\frac{t_1-t}{2}\right).$$

故有

$$|T_{n-1}^{(2)}(t)|\leqslant O\left(\frac{1}{n}\cot\frac{t_1-t}{2}\right)=O(1).$$

根据(3.26) ~ (3.29),得到

$$W_n(t)=\frac{\Gamma(1-r)\sin\frac{\pi}{2}(r+\alpha)}{2\sin(\beta\pi-nt)}t^{r-1}+O(n^{1-r})$$

$$= \frac{\Gamma(1-r)\sin\frac{\pi}{2}(r+\alpha)}{2\sin\beta\pi}\{t^{r-1}+O(n^{1-r})\}+$$

$$\frac{\Gamma(1-r)\sin\frac{\pi}{2}(r+\alpha)\cos\left(\beta\pi-\frac{nt}{2}\right)\sin\frac{nt}{2}}{2\sin\beta\pi\sin(\beta\pi-nt)}\{t^{r-1}+O(n^{1-r})\}$$

$$= \frac{\Gamma(1-r)\sin\frac{\pi}{2}(r+\alpha)}{2\sin\beta\pi}\{t^{r-1}+O(n^{1-r})\}+$$

$$O\left(\frac{\delta_1}{t^{1-r}}\right)+O(\delta_1\cdot n^{1-r})$$

$$= \frac{\Gamma(1-r)\sin\frac{\pi}{2}(r+\alpha)}{2\sin\beta\pi}\{t^{r-1}[1+O(\delta_1)+$$

$$O(\delta_1^{1-r})+O(\delta_1^{2-r})]\}$$

$$= \frac{\Gamma(1-r)\sin\frac{\pi}{2}(r+\alpha)}{2\sin\beta\pi}\{t^{r-1}[1+o(1)]\},$$

此处 $o(1)\to 0$ 当 $\delta_1\to 0$ 时，并且此一极限过渡对于一切适当大的 n 均匀成立.(3.21) 证毕. 同法可证(3.22)，但需利用(3.16).

推论 存在一个适当大的正数 n_4 及一适当小的正数 δ_1，在 $0\leqslant|nt|\leqslant\delta_1$ 及 $n>n_4$ 时

$$W_n(t)>0.$$

引理 9 设 $0<\alpha<r<1$，则存在一个适当大的正数 n_5，对一切 $n>n_5$ 有

$$W_n(t)>0,\quad 当 0\leqslant t\leqslant 2\pi 时.$$

证明 根据引理 4,5,6,8 立得该引理之证.

为了把引理 9 中不等式推广到一切 n，我们用得着下面一个恒等式，它的一个特殊情形($0<\alpha=r<1$) 最初由 V.K.Dzjadyk 证明[2].

引理 10 对于任何二自然数 m 及N，有

$$W_m(t)=\frac{1}{(2N)^{1-r}}\sum_{\nu=0}^{2N-1}W_{2Nm}\left(\frac{t}{2N}+\frac{\nu\pi}{N}\right),\tag{3.30}$$

$0\leqslant t\leqslant 2\pi$. 此处 $r>0$, α 是任何实数.

(3.30) 之成立可以直接验证之. 关于 $r>1$, α 是任何实数的情形，见[7][8].

由引理 9,10 立得

引理 11 设 $0<\alpha<r<1$, 则

$$W_n(t)>0,\quad 0\leqslant t\leqslant 2\pi,$$

对一切自然数成立.

剩下的是 $0<\alpha<r=1$ 的情形.

引理 12 设 $\Psi_{1,\alpha}(t)=\sum_{k=1}^{\infty}k^{-1}\cos\left(kt-\frac{\pi\alpha}{2}\right)$, α 是任意实数, $T_{n-1}(t)$ 是任意给定的次数不高于 $n-1$ 的三角多项式, 则 $\Psi_{1,\alpha}(t)-T_{n-1}(t)$ 在 $0<t<2\pi$ 内至多有 $2n$ 个根(每一个重根应计算其重次).

证 $\alpha\equiv 0(\mathrm{mod}\ 2)$ 及 $\alpha\equiv 1(\mathrm{mod}\ 2)$ 的情形 N. I. Achiezer 曾经考察过[1]. 我们仍用他的方法来考虑一般情形. 假定存在某一次数不高于 $n-1$ 的三角多项式 $T_{n-1}(t)$, 使 $\Psi_{1,\alpha}(t)-T_{n-1}(t)$ 在 $0<t<2\pi$ 内至少有 $2n+1$ 个根, 则由 Rolle 定理知, $\Psi'_{1,\alpha}(t)-T'_{n-1}(t)$ 在 $0<t<2\pi$ 内至少有 $2n$ 个根. 但是由于

$$\sum_{k=1}^{\infty}k^{-1}\cos kt=\ln\frac{1}{2\sin\frac{t}{2}},$$

$$\sum_{k=1}^{\infty}k^{-1}\sin kt=\frac{1}{2}(\pi-t),0<t<2\pi,$$

我们得到 $\Psi'_{1,\alpha}(t)-T'_{n-1}(t)=-\frac{1}{2}\left(\cos\frac{\pi\alpha}{2}\cot\frac{t}{2}+\sin\frac{\pi\alpha}{2}\right)-T'_{n-1}(t)$ 在 $0<t<2\pi$ 内至少有 $2n$ 个根. 把上式用 $1-\cos t$ 乘之, 得

$$2T'_{n-1}(t)(1-\cos t)+\cos\frac{\pi\alpha}{2}\sin t+\sin\frac{\pi\alpha}{2}(1-\cos t)$$

在 $0\leqslant t<2\pi$ 上至少有 $2n+1$ 个根. 上面是一个次数不高于 n 的三角多项式. 由此得

$$2T'_{n-1}(t)(1-\cos t)+\cos\frac{\pi\alpha}{2}\sin t+\sin\frac{\pi\alpha}{2}(1-\cos t)\equiv 0.$$

此不可能. 引理证毕.

由此引理, 我们得到: 对一切 n, 有

$$\mathrm{sign}\{\Psi_{1,\alpha}(t)-U^*_{n-1}(t)\}=\varepsilon_n\mathrm{sign}\sin(nt-\beta\pi),$$

此处 $|\varepsilon_n|=1$. 然后根据引理 3 及引理 10, 得到

推论 设 $0<\alpha<r=1$, 则对一切 n, 有

$$W_n(t)>0,0\leqslant t\leqslant 2\pi.$$

由此得 $\varepsilon_n = -1, n = 1,2,3,\cdots$

§4. 定理 1, 1′, 2 之证

1° 定理 1 之证明.

由引理 11, 12 得, 当 $0<\alpha<r\leqslant 1$ 时, 对于一切 n 成立着

$$\operatorname{sign}\{\Psi_{r,\alpha}(t)-U_{n-1}^{*}(t)\}=-\operatorname{sign}\sin(nt-\beta\pi). \tag{4.1}$$

由此根据 S. M. Nikolskii 的一个定理[4] (或根据 A. A. Markov, 见[1, 第 96 页]), 得到

$$\begin{aligned}\sup_{f\in W^{(r)}(\alpha)} E_n(f) &= \sup_{\substack{f\in W^{(r)}(\alpha)\\ f\perp T_{n-1}}} \|f\|_c=\frac{1}{\pi}E_n(\Psi_{r,\alpha})_L\\ &=\frac{1}{\pi}\int_0^{2\pi}\left|\Psi_{r,\alpha}(t)-U_{n-1}^{*}(t)\right|\mathrm{d}t\\ &=\frac{1}{\pi}\left|\int_0^{2\pi}\{\Psi_{r,\alpha}(t)-U_{n-1}^{*}(t)\}\operatorname{sign}\sin(nt-\beta\pi)\mathrm{d}t\right|\\ &=\frac{1}{\pi}\left|\int_0^{2\pi}\Psi_{r,\alpha}(t)\operatorname{sign}\sin(nt-\beta\pi)\mathrm{d}t\right|.\end{aligned}$$

注意

$$\operatorname{sign}\sin(nt-\beta\pi)=\frac{4}{\pi}\sum_{\nu=0}^{\infty}\frac{\sin(2\nu+1)(nt-\beta\pi)}{2\nu+1},$$

代入上式中, 利用拓广的 Parseval 等式[10] 即得(1.3), (1.4). 定理 1 证毕.

2° 定理 1′ 之证明.

定理 1′ 之证可以归结为前者. 事实上, 设 $2-r<\alpha<2, 0<r\leqslant 1$, 令 $\alpha'=2-\alpha$, 则 $0<\alpha'<r\leqslant 1$. 而

$$\begin{aligned}\Psi_{r,\alpha}(t)&=\sum_{k=1}^{\infty}k^{-r}\cos\left(kt-\frac{\pi\alpha}{2}\right)\\ &=\sum_{k=1}^{\infty}k^{-r}\cos\left[kt-\frac{\pi(2-\alpha')}{2}\right]\\ &=-\sum_{k=1}^{\infty}k^{-r}\cos\left[k(-t)-\frac{\pi\alpha'}{2}\right]\\ &=-\Psi_{r,\alpha'}(-t),\end{aligned}$$

故得

$$\sup_{f\in W^{(r)}(\alpha)} E_n(f)=\sup_{\substack{f\in W^{(r)}(\alpha)\\ f\perp T_{n-1}}}\|f\|_c=\frac{1}{\pi}E_n[\Psi_{r,\alpha'}(-t)]_L$$

$$= \frac{1}{\pi} E_n(\Psi_{r,\alpha'}(t))_L = \frac{4}{\pi} \cdot \frac{K_{r,\alpha'}}{n^r}, n = 1,2,\cdots$$

此处

$$K_{r,\alpha'} = \left| \sum_{\nu=0}^{\infty} \frac{\sin\left[(2\nu+1)\beta'\pi - \frac{\pi\alpha'}{2}\right]}{(2\nu+1)^{r+1}} \right|,$$

$\frac{1}{2} < \beta' < 1$, $\beta'\pi$ 满足方程

$$\sum_{\nu=0}^{\infty} \frac{\cos\left[(2\nu+1)\beta'\pi - \frac{\pi\alpha'}{2}\right]}{(2\nu+1)^r} = 0.$$

令 $\beta = 1 - \beta'$, $\alpha = 2 - \alpha'$, 代入上式即得(1.6)(1.7).

3° 定理 2 之证.

定理 2 可以归结为定理 1′. 事实上, 此时 $\frac{1}{2} < r < 1$, 而

$$\begin{aligned} \widetilde{K}_r(t) &= \sum_{k=1}^{\infty} k^{-r} \sin\left(kt + \frac{\pi r}{2}\right) \\ &= -\sum_{k=1}^{\infty} k^{-r} \cos\left[kt + \frac{\pi(r+1)}{2}\right] \\ &= -\Psi_{r,r+1}(-t). \end{aligned}$$

令 $\alpha = r + 1$, 则见 $2 - r < \alpha = r + 1 < 2$. 故由此得

$$\sup_{f \in \widetilde{W}^{(r)}} E_n(f) = \sup_{\substack{f \in \widetilde{W}^{(r)} \\ f \perp T_{n-1}}} \| f \|_c = \frac{1}{\pi} E_n(\widetilde{K}_r)_L = \frac{4}{\pi} \cdot \frac{K_{r,r+1}}{n^r}, n = 1,2,\cdots$$

(1.10) 证毕.

最后我们指出: 根据 S. M. Nikolskii 的一个定理[4], 定理 1, 1′ 及 2 可以推广到 L_1 中平均近迫的情形. 又对于类 $W^{(r)}(\alpha)$ 及 $\widetilde{W}^{(r)}$ 可以得到它们的最佳线性方法.

参考文献

[1] 阿赫叶惹尔. 逼近论讲义. 北京: 科学出版社, 1957. 96 ~ 97

[2] Dzjadyk, V. K. Best approximation on the class of periodic functions with bounded order s derivative ($0 < s < 1$). Izv. Akademii Nauk USSR, 1953, matem. series(In Russian)(17): 135 ~ 162

[3] Krein, M. G. Concerning theory of best approximation of periodic functions. Dokl. AH USSR, 1938, 13 No. 4 ~ 5: 245 ~249 (In Russian)
[4] Nikolskii, S. M. Approximation of functions in mean by trigonometric polynomials. Izv. Akademii Nauk USSR, 1946, matem. series(10): 207 ~ 256 (In Russian)
[5] Polya. Szego. *Problems and Theorems from Analysis* V. I. M.: GOSIZDAT, 1956. 60 (In Russian)
[6] Stechkin, S. B. On the best approximation of some classes of periodic functions by trigonometric polynomials. Izv. Akademii Nauk USSR, 1956, matem. series(20): 643 ~ 648(In Russian)
[7] Sun Yongsheng. On the best approximation of periodic differentiable funetions by trigonometric polynomials. Izv. Akademii Nauk USSR, 1959, matem. series(23): 67 ~ 92 (In Russian)
[8] Sun Yongsheng. On the best approximation of periodic differentiable functions by trigonometric polynomials (the second issue). Izv. Akademii Nauk USSR, 1961, matem. series(25): 143 ~ 152(In Russian)
[9] Fejér L. Trigonometrische Reihen und Potenzreihen mit mehrfach monotoner Koeffizientenfolge. Trans. Am. Math. Soc. 1936, (39): 18 ~ 59
[10] Zygmund A. *Trigonometrical Series*. London: Cambridge Univ. Press, 1959

数学学报

1964,7,14(4)

一对共轭周期函数的最佳逼近的渐近性质*

Asymptotic Behavior of the Best Approximation of Two Conjugate Periodic Functions

§1. 引言

设 $W^{(r)}$ 表示以 2π 为周期的函数类，其中每一函数 $f(x)$ 具有 r 阶 $(r>0)$ 在 Wyel 意义下的导数 $f^{(r)}(x)$，满足条件

$$\operatorname*{ess\,sup}_{0\leqslant x\leqslant 2\pi}|f^{(r)}(x)|\leqslant 1.$$

我们知道，$f(x)$ 的解析表示式是

$$f(x)=\frac{a_0}{2}+\frac{1}{\pi}\int_0^{2\pi}K_r(x-t)\varphi(t)\mathrm{d}t, \tag{1.1}$$

此处

$$K_r(t)=\sum_{k=1}^{\infty}k^{-r}\cos\left(kt-\frac{\pi r}{2}\right), \tag{1.2}$$

而 $\varphi(t)$ 满足条件

$$\operatorname*{ess\,sup}_{0\leqslant t\leqslant 2\pi}|\varphi(t)|\leqslant 1,\int_0^{2\pi}\varphi(t)\mathrm{d}t=0. \tag{1.3}$$

又设 $\widetilde{W}^{(r)}$ 表示上述函数类之共轭函数类. 这里任何函数 $f(x)\in\widetilde{W}^{(r)}$ 的解析表示式是

$$f(x)=\frac{1}{\pi}\int_0^{2\pi}\widetilde{K}_r(x-t)\varphi(t)\mathrm{d}t, \tag{1.4}$$

其中

---

* 1963 年 6 月 1 日收到，1963 年 9 月 25 日收到修改稿.
本文与张阳春合作.

$$\widetilde{K}_r(t) = \sum_{k=1}^{\infty} k^{-r}\sin\left(kt - \frac{\pi r}{2}\right), \tag{1.5}$$

$\varphi(t)$ 满足(1.3). 采用记号 $E_n(f) = \min\limits_{T_{n-1}} \max\limits_{x} | f(x) - T_{n-1}(x) |$, 此处 $T_{n-1}(x)$ 表示任何次数不超过 $n-1(n \geqslant 1)$ 的三角多项式. 关于类 $W^{(r)}$, $\widetilde{W}^{(r)}$ 的最佳逼近问题, 已由 B. K. Дзядык[1], C. Б. Стечкин[2] 及作者之一[3,4] 等人完全解决. 证明了

$$E_n(W^{(r)}) = \sup_{f\in W^{(r)}} E_n(f) = \frac{A_r}{n^r}, n = 1,2,\cdots \tag{1.6}$$

$$E_n(\widetilde{W}^{(r)}) = \sup_{f\in \widetilde{W}^{(r)}} E_n(f) = \frac{\widetilde{A}_r}{n^r}, n = 1,2,\cdots \tag{1.7}$$

(上面特别地只是指 r 是非整数的情形, r 取整数的情形, 众所周知, 同样的结果是由 J. Favard, H. И. Ахиезер, M. Г. Крейн 等人得到的)[7], 给出了 A_r, $\widetilde{A}_r$ 的解析表示式. 同时又证明: 对于固定的 n, 在类 $W^{(r)}$, $\widetilde{W}^{(r)}$ 中存在着使(1.6), (1.7) 中上确界达到的函数, 它们各是

$$F_{nr}(x) = \frac{1}{\pi}\int_0^{2\pi} K_r(x-t)\cdot \operatorname{sgn}\sin(nt+\alpha\pi)\mathrm{d}t$$

$$= \frac{4}{\pi n^r}\sum_{\nu=0}^{\infty}\frac{\sin\left[(2\nu+1)(nx+\alpha\pi)-\frac{\pi r}{2}\right]}{(2\nu+1)^{r+1}}, \tag{1.8}$$

和

$$\widetilde{F}_{nr}(x) = \frac{1}{\pi}\int_0^{2\pi} \widetilde{K}_r(x-t)\cdot \operatorname{sgn}\sin(nt+\alpha\pi)\mathrm{d}t$$

$$= \frac{4}{\pi n^r}\sum_{\nu=0}^{\infty}\frac{\cos\left[(2\nu+1)(nx+\alpha\pi)-\frac{\pi r}{2}\right]}{(2\nu+1)^{r+1}}. \tag{1.9}$$

此处 α 是任意实数.

还在 1946 年, C. M. Никольский 就曾指出[5]: 对于整数的 r, 如果对于固定的 m 以及不同的 n 考查量 $E_n(F_{mr})$, 则会发现: 当 n 无限增大时, $n^rE_n(F_{mr})$ 的值将变得比 A_r 小得多. 于是发生下列问题: 在 $W^{(r)}$ 中是否存在函数 $f(x)$, 对于它, 当 n 无限增大时, 量 $n^rE_n(f)$ 的上限任意程度地接近 A_r 或者甚至等于 A_r ①? 在 r 取正整数的条件下, C. M. Никольский

① C. M. Никольский 指出[5], 这个问题是由 C. H. Бернштейн 提出的.

给该问题以肯定的解决.

本文目的在于解决以下问题:

(Ⅰ) 对于非整数的 r, С. М. Никольский 的结果是否仍成立?

(Ⅱ) 对于共轭函数类 $\widetilde{W}^{(r)}(r>0)$, 是否亦有类似的结果成立?

将以上两问题结合起来研究, 可以提出:

(Ⅲ) 对于任何 $f(x)\in W^{(r)}$, 显然有

$$\overline{\lim_{n\to\infty}}\, n^r E_n(f)\leqslant A_r,\ \overline{\lim_{n\to\infty}}\, n^r E_n(\tilde f)\leqslant \widetilde{A}_r. \tag{1.10}$$

是否存在一个 $f(x)\in W^{(r)}$, 对于它和它的共轭函数 $\tilde f(x)$, (1.10) 中二式同时有等号成立?

下面定理给予这些问题以肯定解决.

定理 在 $W^{(r)}(r>0)$ 中存在函数 $f(x)$, 对于它和它的共轭函数 $\tilde f(x)$, 成立着:

$$\overline{\lim_{n\to\infty}}\, n^r E_n(f)=A_r,\ \overline{\lim_{n\to\infty}}\, n^r E_n(\tilde f)=\widetilde{A}_r. \tag{1.11}$$

定理之证归结为证明下列

基本引理 设 $r>0, 0<a<1$. 对于任意小数 $\varepsilon>0$, 存在自然数 m_0, 对于每一 $n>m_0$, 有一函数 $\Phi_n^*(t)$, 满足条件:

1° $\Phi_n^*(t)$ 是以 2π 为周期的连续函数;

2° $|\Phi_n^*(t)|\leqslant 1, \int_0^{2\pi}\Phi_n^*(t)\mathrm{d}t=0$;

3° $\Phi_n^*(t)\equiv 0, a\leqslant t\leqslant 2\pi-a, |t|\leqslant a_n'<a$, 此处 $a_n'>0$ 是一与 n(及 r) 有关的数;

4° 对于

$$F_{nr}^*(x)=\frac{1}{\pi}\int_0^{2\pi}K_r(x-t)\Phi_n^*(t)\mathrm{d}t,$$

$$\widetilde{F}_{nr}^*(x)=\frac{1}{\pi}\int_0^{2\pi}\widetilde{K}_r(x-t)\Phi_n^*(t)\mathrm{d}t,$$

成立着

$$E_n(F_{nr}^*)\geqslant\frac{A_r}{n^r}(1-\varepsilon),\ E_n(\widetilde{F}_{nr}^*)\geqslant\frac{\widetilde{A}_r}{n^r}(1-\varepsilon). \tag{1.12}$$

以下在 §2 中给出几条预备事项, 在 §3,4 中给出基本引理之证; 在 §5 中给出定理的证明. 最后在 §6 中对 В. К. Дзядык[1] 研究过的一类函数在 L 中平均逼近问题给一点补充.

§ 2. 某些预备事项

1° 关于函数 $K_r(t)$. 当 $0 < r < 1$ 时, 在 $0 < t < 2\pi$ 内, 已知[6] 有

$$K_r(t) = \frac{\pi}{\Gamma(r)} \cdot \lim_{n\to\infty} \Big\{ t^{r-1} + (t+2\pi)^{r-1} + \cdots + (t+2n\pi)^{r-1} - (2\pi)^{r-1}\frac{n^r}{r} \Big\}, \tag{2.1}$$

记 $(0 < t < 2\pi)$

$$\Phi_r(t) = \lim_{n\to\infty} \frac{\pi}{\Gamma(r)} \Big\{ (t+2\pi)^{r-1} + \cdots + (t+2n\pi)^{r-1} - (2\pi)^{r-1}\frac{n^r}{r} \Big\}, \tag{2.2}$$

В. К. Дзядык 曾指出: $\Phi_r(t)$ 在 $0 \leqslant t \leqslant 2\pi$ 上是连续单调减函数. 在 $[0, 2\pi]$ 之外, $\Phi_r(t)$ 用周期延拓方法定义之, 并且不妨规定 $\Phi_r(0) = \Phi_r(2\pi) = 0$, 于是任何 $f(x) \in W^{(r)}$ 可以表示为:

$$f(x) = \frac{a_0}{2} + \frac{1}{\Gamma(r)} \Big\{ \int_0^x (x-t)^{r-1}\varphi(t)\mathrm{d}t + \int_x^{2\pi} (x-t+2\pi)^{r-1}\varphi(t)\mathrm{d}t \Big\} + \frac{1}{\pi}\int_0^{2\pi} \Phi_r(x-t)\varphi(t)\mathrm{d}t \quad (0 \leqslant x < 2\pi), \tag{2.3}$$

$\varphi(t)$ 满足(1.3).

如果 $r = [r] + r'$, $[r] \geqslant 1$, $0 < r' < 1$, 令 $s = [r]$, 并且

$$f_s(x) = \frac{1}{\pi}\int_0^{2\pi} K_s(x-t)\varphi(t)\mathrm{d}t,$$

易见于 $0 \leqslant x < 2\pi$ 时, 有

$$f(x) = \frac{a_0}{2} + \frac{1}{\Gamma(r')} \Big\{ \int_0^x (x-t)^{r'-1} f_s(t)\mathrm{d}t + \int_x^{2\pi} (x-t+2\pi)^{r'-1} f_s(t)\mathrm{d}t \Big\} + \frac{1}{\pi}\int_0^{2\pi} \Phi_{r'}(x-t) f_s(t)\mathrm{d}t. \tag{2.4}$$

2° 关于函数 $\widetilde{K}_r(t)$. 当 $0 < r < 1$ 时, 在 $0 < t < 2\pi$ 上, 有(见[4], 第 195 页)

$$\widetilde{K}_r(t) = \widetilde{K}_r^{(1)}(t) + \widetilde{K}_r^{(2)}(t) + \widetilde{K}_r^{(3)}(t),$$

此处

$$\widetilde{K}_r^{(1)}(t) = \begin{cases} \dfrac{\Gamma(1-r)\cos \pi r}{t^{1-r}}, & 0 < t \leqslant \pi, \\ 0, & \pi < t < 2\pi. \end{cases}$$

$$\widetilde{K}_r^{(2)}(t)=\begin{cases}0, & 0\leqslant t\leqslant\pi,\\ \dfrac{\Gamma(1-r)}{(2\pi-t)^{1-r}}, & \pi<t<2\pi.\end{cases}$$

至于 $\widetilde{K}_r^{(3)}(t)$，它是$[0,2\pi]$上的有界变差函数，我们没有必要写出其解析表示式. 在基本区间$[0,2\pi]$之外上列各函数皆用周期延拓方法定义之，并且不妨规定它们在 $t=2n\pi$ 时取零值. 于是任何 $\tilde{f}(x)\in\widetilde{W}^{(r)}$ 可以表示为

(a) 当 $0\leqslant x<\pi$ 时，

$$\tilde{f}(x)=\frac{\Gamma(1-r)}{\pi}\left\{\int_0^x\frac{\cos\pi r\cdot\varphi(t)}{(x-t)^{1-r}}\mathrm{d}t+\int_{x+\pi}^{2\pi}\frac{\cos\pi r\varphi(t)}{(2\pi+x-t)^{1-r}}\mathrm{d}t-\int_x^{x+\pi}\frac{\varphi(t)}{(t-x)^{1-r}}\mathrm{d}t\right\}+\frac{1}{\pi}\int_0^{2\pi}\widetilde{K}_r^{(3)}(x-t)\varphi(t)\mathrm{d}t; \tag{2.5}$$

(b) $\pi\leqslant x<2\pi$ 时，

$$\tilde{f}(x)=\frac{\Gamma(1-r)}{\pi}\left\{\int_{x-\pi}^x\frac{\cos\pi r\varphi(t)}{(x-t)^{1-r}}\mathrm{d}t-\int_0^{x-\pi}\frac{\varphi(t)\mathrm{d}t}{(2\pi-x+t)^{1-r}}-\int_x^{2\pi}\frac{\varphi(t)\mathrm{d}t}{(t-x)^{1-r}}\right\}+\frac{1}{\pi}\int_0^{2\pi}\widetilde{K}_r^{(3)}(x-t)\varphi(t)\mathrm{d}t. \tag{2.5$'$}$$

当 $r=s+r'$ 时，只须以 $f_s(t)$ 代换(2.5)，(2.5′)中之 $\varphi(t)$ 以 r' 代换其中的 r，即得 $\tilde{f}(x)$ 之表示式.

3° 当 r 取整数时，我们利用熟知的公式

$$\tilde{f}(x)=-\frac{1}{\pi}\int_0^\pi\frac{f(x+t)-f(x-t)}{2\tan\frac{t}{2}}\mathrm{d}t.$$

令

$$\widetilde{K}_1^{(1)}(t)=\begin{cases}\dfrac{1}{t}, & 0<t<\pi,\\ \dfrac{1}{t-2\pi}, & \pi<t<2\pi,\end{cases}$$

在$(0,2\pi)$之外用周期延拓方法定义之，同时又规定 $\widetilde{K}_1^{(1)}(n\pi)=0$. 令

$$\widetilde{K}_1^{(2)}(t)=\frac{1}{2}\cot\frac{t}{2}-\widetilde{K}_1^{(1)}(t).$$

不难看出，$\widetilde{K}_1^{(2)}(t)$ 是一个以 2π 为周期的有界变差函数.

(a) 如果 $0<x<\pi$，$f(t)$ 在 x 的某邻域内恒等于零，则易见

$$\tilde{f}(x)=-\frac{1}{\pi}\left\{\int_x^{x+\pi}\frac{f(t)}{t-x}\mathrm{d}t-\int_{x+\pi}^{2\pi}\frac{f(t)}{2\pi+x-t}\mathrm{d}t-\int_0^x\frac{f(t)}{x-t}\mathrm{d}t\right\}-$$

$$\frac{1}{\pi}\int_0^{\pi}\widetilde{K}_1^{(2)}(t)[f(x+t)-f(x-t)]\mathrm{d}t. \tag{2.6}$$

(b) 若 $\pi < x < 2\pi$, $f(t)$ 在 x 的某邻域内恒为零, 则易见

$$\tilde{f}(x) = -\frac{1}{\pi}\left\{\int_{x-\pi}^{x}\frac{f(t)}{x-t}\mathrm{d}t + \int_x^{2\pi}\frac{f(t)}{t-x}\mathrm{d}t + \int_0^{x-\pi}\frac{f(t)}{2\pi+t-x}\mathrm{d}t\right\} - \frac{1}{\pi}\int_0^{\pi}\widetilde{K}_1^{(2)}(t)[f(x+t)-f(x-t)]\mathrm{d}t. \tag{2.6'}$$

4° $F_{nr}(x)$, $\widetilde{F}_{nr}(x)$ 的切彼晓夫交错点组. 为后面之需要, 我们须适当择定参数 α. 令

$$H(x) = \sum_{\nu=0}^{\infty}\frac{\sin\left[(2\nu+1)(x+\alpha\pi)-\frac{\pi r}{2}\right]}{(2\nu+1)^{r+1}},$$

$$\widetilde{H}(x) = \sum_{\nu=0}^{\infty}\frac{\cos\left[(2\nu+1)(x+\alpha\pi)-\frac{\pi r}{2}\right]}{(2\nu+1)^{r+1}}.$$

注意在$[0,2\pi)$上, $|H(x)|$, $|\widetilde{H}(x)|$ 在两个点上达到其极大值, 且使极大值达到的两点距离在每一情形下都等于 π, 而 $H(x)$, $\widetilde{H}(x)$ 在各该二点上的值异号. 兹选定 α, 使 $0 < \alpha < 1$, 且 $|H(x)|$, $|\widetilde{H}(x)|$ 在 $x = 0$ 上不达到极大值. 固定这样的一个 α(α 与 r 有关), 在以后各节中凡出现 $F_{nr}(x)$, $\widetilde{F}_{nr}(x)$, 其中参数 α 均如是选定, 不再说明. 此时, 假定在$[0,2\pi)$中使 $|H(x)|$, $|\widetilde{H}(x)|$ 达到极大值之点各为 $\beta\pi$, $(\beta+1)\pi$, $\bar{\beta}\pi$, $(\bar{\beta}+1)\pi$, 此处 $0 < \beta, \bar{\beta} < 1$, 则 $F_{nr}(x)$, $\widetilde{F}_{nr}(x)$ 的 n 阶切彼晓夫交错点组各是:

$$x_k = \frac{k-1}{n}\pi + \frac{\beta\pi}{n}, \quad k = 1,2,\cdots,2n,$$

$$\tilde{x}_k = \frac{k-1}{n}\pi + \frac{\bar{\beta}\pi}{n}, \quad k = 1,2,\cdots,2n. \tag{2.7}$$

5° M.Riesz 的插值公式(见[7], 191 页). 若 $f(x)$ 是周期 2π 的连续函数, 令

$$U_n(f,x) = \frac{\sin(nx-\beta\pi)}{2n}\sum_{k=1}^{2n}(-1)^{k-1}f(x_k)\cot\frac{x-x_k}{2}, \tag{2.8}$$

容易验证: $U_n(f,x)$ 是一个次数 $\leqslant n$ 的三角多项式, 满足条件

$$U_n(f,x_k) = f(x_k), \quad k = 1,2,\cdots,2n.$$

事实上, 当 $x \neq x_k$ 时, 有

$$(-1)^{k-1}\sin(nx-\beta\pi)\cot\frac{x-x_k}{2}$$
$$=\sin n(x-x_k)\cot\frac{x-x_k}{2}$$
$$=\frac{\sin\frac{2n+1}{2}(x-x_k)}{\sin\frac{x-x_k}{2}}-\cos n(x-x_k)$$
$$=\left\{1+2\sum_{m=1}^{n}\cos m(x-x_k)\right\}-\cos n(x-x_k).$$

而 x_k 是 $U_n(f,x)$ 的可除间断点：

$$\lim_{x\to x_k}U_n(f,x)=f(x_k),\quad k=1,2,\cdots,2n.$$

对于任何次数小于等于 $n-1$ 的三角多项式 $T_{n-1}(x)$，有

$$U_n(T_{n-1},x)\equiv T_{n-1}(x).$$

我们令 $\|U_n\|=\sup_{\|f\|_c\leqslant 1}|U_n(f,0)|$，以下给出 $\|U_n\|$ 的渐近值.

引理 1　$\|U_n\|=\frac{2}{\pi}\sin\beta\pi\cdot\ln n+O(1)$.　(2.9)

证明　事实上有

$$\|U_n\|=\frac{\sin\beta\pi}{2n}\sum_{k=1}^{2n}\left|\cot\frac{x_k}{2}\right|.$$

由于当 $1\leqslant k\leqslant n$ 时，$\cot\frac{x_k}{2}>0$，而当 $n+1\leqslant k\leqslant 2n$ 时，$\cot\frac{x_k}{2}<0$，并且

$$\cot\frac{t}{2}-\frac{2}{t}=O(t),0\leqslant t\leqslant\frac{\pi}{2}.$$

故得

$$\sum_{k=1}^{n}\cot\frac{x_k}{2}=\sum_{k=1}^{n}\frac{2}{x_k}+O(n),\tag{2.10}$$

以及

$$-\sum_{k=n+1}^{2n}\cot\frac{x_k}{2}=\sum_{k=1}^{n}\cot\left(\frac{\pi}{2}-\frac{k-1}{2n}\pi-\frac{\beta\pi}{2n}\right)$$
$$=\sum_{k=1}^{n}\cot\left(\frac{\pi}{2}-\frac{x_k}{2}\right)$$
$$=\sum_{k=1}^{n}\frac{2}{\pi-x_k}+O(n),\tag{2.11}$$

合并(2.10),(2.11),得

$$\sum_{k=1}^{2n}\left|\cot\frac{x_k}{2}\right| = \frac{4}{\pi}n\cdot\ln n + O(n),$$ 由此即得(2.9).

引理证毕.

利用 M. Riesz 插值多项式可以估计 $f(x)$ 的最佳一致逼近. 令

$$E_n(f) = \max_x\left|f(x) - T_{n-1}^*(x)\right|.$$

$T_{n-1}^*(x)$ 是 $f(x)$ 的次数 $\leqslant n-1$ 的最佳逼近多项式. 则

$$f(x) - U_n(f,x) = [f(x) - T_{n-1}^*(x)] - U_n(f - T_{n-1}^*, x).$$

故得

$$\begin{aligned}|f(0) - U_n(f,0)| &\leqslant |f(0) - T_{n-1}^*(0)| + |U_n(f - T_{n-1}^*, 0)|\\ &\leqslant (1 + \|U_n\|)E_n(f).\end{aligned}$$

由此得

$$E_n(f) \geqslant \frac{|f(0) - U_n(f,0)|}{1 + \|U_n\|}. \tag{2.12}$$

6° 为后面需要, 此处给出一条由 С. М. Никольский 证明的引理.

引理 2 (С. М. Никольский) 设 r 是正整数, $0 < a < 1$. 存在一适当大的自然数 N 及正数 A, C (它们只与 r 有关), 对于每一 $n > N$, 存在正数 $X_n', X_n'' < \frac{A}{n}$, 及 $[0, a]$ 上的函数 $\psi_n(t)$, 满足条件:

1° $\psi_n(t)$ 在 $[0, a]$ 上殆遍有 r 阶导数, 且 $\operatorname*{ess\,sup}_{0\leqslant t\leqslant a}|\psi_n^{(r)}(t)| \leqslant 1$;

2° $\psi_n(t) \equiv F_{nr}(t), X_n' \leqslant t \leqslant a - X_n''$;

3° $\psi_n(0) = \psi_n'(0) = \cdots = \psi_n^{(r)}(0) = 0,$

$\psi_n(a) = \psi_n'(a) = \cdots = \psi_n^{(r)}(a) = 0;$

4° $|\psi_n^{(k)}(t)| < \frac{C}{n^{r-k}}(k = 0, 1, \cdots, r-1).$

(见[5])

据 С. М. Никольский 的结果, 在 $[0, X_n']$, $[a - X_n'', a]$ 上, 函数 $\psi_n(t)$ 可以构造得恒等于代数多项式. 根据这条引理, 当 r 是正整数时, 我们可以如下地"修正" $F_{nr}(x)$: 设 $0 < a < 1$, 存在适当大的自然数 N, 对每一 $n > N$, 有 $0 < X_n', X_n'', X_n^*, X_n^{**} \leqslant \frac{A}{n}$, $A > 0$ 只与 r 有关, 以及函数 $\Phi_{nr}(x) \in W^{(r)}$, 满足条件:

1° $\Phi_{nr}(x)\equiv 0, a\leqslant x\leqslant 2\pi-a$;

2° $\Phi_{nr}(x)\equiv F_{nr}(x), X_n'\leqslant x\leqslant a-X_n''$,

$\Phi_{nr}(x)\equiv -F_{nr}(x), 2\pi-a+X_n^{**}\leqslant x\leqslant 2\pi-X_n^*$;

3° $\Phi_{nr}(0)=\cdots=\Phi_{nr}^{(r)}(0)=0$,

$$\Phi_{nr}(2\pi)=\cdots=\Phi_{nr}^{(r)}(2\pi)=0. \tag{2.13}$$

"修正" 函数 $\Phi_{nr}(x)$ 对于以后的证明起很大作用.

最后附带指出, 若令 $K(t)$ 表示 $\widetilde{K}_{r'}^{(3)}(t)(0<r'<1)$, $\Phi_{r'}(t)(0<r'<1)$ 或 $\widetilde{K}_1^{(2)}(t)$ 三者之一, 对任二实数 $0\leqslant a<b\leqslant 2\pi$, 有

$$\int_a^b K(x-t)F_{nr}(t)\mathrm{d}t=O\left(\frac{1}{n^{r+1}}\right), \tag{2.14}$$

$$\int_a^b K(x-t)\Phi_{nr}(t)\mathrm{d}t=O\left(\frac{1}{n^{r+1}}\right), \tag{2.15}$$

$$\int_a^b K(x-t)\cdot\operatorname{sgn}\sin(nt+a\pi)\mathrm{d}t=O\left(\frac{1}{n}\right). \tag{2.16}$$

估计对于 x, a, b 是均匀的.

§3. 基本引理之证 (r 取正整数)

设 $0<a<1$. 设 $n>N$, 取前节中所定义的函数 $\Phi_{nr}(x)$ 及其共轭函数:

$$\widetilde{\Phi}_{nr}(x)=-\frac{1}{\pi}\int_0^\pi\frac{\Phi_{nr}(x+t)-\Phi_{nr}(x-t)}{2\tan\frac{t}{2}}\mathrm{d}t. \tag{3.1}$$

同时取

$$\widetilde{F}_{nr}(x)=-\frac{1}{\pi}\int_0^\pi\frac{F_{nr}(x+t)-F_{nr}(x-t)}{2\tan\frac{t}{2}}\mathrm{d}t. \tag{3.2}$$

注意, 由(2.14), (2.15) 知

$$\int_0^\pi\widetilde{K}_1^{(2)}(t)[F_{nr}(x+t)-F_{nr}(x-t)]\mathrm{d}t=O\left(\frac{1}{n^{r+1}}\right),$$

$$\int_0^\pi\widetilde{K}_1^{(2)}(t)[\Phi_{nr}(x+t)-\Phi_{nr}(x-t)]\mathrm{d}t=O\left(\frac{1}{n^{r+1}}\right), \tag{3.3}$$

由(2.6) 推知, 当 $X'_n<x<a-X''_n$ 时,

$$\widetilde{F}_{nr}(x)-\widetilde{\Phi}_{nr}(x)=-\frac{1}{\pi}\left\{\int_{x}^{x+\pi}\frac{F_{nr}(t)-\Phi_{nr}(t)}{t-x}\mathrm{d}t-\right.$$
$$\int_{x+\pi}^{2\pi}\frac{F_{nr}(t)-\Phi_{nr}(t)}{2\pi+x-t}\mathrm{d}t-$$
$$\left.\int_{0}^{x}\frac{F_{nr}(t)-\Phi_{nr}(t)}{x-t}\mathrm{d}t\right\}+O\left(\frac{1}{n^{r+1}}\right). \tag{3.4}$$

假定$\frac{B_1}{n}\leqslant x\leqslant\frac{a}{2}$，$B_1>0$以后择定之，则

$$\int_{x}^{x+\pi}\frac{F_{nr}(t)-\Phi_{nr}(t)}{t-x}\mathrm{d}t=\int_{a-X''_n}^{x+\pi}\frac{F_{nr}(t)-\Phi_{nr}(t)}{t-x}\mathrm{d}t$$
$$=O\left(\frac{1}{(a-x-X''_n)n^{r+1}}\right)=O\left(\frac{1}{an^{r+1}}\right), \tag{3.5}$$

$$\int_{x+\pi}^{2\pi}\frac{F_{nr}(t)-\Phi_{nr}(t)}{2\pi+x-t}\mathrm{d}t=O\left(\frac{1}{xn^{r+1}}\right)=O\left(\frac{1}{B_1n^{r}}\right), \tag{3.6}$$

$$\int_{0}^{x}\frac{F_{nr}(t)-\Phi_{nr}(t)}{x-t}\mathrm{d}t=\int_{0}^{X'_n}\frac{F_{nr}(t)-\Phi_{nr}(t)}{x-t}\mathrm{d}t$$
$$=O\left(\frac{1}{(x-X'_n)n^{r+1}}\right)=O\left(\frac{1}{B_1n^{r}}\right). \tag{3.7}$$

欲使(3.7)成立，只须$B_1>0$选择得适当大，例如$B_1>2A$即可．由(3.4)～(3.7)得：对于任意小的正数η，有$B_1>0$及$N_1>0$，对于每一个$n>N_1$，在$\frac{B_1}{n}\leqslant x\leqslant\frac{a}{2}<a-X''_n$上，

$$\left|n^r\{\widetilde{F}_{nr}(x)-\widetilde{\Phi}_{nr}(x)\}\right|<\eta/2 \tag{3.8}$$

成立．

其次，当$n>N$，$2\pi-a+X_n^{**}<x<2\pi-X_n^{*}$时，有

$$\widetilde{F}_{nr}(x)+\widetilde{\Phi}_{nr}(x)=-\frac{1}{\pi}\left\{\int_{x-\pi}^{x}\frac{F_{nr}(t)+\Phi_{nr}(t)}{t-x}\mathrm{d}t+\right.$$
$$\int_{x}^{2\pi}\frac{F_{nr}(t)+\Phi_{nr}(t)}{t-x}\mathrm{d}t+$$
$$\left.\int_{0}^{x-\pi}\frac{F_{nr}(t)+\Phi_{nr}(t)}{2\pi+t-x}\mathrm{d}t\right\}+O\left(\frac{1}{n^{r+1}}\right). \tag{3.9}$$

令$2\pi-\frac{a}{2}\leqslant x\leqslant 2\pi-\frac{B_2}{n}$，$B_2>2A$，则

$$\int_{x-\pi}^{x}\frac{F_{nr}(t)+\Phi_{nr}(t)}{x-t}\mathrm{d}t=\int_{x-\pi}^{2\pi-a+X_n^{**}}\frac{F_{nr}(t)+\Phi_{nr}(t)}{x-t}\mathrm{d}t$$

$$= O\left(\frac{1}{(x + a - X_n^{**} - 2\pi)n^{r+1}}\right)$$

$$= O\left(\frac{1}{an^{r+1}}\right), \tag{3.10}$$

$$\int_x^{2\pi}\frac{F_{nr}(t) + \Phi_{nr}(t)}{t - x}\mathrm{d}t = \int_{2\pi-X_n^*}^{2\pi}\frac{F_{nr}(t) + \Phi_{nr}(t)}{t - x}\mathrm{d}t$$

$$= O\left(\frac{1}{(2\pi - X_n^* - x)n^{r+1}}\right)$$

$$= O\left(\frac{1}{B_2 n^r}\right), \tag{3.11}$$

$$\int_0^{x-\pi}\frac{F_{nr}(t) + \Phi_{nr}(t)}{2\pi + t - x}\mathrm{d}t = O\left(\frac{1}{(2\pi - x)n^{r+1}}\right) = O\left(\frac{1}{B_2 n^r}\right). \tag{3.12}$$

从(3.9) ~ (3.12) 得:对任意小的 $\eta > 0$,有适当大的正数 B_2 及 N_2,对于每一 $n > N_2$,在 $2\pi - \frac{a}{2} \leqslant x \leqslant 2\pi - \frac{B_2}{n}$ 上成立

$$\left|n^r\{\tilde{F}_{nr}(x) + \tilde{\Phi}_{nr}(x)\}\right| < \eta/2. \tag{3.13}$$

不难估计 $\tilde{\Phi}_{nr}(x)$ 在其他部分区间上的值. 事实上, 由

$$\left|\int_0^{\frac{\pi}{n}}\frac{\Phi_{nr}(x + t) - \Phi_{nr}(x - t)}{2\tan\frac{t}{2}}\mathrm{d}t\right| \leqslant \frac{\pi}{n}\cdot\max_x\left|\Phi_{nr}'(x)\right| = O\left(\frac{1}{n^r}\right) \tag{3.14}$$

以及

$$\int_{\frac{\pi}{n}}^{\pi}\frac{\Phi_{nr}(x + t) - \Phi_{nr}(x - t)}{2\tan\frac{t}{2}}\mathrm{d}t = O\left[\frac{1}{\left(\tan\frac{\pi}{n}\right)\cdot n^{r+1}}\right] = O\left(\frac{1}{n^r}\right) \tag{3.15}$$

知:在 $0 \leqslant t \leqslant 2\pi$ 上一致成立

$$\tilde{\Phi}_{nr}(x) = O\left(\frac{1}{n^r}\right). \tag{3.16}$$

对 $\Phi_{nr}(x)$, $\tilde{\Phi}_{nr}(x)$ 再"修正"一下. 注意, 对每一个 n, 存在正数 $\delta_n > 0$, 对于任一可测集 $e \subset [0,2\pi]$, 只须 $\mathrm{mes}(e) < \delta_n$, 即有

$$\int_e\left|K_r(x - t)\right|\mathrm{d}t < \frac{1}{n^{r+1}}, \quad \int_e\left|\tilde{K}_r(x - t)\right|\mathrm{d}t < \frac{1}{n^{r+1}},$$

对 x 一致成立. 明乎此, 我们取 $\Phi_{nr}^{(r)}(x)$, 则见

$$\operatorname{ess\,sup}\left|\Phi_{nr}^{(r)}(x)\right|\leqslant 1,\int_0^{2\pi}\Phi_{nr}^{(r)}(x)\mathrm{d}x = 0.$$

并且 $\Phi_{nr}^{(r)}(x)\equiv 0$，当 $a<x<2\pi-a$ 时 $(n>\max(N,N_1,N_2))$. 不难了解①，存在一个 2π 周期的连续函数 $\Phi_n^*(x)$，满足条件：

1° $\left|\Phi_n^*(t)\right|\leqslant 1,\int_0^{2\pi}\Phi_n^*(t)\mathrm{d}t = 0$;

2° $\Phi_n^*(t)\equiv 0, a\leqslant t\leqslant 2\pi-a$，$|t|\leqslant a'_n< a, a'_n>0$ 是一适当小的数；

3° $\operatorname{mes}\mathscr{E}\{x:0\leqslant x\leqslant 2\pi,\Phi_{nr}^{(r)}(x)\neq\Phi_n^*(x)\}<\delta_n$.

令

$$F_{nr}^*(x)=\frac{1}{\pi}\int_0^{2\pi}K_r(x-t)\Phi_n^*(t)\mathrm{d}t, \tag{3.17}$$

$$\widetilde{F}_{nr}^*(x)=\frac{1}{\pi}\int_0^{2\pi}\widetilde{K}_r(x-t)\Phi_n^*(t)\mathrm{d}t, \tag{3.18}$$

则有

$$\left|\Phi_{nr}(x)-F_{nr}^*(x)\right|<\frac{2}{\pi n^{r+1}},$$

$$\left|\widetilde{\Phi}_{nr}(x)-\widetilde{F}_{nr}^*(x)\right|<\frac{2}{\pi n^{r+1}}.$$

由此，对于任意小的数 $\eta>0$，有适当大的自然数 $N_3>N$，对于每一 $n>N_3$，在 $0\leqslant x\leqslant 2\pi$ 上一致成立：

$$\left|n^r\{\Phi_{nr}(x)-F_{nr}^*(x)\}\right|<\frac{\eta}{2},$$

$$\left|n^r\{\widetilde{\Phi}_{nr}(x)-\widetilde{F}_{nr}^*(x)\}\right|<\frac{\eta}{2}.$$

把这些结果与(3.8)，(3.13)，(3.16) 合并，即可得到下述结果：

对于任意小的数 $\eta>0$，有适当大的正数 $B>0$ 和 $N^*>0$，对于每一个 $n>N^*$，成立：

1° 在 $\frac{B}{n}\leqslant x\leqslant\frac{a}{2}$ 上，

$$|n^r\{F_{nr}(x)-F_{nr}^*(x)\}|<\eta,$$

$$\left|n^r\{\widetilde{F}_{nr}(x)-\widetilde{F}_{nr}^*(x)\}\right|<\eta; \tag{3.19}$$

① 可以认为，$\Phi_{nr}^{(r)}(x)$ 只有有限个第一类间断点，见 §2，6°.

2° 在 $2\pi-\frac{a}{2}\leqslant x\leqslant 2\pi-\frac{B}{n}$ 上,

$$\left|n^r\{F_{nr}(x)+F_{nr}^*(x)\}\right|<\eta,$$

$$\left|n^r\{\widetilde{F}_{nr}(x)+\widetilde{F}_{nr}^*(x)\}\right|<\eta; \tag{3.20}$$

3° 在 $0\leqslant x\leqslant 2\pi$ 上,

$$F_{nr}^*(x)=O\left(\frac{1}{n^r}\right),\quad \widetilde{F}_{nr}^*(x)=O\left(\frac{1}{n^r}\right). \tag{3.21}$$

利用(2.12) 估计 $E_n(F_{nr}^*),E_n(\widetilde{F}_{nr}^*)$. 注意

$$U_n(F_{nr}^*,0)=\frac{\sin\beta\pi}{2n}\sum_{k=1}^{2n}(-1)^{k-1}F_{nr}^*(x_k)\cot\frac{x_k}{2}, \tag{3.22}$$

$$U_n(\widetilde{F}_{nr}^*,0)=\frac{\sin\widetilde{\beta}\,\pi}{2n}\sum_{k=1}^{2n}(-1)^{k-1}\widetilde{F}_{nr}^*(\widetilde{x}_k)\cot\frac{\widetilde{x}_k}{2}, \tag{3.22$'$}$$

$0<\beta,\widetilde{\beta}<1$. 估计方法一样, 我们只计算 $E_n(F_{nr}^*)$. 假定 $n>N^*$, 而

$$U_n(F_{nr}^*,0)=\frac{\sin\beta\pi}{2n}\left\{\sum_{x_k\in\left(0,\frac{B}{n}\right]}+\sum_{x_k\in\left[\frac{B}{n},\frac{a}{2}\right)}+\sum_{x_k\in\left[\frac{a}{2},2\pi-\frac{a}{2}\right]}+\sum_{x_k\in\left(2\pi-\frac{a}{2},2\pi-\frac{B}{n}\right]}+\sum_{x_k\in\left(2\pi-\frac{B}{n},2\pi\right)}\right\}.$$

分别估计以上五个和:

$$\left|\sum_{x_k\in\left(0,\frac{B}{n}\right]}(-1)^{k-1}F_{nr}^*(x_k)\cot\frac{x_k}{2}\right|=O\left(\frac{1}{n^r}\right)\left|\sum\cot\frac{x_k}{2}\right|=O\left(\frac{|\ln B|}{n^{r-1}}\right), \tag{3.23}$$

$$\left|\sum_{x_k\in\left(2\pi-\frac{B}{n},2\pi\right)}(-1)^{k-1}F_{nr}^*(x_k)\cot\frac{x_k}{2}\right|=O\left(\frac{1}{n^r}\right)\sum\left|\cot\frac{x_k}{2}\right|=O\left(\frac{|\ln B|}{n^{r-1}}\right), \tag{3.24}$$

$$\left|\sum_{x_k\in\left[\frac{a}{2},2\pi-\frac{a}{2}\right]}(-1)^{k-1}F_{nr}^*(x_k)\cot\frac{x_k}{2}\right|=O\left(\frac{1}{n^r}\right)\sum\left|\cot\frac{x_k}{2}\right|=O\left(\frac{1}{an^{r-1}}\right). \tag{3.25}$$

假定 $0<\eta<1$, 则于 $x_k\in\left[\frac{B}{n},\frac{a}{2}\right]$时, $F_{nr}^*(x_k)$ 与 $F_{nr}(x_k)$ 同号: $(A_r>1)$

$$F_{nr}(x_k)-\frac{\eta}{n^r}<F_{nr}^*(x_k)<F_{nr}(x_k)+\frac{\eta}{n^r}.$$

而当 $x_k\in\left[2\pi-\frac{a}{2},2\pi-\frac{B}{n}\right]$时, $F_{nr}^*(x_k)$ 与 $F_{nr}(x_k)$ 异号:

$$-F_{nr}(x_k)-\frac{\eta}{n^r}<F_{nr}^*(x_k)<-F_{nr}(x_k)+\frac{\eta}{n^r}.$$

注意 $F_{nr}(x_k)=\mathscr{E}(-1)^k\frac{A_r}{n^r}$，$|\mathscr{E}|=1$，以及当 $x_k\in\left[\frac{B}{n},\frac{a}{2}\right]$时，$\cot\frac{x_k}{2}>0$，当 $x_k\in\left[2\pi-\frac{a}{2},2\pi-\frac{B}{n}\right]$时，$\cot\frac{x_k}{2}<0$. 我们得到

$$\left|\sum_{x_k\in\left[\frac{B}{n},\frac{a}{2}\right]}+\sum_{x_k\in\left[2\pi-\frac{a}{2},2\pi-\frac{B}{n}\right]}\right|$$
$$\geqslant\frac{A_r-\eta}{n^r}\left|\sum_{x_k\in\left[\frac{B}{n},\frac{a}{2}\right]}\cot\frac{x_k}{2}+\sum_{x_k\in\left[2\pi-\frac{a}{2},2\pi-\frac{B}{n}\right]}\left|\cot\frac{x_k}{2}\right|\right|.$$

但

$$\sum_{x_k\in\left[\frac{B}{n},\frac{a}{2}\right]}\cot\frac{x_k}{2}=\sum_{x_k\in\left[\frac{B}{n},\frac{a}{2}\right]}\frac{2}{x_k}+O(n)=\frac{2n}{\pi}\sum_{x_k}\frac{1}{k}+O(n)$$
$$=\frac{2n}{\pi}\left\{\ln n-\left|\ln\frac{B}{a}\right|\right\}+O(n). \tag{3.26}$$

同样，有

$$\left|\sum_{x_k\in\left[2\pi-\frac{a}{2},2\pi-\frac{B}{n}\right]}\cot\frac{x_k}{2}\right|=\frac{2n}{\pi}\left\{\ln n-\left|\ln\frac{B}{a}\right|\right\}+O(n). \tag{3.27}$$

综合以上各式，得

$$|U_n(F_{nr}^*,0)|\geqslant\frac{\sin\beta\pi}{2n}\left\{\frac{A_r-\eta}{n^r}\cdot\frac{4n}{\pi}\left[\ln n-\left|\ln\frac{B}{a}\right|\right]-\right.$$
$$\left.O\left(\frac{1}{n^{r-1}}\right)-O\left(\frac{\ln B}{n^{r-1}}\right)-O\left(\frac{1}{an^{r-1}}\right)\right\}$$
$$=\frac{2\sin\beta\pi}{\pi}\frac{\ln n}{n^r}(A_r-\eta)-$$
$$\left\{O\left(\frac{|\ln B|}{an^r}\right)+O\left(\frac{|\ln B|}{n^r}\right)\right\}.$$

注意到

$$F_{nr}^*(0)=O\left(\frac{1}{n^r}\right),\ \|U_n\|\approx\frac{2}{\pi}\sin\beta\pi\ln n,$$

得

$$E_n(F_{nr}^*)\geqslant\frac{A_r-\eta}{n^r}-\frac{1}{n^r\ln n}\left\{O\left(\frac{|\ln B|}{a}\right)+O\left(\frac{1}{a}\right)\right\}.$$

对于 $\eta>0$，可以选取 $m_0'>N^*$，使

$$\frac{1}{\ln n}\left\{\left|O\left(\frac{|\ln B|}{a}\right)\right|+\left|O\left(\frac{1}{a}\right)\right|\right\}<\eta, n>m'_0.$$

则有

$$E_n(F^*_{nr})\geqslant\frac{A_r-2\eta}{n^r}>\frac{A_r}{n^r}(1-2\eta).$$

令 $\varepsilon=2\eta$，即得所求. 至于 $E_n(\widetilde{F}^*_{nr})$，可用同法估计之，得：于 $n>m''_0$ 时 (m''_0 是与 ε 有关的正数)，有

$$E_n(\widetilde{F}^*_{nr})\geqslant\frac{\widetilde{A}_r}{n^r}(1-\varepsilon).$$

最后令 $m_0=\max(m'_0, m''_0)$，即得基本引理之证.

§4. 基本引理之证（r 取非整数时）

(Ⅰ) $0<r<1$. 存在适当大的正数 N，对于每一自然数 $n>N$，有

$$\Phi_n(t)=\begin{cases}\operatorname{sgn}\sin(nt+\alpha\pi), & 0\leqslant t\leqslant a-X_n,\\ 0, & a-X_n\leqslant t\leqslant 2\pi-a+X^*_n,\\ -\operatorname{sgn}\sin(nt+\alpha\pi), & 2\pi-a+X^*_n\leqslant t\leqslant 2\pi,\end{cases}$$

此处 $0<X_n, X^*_n\leqslant\frac{A}{n}$，($A>0$ 只与 r 有关)，且 $\int_0^{2\pi}\Phi_n(t)\mathrm{d}t=0$. 令

$$\Phi_{nr}(x)=\frac{1}{\Gamma(r)}\left\{\int_0^x(x-t)^{r-1}\Phi_n(t)\mathrm{d}t+\int_x^{2\pi}(x-t+2\pi)^{r-1}\Phi_n(t)\mathrm{d}t\right\}+\frac{1}{\pi}\int_0^{2\pi}\Phi_r(x-t)\Phi_n(t)\mathrm{d}t, 0\leqslant x\leqslant 2\pi. \tag{4.1}$$

由于

$$\int_0^{2\pi}\Phi_r(x-t)\operatorname{sgn}\sin(nt+\alpha\pi)\mathrm{d}t=O\left(\frac{1}{n}\right),$$

$$\int_0^{2\pi}\Phi_r(x-t)\Phi_n(t)\mathrm{d}t=O\left(\frac{1}{n}\right),$$

得到

$$\begin{aligned}&F_{nr}(x)-\Phi_{nr}(x)\\&=\frac{1}{\Gamma(r)}\left\{\int_0^x(x-t)^{r-1}[\operatorname{sgn}\sin(nt+\alpha\pi)-\Phi_n(t)]\mathrm{d}t+\right.\\&\left.\int_x^{2\pi}(x-t+2\pi)^{r-1}[\operatorname{sgn}\sin(nt+\alpha\pi)-\Phi_n(t)]\mathrm{d}t\right\}+O\left(\frac{1}{n}\right).\end{aligned} \tag{4.2}$$

如果 $0 < x < a - X_n$，则

$$\int_0^x (x-t)^{r-1}[\operatorname{sgn}\sin(nt+\alpha\pi) - \Phi_n(t)]\mathrm{d}t = 0,$$

$$\int_x^{2\pi} (x-t+2\pi)^{r-1}[\operatorname{sgn}\sin(nt+\alpha\pi) - \Phi_n(t)]\mathrm{d}t$$

$$= \int_{a-X_n}^{2\pi} (x-t+2\pi)^{r-1}[\operatorname{sgn}\sin(nt+\alpha\pi) - \Phi_n(t)]\mathrm{d}t$$

$$= O\left(\frac{1}{x^{1-r}n}\right).$$

故若 $0 < B_1 \leqslant nx$，则

$$\int_0^x (x-t+2\pi)^{r-1}[\operatorname{sgn}\sin(nt+\alpha\pi) - \Phi_n(t)]\mathrm{d}t = O\left(\frac{1}{B_1^{1-r}n^r}\right). \tag{4.3}$$

由(4.2), (4.3) 得：对于任意小的数 $\eta > 0$，有适当大的 $B_1 > 0$ 与 $N_1 > 0$，对每一 $n > N_1$，在 $\frac{B_1}{n} \leqslant x \leqslant a - X_n$ 上，成立着

$$\left| n^r\{F_{nr}(x) - \Phi_{nr}(x)\}\right| < \frac{\eta}{2}. \tag{4.4}$$

然后在 $2\pi - \frac{a}{2} < x < 2\pi$ 上考查

$$F_{nr}(x) + \Phi_{nr}(x)$$

$$= \frac{1}{\Gamma(r)}\left\{\int_0^x (x-t)^{r-1}[\operatorname{sgn}\sin(nt+\alpha\pi) + \Phi_n(t)]\mathrm{d}t + \right.$$

$$\left.\int_x^{2\pi} (x-t+2\pi)^{r-1}[\operatorname{sgn}\sin(nt+\alpha\pi) + \Phi_n(t)]\mathrm{d}t\right\} + O\left(\frac{1}{n}\right). \tag{4.5}$$

注意，此时

$$\int_x^{2\pi} (2\pi+x-t)^{r-1}[\operatorname{sgn}\sin(nt+\alpha\pi) + \Phi_n(t)]\mathrm{d}t = 0, \tag{4.6}$$

并且

$$\int_0^x (x-t)^{r-1}[\operatorname{sgn}\sin(nt+\alpha\pi) + \Phi_n(t)]\mathrm{d}t$$

$$= \int_0^{2\pi-a+X_n^*} (x-t)^{r-1}[\operatorname{sgn}\sin(nt+\alpha\pi) + \Phi_n(t)]\mathrm{d}t$$

$$= O\left(\frac{1}{a^{1-r}n}\right). \tag{4.7}$$

$\left(\text{此事对一切充分大的 } n \text{ 成立:只须 } X_n^* < \frac{a}{2} \text{ 即可}\right)$由此知:对于一切充分大的 n,在 $2\pi - \frac{a}{2} \leqslant x \leqslant 2\pi$ 上成立着

$$n^r\{F_{nr}(x) + \Phi_{nr}(x)\} = O\left(\frac{1}{(an)^{1-r}}\right). \tag{4.8}$$

故对于任意的小数 $\eta > 0$,存在一适当大的正数 N'_1,对每一个 $n > N'_1$,有

$$|n^r\{F_{nr}(x) + \Phi_{nr}(x)\}| < \frac{\eta}{2} \tag{4.9}$$

在 $2\pi - \frac{a}{2} \leqslant x \leqslant 2\pi$ 上成立.

至于在其他区间上,不难证明

$$\Phi_{nr}(x) = O\left(\frac{1}{n^r}\right). \tag{4.10}$$

以下利用(2.5),(2.5′) 估计 $\tilde{\Phi}_{nr}(x)$.

对于充分大的 n,当 $0 < x < a - X_n$ 时,有

$$\begin{aligned}
&\tilde{F}_{nr}(x) - \tilde{\Phi}_{nr}(x) \\
&= \frac{\Gamma(1-r)}{\pi}\left\{\int_0^x \frac{\cos \pi r}{(x-t)^{1-r}}[\operatorname{sgn} \sin (nt + \alpha\pi) - \Phi_n(t)]\mathrm{d}t + \right. \\
&\int_{x+\pi}^{2\pi} \frac{\cos \pi r}{(2\pi + x - t)^{1-r}}[\operatorname{sgn} \sin (nt + \alpha\pi) - \Phi_n(t)]\mathrm{d}t - \\
&\left.\int_x^{x+\pi} \frac{\operatorname{sgn} \sin (nt + \alpha\pi) - \Phi_n(t)}{(t-x)^{1-r}}\mathrm{d}t\right\} + O\left(\frac{1}{n}\right).
\end{aligned} \tag{4.11}$$

$$\int_0^x \frac{\operatorname{sgn} \sin(nt + \alpha\pi) - \Phi_n(t)}{(x-t)^{1-r}} \cos \pi r \mathrm{d}t = 0, \tag{4.12}$$

$$\begin{aligned}
&\int_x^{x+\pi} \frac{\operatorname{sgn} \sin(nt + \alpha\pi) - \Phi_n(t)}{(t-x)^{1-r}}\mathrm{d}t \\
&= \int_{a-X_n}^{x+\pi} \frac{\operatorname{sgn} \sin(nt + \alpha\pi) - \Phi_n(t)}{(t-x)^{1-r}}\mathrm{d}t \\
&= O\left(\frac{1}{a^{1-r}n}\right),
\end{aligned} \tag{4.13}$$

当 $x < a - X_n, x \leqslant \frac{a}{2}$ 时(此事只须 n 充分大即成立).最后对任意正数

B_2, 当$\frac{B_2}{n} \leqslant x \leqslant \frac{a}{2}$时, 成立着

$$\int_{\pi+x}^{2\pi} \frac{\operatorname{sgn}\sin(nt+\alpha\pi) - \Phi_n(t)}{(2\pi+x-t)^{1-r}} \mathrm{d}t$$

$$= O\left(\frac{1}{x^{1-r}n}\right) = O\left(\frac{1}{B_2^{1-r}n^r}\right). \tag{4.14}$$

由(4.11) ~ (4.14) 得: 对于任意小的数 $\eta > 0$, 存在一个适当大的数 $N_2 > 0$, 与 $B_2 > 0$, 对每一个 $n > N_2$, 在$\frac{B_2}{n} \leqslant x \leqslant \frac{a}{2} < a - X_n$ 上, 成立

$$\left| n^r \{\widetilde{F}_{nr}(x) - \widetilde{\Phi}_{nr}(x)\} \right| < \frac{\eta}{2}. \tag{4.15}$$

当$2\pi - \frac{a}{2} \leqslant x \leqslant 2\pi$时, 我们利用(2.5′). 用相同的方法可以证明: 对任意小的数 $\eta > 0$ 有正数 $N_2' > 0, B_2' > 0$, 对每一个 $n > N_2'$, 在 $2\pi - \frac{a}{2} \leqslant x \leqslant 2\pi - \frac{B_2'}{n}$ 上, 成立

$$\left| n^r \{\widetilde{F}_{nr}(x) + \widetilde{\Phi}_{nr}(x)\} \right| < \frac{\eta}{2}. \tag{4.16}$$

另外, 容易证明, 在 $0 \leqslant x \leqslant 2\pi$ 上, 一致成立

$$\widetilde{\Phi}_{nr}(x) = O\left(\frac{1}{n^r}\right). \tag{4.17}$$

现在仿照 §3 中的方法"修正"$\Phi_{nr}(x), \widetilde{\Phi}_{nr}(x)$. 对于每一个自然数 n, 有 $\delta_n > 0$, 对任一可测集 $e \subset [0, 2\pi]$, 只须 $\operatorname{mes}(e) < \delta_n$, 即有(对任意的 x)

$$\int_e |K_r(x-t)| \mathrm{d}t < n^{-1}, \int_e |\widetilde{K}_r(x-t)| \mathrm{d}t < n^{-1}. \tag{4.18}$$

据此, 对于一切充分大的 n(例如, $n > \max(N, N_1, N_1', N_2, N_2')$), 存在一个 2π 周期的连续函数 $\Phi_n^*(t)$, 使

1° $|\Phi_n^*(t)| \leqslant 1, \int_0^{2\pi} \Phi_n^*(t) \mathrm{d}t = 0$;

2° $\Phi_n^*(t) \equiv 0, a \leqslant t \leqslant 2\pi - a, |t| \leqslant a_n' < a, a_n' > 0$;

3° $\operatorname{mes} \mathscr{E}\{x: 0 \leqslant x \leqslant 2\pi, \Phi_n^*(x) \neq \Phi_n(x)\} < \delta_n$, 此种 $\Phi_n^*(t)$ 无疑存在. 对此, 令

$$F_{nr}^*(x) = \frac{1}{\pi} \int_0^{2\pi} K_r(x-t) \Phi_n^*(t) \mathrm{d}t,$$

$$\widetilde{F}^*_{nr}(x)=\frac{1}{\pi}\int_0^{2\pi}\widetilde{K}_r(x-t)\Phi^*_n(t)\mathrm{d}t.$$

则在 $0\leqslant x\leqslant 2\pi$ 上, 一致成立

$$\begin{aligned}&\left|F^*_{nr}(x)-\Phi_{nr}(x)\right|<\frac{2}{\pi n},\\&\left|\widetilde{F}^*_{nr}(x)-\widetilde{\Phi}_{nr}(x)\right|<\frac{2}{\pi n}.\end{aligned}\tag{4.19}$$

从(4.16),(4.17),(4.19) 不难引出以下结论:

对任意小的数 $\eta>0$, 有适当大的正数 $N>0$ 与 $B>0$, 对于每一个 $n>N$, 对 $F^*_{nr}(x),\widetilde{F}^*_{nr}(x)$ 成立着

1° 在 $\frac{B}{n}\leqslant x\leqslant\frac{a}{2}<a-X_n$ 上, 有

$\left|n^r\{F_{nr}(x)-F^*_{nr}(x)\}\right|<\eta,\left|n^r\{\widetilde{F}_{nr}(x)-\widetilde{F}^*_{nr}(x)\}\right|<\eta;$

2° 在 $2\pi-\frac{a}{2}\leqslant x\leqslant 2\pi-\frac{B}{n}$ 上, 有

$\left|n^r\{F_{nr}(x)+F^*_{nr}(x)\}\right|<\eta,\left|n^r\{\widetilde{F}_{nr}(x)+\widetilde{F}^*_{nr}(x)\}\right|<\eta;$

3° 在 $0\leqslant x\leqslant 2\pi$ 上, 有

$$F^*_{nr}(x)=O\left(\frac{1}{n^r}\right),\widetilde{F}^*_{nr}(x)=O\left(\frac{1}{n^r}\right).\tag{4.20}$$

至此, 使用 §3 中一样的方法估计 $E_n(F^*_{nr}),E_n(\widetilde{F}^*_{nr})$, 即得引理在 $0<r<1$ 时之证. 因步骤与 §3 基本上无差别, 故不赘述.

(Ⅱ) $r=s+r',s=[r]\geqslant 1,0<r'<1$. 这种情形基本上和 §3 以及 $0<r<1$ 的情形没有差别, 为节省篇幅, 我们只将证明的主要步骤和结论给出来. 在 §2 的公式(2.4) 中的 $f_s(t)$ 换以 $F_{ns}(t)$ 及 $\Phi_{ns}(t)$; (2.5),(2.5′) 中的 $\varphi(t)$, 都各用 $F_{ns}(t)$, 以及 $\Phi_{ns}(t)$ 代换且把 r 换为 r'(关于 $\Phi_{ns}(t)$, 见 §2,6°), 所得函数各表以 $F_{nr}(x)(0\leqslant x\leqslant 2\pi)$, $\widetilde{F}_{nr}(x)(0\leqslant x<\pi,\pi\leqslant x<2\pi$ 由二式给出) 和 $\Phi_{nr}(x)(0\leqslant x\leqslant 2\pi)$, $\widetilde{\Phi}_{nr}(x)(0\leqslant x<\pi,\pi\leqslant x<2\pi$ 由二式给出). 显然, $\Phi_{nr}(x)\in W^{(r)}$, $\widetilde{\Phi}_{nr}\in\widetilde{W}^{(r)}$. 此外自然假定 n 适当大(至少 $n>N$, N 是 §2,6° 中给出的) 用 §3, 和 $0<r<1$ 中一样的方法和步骤, 可以证明下述结论: 对于任意小的数 $\eta>0$, 有适当大的正数 N 及 B, 对于每一 $n>N$, 成立着

$$\left.\begin{aligned}
&1^\circ\quad 在\frac{B}{n}\leqslant x\leqslant\frac{a}{2}<a-X_n''上\\
&\qquad\qquad |n^r\{F_{nr}(x)-\Phi_{nr}(x)\}|<\frac{\eta}{2},\\
&\qquad\qquad |n^r\{\widetilde{F}_{nr}(x)-\widetilde{\Phi}_{nr}(x)\}|<\frac{\eta}{2};\\
&2^\circ\quad 在 2\pi-\frac{a}{2}\leqslant x\leqslant 2\pi-\frac{B}{n}上\\
&\qquad\qquad |n^r\{F_{nr}(x)+\Phi_{nr}(x)\}|<\frac{\eta}{2},\\
&\qquad\qquad |n^r\{\widetilde{F}_{nr}(x)+\widetilde{\Phi}_{nr}(x)\}|<\frac{\eta}{2};\\
&3^\circ\quad \Phi_{nr}(x)=O\left(\frac{1}{n^r}\right),\widetilde{\Phi}_{nr}(x)=O\left(\frac{1}{n^r}\right).
\end{aligned}\right\}\tag{4.21}$$

对 $\Phi_{nr}(x)$, $\widetilde{\Phi}_{nr}(x)$ 的“修正”可以如下地进行：取 $\Phi_{ns}^{(s)}(x)$，则见 $\operatorname{ess\,sup}|\Phi_{ns}^{(s)}(x)|\leqslant 1,\int_0^{2\pi}\Phi_{ns}^{(s)}(t)\mathrm{d}t=0$，且 $\Phi_{ns}^{(s)}(x)\equiv 0, a<x<2\pi-a$. 对于每一个 n，有 $\delta_n>0$，任取可测集 $e\subset[0,2\pi]$，由 $\mathrm{mes}(e)<\delta_n$ 得到(对任意 x)

$$\int_e|K_r(x-t)|\mathrm{d}t<\frac{1}{n^{s+1}},\int_e|\widetilde{K}_r(x-t)|\mathrm{d}t<\frac{1}{n^{s+1}}.$$

于是，对于一切充分大的 n，存在一个 2π 周期的连续函数 $\Phi_n^*(t)$，满足：

1° $|\Phi_n^*(t)|\leqslant 1,\int_0^{2\pi}\Phi_n^*(t)\mathrm{d}t=0$;

2° $\Phi_n^*(t)\equiv 0, a\leqslant t\leqslant 2\pi-a, |t|\leqslant a'_n<a, a_n>0$;

3° $\mathrm{mes}\,\mathscr{E}\{t:0\leqslant t\leqslant 2\pi,\Phi_n^*(t)\neq\Phi_{ns}^{(s)}(t)\}<\delta_n$.

最后，取

$$\left.\begin{aligned}
F_{nr}^*(x)&=\frac{1}{\pi}\int_0^{2\pi}K_r(x-t)\Phi_n^*(t)\mathrm{d}t,\\
\widetilde{F}_{nr}^*(x)&=\frac{1}{\pi}\int_0^{2\pi}\widetilde{K}_r(x-t)\Phi_n^*(t)\mathrm{d}t.
\end{aligned}\right.\tag{4.22}$$

则有

$$\begin{aligned} &\left|F^*_{nr}(x) - \Phi_{nr}(x)\right| < \frac{2}{\pi n^{s+1}}, \\ &\left|\widetilde{F}^*_{nr}(x) - \widetilde{\Phi}_{nr}(x)\right| < \frac{2}{\pi n^{s+1}}. \end{aligned} \tag{4.23}$$

由此不难得出如下结论：对于任意小数 $\eta > 0$，存在适当大的正数 N 及 B，对每一 $n > N$，成立着

$$\left.\begin{aligned} &1^\circ \quad 在 \frac{B}{n} \leqslant x \leqslant \frac{a}{2} < a - X''_n 上, \\ &\quad \left|n^r\{F_{nr}(x) - F^*_{nr}(x)\}\right| < \eta, \\ &\quad \left|n^r\{\widetilde{F}_{nr}(x) - \widetilde{F}^*_{nr}(x)\}\right| < \eta. \\ &2^\circ \quad 在 2\pi - \frac{a}{2} \leqslant x \leqslant 2\pi - \frac{B}{n} 上, \\ &\quad \left|n^r\{F_{nr}(x) + F^*_{nr}(x)\}\right| < \eta, \\ &\quad \left|n^r\{\widetilde{F}_{nr}(x) + \widetilde{F}^*_{nr}(x)\}\right| < \eta. \\ &3^\circ \quad F^*_{nr}(x) = O\left(\frac{1}{n^r}\right), \widetilde{F}^*_{nr}(x) = O\left(\frac{1}{n^r}\right). \end{aligned}\right\} \tag{4.24}$$

最后，利用 §3 中的方法估计 $E_n(F^*_{nr})$，$E_n(\widetilde{F}^*_{nr})$，即得基本引理之证. 因计算的步骤与 §3 无区别，故不再重复. 基本引理完全证毕.

§5. 定理之证

取二减数列 $\{\varepsilon_k\}$，$\{\eta_k\}$，$\lim\limits_{k\to\infty}\varepsilon_k = \lim\limits_{k\to\infty}\eta_k = 0$. 任取 $0 < a_1 < 1$. 由基本引理，知存在自然数 n_1，对其有函数 $\Phi^*_{n_1}(t)$，满足

1° $\Phi^*_{n_1}(t)$ 是以 2π 为周期的连续函数；

2° $\left|\Phi^*_{n_1}(t)\right| \leqslant 1, \int_0^{2\pi}\Phi^*_{n_1}(t)\mathrm{d}t = 0$；

3° $\Phi^*_{n_1}(t) \equiv 0, a_1 \leqslant t \leqslant 2\pi - a_1, \mid t \mid \leqslant a_2 < a_1, a_2 > 0$；

4° 对于 $F^*_{n_1r}(x)$，$\widetilde{F}^*_{n_1r}(x)$ 成立着：

$$E_{n_1}(F^*_{n_1r}) \geqslant \frac{A_r}{n_1^r}(1 - \varepsilon_1), E_{n_1}(\widetilde{F}^*_{n_1r}) \geqslant \frac{\widetilde{A}_r}{n_1^r}(1 - \varepsilon_1).$$

对于 $a_2 > 0$，有自然数 $n_2 > n_1$，及以 2π 为周期的连续函数 $\Phi^*_{n_2}(t)$，满足

1° $\left|\Phi^*_{n_2}(t)\right| \leqslant 1, \int_0^{2\pi}\Phi^*_{n_2}(t)\mathrm{d}t = 0$；

2° $\Phi_{n_2}^*(t) \equiv 0, a_2 \leqslant t \leqslant 2\pi - a_2, |t| \leqslant a_3 < a_2, a_3 > 0$;

3° 对于 $F_{n_2r}^*(x), \widetilde{F}_{n_2r}^*(x)$, 成立

$$E_{n_2}(F_{n_2r}^*) \geqslant \frac{A_r}{n_2^r}(1 - \varepsilon_2),$$

$$E_{n_2}(\widetilde{F}_{n_2r}^*) \geqslant \frac{\widetilde{A}_r}{n_2^r}(1 - \varepsilon_2);$$

4° $E_{n_2}(F_{n_1r}^*) \leqslant \frac{\eta_1}{n_2^r}, E_{n_2}(\widetilde{F}_{n_1r}^*) \leqslant \frac{\eta_1}{n_2^r}$.

最后这一条是可以做到的, 因为 $\Phi_{n_1}^*(t)$ 是连续函数, 由[8], 对任意自然数 n, 有

$$E_n(F_{n_1r}^*) \leqslant \frac{1}{\pi} E_n(K_r)_L \cdot E_n(\Phi_{n_1r}^*),$$

$$E_n(\widetilde{F}_{n_1r}^*) \leqslant \frac{1}{\pi} E_n(\widetilde{K}_r)_L \cdot E_n(\Phi_{n_1r}^*).$$

故由此得

$$E_n(F_{n_1r}^*) = o\left(\frac{1}{n_r}\right), E_n(\widetilde{F}_{n_1r}^*) = o\left(\frac{1}{n_r}\right).$$

这样用数学归纳法不难证明: 存在一个减数列 $a_1, a_2, \cdots, a_k, \cdots, \lim\limits_{k\to\infty} a_k = 0, a_k > 0$, 以及自然数子列 $n_1 < n_2 < \cdots < n_k < \cdots$, 与以 2π 为周期的连续函数列

$$\Phi_{n_1}^*(t), \Phi_{n_2}^*(t), \cdots, \Phi_{n_k}^*(t), \cdots$$

满足条件:

1° $\left|\Phi_{n_k}^*(t)\right| \leqslant 1, \int_0^{2\pi} \Phi_{n_k}^*(t)\mathrm{d}t = 0, k = 1, 2, \cdots$

2° $\Phi_{n_k}^*(t) \equiv 0, a_k \leqslant t \leqslant 2\pi - a_k, |t| \leqslant a_{k+1}, k = 1, 2, \cdots$

3° $E_{n_k}(F_{n_kr}^*) \geqslant \frac{A_r}{n_k^r}(1 - \varepsilon_k)$,

$E_{n_k}(\widetilde{F}_{n_kr}^*) \geqslant \frac{\widetilde{A}_r}{n_k^r}(1 - \varepsilon_k), k = 1, 2, \cdots$

4° $E_{n_k}(F_{n_1r}^* + \cdots + F_{n_{k-1}r}^*) \leqslant \frac{\eta_{k-1}}{n_k^r}, k = 2, 3, \cdots$

$E_{n_k}(\widetilde{F}_{n_1r}^* + \cdots + \widetilde{F}_{n_{k-1}r}^*) \leqslant \frac{\eta_{k-1}}{n_k^r}, k = 2, 3, \cdots$

5°　$\sum_{s=k+1}^{\infty}\frac{1}{n_s^r}=o\left(\frac{1}{n_k^r}\right), k\to\infty$ 时.

令

$$\Phi(t)=\sum_{k=1}^{\infty}\Phi_{n_k}^*(t), \tag{5.1}$$

显见

$$|\Phi(t)|\leqslant 1, \int_0^{2\pi}\Phi(t)\mathrm{d}t=0.$$

令

$$F(x)=\frac{1}{\pi}\int_0^{2\pi}K_r(x-t)\Phi(t)\mathrm{d}t, \tag{5.2}$$

$$\widetilde{F}(x)=\frac{1}{\pi}\int_0^{2\pi}\widetilde{K}_r(x-t)\Phi(t)\mathrm{d}t. \tag{5.3}$$

成立着

$$F(x)=\sum_{k=1}^{\infty}F_{n_kr}^*(x), \widetilde{F}(x)=\sum_{k=1}^{\infty}\widetilde{F}_{n_kr}^*(x). \tag{5.4}$$

(5.4) 中的级数在$[0,2\pi]$上一致收敛, 由此得

$$\begin{aligned}E_{n_k}(F)&\geqslant E_{n_k}(F_{n_kr}^*)-E_{n_k}(F_{n_1r}^*+\cdots+F_{n_{k-1}r}^*)-\sum_{s=k+1}^{\infty}\|F_{n_sr}^*\|\\&\geqslant\frac{A_r}{n_k^r}(1-\varepsilon_k)-\frac{\eta_{k-1}}{n_k^r}-o\left(\frac{1}{n_k^r}\right),\end{aligned}$$

$$\begin{aligned}E_{n_k}(\widetilde{F})&\geqslant E_{n_k}(\widetilde{F}_{n_kr}^*)-E_{n_k}(\widetilde{F}_{n_1r}^*+\ldots+\widetilde{F}_{n_{k-1}r}^*)-\sum_{s=k+1}^{\infty}\|\widetilde{F}_{n_sr}^*\|\\&\geqslant\frac{\widetilde{A}_r}{n_k^r}(1-\varepsilon_k)-\frac{\eta_{k-1}}{n_k^r}-o\left(\frac{1}{n_k^r}\right),\end{aligned}$$

此处 $\|F_{n_sr}^*\|=\max_x|F_{n_sr}^*(x)|=O\left(\frac{1}{n_s^r}\right)$, 等等. 由此即得

$$\lim_{k\to\infty}n_k^rE_{n_k}(F)=A_r,\quad \lim_{k\to\infty}n_k^rE_{n_k}(\widetilde{F})=\widetilde{A}r.$$

定理证毕.

§6. 一类周期函数及其共轭函数在 L 中平均逼近的渐近性质

B. K. Дзядык 在[1]中(并参见[1]中引的有关文献)给出以下

定理　(B. K. Дзядык) 设 $r>0$, 若2π周期的连续函数$f(x)$具有 r

阶(在 Weyl 意义下)的导函数 $f^{(r)}(x)$,满足条件:

1° $f^{(r)}(t)$ 在$[0,2\pi]$上有界变差;

2° $f^{(r)}(t) = g(t) + h(t)$,$h(t)$ 绝对连续,而 $g(t)$ 是跃度函数;

3° $g(t)$ 在区间$[0,2\pi]$内至少有一个真正间断点,则

$$\lim_{n\to\infty} n^r E_n(f)_L = A_r \sum_k | g(a_k+0) - g(a_k-0) |,$$

此处$\{a_k\}$是$[0,2\pi]$中的 $g(t)$ 的真正间断点集.

对于该定理可补充如下:

$$\lim_{n\to\infty} n^r E_n(\tilde{f})_L = \tilde{A}_r \sum_k | g(a_k+0) - g(a_k-0) |.$$

其证法和 В. К. Дзядык 考虑的情形所用方法一样,故不重复.

参考文献

[1] В.К. Дзядык. О наилучшем приближении на классах периодических функций, определяемых ядрами, являющимися интегралами от абсолютно монотонных функций. Изв. АН СССР, 1959, сер. матем. (23): 933 ~ 950

[2] С. Б. Стечкин. О наилучшем приближении некоторых классов периодических функций тригонометрическими полиномами. Изв. АН СССР, 1956, сер. матем. (20): 643 ~ 648

[3] Сунь Юн-шен. О наилучшем приближении периодических дифференцируемых функций тригонометрическими полиномами. Изв. АН СССР, 1961, сер. матем. (25): 143 ~ 152

[4] 孙永生. 用三角多项式近迫周期可微函数. 数学学报, 1962, (12): 181 ~ 202

[5] С. М. Никольский. Об интерполировании и наилучшем приближении дифференцируемых периодических функций тригонометрическими полиномами. Изв. АН СССР, 1946, сер. матем. (10): 393 ~ 410

[6] A. Zygmund. *Trigonometrical Series*. London: Cambridge Univ. Press, 1959

[7] А. Ф. Тиман. *Теория приближения функций действительного переменного.* Москва: Физматгиз, 1960. 301 ~ 303

[8] Сунь Юн-шен. О наилучшем приближении классов функций, представимых в форме свертки. ДАН СССР, 1958, 118: 2, 247 ~ 250

数学学报

1981, 7, 24(4)

关于 Cesàro 算子的逼近常数*

The Exact Constant of Approximation for Cesàro Operators

§1. 引言

我们用 $L^p_{2\pi}(1 \leqslant p < \infty)$ 表示 2π 周期的 p 次幂可和的函数空间，当 $p=\infty$ 时约定 $L^\infty_{2\pi}=C_{2\pi}$，后者表示 2π 周期的连续函数空间. 对于 $f(x)\in L^p_{2\pi}$:

$$\| f \|_p = \left\{\int_0^{2\pi} | f(x) |^p \mathrm{d}x\right\}^{\frac{1}{p}} (1 \leqslant p < \infty).$$

$$\| f \|_c = \max_x | f(x) |.$$

$$\omega_p(f;t) = \sup_{|u|\leqslant t}\left\{\int_0^{2\pi} | f(x+u) - f(x) |^p \mathrm{d}x\right\}^{\frac{1}{p}} (1 \leqslant p < \infty).$$

$$\omega(f;t)_c = \sup_{|u|\leqslant t} \max_x | f(x+u) - f(x) |.$$

$f(x)$ 的傅里叶级数记作

$$\sigma(f) = \frac{a_0}{2} + \sum_{k=1}^{\infty} (a_k \cos kx + b_k \sin kx), \tag{1.1}$$

$S_n(f;x)$ 为其 n 阶部分和.

置 $A_n^\alpha = (\alpha+1)\cdots(\alpha+n)/n!(\alpha \neq -1, -2, \cdots)$,

$$\sigma_n^\alpha(f;x) = \frac{1}{A_n^\alpha}\sum_{k=0}^{n} A_{n-k}^{\alpha-1} S_k(f;x) \tag{1.2}$$

* 数学学报于 1979 年 10 月 23 日收到本文.

是 $\sigma(f)$ 的 Cesàro 平均，以下称之为 Cesàro 算子. 当 $\alpha = 1$ 时，(1.2) 是 $\sigma(f)$ 的算术平均，这便是熟识的 Fejér 算子，以下简记为 $\sigma_n(f;x)$. 关于 $f(x)$ 用其 Cesàro 算子特别是 Fejér 算子逼近的问题有各种提法，并且已经知道很多结果. 我们要提到的是 П.Л. Ульянов 在 1964 年发表的下述结果[1]：

定理 若 $f(x) \in L_{2\pi}^p (1 \leqslant p \leqslant \infty)$，则对每一整数 $n \geqslant 0$，成立

$$(1)\ \| \sigma_n^\alpha(f;x) - f(x) \|_p \leqslant B(\alpha)\omega_p\left(f;\frac{\ln(n+2)}{n+1}\right), \tag{1.3}$$

当 $p = 1, \infty, \alpha > 0$ 时.

$$(2)\ \| \sigma_n^\alpha(f;x) - f(x) \|_p \leqslant B(\alpha,p)\omega_p\left(f;\frac{1}{n+1}\right), \tag{1.4}$$

当 $1 < p < \infty$，且 $\alpha \geqslant 1$ 时，

Ульянов 指出，在(1.3)，(1.4) 内，不等式在阶的意义下不能再行改进.

本文目的是讨论 Ульянов 不等式(1.3) 的精确常数问题. 引入记号

$$A_n(\alpha)_X = \sup_{\substack{f \in X \\ f \neq \text{const}}} \frac{\| \sigma_n^\alpha(f;x) - f(x) \|_X}{\ln(n+1)\omega\left(f;\frac{2\pi}{n+1}\right)_X} \tag{1.5}$$

$$B_n(\alpha)_X = \sup_{\substack{f \in X \\ f \neq \text{const}}} \frac{\| \sigma_n^\alpha(f;x) - f(x) \|_X}{\omega\left(f;\frac{2\pi}{n+1}\ln(n+1)\right)_X} \tag{1.6}$$

此处 $X = C_{2\pi}$ 及 $L_{2\pi}$($p = 1$ 时 $L_{2\pi}^1$ 简写成 $L_{2\pi}$)，$\omega(f;t)_X$ 表示$f(x) \in X$ 在 X 尺度下的连续模. 很明白，$B_n(\alpha)_X$ 是(1.3) 右端系数之最佳者. 由

$$\omega\left(f;\frac{2\pi}{n+1}\ln(n+1)\right)_X \leqslant C\ln(n+1)\omega\left(f;\frac{2\pi}{n+1}\right)_X$$

得到较(1.3) 弱的不等式

$$\| \sigma_n^\alpha(f;x) - f(x) \|_X \leqslant B'(\alpha)\ln(n+1)\omega\left(f;\frac{2\pi}{n+1}\right)_X, \tag{1.3'}$$

$A_n(\alpha)_X$ 是(1.3′) 右端系数之最佳者. 在系数 $A_n(\alpha)_X$ 与$B_n(\alpha)_X$ 之间存在着简单的关系. 由

$$\begin{aligned}\| \sigma_n^\alpha(f;x) - f(x) \|_X &\leqslant B_n(\alpha)_X\omega\left(f;\frac{2\pi}{n+1}\cdot\ln(n+1)\right)_X \\ &\leqslant \{1+\ln(n+1)\}B_n(\alpha)_X\omega\left(f;\frac{2\pi}{n+1}\right)_X\end{aligned}$$

以及 $A_n(\alpha)_X$ 的定义即得

$$A_n(\alpha)_X \leqslant \left\{1 + \frac{1}{\ln(n+1)}\right\} B_n(\alpha)_X. \quad n = 1,2,\cdots \tag{1.7}$$

П. Л. Ульянов 定理表明 $B_n(\alpha)_X$ 是有界量，从而 $A_n(\alpha)_X$ 也是有界量. 对于一般情况 $\alpha > 0$，量 $A_n(\alpha)$，$B_n(\alpha)$ 很难计算. $\alpha = 1$ 时有一些结果. 例如，В. О. Дудас[2] 于 1974 年证明

$$\sup_{n\in\mathbf{N}} A_n(1)_c = \frac{1}{\ln 2}. \tag{1.8}$$

N 表自然数集. 姚璧芸[3]、李文清[4] 讨论了量 $A_n(1)_c$ 的极限. 至于系数 $B_n(\alpha)_X$，我们只知道李文清[4] 给出了 $\lim B_n(1)_c$ 的上界及下界估计. 在本文内我们将证明

定理 1 若 $\alpha > 0$，且 H_X^1 表示空间 X 内的李普西兹类

$(f(x) \in H_X^1 \Leftrightarrow \| f(x+t) - f(x) \|_X \leqslant | t |)$ 则

$$\lim_{n\to\infty} A_n(\alpha)_X = \lim_{n\to\infty} \sup_{\substack{f\in H_X^1 \\ f\neq \text{const}}} \frac{\| \sigma_n^\alpha(f;x) - f(x) \|_X}{\ln(n+1)\,\omega\left(f;\dfrac{2\pi}{n+1}\right)_X} = \frac{\alpha}{\pi^2}. \tag{1.9}$$

由定理 1 及(1.7) 可以得到 $B_n(\alpha)_X$ 的下限的下界估计：

$$\frac{\alpha}{\pi^2} \leqslant \varliminf_{n\to\infty} B_n(\alpha)_X. \tag{1.10}$$ ①

定理 1 的证明在 §3, §4 给出.

然后我们提出下列问题：设有实数三角阵

$$\Lambda = \left\{ \lambda_k^{(n)} \middle| \begin{array}{l} k = 0,\cdots,n, \\ n = 0,1,\cdots \end{array} \quad \lambda_0^{(n)} = 1 \right\}$$

$f(x) \in X$. 利用 Λ 构作 $\sigma(f)$ 的 Λ 平均：

$$A_n(f;x,\lambda) = \frac{a_0}{2} + \sum_{k=1}^{n} \lambda_k^{(n)} (a_k \cos kx + b_k \sin kx). \tag{1.11}$$

A_n 是线性算子，其积分表达式是

$$A_n(f;x,\lambda) = \frac{1}{\pi}\int_0^{2\pi} f(x+t) K_n^\lambda(t)\,\mathrm{d}t, \tag{1.12}$$

此处

$$K_n^\lambda(t) = \frac{1}{2} + \sum_{k=1}^{n} \lambda_k^{(n)} \cos kt. \tag{1.13}$$

---

① 可证：$\frac{\alpha}{\pi^2} \leqslant \varliminf B_n(\alpha)_X \leqslant \varlimsup B_n(\alpha)_X \leqslant \frac{2\alpha}{\pi^2}$.

置

$$C_n(\lambda)_X = \sup_{\substack{f\in X\\ f\neq \mathrm{const}}} \frac{\| f(x) - A_n(f;x,\lambda) \|_X}{\omega\left(f;\frac{\pi}{n+1}\right)_X}. \tag{1.14}$$

定理 1 表明:当 Λ 阵取为 Cesàro 求和系数时有

$$C_n(\lambda)_L \approx C_n(\lambda)_c\ ^{①}. \tag{1.15}$$

给出使(1.15) 成立的一类 Λ 阵的性质, 并对某些重要的 Λ 阵确定出(1.15) 内的常数将是很有意思的. 在本文最后一节内我们将对满足条件

$$K_n^\lambda(t) \geqslant 0 \tag{1.16}$$

的 Λ 阵给出保证(1.15) 成立的一种充分条件.

§2. 几条引理

熟知(见[5], 第 157 页)

$$K_n^\alpha(t) = \frac{1}{A_n^\alpha}\sum_{k=0}^{n} A_{n-k}^{\alpha-1} D_k(t) \tag{2.1}$$

是 Cesàro 算子的核, 其中 $D_k(t)$ 是狄里赫来核. Kogbetliantz 证明(见[6], 第 477 页): 当 $0<\alpha<1, 0<t\leqslant\pi$ 时, 有

$$K_n^\alpha(t) = \Omega_n(t) + W_n(t), \tag{2.2}$$

此处

$$\Omega_n(t) = \frac{\sin\left[\left(n+\frac{1}{2}+\frac{\alpha}{2}\right)t - \frac{\pi\alpha}{2}\right]}{A_n^\alpha\left(2\sin\frac{t}{2}\right)^{\alpha+1}}, \tag{2.3}$$

而 $W_n(t) = O(n^{-1}t^{-2})$. 在引理 1 内给出 $W_n(t)$ 的一个表示式.

引理 1 若 $0<\alpha<1, 0<t\leqslant\pi$, 则

$$W_n(t) = \frac{1}{A_n^\alpha}\sum_{\nu=0}^{\infty} \left| A_{n+\nu+2}^{\alpha-2} \right| \frac{1-\cos(\nu+1)t}{\left(2\sin\frac{t}{2}\right)^2} \geqslant 0. \tag{2.4}$$

证明 当 $0<\alpha<1, 0<t\leqslant\pi$ 时有

---

① $C_n(\lambda)_L \approx C_n(\lambda)_c$ 含义为 $\lim\limits_{n\to\infty}\frac{C_n(\lambda)_L}{C_n(\lambda)_c} = 1$. 又在 Cesàro 算子情形 $\omega(f;t)_X$ 取在 $t=\frac{2\pi}{n+1}$ 点. 此无关紧要.

$$K_n^\alpha(t) = \operatorname{Im}\left\{\frac{e^{i\left(n+\frac{1}{2}\right)t}}{2A_n^\alpha \sin\frac{t}{2}}\left[\frac{e^{i\left(\frac{\alpha t}{2}-\frac{\pi\alpha}{2}\right)}}{\left(2\sin\frac{t}{2}\right)^\alpha} - \sum_{\nu=n+1}^{\infty} A_\nu^{\alpha-1} e^{-i\nu t}\right]\right\}$$

$$= \Omega_n(t) + \frac{1}{2A_n^\alpha \sin\frac{t}{2}} \sum_{\nu=n+1}^{\infty} A_\nu^{\alpha-1} \sin\left(\nu - n - \frac{1}{2}\right)t.$$

注意到

$$D_{\nu-n-1}(t) = \frac{\sin\left(\nu - n - \frac{1}{2}\right)t}{2\sin\frac{t}{2}} \quad (\nu \geqslant n+1),$$

有

$$W_n(t) = \frac{1}{A_n^\alpha}\sum_{\nu=0}^{\infty} A_{n+\nu+1}^{\alpha-1} D_\nu(t)$$

$$= \frac{1}{A_n^\alpha}\sum_{\nu=0}^{\infty} \Delta A_{n+\nu+1}^{\alpha-1}[D_0(t) + \cdots + D_\nu(t)]$$

$$= \frac{1}{A_n^\alpha}\sum_{\nu=0}^{\infty} \Delta A_{n+\nu+1}^{\alpha-1} \frac{1-\cos(\nu+1)t}{\left(2\sin\frac{t}{2}\right)^2}.$$

由 $A_\mu^{\alpha-1} - A_{\mu+1}^{\alpha-1} = -A_{\mu+1}^{\alpha-2} > 0$ 即得引理 1 之证. 顺便指出, 函数

$$\sum_{\nu=0}^{\infty} \left|A_{n+\nu+2}^{\alpha-2}\right| \{1-\cos(\nu+1)t\} \tag{2.5}$$

是一非负连续函数.

引理 2 (П. Л. Ульянов) 若 $f \in L_{2\pi}^p, 1 < p < \infty, 0 < \alpha < 1$, 则

$$\left\|\int_{\frac{2\pi}{n+1}}^{\pi} [f(x+t) + f(x-t) - 2f(x)]\Omega_n(t)\mathrm{d}t\right\|_X$$

$$\leqslant B(\alpha)\cdot\omega\left(f; \frac{2\pi}{n+1}\right)_X \tag{2.6}$$

此处 $B(\alpha) > 0$ 是一仅与 α 有关的常数.

引理 2 的证明见[1].

引理 3 若 $F_\nu^*(t) = \int_t^\pi \frac{1-\cos(\nu+1)u}{\left(2\sin\frac{u}{2}\right)^2}\mathrm{d}u, j = \left[\frac{n+1}{2}\right]$,

则

$$\sum_{k=1}^{j} F_\nu^*\left(\frac{2k\pi}{n+1}\right) = \begin{cases} \frac{n+1}{2\pi}\ln(n+1) + O(n), \nu \geqslant n \text{ 时}, \\ \frac{n+1}{2\pi}\ln(\nu+1) + O(n), 0 < \nu < n \text{ 时}. \end{cases}$$

此处余项①$|O(n)| \leqslant C_1 n$, C_1 与 n, ν 无关.

证明

$$F_\nu^*(t) = 2\int_t^\pi \left(\frac{\sin\frac{\nu+1}{2}u}{2\sin\frac{u}{2}}\right)^2 du$$

$$= 2\int_t^\pi \left(\frac{\sin\frac{\nu+1}{2}u}{u}\right)^2 du + 2\int_t^\pi \left\{\frac{1}{\left(2\sin\frac{u}{2}\right)^2} - \frac{1}{u^2}\right\}\sin^2\frac{\nu+1}{2}u\,du$$

$$= 2\int_t^\pi \left(\frac{\sin\frac{\nu+1}{2}u}{u}\right)^2 du + O(1),$$

$|O(1)| \leqslant C_2$, C_2 是一绝对常数. 由此得

$$F_\nu^*(t) = (\nu+1)\int_{\frac{\nu+1}{2}t}^{\frac{\nu+1}{2}\pi} \left(\frac{\sin u}{u}\right)^2 du + O(1)$$

$$= (\nu+1)\int_{\frac{\nu+1}{2}t}^{\infty} \left(\frac{\sin u}{u}\right)^2 du - (\nu+1)\int_{\frac{\nu+1}{2}\pi}^{\infty} \left(\frac{\sin u}{u}\right)^2 du + O(1)$$

$$= (\nu+1)\int_{\frac{\nu+1}{2}t}^{\infty} \left(\frac{\sin u}{u}\right)^2 du + O(1).$$

从而有

$$\sum_{k=1}^{j} F_\nu^*(t_k) = (\nu+1)\sum_{k=1}^{j}\int_{\frac{\nu+1}{2}t_k}^{\infty} \left(\frac{\sin u}{u}\right)^2 du + O(n). \quad (2.7)$$

此处 $t_k = \frac{2k\pi}{n+1}$. 令

$$S = (\nu+1)\sum_{k=1}^{j}\int_{\frac{\nu+1}{2}t_k}^{\infty} \left(\frac{\sin u}{u}\right)^2 du. \quad (2.8)$$

---

① 当 ν/n 很小时, 第二个估计式右边的第一项的阶也是 $O(n)$, 这时它不是主项. 但采取这样的表示式是无妨的, 且便于后面的估计.

由

$$\int_t^{\infty} \frac{\sin^2 u}{u^2} \mathrm{d} u+\int_t^{\infty} \frac{\cos^2 u}{u^2} \mathrm{d} u=1 / t, \tag{2.9}$$

以及

$$\begin{aligned}\int_t^{\infty} \frac{\cos^2 u}{u^2} \mathrm{d} u & =\int_t^{t+\frac{\pi}{2}} \frac{\cos^2 u}{u^2} \mathrm{d} u+\int_{t+\frac{\pi}{2}}^{\infty} \frac{\cos^2 u}{u^2} \mathrm{d} u \\ & =\int_t^{t+\frac{\pi}{2}} \frac{1-\sin^2 u}{u^2} \mathrm{d} u+\int_t^{\infty} \frac{\sin^2 u}{\left(u+\frac{\pi}{2}\right)^2} \mathrm{d} u .\end{aligned} \tag{2.10}$$

得到

$$\begin{aligned}& \int_t^{\infty} \frac{\cos^2 u}{u^2} \mathrm{d} u-\int_t^{\infty} \frac{\sin^2 u}{u^2} \mathrm{d} u \\ & =\int_t^{t+\frac{\pi}{2}} \frac{1-\sin^2 u}{u^2} \mathrm{d} u+\int_t^{\infty}\left\{\frac{1}{\left(u+\frac{\pi}{2}\right)^2}-\frac{1}{u^2}\right\} \sin^2 u \mathrm{d} u .\end{aligned} \tag{2.11}$$

由(2.9),(2.11) 得

$$\begin{aligned}\int_t^{\infty} \frac{\sin^2 u}{u^2} \mathrm{d} u= & \frac{1}{2 t}-\frac{\pi}{4 t\left(t+\frac{\pi}{2}\right)}+\frac{1}{2} \int_t^{t+\frac{\pi}{2}} \frac{\sin^2 u}{u^2} \mathrm{d} u+ \\ & \frac{1}{2} \int_t^{\infty} \frac{\left(2 u+\frac{\pi}{2}\right) \frac{\pi}{2}}{u^2\left(u+\frac{\pi}{2}\right)^2} \sin^2 u \mathrm{d} u .\end{aligned} \tag{2.12}$$

记 $S=S_1+S_2+S_3+S_4$, 此处

$$S_1=\frac{\nu+1}{2} \sum_{k=1}^j \frac{1}{\frac{\nu+1}{2} t_k}=\sum_{k=1}^j \frac{1}{t_k}, \tag{2.13}$$

$$S_2=-\frac{\pi(\nu+1)}{4} \sum_{k=1}^j \frac{1}{\frac{\nu+1}{2} t_k\left(\frac{\nu+1}{2} t_k+\frac{\pi}{2}\right)}, \tag{2.14}$$

$$S_3=\frac{\nu+1}{2} \sum_{k=1}^j \int_{\frac{\nu+1}{2} t_k}^{\infty} \frac{\left(2 u+\frac{\pi}{2}\right) \frac{\pi}{2}}{u^2\left(u+\frac{\pi}{2}\right)^2} \sin^2 u \mathrm{d} u, \tag{2.15}$$

$$S_4=\frac{\nu+1}{2} \sum_{k=1}^j \int_{\frac{\nu+1}{2} t_k}^{\frac{\nu+1}{2} t_k+\frac{\pi}{2}} \frac{\sin^2 u}{u^2} \mathrm{d} u . \tag{2.16}$$

先估计 S_3. 注意有

$$\int_0^{\infty}\frac{2u+\frac{\pi}{2}}{u^2\left(u+\frac{\pi}{2}\right)^2}\sin^2 u\,du \leqslant 2\int_0^{\infty}\frac{\sin^2 u}{u^2\left(u+\frac{\pi}{2}\right)}du$$

$$\leqslant \frac{2}{\frac{\pi}{2}}\int_0^{\infty}\left(\frac{\sin u}{u}\right)^2 du = 2,$$

以及

$$\int_{\frac{\nu+1}{2}t_k}^{\infty}\frac{2u+\frac{\pi}{2}}{u^2\left(u+\frac{\pi}{2}\right)^2}\sin^2 u\,du \leqslant 2\int_{\frac{\nu+1}{2}t_k}^{\infty}\frac{du}{u^3} = \frac{1}{\left(\frac{\nu+1}{2}t_k\right)^2}$$

$$= \left(\frac{n+1}{\nu+1}\right)^2\frac{1}{k^2\pi^2}.$$

从而有

$$\int_{\frac{\nu+1}{2}t_k}^{\infty}\frac{2u+\frac{\pi}{2}}{u^2\left(u+\frac{\pi}{2}\right)^2}\sin^2 u\,du \leqslant \min\left\{2,\left(\frac{n+1}{\nu+1}\right)^2\frac{1}{k^2\pi^2}\right\}.$$

故得

$$S_3 \leqslant \frac{\pi}{4}(\nu+1)\sum_{k=1}^{j}\min\left\{2,\left(\frac{n+1}{\nu+1}\right)^2\frac{1}{k^2\pi^2}\right\}. \tag{2.17}$$

当 $\nu \geqslant n$ 时,易见有 $\min\left\{2,\left(\frac{n+1}{\nu+1}\right)^2\frac{1}{k^2\pi^2}\right\} = \left(\frac{n+1}{\nu+1}\right)^2\frac{1}{k^2\pi^2}$,故有

$$S_3 \leqslant \frac{\pi}{4}(\nu+1)\left(\frac{n+1}{\nu+1}\right)^2\sum_{k=1}^{j}\frac{1}{k^2\pi^2} = \left(\frac{n+1}{\nu+1}\right)(n+1)\cdot O(1) = O(n), \tag{2.18}$$

当 $0<\nu<n$ 时,置 $\nu=\theta n, 0<\theta<1$,此时

$$S_3 \leqslant \frac{\pi}{4}(\theta n+1)\left\{\sum_{2<\left(\frac{n+1}{\theta n+1}\right)^2\frac{1}{k^2\pi^2}} + \sum_{2\geqslant\left(\frac{n+1}{\theta n+1}\right)^2\frac{1}{k^2\pi^2}}\right\}.$$

由于 $2<\left(\frac{n+1}{\theta n+1}\right)^2\frac{1}{k^2\pi^2} \Leftrightarrow k<\frac{1}{\sqrt{2}\pi}\frac{n+1}{\theta n+1}$,而 $\frac{n+1}{\theta n+1}<\frac{1}{\theta}$,故得

$$S_3 \leqslant \frac{\pi}{4}(\theta n+1)\left\{\sum_{k<\frac{1}{\sqrt{2}\pi}\cdot\frac{1}{\theta}}2 + \sum_{k\geqslant\frac{1}{\sqrt{2}\pi}\cdot\frac{1}{\theta}}\left(\frac{n+1}{\theta n+1}\right)^2\frac{1}{k^2\pi^2}\right\}.$$

$$\sum_{k<\frac{1}{\sqrt{2}\pi}\cdot\frac{1}{\theta}}2 = 2\left[\frac{1}{\sqrt{2}\pi\theta}\right] \leqslant \frac{\sqrt{2}}{\pi\theta}.$$

$$\sum_{k\geqslant\frac{1}{\sqrt{2}\pi}\cdot\frac{1}{\theta}}\left(\frac{n+1}{\theta n+1}\right)^2\frac{1}{k^2\pi^2}\leqslant\frac{1}{\pi^2\theta^2}\sum_{k\geqslant\frac{1}{\sqrt{2}\pi\theta}}\frac{1}{k^2}\leqslant\frac{1}{\pi^2\theta^2\cdot\left[\frac{1}{\sqrt{2}\pi\theta}\right]}.$$

从而有

$$S_3\leqslant\frac{\pi}{4}\left\{2\theta n\cdot\frac{\sqrt{2}}{\pi\theta}+\frac{2n}{\pi^2\theta}\cdot\frac{1}{\left[\frac{1}{\sqrt{2}\pi\theta}\right]}\right\}=O(n).\tag{2.19}$$

S_4 可用同法估计. 事实上,

$$\int_{\frac{\nu+1}{2}t_k}^{\frac{\nu+1}{2}t_k+\frac{\pi}{2}}\frac{\sin^2u}{u^2}\mathrm{d}u\leqslant\int_{\frac{\nu+1}{2}t_k}^{\frac{\nu+1}{2}t_k+\frac{\pi}{2}}u^{-2}\mathrm{d}u<\frac{\pi}{2}\frac{1}{\left(\frac{\nu+1}{2}t_k\right)^2}$$

$$=\frac{\pi}{2}\left(\frac{n+1}{\nu+1}\right)^2\frac{1}{k^2\pi^2},$$

又有

$$\int_{\frac{\nu+1}{2}t_k}^{\frac{\nu+1}{2}t_k+\frac{\pi}{2}}\frac{\sin^2u}{u^2}\mathrm{d}u<\int_0^{\infty}\frac{\sin^2u}{u^2}\mathrm{d}u=\frac{\pi}{2},$$

从而得

$$\int_{\frac{\nu+1}{2}t_k}^{\frac{\nu+1}{2}t_k+\frac{\pi}{2}}\frac{\sin^2u}{u^2}\mathrm{d}u\leqslant\min\left\{\frac{\pi}{2},\frac{\pi}{2}\left(\frac{n+1}{\nu+1}\right)^2\frac{1}{k^2\pi^2}\right\},$$

$$S_4\leqslant\frac{\nu+1}{2}\sum_{k=1}^{j}\min\left\{\frac{\pi}{2},\frac{\pi}{2}\left(\frac{n+1}{\nu+1}\right)^2\frac{1}{k^2\pi^2}\right\}.\tag{2.20}$$

故用与估计 S_3 一样的方法可得

$$S_4=O(n).\tag{2.21}$$

至于 S_2, 从

$$\frac{1}{t_k\left(\frac{\nu+1}{2}t_k+\frac{\pi}{2}\right)}=\frac{1}{2\pi^2}\frac{n+1}{k\left[\left(\frac{\nu+1}{n+1}\right)k+\frac{1}{2}\right]}$$

得

$$-S_2=(\nu+1)\sum_{k=1}^{j}\frac{\frac{\pi}{4}}{\frac{\nu+1}{2}t_k\left(\frac{\nu+1}{2}t_k+\frac{\pi}{2}\right)}$$

$$=\frac{n+1}{4\pi}\sum_{k=1}^{j}\frac{1}{k\left[\left(\frac{\nu+1}{n+1}\right)k+\frac{1}{2}\right]}.$$

当 $\nu\geqslant n$ 时, 由于

$$\sum_{k=1}^{j} \frac{1}{k\left[\left(\frac{\nu+1}{n+1}\right)k+\frac{1}{2}\right]} \leqslant \sum_{k=1}^{j} \frac{1}{k\left(k+\frac{1}{2}\right)} \leqslant C_3,$$

C_3 为一绝对常数, 知道有 $S_2 = O(n)$. 当 $0 < \nu < n$ 时, 令 $\theta = \frac{\nu+1}{n+1}$, $0 < \theta < 1$, 则有

$$\begin{aligned}
&\frac{n+1}{4\pi}\sum_{k=1}^{j}\frac{1}{k(\theta k+1/2)} \\
&= \frac{n+1}{2\pi}\left[\sum_{k=1}^{j}\frac{1}{k} - \sum_{k=1}^{j}\frac{1}{k+\frac{1}{2\theta}}\right] \\
&= \frac{n+1}{2\pi}\left[\ln j + \ln\left(1+\frac{1}{2\theta}\right) - \ln\left(j+\frac{1}{2\theta}\right) + O(1)\right] \\
&= \frac{n+1}{2\pi}\left[\ln\left(1+\frac{1}{2\theta}\right) - \ln\left(1+\frac{1}{2\theta j}\right) + O(1)\right] \\
&= \frac{n+1}{2\pi}\left[\ln\left(1+\frac{1}{2\theta}\right) + O(1)\right] \\
&= \frac{n+1}{2\pi}\left[\ln\frac{1}{\theta} + O(1)\right] \\
&= \frac{n+1}{2\pi}\left[\ln\frac{n+1}{\nu+1} + O(1)\right].
\end{aligned}$$

结果是

$$S_2 = \begin{cases} O(n), & \text{当 } \nu \geqslant n \text{ 时}, \\ -\frac{n+1}{2\pi}\left[\ln\frac{n+1}{\nu+1} + O(1)\right], & 0 < \nu < n \text{ 时}. \end{cases} \tag{2.22}$$

最后

$$S_1 = \sum_{k=1}^{j}\frac{1}{t_k} = \frac{n+1}{2\pi}\sum_{k=1}^{j}\frac{1}{k} = \frac{n+1}{2\pi}[\ln(n+1)+O(1)]. \tag{2.23}$$

综合(2.19),(2.21),(2.22),(2.23) 得: 当 $\nu \geqslant n$ 时,

$$S = \frac{n+1}{2\pi}\ln(n+1) + O(n); \tag{2.24}$$

而当 $0 < \nu < n$ 时, 有

$$\begin{aligned}
S &= \frac{n+1}{2\pi}\ln(n+1) - \frac{n+1}{2\pi}\ln\frac{n+1}{\nu+1} + O(n) \\
&= \frac{n+1}{2\pi}\ln(\nu+1) + O(n).
\end{aligned} \tag{2.25}$$

引理 3 得证.

后面需要函数 $\int_t^\pi K_n^\alpha(u)\mathrm{d}u = \Psi(t)$. 当 $0 < \alpha < 1$ 时, Zygmund(见[5], 第 184 页的(11.5)) 指出:

$$\Psi(t) = \frac{\alpha}{n+1}\cdot\frac{1}{2}\cot\frac{t}{2} + \frac{2\theta_1}{nA_n^\alpha\left(2\sin\frac{t}{2}\right)^{\alpha+1}} + \frac{B}{n^2t^2}, \quad (2.26)$$

此处 $|\theta_1|\leqslant 1$, $|B|\leqslant C_4$, C_4 为一仅与 α 有关的常数, 而 $0 < t \leqslant \pi$. 下面引理给出 $\Psi(t)$ 在 $\alpha > 1$ 时的估计式.

引理 4 当 $\alpha > 1$ 时, 若令 $r = [\alpha]$, $0 < t \leqslant \pi$, 则有

$$\left|\Psi(t) - \frac{\alpha}{\alpha+n-1}\frac{1}{2}\cot\frac{t}{2}\right| \leqslant \frac{A_1}{(nt)^{\alpha+1}} + \sum_{j=2}^{r+1}\frac{B_{j-1}}{(nt)_j}.$$

此处 $A_1, B_1, \cdots, B_r$ 是仅与 α 有关的常数.

证明 由文献[7] 知, 当 $[\alpha] = r = 1$ 时, 有

$$K_n^\alpha(t) = \Omega_n(t) + \frac{\alpha}{\alpha+n-1}\frac{1}{\left(2\sin\frac{t}{2}\right)^2} - \frac{1}{A_n^\alpha}\mathrm{Im}\left\{\frac{\sum\limits_{\nu=n+1}^{\infty}A_\nu^{\alpha-2}\mathrm{e}^{-i\left(\nu-n-\frac{1}{2}\right)t}}{(1-\mathrm{e}^{-it})2\sin\frac{t}{2}}\right\}.$$

$[\alpha] = r = 2$ 时, 有

$$K_n^\alpha(t) = \Omega_n(t) + \frac{\alpha}{\alpha+n-1}\frac{1}{\left(2\sin\frac{t}{2}\right)^2} + \frac{\alpha(\alpha-1)}{8(\alpha+n)(\alpha+n-1)}\frac{1}{\left(\sin\frac{t}{2}\right)^2} - \frac{1}{A_n^\alpha}\mathrm{Im}\left\{\frac{\sum\limits_{\nu=n+1}^{\infty}A_\nu^{\alpha-3}\mathrm{e}^{-i\left(\nu-n-\frac{1}{2}\right)t}}{(1-\mathrm{e}^{-it})^2\cdot 2\sin\frac{t}{2}}\right\}.$$

$[\alpha] = r \geqslant 3$ 时, 有

$$K_n^\alpha(t) = \Omega_n(t) + \frac{\alpha}{\alpha+n-1}\frac{1}{\left(2\sin\frac{t}{2}\right)^2} + \frac{\alpha(\alpha-1)}{8(\alpha+n)(\alpha+n-1)}\frac{1}{\left(\sin\frac{t}{2}\right)^2} +$$

$$\frac{1}{A_n^\alpha}\operatorname{Im}\left\{\frac{-\mathrm{e}^{-i/2}}{2\sin t/2}\sum_{j=3}^{r}\frac{A_n^{\alpha-j}}{(1-\mathrm{e}^{-it})^j}-\right.$$
$$\left.\sum_{\nu=n+1}^{\infty}A_\nu^{\alpha-r-1}\mathrm{e}^{-i\left(\nu-n-\frac{1}{2}\right)t}/(1-\mathrm{e}^{-it})^r\left(2\sin\frac{t}{2}\right)\right\}.$$

注意到

$$\frac{\alpha}{\alpha+n-1}\int_t^\pi\frac{\mathrm{d}u}{\left(2\sin\frac{u}{2}\right)^2}=\frac{\alpha}{\alpha+n-1}\frac{1}{2}\cot\frac{t}{2},\tag{2.27}$$

$$\left|\frac{1}{A_n^\alpha}\int_t^\pi\frac{\sin\left[\left(n+\frac{1}{2}+\frac{\alpha}{2}\right)u-\frac{\pi\alpha}{2}\right]}{\left(2\sin\frac{u}{2}\right)^{\alpha+1}}\mathrm{d}u\right|$$
$$=\left|\frac{1}{A_n^\alpha\left(2\sin\frac{t}{2}\right)^{\alpha+1}}\int_t^\xi\sin\left[\left(n+\frac{1}{2}+\frac{\alpha}{2}\right)u-\frac{\pi\alpha}{2}\right]\mathrm{d}u\right|$$
$$=\frac{2\mid\theta_1\mid}{\left(n+\frac{1}{2}+\frac{\alpha}{2}\right)A_n^\alpha\left(2\sin\frac{t}{2}\right)^{\alpha+1}}\leqslant\frac{A_1}{(nt)^{\alpha+1}}.\tag{2.28}$$

此处：

$$t<\xi<\pi,\ \mid\theta_1\mid\leqslant 1,0<t\leqslant\pi.$$

$$\left|\frac{\alpha(\alpha-1)}{8(\alpha+n)(\alpha+n-1)}\int_t^\pi\frac{\mathrm{d}u}{\left(\sin\frac{u}{2}\right)^2}\right|\leqslant\frac{B_1}{(nt)^2}.\tag{2.29}$$

再由

$$\frac{1}{A_n^\alpha}\left|\operatorname{Im}\left\{\frac{\sum_{\nu=n+1}^{\infty}A_\nu^{\alpha-r-1}\mathrm{e}^{-i\left(\nu-n-\frac{1}{2}\right)t}}{(1-\mathrm{e}^{-it})^r\left(2\sin\frac{t}{2}\right)}\right\}\right|\leqslant\frac{A_{n+1}^{\alpha-r-1}}{2^{r+1}A_n^\alpha\left(\sin\frac{t}{2}\right)^{r+2}},$$

以及

$$\frac{1}{A_n^\alpha}\left|\operatorname{Im}\left\{\frac{-\mathrm{e}^{-i/2}}{2\sin\frac{t}{2}}\sum_{j=3}^{r}\frac{A_n^{\alpha-j}}{(1-\mathrm{e}^{-it})^j}\right\}\right|\leqslant\sum_{j=3}^{r}\frac{A_n^{\alpha-j}}{A_n^\alpha\left(2\sin\frac{t}{2}\right)^{j+1}}$$

得到

$$\left|\frac{1}{A_n^\alpha}\int_t^\pi\operatorname{Im}\left\{\frac{\sum_{\nu=n+1}^{\infty}A_\nu^{\alpha-r-1}\mathrm{e}^{-i\left(\nu-n-\frac{1}{2}\right)u}}{(1-\mathrm{e}^{-iu})^r\left(2\sin\frac{u}{2}\right)}\right\}\mathrm{d}u\right|$$

$$\leqslant \frac{A_{n+1}^{\alpha-r-1}}{2^{r+1}A_n^\alpha}\int_t^\pi \frac{\mathrm{d}u}{\left(\sin\frac{u}{2}\right)^{r+2}} \leqslant \frac{B_r}{(nt)^{r+1}}. \tag{2.30}$$

$$\left|\frac{1}{A_n^\alpha}\int_t^\pi \mathrm{Im}\left\{\frac{-\mathrm{e}^{-i/2}}{2\sin\frac{u}{2}}\sum_{j=3}^{r}\frac{A_n^{\alpha-j}}{(1-\mathrm{e}^{-iu})^j}\right\}\mathrm{d}u\right|$$

$$\leqslant \frac{1}{A_n^\alpha}\sum_{j=3}^{r}\int_t^\pi \frac{A_n^{\alpha-j}\mathrm{d}u}{\left(2\sin\frac{u}{2}\right)^{j+1}} \leqslant \sum_{j=3}^{r}\frac{B_{j-1}}{(nt)^j}. \tag{2.31}$$

利用(2.27) ~ (2.31) 诸式即得引理 4 之证.

引理 5 若 $f(x)\in X(C,L), 0<\alpha<1$, 则

$$\| f(x)-\sigma_n^\alpha(f;x)\|_X$$

$$\leqslant \left\{\frac{2}{\pi}\int_{\frac{2\pi}{n+1}}^{\pi}\left[\frac{n+1}{2\pi}t\right]W_n(t)\mathrm{d}t\right\}\omega\left(f;\frac{2\pi}{n+1}\right)_X +$$

$$O\left(\omega\left(f;\frac{2\pi}{n+1}\right)_X\right). \tag{2.32}$$

证明 记 $\Delta_t^2(f;x)=f(x+t)+f(x-t)-2f(x)$. 有

$$\begin{aligned}\sigma_n^\alpha(f;x)-f(x) &= \frac{1}{\pi}\int_0^\pi \Delta_t^2(f;x)K_n^\alpha(t)\mathrm{d}t\\ &= \frac{1}{\pi}\int_0^{\frac{2\pi}{n+1}}\Delta_t^2(f;x)K_n^\alpha(t)\mathrm{d}t+\frac{1}{\pi}\int_{\frac{2\pi}{n+1}}^{\pi}\Delta_t^2(f;x)\Omega_n(t)\mathrm{d}t+\\ &\quad \frac{1}{\pi}\int_{\frac{2\pi}{n+1}}^{\pi}\Delta_t^2(f;x)W_n(t)\mathrm{d}t\\ &= I_1+I_2+I_3.\end{aligned}$$

由于

$$\begin{aligned}\| I_1(x)\|_X &\leqslant \frac{1}{\pi}\int_0^{\frac{2\pi}{n+1}}\|\Delta_t^2(f;\cdot)\|_X\,|K_n^\alpha(t)|\,\mathrm{d}t\\ &\leqslant B(\alpha)\omega\left(f;\frac{2\pi}{n+1}\right)_X,\end{aligned}$$

$$\begin{aligned}\| I_2(x)\|_X &\leqslant \frac{1}{\pi}\left\|\int_{\frac{2\pi}{n+1}}^{\pi}\Delta_t^2(f;x)\Omega_n(t)\mathrm{d}t\right\|_X\\ &\leqslant B_1(\alpha)\omega\left(f;\frac{2\pi}{n+1}\right)_X,\end{aligned}$$

(见引理 2)

$$\| I_3(x)\|_X \leqslant \frac{1}{\pi}\int_{\frac{2\pi}{n+1}}^{\pi}\|\Delta_t^2(f;\cdot)\|_X W_n(t)\mathrm{d}t$$

$$\leqslant \frac{2}{\pi}\int_{\frac{2\pi}{n+1}}^{\pi}\omega(f;t)_X\cdot W_n(t)\mathrm{d}t,$$

由 $\omega(f;t)_X=\omega\left(f;\frac{2\pi}{n+1}\cdot\frac{n+1}{2\pi}t\right)\leqslant\left\{1+\left[\frac{n+1}{2\pi}t\right]\right\}\omega\left(f;\frac{2\pi}{n+1}\right)_X$

得到

$$\|I_3(x)\|_X\leqslant\frac{2\omega\left(f;\frac{2\pi}{n+1}\right)_X}{\pi}\int_{\frac{2\pi}{n+1}}^{\pi}\left\{1+\left[\frac{n+1}{2\pi}t\right]\right\}W_n(t)\mathrm{d}t.$$

由引理 1 得

$$\begin{aligned}\int_{\frac{2\pi}{n+1}}^{\pi}W_n(t)\mathrm{d}t&=\frac{1}{A_n^{\alpha}}\sum_{\nu=0}^{\infty}\Delta A_{n+\nu+1}^{\alpha-1}\int_{\frac{2\pi}{n+1}}^{\pi}\frac{1-\cos(\nu+1)t}{\left(2\sin\frac{t}{2}\right)^2}\mathrm{d}t\\&\leqslant\frac{1}{2A_n^{\alpha}}\sum_{\nu=0}^{\infty}\Delta A_{n+\nu+1}^{\alpha-1}\int_{\frac{2\pi}{n+1}}^{\pi}\frac{\mathrm{d}t}{\left(\sin\frac{t}{2}\right)^2}\\&\leqslant\frac{\pi}{4}(n+1)\cdot\frac{A_{n+1}^{\alpha-1}}{A_n^{\alpha}}\\&\leqslant C_5.\end{aligned}$$

合并之得

$$\begin{aligned}\|\sigma_n^{\alpha}(f;x)-f(x)\|_X&\leqslant\|I_1(x)\|_X+\|I_2(x)\|_X+\|I_3(x)\|_X\\&\leqslant\left\{\frac{2}{\pi}\int_{\frac{2\pi}{n+1}}^{\pi}\left[\frac{n+1}{2\pi}t\right]W_n(t)\mathrm{d}t\right\}\omega\left(f;\frac{2\pi}{n+1}\right)_X+\\&\qquad O\left(\omega\left(f;\frac{2\pi}{n+1}\right)_X\right).\end{aligned}$$

引理 5 获证.

§3. 定理 1 的证明

$X=C_{2\pi}$ 的情形

(1) $0<\alpha<1$.

由引理 5, 对任取的 $f(x)\in C_{2\pi}$ 有

$$\|\sigma_n^{\alpha}(f;x)-f(x)\|_c\leqslant\frac{2\omega\left(f;\frac{2\pi}{n+1}\right)_c}{\pi}\int_{\frac{2\pi}{n+1}}^{\pi}\left[\frac{n+1}{2\pi}t\right]W_n(t)\mathrm{d}t+ O\left(\omega\left(f;\frac{2\pi}{n+1}\right)_c\right).\tag{3.1}$$

仿照文献[8]，定义一个 2π 周期的连续偶函数 $f_{\varepsilon,n}(x)$ 如下：取充分小的 $\varepsilon(0<\varepsilon<1)$，令

$$f_{\varepsilon,n}(x)=\begin{cases} k+\varepsilon^{-1}\left(x-\dfrac{2k\pi}{n+1}\right), & \dfrac{2k\pi}{n+1}\leqslant x\leqslant\dfrac{2k\pi}{n+1}+\varepsilon, \\ k+1, & \dfrac{2k\pi}{n+1}+\varepsilon\leqslant x\leqslant\dfrac{2(k+1)\pi}{n+1}, \\ & \left(k=0,\cdots,\left[\dfrac{n}{2}\right]-1\right) \\ \left[\dfrac{n}{2}\right]+\varepsilon^{-1}\left(x-\dfrac{2\left[\frac{n}{2}\right]\pi}{n+1}\right), & \dfrac{2\left[\frac{n}{2}\right]\pi}{n+1}\leqslant x\leqslant\dfrac{2\left[\frac{n}{2}\right]\pi}{n+1}+\varepsilon, \\ \left[\dfrac{n}{2}\right]+1, & \dfrac{2\left[\frac{n}{2}\right]\pi}{n+1}+\varepsilon\leqslant x\leqslant\pi. \end{cases}$$

易见 $\omega\left(f_{\varepsilon,n};\dfrac{2\pi}{n+1}\right)_c=1,f_{\varepsilon,n}(0)=0$，从而有

$$\begin{aligned}\sigma_n^\alpha(f_{\varepsilon,n};0)-f_{\varepsilon,n}(0)&=\sigma_n^\alpha(f_{\varepsilon,n};0)\\&=\frac{2}{\pi}\int_0^{\frac{2\pi}{n+1}}f_{\varepsilon,n}(t)K_n^\alpha(t)\mathrm{d}t+\frac{2}{\pi}\int_{\frac{2\pi}{n+1}}^{\pi}f_{\varepsilon,n}(t)\Omega_n(t)\mathrm{d}t+\\&\quad\frac{2}{\pi}\int_{\frac{2\pi}{n+1}}^{\pi}f_{\varepsilon,n}(t)W_n(t)\mathrm{d}t\\&=J_1(\varepsilon)+J_2(\varepsilon)+J_3(\varepsilon).\end{aligned}$$

$$|J_1(\varepsilon)|\leqslant\frac{2}{\pi}\int_0^{\frac{2\pi}{n+1}}|K_n^\alpha(t)|\mathrm{d}t\leqslant 4;$$

$|J_2(\varepsilon)|\leqslant B(\alpha)$，$B(\alpha)$ 只与 α 有关. 此事之证明可仿照文献[1]的引理1. 所以有

$$\begin{aligned}\|\sigma_n^\alpha(f_{\varepsilon,n};x)-f_{\varepsilon,n}(x)\|_c&\geqslant|\sigma_n^\alpha(f_{\varepsilon,n};0)|\\&\geqslant J_3(\varepsilon)-(4+B(\alpha)).\end{aligned}$$

从而有

$$\begin{aligned}\sup_{\substack{f\in C_{2\pi}\\ f\neq\mathrm{const}}}\frac{\|\sigma_n^\alpha(f;x)-f(x)\|_c}{\ln(n+1)\omega\left(f;\dfrac{2\pi}{n+1}\right)_c}&\geqslant\frac{|\sigma_n^\alpha(f_{\varepsilon,n};0)|}{\ln(n+1)}\\&\geqslant\frac{J_3(\varepsilon)}{\ln(n+1)}-\frac{4+B(\alpha)}{\ln(n+1)}.\end{aligned}$$

当 $\varepsilon\to 0+$ 时，有

$$\lim_{\varepsilon\to 0+} J_3(\varepsilon) = \frac{2}{\pi}\int_{\frac{2\pi}{n+1}}^{\pi}\left\{1+\left[\frac{n+1}{2\pi}t\right]\right\}W_n(t)\mathrm{d}t.$$

从而得

$$A_n(\alpha)_c \geqslant \frac{2}{\pi\ln(n+1)}\int_{\frac{2\pi}{n+1}}^{\pi}\left[\frac{n+1}{2\pi}t\right]W_n(t)\mathrm{d}t + O\left(\frac{1}{\ln n}\right).$$

此式与(3.1) 比较即得

$$\lim_{n\to\infty} A_n(\alpha)_c = \lim_{n\to\infty}\frac{2}{\pi\ln(n+1)}\int_{\frac{2\pi}{n+1}}^{\pi}\left[\frac{n+1}{2\pi}t\right]W_n(t)\mathrm{d}t. \tag{3.2}$$

下面计算(3.2).

$$\int_{\frac{2\pi}{n+1}}^{\pi}\left[\frac{n+1}{2\pi}t\right]W_n(t)\mathrm{d}t$$
$$= \frac{1}{A_n^{\alpha}}\sum_{\nu=0}^{\infty}\Delta A_{n+\nu+1}^{\alpha-1}\int_{\frac{2\pi}{n+1}}^{\pi}\left[\frac{n+1}{2\pi}t\right]\frac{1-\cos(\nu+1)t}{\left(2\sin\dfrac{t}{2}\right)^2}\mathrm{d}t.$$

但因

$$\int_{\frac{2\pi}{n+1}}^{\pi}\left[\frac{n+1}{2\pi}t\right]\frac{1-\cos(\nu+1)t}{\left(2\sin\dfrac{t}{2}\right)^2}\mathrm{d}t$$
$$= -\int_{\frac{2\pi}{n+1}}^{\pi}\left[\frac{n+1}{2\pi}t\right]\mathrm{d}F_{\nu}^{*}(t)$$
$$= -\left[\frac{n+1}{2\pi}t\right]F_{\nu}^{*}(t)\Bigg|_{\frac{2\pi}{n+1}}^{\pi} + \int_{\frac{2\pi}{n+1}}^{\pi}F_{\nu}^{*}(t)\mathrm{d}\left[\frac{n+1}{2\pi}t\right]$$
$$= \sum_{k=1}^{j}F_{\nu}^{*}(t_k). \tag{3.3}$$

故有

$$\int_{\frac{2\pi}{n+1}}^{\pi}\left[\frac{n+1}{2\pi}t\right]W_n(t)\mathrm{d}t = \frac{1}{A_n^{\alpha}}\sum_{\nu=0}^{\infty}\Delta A_{n+\nu+1}^{\alpha-1}\sum_{k=1}^{j}F_{\nu}^{*}(t_k).$$

由引理 3 得

$$\int_{\frac{2\pi}{n+1}}^{\pi}\left[\frac{n+1}{2\pi}t\right]W_n(t)\mathrm{d}t$$
$$= \frac{n+1}{2\pi A_n^{\alpha}}\sum_{\nu=0}^{n-1}\Delta A_{n+\nu+1}^{\alpha-1}\ln(\nu+1) +$$
$$\frac{(n+1)\ln(n+1)}{2\pi\cdot A_n^{\alpha}}\sum_{\nu=n}^{\infty}\Delta A_{n+\nu+1}^{\alpha-1} + O(1)$$

$$= \frac{n+1}{2\pi A_n^{\alpha}} \sum_{\nu=0}^{n-1} \Delta A_{n+\nu+1}^{\alpha-1}\left(1+\cdots+\frac{1}{\nu+1}\right)+$$

$$\frac{(n+1)\ln(n+1)}{2\pi} \frac{A_{2n+1}^{\alpha-1}}{A_n^{\alpha}}+O(1)$$

$$= \frac{n+1}{2\pi A_n^{\alpha}} \sum_{k=1}^{n} A_{n+k}^{\alpha-1} \frac{1}{k}+O(1). \tag{3.4}$$

由于

$$\frac{n+1}{A_n^{\alpha}} \sum_{k=1}^{n} A_{n+k}^{\alpha-1} \frac{1}{k} \approx (n+1) \frac{\Gamma(\alpha+1)}{n^{\alpha}} \sum_{k=1}^{n} \frac{(n+k)^{\alpha-1}}{\Gamma(\alpha)} \cdot \frac{1}{k}$$

$$= \frac{n+1}{n} \cdot \alpha \cdot \sum_{k=1}^{n}\left(1+\frac{k}{n}\right)^{\alpha-1} \frac{1}{k},$$

并且

$$\sum_{k=1}^{n}\left(1+\frac{k}{n}\right)^{\alpha-1} \frac{1}{k}=\sum_{k=1}^{\left[\frac{n}{2}\right]}\left(1+\frac{k}{n}\right)^{\alpha-1} \frac{1}{k}+\sum_{k=\left[\frac{n}{2}\right]+1}^{n}\left(1+\frac{k}{n}\right)^{\alpha-1} \frac{1}{k},$$

而

$$\sum_{k=\left[\frac{n}{2}\right]+1}^{n}\left(1+\frac{k}{n}\right)^{\alpha-1} \frac{1}{k} \leqslant c_6 \sum_{k=\left[\frac{n}{2}\right]+1}^{n} \frac{1}{k}=O(1),$$

$$\sum_{k=1}^{\left[\frac{n}{2}\right]}\left(1+\frac{k}{n}\right)^{\alpha-1} \frac{1}{k} \approx \sum_{k=1}^{\left[\frac{n}{2}\right]}\left\{1-(1-\alpha) \frac{k}{n}\right\} \frac{1}{k}$$

$$= \sum_{k=1}^{\left[\frac{n}{2}\right]} \frac{1}{k}+O(1)=\ln(n+1)+O(1),$$

故有

$$\frac{n+1}{A_n^{\alpha}} \sum_{k=1}^{n} A_{n+k}^{\alpha-1} \frac{1}{k}=\alpha \ln(n+1)+O(1), \tag{3.5}$$

$$\int_{\frac{2\pi}{n+1}}^{\pi}\left[\frac{n+1}{2\pi} t\right] W_n(t) \mathrm{d} t=\frac{\alpha}{2\pi} \ln(n+1)+O(1). \tag{3.6}$$

把(3.6) 代入(3.2) 即得

$\lim\limits_{n \to \infty} A_n(\alpha)_c = \dfrac{\alpha}{\pi^2}$. $0 < \alpha < 1$ 的情形证完.

(2)$\alpha = 1$ 时, 记

$$\sigma_n(f;x)-f(x)=\frac{1}{2(n+1)\pi} \int_0^{\pi} \Delta_t^2(f;x)\left(\frac{\sin \frac{n+1}{2} t}{\sin \frac{t}{2}}\right)^2 \mathrm{d} t, \tag{3.7}$$

从而有

$$\| \sigma_n(f;x) - f(x) \|_X \leqslant \left\{ \frac{1}{(n+1)\pi} \int_0^{\pi} \left[\frac{n+1}{2\pi} t \right] \left(\frac{\sin \frac{n+1}{2} t}{\sin \frac{t}{2}} \right)^2 \mathrm{d}t \right\}$$

$$\omega\left(f; \frac{2\pi}{n+1} \right)_X + O\left(\omega\left(f; \frac{2\pi}{n+1} \right)_X \right). \quad (3.8)$$

与(1) 一样方法可证

$$A_n(1)_c = \frac{1}{(n+1)\pi \ln(n+1)} \int_0^{\pi} \left[\frac{n+1}{2\pi} t \right] \left(\frac{\sin \frac{n+1}{2} t}{\sin \frac{t}{2}} \right)^2 \mathrm{d}t + O\left(\frac{1}{\ln n} \right)$$

$$= \frac{2}{(n+1)\pi \ln(n+1)} \sum_{k=1}^{j} F_n^*(t_k) + O\left(\frac{1}{\ln n} \right)$$

$$= \frac{1}{\pi^2} + O\left(\frac{1}{\ln n} \right). \quad (3.9)$$

关于这个常数可参看[3],[4].

(3)$\alpha > 1$ 时,记

$$K_n^{\alpha}(t) = \frac{1}{A_n^{\alpha}} \sum_{k=0}^{n} A_{n-k}^{\alpha-2} \frac{\sin^2 \frac{k+1}{2} t}{2\sin^2 \frac{t}{2}} \geqslant 0.$$

用(1) 的方法即得

$$\sup_{\substack{f \in C_{2\pi} \\ f \neq \text{const}}} \frac{\| \sigma_n^{\alpha}(f;x) - f(x) \|_c}{\omega\left(f; \frac{2\pi}{n+1} \right)_c} = \frac{2}{\pi} \int_0^{\pi} \left\{ 1 + \left[\frac{n+1}{2\pi} t \right] \right\} K_n^{\alpha}(t) \mathrm{d}t$$

$$= 1 + \frac{2}{\pi} \int_0^{\pi} \left[\frac{n+1}{2\pi} t \right] K_n^{\alpha}(t) \mathrm{d}t. \quad (3.10)$$

$$\frac{2}{\pi} \int_0^{\pi} \left[\frac{n+1}{2\pi} t \right] K_n^{\alpha}(t) \mathrm{d}t = \frac{2}{\pi A_n^{\alpha}} \sum_{k=0}^{n} A_{n-k}^{\alpha-2} \int_0^{\pi} \left[\frac{n+1}{2\pi} t \right] \frac{\sin^2 \frac{k+1}{2} t}{2\sin^2 \frac{t}{2}} \mathrm{d}t$$

$$= \frac{2}{\pi A_n^{\alpha}} \sum_{k=0}^{n} A_{n-k}^{\alpha-2} \sum_{\nu=1}^{j} F_k^*(t_{\nu})$$

$$= \frac{1}{\pi^2} \frac{n+1}{A_n^{\alpha}} \sum_{k=0}^{n} A_{n-k}^{\alpha-2} \ln(k+1) + O(1)$$

$$= \frac{1}{\pi^2} \frac{n+1}{A_n^{\alpha}} \sum_{k=0}^{n} A_{n-k}^{\alpha-2} \left(1 + \cdots + \frac{1}{k+1} \right) + O(1)$$

$$= \frac{1}{\pi^2}\frac{n+1}{A_n^\alpha}\sum_{k=0}^{n}A_{n-k}^{\alpha-1}\frac{1}{k+1}+O(1). \tag{3.11}$$

但因

$$\frac{n+1}{A_n^\alpha}\sum_{k=0}^{n}A_{n-k}^{\alpha-1}\frac{1}{k+1}=\frac{n+1}{A_n^\alpha}\left\{\sum_{k=0}^{\left[\frac{n}{2}\right]}+\sum_{k=\left[\frac{n}{2}\right]+1}^{n}\right\},$$

$$\frac{n+1}{A_n^\alpha}\sum_{k=\left[\frac{n}{2}\right]+1}^{n}A_{n-k}^{\alpha-1}\frac{1}{k+1}=O(1), \tag{3.12}$$

$$\frac{n+1}{A_n^\alpha}\sum_{k=0}^{\left[\frac{n}{2}\right]}A_{n-k}^{\alpha-1}\frac{1}{k+1}\approx\frac{\alpha(n+1)}{n^\alpha}\sum_{k=0}^{\left[\frac{n}{2}\right]}(n-k)^{\alpha-1}\frac{1}{k+1}$$

$$=\frac{\alpha(n+1)}{n}\sum_{k=0}^{\left[\frac{n}{2}\right]}\left(1-\frac{k}{n}\right)^{\alpha-1}\frac{1}{k+1}$$

$$\approx\alpha\sum_{k=0}^{\left[\frac{n}{2}\right]}\left\{1-(\alpha-1)\frac{k}{n}\right\}\frac{1}{k+1}$$

$$=\alpha\ln(n+1)+O(1), \tag{3.13}$$

故由(3.10) ~ (3.13) 即得

$$\sup_{\substack{f\in C_{2\pi}\\ f\neq\text{const}}}\frac{\|\sigma_n^\alpha(f;x)-f(x)\|_c}{\ln(n+1)\omega\left(f;\frac{2\pi}{n+1}\right)_c}=\frac{\alpha}{\pi^2}+O\left(\frac{1}{\ln n}\right). \tag{3.14}$$

定理 1 关于 $X=C_{2\pi}$ 的情形全证.

下面给出几条推论. 用 W_M^1 表示 2π 周期的绝对连续函数类, $f(x)$ 的导数满足条件 $\operatorname{ess\,sup}|f'(x)|\leqslant 1$ 者.

推论 1 $\alpha>0$. 成立着

$$\sup_{f\in W_M^1}\|\sigma_n^\alpha(f;x)-f(x)\|_c=\frac{2\alpha}{\pi}\frac{\ln(n+1)}{n+1}+O\left(\frac{1}{n}\right). \tag{3.15}$$

$\alpha=1$ 的情形见[9]. $\alpha=2,3,\cdots$ 情形见[13].

证明 $f(x)\in W_M^1$ 含有 $\omega(f;t)_c\leqslant t$, 故由定理 1 推知

$$\|\sigma_n^\alpha(f;x)-f(x)\|_c\leqslant\frac{\alpha}{\pi^2}\ln(n+1)\omega\left(f;\frac{2\pi}{n+1}\right)_c+O\left(\omega\left(f;\frac{2\pi}{n+1}\right)_c\right)$$

$$\leqslant\frac{\alpha}{\pi^2}\ln(n+1)\cdot\frac{2\pi}{n+1}+O\left(\frac{1}{n}\right)$$

$$=\frac{2\alpha}{\pi}\frac{\ln(n+1)}{n+1}+O\left(\frac{1}{n}\right). \tag{3.16}$$

取 $f_*(x)=|x|$, $-\pi\leqslant x\leqslant\pi$, $f_*(x+2\pi)=f_*(x)$, 则

$f_*(x) \in W_M^1$,并且

$$\sup_{f\in W_M^1} \| \sigma_n^\alpha(f;x) - f(x) \|_c \geqslant | \sigma_n^\alpha(f_*;0) |.$$

完全重复前面定理中采用的方法可得

$$| \sigma_n^\alpha(f_*;0) | = \frac{2\alpha}{\pi} \frac{\ln(n+1)}{n+1} + O\left(\frac{1}{n}\right).$$

推论 1 证完.

推论 2 若 $\alpha > 0$,则

$$\begin{aligned}\lim_{n\to\infty} A_n(\alpha)_c &= \lim_{n\to\infty} \sup_{\omega\in\Omega} \sup_{\substack{f\in H_M^\omega \\ f\neq \mathrm{const}}} \frac{\| \sigma_n^\alpha(f;x) - f(x) \|_c}{\ln(n+1)\omega\left(f;\frac{2\pi}{n+1}\right)_c} \\ &= \lim_{n\to\infty} \sup_{\substack{f\in H_M^1 \\ f\neq \mathrm{const}}} \frac{\| \sigma_n^\alpha(f;x) - f(x) \|_c}{\ln(n+1)\omega\left(f;\frac{2\pi}{n+1}\right)_c} \\ &= \frac{\alpha}{\pi^2}.\end{aligned} \tag{3.17}$$

此处:$\omega(t)$ 是连续模,Ω 是连续模类,H_M^ω 表示 2π 周期连续函数 $f(x)$ 之类,$f(x)$ 满足条件 $\omega(f;t) \leqslant \omega(t)$ 者. 当 $\omega(t) = t$ 时,H_M^ω 简记作 H_M^1. 显然 $H_M^1 = W_M^1$. 推论 2 表明,(1.5) 中的极值当 $X = C_{2\pi}$ 时渐近地等于在函数类 H_M^1 上的极值.

证明 我们有

$$\sup_{\substack{f\in H_M^1 \\ f\neq \mathrm{const}}} \frac{\| \sigma_n^\alpha(f;x) - f(x) \|_c}{\ln(n+1)\omega\left(f;\frac{2\pi}{n+1}\right)_c} \leqslant A_n(\alpha)_c = \frac{\alpha}{\pi^2} + O\left(\frac{1}{\ln n}\right).$$

由 $\omega\left(f;\frac{2\pi}{n+1}\right)_c \leqslant \frac{2\pi}{n+1}$ 知道

$$\begin{aligned}\sup_{\substack{f\in H_M^1 \\ f\neq \mathrm{const}}} \frac{\| \sigma_n^\alpha(f;x) - f(x) \|_c}{\ln(n+1)\omega\left(f;\frac{2\pi}{n+1}\right)_c} &\geqslant \sup_{\substack{f\in H_M^1 \\ f\neq \mathrm{const}}} \frac{\| \sigma_n^\alpha(f;x) - f(x) \|_c}{\ln(n+1)\cdot\frac{2\pi}{n+1}} \\ &= \frac{\alpha}{\pi^2} + O\left(\frac{1}{\ln n}\right).\end{aligned}$$

推论 2 得证.

§4. 定理 1 的证明(续)

$X = L_{2\pi}$ 的情形

引入函数类 H_L^1 及 W_L^1. 前者是 2π 周期 L 可和函数类, 满足条件

$$\omega(f;t)_L = \sup_{|u|\leqslant t} \| f(\cdot + u) - f(\cdot) \|_L \leqslant t \tag{4.1}$$

者. W_L^1 是 2π 周期绝对连续函数类, 满足条件 $\| f'(x) \|_L \leqslant 1$ 者. 自然是 $W_L^1 \subset H_L^1$. 为证明定理 1 在 $X = L_{2\pi}$ 时成立, 需要先证明下面定理 2, 其特殊情形($\alpha = 1$) 是 С. М. Никольский 得到的(见[10]).

定理 2 若 $\alpha > 0$, 则

$$\begin{aligned}\sup_{f\in H_L^1} \| \sigma_n^\alpha(f;x) - f(x) \|_L &\approx \sup_{f\in W_L^1} \| \sigma_n^\alpha(f;x) - f(x) \|_L \\ &\approx \sup_{f\in W_M^1} \| \sigma_n^\alpha(f;x) - f(x) \|_c \\ &= \frac{2\alpha}{\pi} \frac{\ln(n+1)}{n+1} + O\left(\frac{1}{n}\right).\end{aligned} \tag{4.2}$$

证明 先证当 $\alpha > 0$ 时有

$$\sup_{f\in H_L^1} \| \sigma_n^\alpha(f;x) - f(x) \|_L \leqslant \frac{2\alpha}{\pi} \frac{\ln(n+1)}{n+1} + O\left(\frac{1}{n}\right). \tag{4.3}$$

当 $0 < \alpha < 1$ 时, 由(2.32), (3.6) 即得(4.3). 当 $\alpha > 1$ 时, 与(3.10) 类似的有

$$\sup_{\substack{f\in L\\ f\neq \text{const}}} \frac{\| \sigma_n^\alpha(f;x) - f(x) \|_L}{\omega\left(f;\frac{2\pi}{n+1}\right)_L} \leqslant 1 + \frac{2}{\pi}\int_0^\pi \left[\frac{n+1}{2\pi}t\right] K_n^\alpha(t)\mathrm{d}t. \tag{4.4}$$

由(3.11), (3.12), (3.13) 即得(4.3). 然后由 $W_L^1 \subset H_L^1$ 得

$$\sup_{f\in W_L^1} \| \sigma_n^\alpha(f;x) - f(x) \|_L \leqslant \sup_{f\in H_L^1} \| \sigma_n^\alpha(f;x) - f(x) \|_L. \tag{4.5}$$

С. М. Никольский 在 $\alpha = 1$ 时算得(见[10])

$$\sup_{f\in W_L^1} \| \sigma_n(f;x) - f(x) \|_L = \frac{2}{\pi} \frac{\ln(n+1)}{n+1} + O\left(\frac{1}{n}\right). \tag{4.6}$$

我们用[10] 的方法来处理 $\alpha > 0$ 的情形.

(1)$0 < \alpha < 1$ 时, 由于

$$f(x) = \frac{a_0}{2} + \frac{1}{\pi}\int_0^{2\pi} D_1^{(1)}(t) f'(x+t)\mathrm{d}t, \tag{4.7}$$

此处

$$D_1^{(1)}(t)=\frac{\pi-t}{2}(0<t<2\pi),\quad D_1^{(1)}(t+2\pi)=D_1^{(1)}(t),$$

以及

$$\sigma_n^\alpha(f;x)=\frac{a_0}{2}+\frac{1}{\pi}\int_0^{2\pi}K_{n*}^\alpha(t)f'(x+t)\mathrm{d}t,\tag{4.8}$$

此处

$$K_{n*}^\alpha(t)=\sum_{k=1}^{n}\frac{A_{n-k}^\alpha}{A_n^\alpha}\frac{\sin kt}{k},\tag{4.9}$$

故有

$$f(x)-\sigma_n^\alpha(f;x)=\frac{1}{\pi}\int_0^{2\pi}\{D_1^{(1)}(t)-K_{n*}^\alpha(t)\}f'(x+t)\mathrm{d}t.$$

Zygmund([5] 的 p.184 的(11.5)) 指出,如果 $0<\alpha<1$,则当 $0<t\leqslant\pi$ 时,有

$$\begin{aligned}D_1^{(1)}(t)-K_{n*}^\alpha(t)=\Psi(t)&=\int_t^\pi K_n^\alpha(t)\mathrm{d}t\\&=\frac{\alpha}{2(n+1)}\cot\frac{t}{2}-\frac{2\theta_1}{nA_n^\alpha\left(2\sin\frac{t}{2}\right)^{\alpha+1}}+\frac{B}{n^2t^2}.\end{aligned}\tag{4.10}$$

当 $0\leqslant t<\pi$ 时,易见有

$$\begin{aligned}D_1^{(1)}(\pi+t)-K_{n*}^\alpha(\pi+t)&=-\{D_1(\pi-t)-K_{n*}^\alpha(\pi-t)\}\\&=-\Psi(\pi-t).\end{aligned}\tag{4.11}$$

根据[10](第 216 ~ 217 页),有

$$\begin{aligned}&\sup_{f\in W_L^1}\|\sigma_n^\alpha(f;x)-f(x)\|_L\\&=\frac{1}{2\pi}\max_t\int_0^{2\pi}|\Psi(x)-\Psi(x+t)|\mathrm{d}x\\&\geqslant\frac{1}{2\pi}\left\{\int_{\frac{1}{n}}^{\pi-\frac{1}{n}}|\Psi(x)-\Psi(x+\pi)|\mathrm{d}x+\right.\\&\left.\int_{\pi+\frac{1}{n}}^{2\pi-\frac{1}{n}}|\Psi(x)-\Psi(x+\pi)|\mathrm{d}x\right\}\\&\geqslant\frac{\alpha}{4\pi(n+1)}\left\{\int_{\frac{1}{n}}^{\pi-\frac{1}{n}}\left|\cot\frac{x}{2}-\cot\frac{x+\pi}{2}\right|\mathrm{d}x+\right.\end{aligned}$$

$$\int_{\pi+\frac{1}{n}}^{2\pi-\frac{1}{n}}\left|\cot\frac{x}{2}-\cot\frac{x+\pi}{2}\right|\mathrm{d}x\Bigg\}-$$

$$\left\{O\left(\int_{\frac{1}{n}}^{\pi}\frac{\mathrm{d}x}{(nx)^{\alpha+1}}\right)+O\left(\int_{\frac{1}{n}}^{\pi}\frac{\mathrm{d}x}{(nx)^{2}}\right)\right\}$$

$$=\frac{2\alpha}{\pi}\frac{\ln(n+1)}{n+1}+O\left(\frac{1}{n}\right). \tag{4.12}$$

综合(4.3),(4.5),(4.12) 即得(4.2). $0<\alpha<1$ 情形证完.

(2) $\alpha>1$ 时. 仍根据 С. М. Никольский 得

$$\sup_{f\in W_L^1}\|\sigma_n^{\alpha}(f;x)-f(x)\|_L=\frac{1}{2\pi}\max_t\int_0^{2\pi}|\Psi(x+t)-\Psi(x)|\mathrm{d}x$$

$$\geqslant\frac{1}{2\pi}\Bigg\{\int_{\frac{1}{n}}^{\pi-\frac{1}{n}}|\Psi(x)-\Psi(x+\pi)|\mathrm{d}x+$$

$$\int_{\pi+\frac{1}{n}}^{2\pi-\frac{1}{n}}|\Psi(x+\pi)-\Psi(x)|\mathrm{d}x\Bigg\}.$$

利用引理 4 即得: 当 $[\alpha]=r=1$ 时,

$$\sup_{f\in W_L^1}\|\sigma_n^{\alpha}(f;x)-f(x)\|_L\geqslant\frac{\alpha}{4\pi(\alpha+n-1)}\Bigg\{\int_{\frac{1}{n}}^{\pi-\frac{1}{n}}\left|\cot\frac{x}{2}-\cot\frac{x+\pi}{2}\right|\mathrm{d}x+$$

$$\int_{\pi+\frac{1}{n}}^{2\pi-\frac{1}{n}}\left|\cot\frac{x}{2}-\cot\frac{x+\pi}{2}\right|\mathrm{d}x\Bigg\}-$$

$$O\left(\int_{\frac{1}{n}}^{\pi}\frac{\mathrm{d}x}{(nx)^{\alpha+1}}\right)-O\left(\int_{\frac{1}{n}}^{\pi}\frac{\mathrm{d}x}{(nx)^{2}}\right)$$

$$=\frac{2\alpha}{\pi}\frac{\ln(n+1)}{n+1}+O\left(\frac{1}{n}\right). \tag{4.13}$$

当 $[\alpha]=r\geqslant 2$ 时, 有

$$\sup_{f\in W_L^1}\|\sigma_n^{\alpha}(f;x)-f(x)\|_L\geqslant\frac{\alpha}{4\pi(\alpha+n-1)}\Bigg\{\int_{\frac{1}{n}}^{\pi-\frac{1}{n}}\left|\cot\frac{x}{2}-\cot\frac{x+\pi}{2}\right|\mathrm{d}x+$$

$$\int_{\pi+\frac{1}{n}}^{2\pi-\frac{1}{n}}\left|\cot\frac{x}{2}-\cot\frac{x+\pi}{2}\right|\mathrm{d}x\Bigg\}-$$

$$O\left(\int_{\frac{1}{n}}^{\pi}\frac{\mathrm{d}x}{(nx)^{\alpha+1}}\right)-O\left(\int_{\frac{1}{n}}^{\pi}\sum_{j=2}^{r+1}\frac{1}{(nx)^{j}}\mathrm{d}x\right)$$

$$=\frac{2\alpha}{\pi}\frac{\ln(n+1)}{n+1}+O\left(\frac{1}{n}\right). \tag{4.14}$$

综合(4.3),(4.5),(4.13),(4.14) 即得 $\alpha>1$ 时(4.2)之证. 定理 2 证毕.

当 $X = L_{2\pi}$ 时定理 1 之证

已经知道有

$$\overline{\lim_{n\to\infty}} A_n(\alpha)_L \leqslant \frac{\alpha}{\pi^2}. \tag{4.15}$$

假定有 $\underline{\lim}_{n\to\infty} A_n(\alpha)_L < \frac{\alpha}{\pi^2}$. 对适当小的 $\varepsilon > 0$, 有自然数子序列 $n_1 < n_2 < \cdots$ 使

$$A_{n_i}(\alpha)_L \leqslant \frac{\alpha}{\pi^2} - \varepsilon(>0). \tag{4.16}$$

这样一来, 对于 $f(x) \in H_L^1$, $f(x) \neq \mathrm{const}$ 者, 必有

$$\frac{\| \sigma_{n_i}^{\alpha}(f;x) - f(x) \|_L}{\ln(n_i+1)\omega\left(f;\frac{2\pi}{n_i+1}\right)_L} \leqslant \frac{\alpha}{\pi^2} - \varepsilon.$$

由 $\omega\left(f;\frac{2\pi}{n_i+1}\right)_L \leqslant \frac{2\pi}{n_i+1}$, 更会有

$$\frac{\| \sigma_{n_i}^{\alpha}(f;x) - f(x) \|_L}{\ln(n_i+1)\cdot\frac{2\pi}{n_i+1}} \leqslant \frac{\alpha}{\pi^2} - \varepsilon$$

对一切 $f(x) \in H_L^1$ 成立. 从而有

$$\sup_{f\in H_L^1} \| \sigma_{n_i}^{\alpha}(f;x) - f(x) \|_L \leqslant \left(\frac{\alpha}{\pi^2} - \varepsilon\right)\frac{2\pi\ln(n_i+1)}{n_i+1}.$$

此与定理 2 矛盾. 故得

$$\lim_{n\to\infty} A_n(\alpha)_L = \frac{\alpha}{\pi^2}. \tag{4.17}$$

定理 1 对 $X = L_{2\pi}$ 情形证完.

推论 3 若 $\alpha > 0$, 则

$$\begin{aligned}\lim_{n\to\infty} A_n(\alpha)_L &= \lim_{n\to\infty} \sup_{\omega\in\Omega_1} \sup_{\substack{f\in H_L^{\omega}\\ f\neq\mathrm{const}}} \frac{\| \sigma_n^{\alpha}(f;x) - f(x) \|_L}{\ln(n+1)\omega\left(f;\frac{2\pi}{n+1}\right)_L} \\ &= \lim_{n\to\infty} \sup_{\substack{f\in H_L^1\\ f\neq\mathrm{const}}} \frac{\| \sigma_n^{\alpha}(f;x) - f(x) \|_L}{\ln(n+1)\omega\left(f;\frac{2\pi}{n+1}\right)_L} = \frac{\alpha}{\pi^2}.\end{aligned}$$

此处: Ω_1 表示 $L_{2\pi}$ 类的函数的连续积分模类, H_L^{ω} 表示 2π 周期 L 可和函数, 其连续积分模满足条件 $\omega(f;t)_L \leqslant \omega(t)$ 者. 推论 3 表明, (1.5) 的极值当 $X = L_{2\pi}$ 时在 H_L^1 类上渐近地达到, 恰如当 $X = C_{2\pi}$ 时该极值在 H_M^1

类上渐近地达到一样.

§5. 关于线性多项式型正算子逼近常数的注记

定理1表明,对于正阶 Cesàro 算子的逼近,成立着以下关系式:

$$\sup_{\substack{f\in C_{2\pi}\\ f\neq\text{const}}}\frac{\|\sigma_n^\alpha(f;x)-f(x)\|_c}{\omega\left(f;\frac{2\pi}{n+1}\right)_c}\approx\sup_{\substack{f\in L_{2\pi}\\ f\neq\text{const}}}\frac{\|\sigma_n^\alpha(f;x)-f(x)\|_L}{\omega\left(f;\frac{2\pi}{n+1}\right)_L}. \tag{5.1}$$

本节目的是指出(5.1)类型的渐近等式可以容易地推广到用任意正算子来逼近的情形. 仍用 X 代表 $C_{2\pi}$ 及 $L_{2\pi}$, Λ 是 §1 内给出的实数三角阵,满足条件

$$K_n^\lambda(t)=\frac{1}{2}+\sum_{k=1}^n\lambda_k^{(n)}\cos kt\geqslant 0. \tag{5.2}$$

算子 $A_n(f;x,\lambda)$ 的定义已在 §1 内给出. 置

$$C_n(\lambda)_X=\sup_{\substack{f\in X\\ f\neq\text{const}}}\frac{\|f(x)-A_n(f;x,\lambda)\|_X}{\omega\left(f;\frac{\pi}{n+1}\right)_X}. \tag{5.3}$$

王兴华[11] 指出

$$C_n(\lambda)_c=1+\frac{1}{\pi}\int_0^\pi\left[\frac{n+1}{\pi}t\right]K_n^\lambda(t)\mathrm{d}t\overset{\text{def}}{=}\alpha_n. \tag{5.4}$$

利用前节中用过的论证方法可得

$$C_n(\lambda)_L\leqslant C_n(\lambda)_c. \tag{5.5}$$

对于类 $W_M^1=H_M^1$ 有

$$\sup_{f\in H_M^1}\|f(x)-A_n(f;x,\lambda)\|_c\leqslant\frac{\alpha_n\pi}{n+1}. \tag{5.6}$$

定理3　若 Λ 阵满足(5.2),并且有

$$\sup_{f\in H_M^1}\|f(x)-A_n(f;x,\lambda)\|_c\approx\frac{\alpha_n\pi}{n+1}, \tag{5.7}$$

则成立

$$\lim_{n\to\infty}\frac{C_n(\lambda)_L}{C_n(\lambda)_c}=\lim_{n\to\infty}\sup_{\substack{f\in L\\ f\neq\text{const}}}\frac{\|f(x)-A_n(f;x,\lambda)\|_L}{\alpha_n\omega\left(f;\frac{\pi}{n+1}\right)_L}=1.$$

证明　基本上是重复上节中的论证. 首先有

$$\sup_{f\in W_L^1}\|f(x)-A_n(f;x,\lambda)\|_L\leqslant\sup_{f\in H_L^1}\|f(x)-A_n(f;x,\lambda)\|_L\leqslant\frac{\alpha_n\pi}{n+1}.\tag{5.8}$$

根据[12],对于线性正算子有

$$\sup_{f\in W_L^1}\|f(x)-A_n(f;x,\lambda)\|_L=\sup_{f\in W_M^1}\|f(x)-A_n(f;x,\lambda)\|_c.\tag{5.9}$$

由此得

$$\sup_{f\in W_L^1}\|f(x)-A_n(f;x,\lambda)\|_L\approx\frac{\alpha_n\pi}{n+1}.\tag{5.10}$$

由(5.5)已知有$\overline{\lim_{n\to\infty}}\frac{C_n(\lambda)_L}{C_n(\lambda)_c}\leqslant 1$.设有

$$\underline{\lim_{n\to\infty}}\frac{C_n(\lambda)_L}{C_n(\lambda)_c}<1.$$

则存在充分小的$\varepsilon>0$及自然数子序列$n_1<n_2<\cdots$使

$$\sup_{\substack{f\in L\\ f\neq\mathrm{const}}}\frac{\|f(x)-A_{n_i}(f;x,\lambda)\|_L}{\alpha_{n_i}\omega\left(f;\frac{\pi}{n_i+1}\right)_L}\leqslant 1-\varepsilon(>0).\tag{5.11}$$

从而对于$f(x)\in H_L^1$也有(5.11).当$f(x)\in H_L^1$时有$\omega\left(f;\frac{\pi}{n+1}\right)_L\leqslant\frac{\pi}{n+1}$,所以得到

$$\sup_{f\in H_L^1}\frac{\|f(x)-A_{n_i}(f;x,\lambda)\|_L}{\alpha_{n_i}\cdot\frac{\pi}{n_i+1}}\leqslant 1-\varepsilon.$$

此与

$$\sup_{f\in H_L^1}\|f(x)-A_n(f;x,\lambda)\|_L\approx\frac{\alpha_n\pi}{n+1}\tag{5.12}$$

矛盾.定理3证完.

推论 在定理3条件下成立着

$$\sup_{\substack{f\in C\\ f\neq\mathrm{const}}}\frac{\|f(x)-A_n(f;x,\lambda)\|_c}{\omega\left(f;\frac{\pi}{n+1}\right)_c}\approx\sup_{\substack{f\in H_M^1\\ f\neq\mathrm{const}}}\frac{\|f(x)-A_n(f;x,\lambda)\|_c}{\omega\left(f;\frac{\pi}{n+1}\right)_c}$$

$$\approx\sup_{\substack{f\in L\\ f\neq\mathrm{const}}}\frac{\|f(x)-A_n(f;x,\lambda)\|_L}{\omega\left(f;\frac{\pi}{n+1}\right)_L}.$$

$$\approx \sup_{\substack{f\in H_L^1\\ f\neq \text{const}}} \frac{\| f(x) - A_n(f;x,\lambda) \|_L}{\omega\left(f;\dfrac{\pi}{n+1}\right)_L}$$

$$\approx \alpha_n. \tag{5.13}$$

参考文献

[1] Ульянов П. Л. О приближении функций. Сибирский матем. журн., 1964, 5(2): 418 ~ 437

[2] Дудас В. О. Приближения непрерывной периодической функции суммами фейера. 《Вопросы теории прибл. функц. и её прил.》Киев АН УССР, 1976, 109 ~ 115

[3] 姚璧芸. 逼近算子的极限常数. 杭州大学学报, 1963, 自然科学版(1): 111 ~ 115

[4] 李文清. 关于费叶尔算子的逼近度. 厦门大学学报, 1977, 自然科学版, (2): 5 ~ 15

[5] Зигмунд А. *Тригонометрические ряды*. Т. 1, Москва: Издат. 《Мир》, 1965

[6] Харди Г. *Расходащиеся ряды*. Москва: Издат. иностранной-литературы, 1951

[7] 郭竹瑞. 连续函数用它的富理埃级数的蔡查罗平均数来逼近. 数学学报, 1962, 12(3): 320 ~ 329

[8] 王兴华. 爵克松奇异积分对连续函数逼近的准确常数. 数学学报, 1964, 14(2): 231 ~ 237

[9] Никольский С. М. Об асимптотическом поведении остатка при приближении функций, удовлетворяющих условием Липшица суммами Фейера. Изв. АН СССР, 1940, сер. мамем. (4): 501 ~ 507

[10] Никольский С. М. Приближение функций тригонометрическими полиномами в среднем, Изв. АН СССР, 1946, сер. мемем. (10): 207 ~ 256

[11] 王兴华. 用富理埃级数的正线性和逼近连续函数. 数学进展, 1965, 8(3): 322 ~ 325

[12] Моторный В. П. Приближение периодических функций тригонометрическими полиномами в среднем. Матем. замемкu. 1974, 16(1): 15 ~ 26

[13] Фалалеев Л. Approximation of functions by(C, α)means of the Fourier series. Изв. АН Казах. ССР, 1972, сер. физ-мам. (1)

数学学报
1982,9,25(5)

关于周期函数用线性算子的平均逼近*

On the Approximation of Periodic Functions in Mean by Linear Operators

§1.

设 C,L 各表示 2π 周期的连续函数空间及 L 可和函数空间，其范数分别是：

对 $f\in C$: $\|f\|_c=\max|f(x)|$;

对 $f\in L$: $\|f\|_L=\int_0^{2\pi}|f(x)|\mathrm{d}x$.

令 M 表示本性有界的 2π 周期可测函数空间，范数为

$$\|f\|_M=\operatorname{ess\,sup}|f(x)|.$$

引入函数类

$$H_X^0=\{\varphi:\varphi\in X,\|\varphi\|_X\leqslant 1,\int_0^{2\pi}\varphi(t)\mathrm{d}t=0\},\qquad(1.1)$$

$X=M,L$.

$$H_X^\omega=\{\varphi:\varphi\in X,\|\varphi(\cdot+t)-\varphi(\cdot)\|_X\leqslant\omega(|t|)\},\qquad(1.2)$$

$X=C,L,\omega(t)$ 为给定的连续模.

给定 $K(t)\in L$，由 K 对 H_X^0,H_X^ω 的函数进行卷积变换，给出以下函数类：

$$\mathfrak{M}_{X(C,L)}=\left\{f:f=\frac{a_0}{2}+\pi^{-1}K*\varphi,\varphi\in H_X^0\right\},\qquad(1.3)$$

---

* 数学学报于1980年4月8日收到本文.

$$\mathfrak{M}_{X(C,L)}^{\omega} = \left\{ f: f = \frac{a_0}{2} + \pi^{-1} K * \varphi, \varphi \in H_X^{\omega} \right\}, \tag{1.4}$$

$\mathfrak{M}_C, \mathfrak{M}_C^{\omega}$，以及 $\mathfrak{M}_L, \mathfrak{M}_L^{\omega}$ 各是 C, L 空间内的局部紧致集. 当 $K(t)$ 取具体函数时，它们可以给出一些在逼近论中经常讨论的函数类. 例如，当

$$K(t) = K_r(t) = \sum_{k=1}^{\infty} k^{-r} \cos\left(kt + \frac{\pi r}{2}\right) (r = 1, 2, \cdots),$$

以及

$$K(t) = \widetilde{K}_r(t) = \sum_{k=1}^{\infty} k^{-r} \sin\left(kt + \frac{\pi r}{2}\right) (r = 1, 2, \cdots)$$

时，所得到的函数类按习惯分别表示为 W_X^r, $W^r H_X^{\omega}$ 以及 $\widetilde{W}_X^r$, $\widetilde{W}^r H_X^{\omega}$(在共轭情形下，$a_0 = 0$).

为了构作线性算子列，我们给定可测函数列 $K_n(t) \in M$，约定满足以下几个条件：

(1) $K_n(-t) = K_n(t), n = 1, 2, \cdots$

(2) $\operatorname{ess\,sup} | K_n(t) | \leqslant M_n, n = 1, 2, \cdots$

(3) $\pi^{-1} \int_{-\pi}^{\pi} K_n(t) \mathrm{d}t = 1. n = 1, 2, \cdots$ (1.5)

$K_n(t)$ 的 Fourier 展开

$$K_n(t) \sim \frac{1}{2} + \sum_{k=1}^{\infty} \lambda_k^{(n)} \cos kt. \tag{1.6}$$

$\{\lambda_k^{(n)}\}$ 是一实数阵. 有时 $K_n(t)$ 是由预先给出的 λ 数阵来确定的(例如下三角阵)，这时为强调 $K_n(t)$ 对 λ 数阵的依赖关系就记作 $K_n^{\lambda}(t)$. 利用 $K_n(t)$ 构作线性算子列

$$L_n(f) = \pi^{-1} K_n * f = \frac{1}{\pi} \int_{-\pi}^{\pi} K_n(t - x) f(t) \mathrm{d}t. \tag{1.7}$$

$L_n(f)$ 是 L 空间上的线性有界算子.

算子 $L_n(f)$ 对类 $\mathfrak{M}_X, \mathfrak{M}_X^{\omega} (X = C, L)$ 的逼近度由下面的量来表征：

$$\mathscr{E}[\mathfrak{M}_X; L_n]_X = \sup_{f \in \mathfrak{M}_X} \| f - L_n(f) \|_X, \tag{1.8}$$

$$\mathscr{E}[\mathfrak{M}_X^{\omega}; L_n]_X = \sup_{f \in \mathfrak{M}_X^{\omega}} \| f - L_n(f) \|_X. \tag{1.9}$$

$X = C, L$.

对于量(1.8)，(1.9) 的研究肇端于 С. М. Никольский. 1946 年 С. М.

Никольский[1] 证明:当 $K(t) = K_r(t), K_n^\lambda(t) = \frac{1}{2} + \sum_{k=1}^{n} \lambda_k^{(n)} \cos kt$ 时,有

$$\mathscr{E}[W_L^r; L_n]_L \leqslant \mathscr{E}[W_M^r; L_n]_C. \tag{1.10}$$

并且指出当 $L_n(f)$ 为 Fejér 算子及 Fourier 算子时,(1.10) 转化为渐近等式.其后出现了一系列研究成果,主要涉及两方面的问题:

(1) 对广泛的函数类 $\mathfrak{M}_X, \mathfrak{M}_X^\omega$ 及算子 $L_n(f)$ 建立 Никольский 型不等式(1.10),并且研究不等式转化为等式或渐近等式的条件.

(2) 对一些重要的函数类和具体的算子算出量(1.8),(1.9) 的渐近展开式的主项.

1967 年 С.В.Стечкин 与 С.А.Теляковский[2] 在 $K(t)$ 取

$$K_r^\alpha(t) = \sum_{k=1}^{\infty} k^{-r} \cos\left(kt + \frac{\pi\alpha}{2}\right) (r > 0, -\infty < \alpha < +\infty)$$

的情况下证明

$$\mathscr{E}[\mathfrak{M}_L; L_n]_L \leqslant \mathscr{E}[\mathfrak{M}_C; L_n]_C \tag{1.11}$$

对任意的 $L_n(f) = \pi^{-1} K_n^\lambda * f$ 成立,此处 $K_n^\lambda(t) = \frac{1}{2} + \sum_{k=1}^{n} \lambda_k^{(n)} \cos kt$.他们指出,对很多具体算子(1.11) 中有等式或渐近等式成立,并且给出了使(1.11) 中没有渐近等式成立的线性算子之例.

1974 年 В.П.Моторный[3] 应用卷积变换方法证明了对于类 $K_r^\alpha * H_X^0 (X = M, L), K_r^\alpha * H_X^\omega (X = C, L)$ 和任意线性算子可以建立 (1.11) 型不等式.他特别在 $K_n^\lambda(t) = \frac{1}{2} + \sum_{k=1}^{n} \lambda_k^{(n)} \cos kt \geqslant 0, r = 1, 2, \cdots, \alpha = r$ 或 $\alpha = r + 1, \omega(t)$ 为上凸连续模的条件下,讨论不等式转化成等式的问题,得到了一系列相当广泛的结果.

本文目的是要在文章开始列举的情况下讨论 Никольский 不等式转化为等式的问题.我们给的条件足够广泛,能够概括很多具体情况.文章的大致内容如下:

在 §2 内我们给出基本不等式

$$\mathscr{E}[\mathfrak{M}_L; L_n]_L \leqslant \mathscr{E}[\mathfrak{M}_C; L_n]_C \tag{1.12}$$

转化成等式的充分必要条件.这个条件的一个特殊形式(N 性质) 对周期可微函数类及其共轭类有直接应用,是 §3,4 中定理论证的基础.

在 §3 内,我们引入条件

$$G_n(t)=\int_t^{\pi}K_n(u)\mathrm{d}u\geqslant 0,\quad 0\leqslant t\leqslant\pi \tag{1.13}$$

取代 В. П. Моторный 在[3] 内较苛刻的条件

$$K_n^{\lambda}(t)=\frac{1}{2}+\sum_{k=1}^{n}\lambda_k^{(n)}\cos kt\geqslant 0. \tag{1.14}$$

证明在条件(1.13) 之下有

$$\mathscr{E}[W_L^r,L_n]_L=\mathscr{E}[W_M^r,L_n]_C,\quad r=1,2,\cdots \tag{1.15}$$

$$\mathscr{E}[\widetilde{W}_L^r,L_n]_L=\mathscr{E}[\widetilde{W}_M^r,L_n]_C,\ r=2,3,\cdots \tag{1.16}$$

在 §4 内,仍在条件(1.13) 之下证明

$$\mathscr{E}[W^rH_L^{\omega},L_n]_L=\mathscr{E}[W^rH_C^{\omega},L_n]_C,\quad r=1,3,5,\cdots \tag{1.17}$$

$$\mathscr{E}[\widetilde{W}^rH_L^{\omega},L_n]_L=\mathscr{E}[\widetilde{W}^rH_C^{\omega},L_n]_C,\ r=2,4,6,\cdots \tag{1.18}$$

$\omega(t)$ 是上凸连续模.

给出了量(1.15) ~ (1.18) 的精确的解析表达式.利用这些解析表达式,当 λ 数阵具体给出时,就可以计算逼近度的渐近展开的主项.

§2. 关于量$\mathscr{E}[\mathfrak{M}_X,L_n]_X(X=C,L)$的一般定理

记

$$\mathfrak{M}_X=\left\{f:f=\frac{a_0}{2}+\pi^{-1}K*\varphi,\varphi\in H_X^0\right\}, \tag{2.1}$$

此处

$$K*\varphi=\int_0^{2\pi}K(t-x)\varphi(t)\mathrm{d}t.$$

又

$$L_n(f)=\pi^{-1}K_n*f. \tag{2.2}$$

对 $f\in\mathfrak{M}_X$,有

$$\begin{aligned}L_n(f)&=\pi^{-1}\int_0^{2\pi}K_n(t-x)f(t)\mathrm{d}t\\&=\frac{1}{\pi}\int_0^{2\pi}K_n(t)\left\{\frac{a_0}{2}+\frac{1}{\pi}\int_0^{2\pi}K(u-x-t)\varphi(u)\mathrm{d}u\right\}\mathrm{d}t\\&=\frac{a_0}{2}+\frac{1}{\pi}\int_0^{2\pi}\varphi(u)\left\{\frac{1}{\pi}\int_0^{2\pi}K_n(t)K(u-x-t)\mathrm{d}t\right\}\mathrm{d}u.\end{aligned}$$

由于

$$\int_0^{2\pi} K_n(t)K(s-t)\mathrm{d}t = \int_0^{2\pi} K_n(t-s)K(t)\mathrm{d}t,$$

若置

$$\Phi_n(s) = \int_0^{2\pi} K_n(t-s)K(t)\mathrm{d}t,$$

则有

$$L_n(f;x) - f(x) = \frac{1}{\pi}\int_0^{2\pi} \{\Phi_n(u) - K(u)\}\varphi(u+x)\mathrm{d}u$$

$$= \frac{1}{\pi}\int_0^{2\pi} K_n^*(u)\varphi(u+x)\mathrm{d}u.$$

此处

$$K_n^*(u) = \Phi_n(u) - K(u). \tag{2.3}$$

当 $\varphi \in H_M^0$ 时, 有

$$\begin{aligned}\mathscr{E}[\mathfrak{M}_C; L_n]_C &= \sup_{f\in\mathfrak{M}_C} \| f - L_n(f) \|_C \\ &= \sup_{f\in\mathfrak{M}_C} | f(0) - L_n(f;0) | \\ &= \frac{1}{\pi}\sup_{\varphi\in H_M^0}\left|\int_0^{2\pi} K_n^*(u)\varphi(u)\mathrm{d}u\right| \\ &= \frac{1}{\pi}E_1(K_n^*)_L.\end{aligned} \tag{2.4}$$

此处

$$E_1(K_n^*)_L = \min_C \int_0^{2\pi} | K_n^*(u) - C | \mathrm{d}u.$$

当 $\varphi \in H_L^0$ 时, 根据 C. M. Никольский 的一个结果(见[4], 第 183 页) 有

$$\begin{aligned}\| f - L_n(f) \|_L &= \frac{1}{\pi}\int_0^{2\pi}\left|\int_0^{2\pi} K_n^*(u)\varphi(u+x)\mathrm{d}u\right|\mathrm{d}x \\ &\leqslant \frac{1}{2\pi}\max_t \int_0^{2\pi} | K_n^*(u+t) - K_n^*(u) | \mathrm{d}u \\ &\leqslant \frac{1}{\pi}E_1(K_n^*)_L.\end{aligned}$$

从而得

$$\mathscr{E}[\mathfrak{M}_L; L_n]_L \leqslant \mathscr{E}[\mathfrak{M}_C, L_n]_C = \frac{1}{\pi}E_1(K_n^*)_L. \tag{2.5}$$

进而研究(2.5) 内有等号成立的条件. 为此, 先推导下列关系式:

$\mathscr{E}[\mathfrak{M}_L; L_n]_L$

$$= \frac{1}{\pi}\sup_{\varphi\in H_L^0}\int_0^{2\pi}\left|\int_0^{2\pi}K_n^*(u)\varphi(u+x)\mathrm{d}u\right|\mathrm{d}x$$

$$= \frac{1}{\pi}\sup_{\varphi\in H_L^0}\sup_{\psi\in H_M}\int_0^{2\pi}\int_0^{2\pi}K_n^*(u)\varphi(u+x)\psi(x)\mathrm{d}u\mathrm{d}x$$

$$= \frac{1}{\pi}\sup_{\varphi\in H_L^0}\sup_{\psi\in H_M}\int_0^{2\pi}\int_0^{2\pi}\{K_n^*(u)-C\}\varphi(u+x)\psi(x)\mathrm{d}u\mathrm{d}x$$

$$= \frac{1}{\pi}\sup_{\varphi\in H_L^0}\sup_{\psi\in H_M}\int_0^{2\pi}\{K_n^*(u)-C\}\left\{\int_0^{2\pi}\varphi(u+x)\psi(x)\mathrm{d}x\right\}\mathrm{d}u. \tag{2.6}$$

此处

$$H_M = \{\psi:\psi\in M, \|\psi\|_M \leqslant 1\},$$

而 C 为任意实数.注意对任取的 $\varphi\in H_L^0$ 及 $\psi\in H_M$,有

$$\int_0^{2\pi}\varphi(u+x)\psi(x)\mathrm{d}x \in H_M^0.$$

记
$$\chi(u,C) = \mathrm{sgn}\{K_n^*(u)-C\}. \tag{2.7}$$

命题1 若存在实数 C^* 满足条件:对任意小的数 $\varepsilon>0,\delta>0$,有 $\varphi_\varepsilon\in H_L^0,\psi_\varepsilon\in H_M$,使

$$\chi_\varepsilon(x) = \psi_\varepsilon * \varphi_\varepsilon = \int_0^{2\pi}\varphi_\varepsilon(u+x)\psi_\varepsilon(u)\mathrm{d}u$$

有

$$\mathrm{mes}E\{x: |\chi_\varepsilon(x)-\chi(x,C^*)|\geqslant\delta, 0\leqslant x\leqslant 2\pi\} < \varepsilon, \tag{2.8}$$

则有

$$\mathscr{E}[\mathfrak{M}_L;L_n]_L = \mathscr{E}[\mathfrak{M}_C,L_n]_C = \frac{1}{\pi}E_1(K_n^*)_L$$
$$= \frac{1}{\pi}\int_0^{2\pi}|K_n^*(u)-C^*|\mathrm{d}u. \tag{2.9}$$

证明 由(2.6)得

$$\mathscr{E}[\mathfrak{M}_L;L_n]_L \geqslant \frac{1}{\pi}\int_0^{2\pi}\{K_n^*(u)-C^*\}\chi_\varepsilon(u)\mathrm{d}u$$
$$= \frac{1}{\pi}\int_0^{2\pi}|K_n^*(u)-C^*|\mathrm{d}u +$$
$$\frac{1}{\pi}\int_0^{2\pi}\{K_n^*(u)-C^*\}\{\chi_\varepsilon(u)-\chi(u,C^*)\}\mathrm{d}u.$$

令

$$e_1 = \mathscr{E}\{x : |\chi_\varepsilon(x) - \chi(x, C^*)| < \delta, 0 \leqslant x \leqslant 2\pi\},$$

$$e_2 = [0, 2\pi] \setminus e_1.$$

则由

$$\int_0^{2\pi} \{K_n^*(u) - C^*\}\{\chi_\varepsilon(u) - \chi(u, C^*)\} du = \int_{e_1} + \int_{e_2},$$

$$\left| \int_{e_1} \{K_n^*(u) - C^*\}\{\chi_\varepsilon(u) - \chi(u, C^*)\} du \right| < \delta \int_0^{2\pi} |K_n^*(u) - C^*| du,$$

$$\left| \int_{e_2} \{K_n^*(u) - C^*\}\{\chi_\varepsilon(u) - \chi(u, C^*)\} du \right| \leqslant 2 \int_{e_2} |K_n^*(u) - C^*| du,$$

对任意小的 $\eta > 0$, 有充分小的 $\varepsilon, \delta > 0$, 使

$$\pi^{-1} \left\{ \delta \int_0^{2\pi} |K_n^*(u) - C^*| du + 2 \int_{e_2} |K_n^*(u) - C^*| du \right\} < \eta.$$

由此得

$$\mathscr{E}[\mathfrak{M}_L; L_n]_L \geqslant \pi^{-1} \int_0^{2\pi} |K_n^*(u) - C^*| du - \eta.$$

(2.9) 得证.

进而讨论命题 1 的条件对(2.9) 的成立的必要性问题.

假定有实数 C^*, 使

$$\begin{aligned}
&\mathscr{E}[\mathfrak{M}_L; L_n]_L \\
&= \frac{1}{\pi} \sup_{\varphi \in H_L^0} \sup_{\psi \in H_M} \int_0^{2\pi} \{K_n^*(u) - C^*\} \left\{ \int_0^{2\pi} \varphi(u + x) \psi(x) dx \right\} du. \\
&= \mathscr{E}[\mathfrak{M}_c, L_n]_C = \frac{1}{\pi} \int_0^{2\pi} |K_n^*(u) - C^*| du. \qquad (2.10)
\end{aligned}$$

给 $K_n^*(u)$ 加上限制条件

$$K_n^*(u) - C^* \overset{\text{a.e.}}{\neq} 0. \qquad (2.11)$$

命题2 设$\mathscr{E}[\mathfrak{M}_L; L_n]_L = \mathscr{E}[\mathfrak{M}_C, L_n]_C$, 且出现在(2.10) 内的 C^* 满足(2.11). 则任取 $\delta > 0, \eta > 0$ 有 $\xi > 0$, 对任一 $\varepsilon \in (0, \xi)$, 只要 $\varphi_\varepsilon \in H_L^0, \psi_\varepsilon \in H_M$ 的卷积 $\chi_\varepsilon = \psi_\varepsilon * \varphi_\varepsilon$ 能使

$$\int_0^{2\pi} \{K_n^*(u) - C^*\} \chi_\varepsilon(u) du > \mathscr{E}[\mathfrak{M}_L, L_n]_L - \varepsilon,$$

就有

$$\text{mes}\, \mathscr{E}\{x : |\chi_\varepsilon(x) - \chi(x, C^*)| \geqslant \delta, 0 \leqslant x \leqslant 2\pi\} < \eta.$$

换言之，任何能使量 $\pi^{-1}\int_0^{2\pi}\{K_n^*(u)-C^*\}\chi_\varepsilon(u)\mathrm{d}u$ 充分接近于量 $\mathscr{E}[\mathfrak{M}_L;L_n]_L$ 的函数 $\varphi_\varepsilon\in H_L^0$ 与 $\psi_\varepsilon\in H_M$，其卷积 $\chi_\varepsilon(u)$ 必在测度收敛意义下任意程度地逼近于 $\chi(u,C^*)$.

证明 假定命题2不成立，则有 $\delta_0>0,\eta_0>0$，对它们有数列 $\varepsilon_1,\cdots,\varepsilon_j,\cdots,\varepsilon_j\to 0$，以及对应的函数列 $\varphi_{\varepsilon_j}(x)\in H_L^0,\psi_{\varepsilon_j}(x)\in H_M,\chi_{\varepsilon_j}=\psi_{\varepsilon_j}*\varphi_{\varepsilon_j}$，能使

$$\lim_{j\to\infty}\int_0^{2\pi}\{K_n^*(u)-C^*\}\chi_{\varepsilon_j}(u)\mathrm{d}u=\int_0^{2\pi}|K_n^*(u)-C^*|\mathrm{d}u.$$

但

$$\operatorname{mes}E\{u:|\chi_{\varepsilon_j}(u)-\chi(u,C^*)|\geqslant\delta_0,0\leqslant u\leqslant 2\pi\}\geqslant\eta_0$$

$(j=1,2,\cdots)$. 由于

$$\{K_n^*(u)-C^*\}\chi_{\varepsilon_j}(u)\leqslant|K_n^*(u)-C^*|,$$

我们有

$$\begin{aligned}&|K_n^*(u)-C^*|-\{K_n^*(u)-C^*\}\chi_{\varepsilon_j}(u)\\&=\{K_n^*(u)-C^*\}\{\chi(u,C^*)-\chi_{\varepsilon_j}(u)\}\geqslant 0.\end{aligned}$$

从而得

$$\int_0^{2\pi}|K_n^*(u)-C^*||\chi_{\varepsilon_j}(u)-\chi(u,C^*)|\mathrm{d}u\to 0.$$

对任一自然数 m，定义

$$E_m=\{u:|K_n^*(u)-C^*|\geqslant m^{-1},0\leqslant u\leqslant 2\pi\}.$$

由于条件(2.11)，故显然有 $\lim\limits_{m\to\infty}\operatorname{mes}(E_m)=2\pi$. 对每一个 m，当 $j\to\infty$ 时，有

$$\begin{aligned}&m^{-1}\int_{E_m}|\chi_{\varepsilon_j}(u)-\chi(u,C^*)|\mathrm{d}u\\&\leqslant\int_0^{2\pi}|K_n^*(u)-C^*|\cdot|\chi_{\varepsilon_j}(u)-\chi(u,C^*)|\mathrm{d}u\to 0.\end{aligned}$$

故对每一个 m，有

$$\lim_{j\to\infty}\int_{E_m}|\chi_{\varepsilon_j}(u)-\chi(u,C^*)|\mathrm{d}u=0.$$

注意到

$$\int_0^{2\pi} |\chi_{\varepsilon_j}(u)-\chi(u,C^*)|\,du=\int_{E_m}+\int_{[0,2\pi]\setminus E_m},$$

对任取的 $\varepsilon>0$, 有 m 可使

$$\int_{[0,2\pi]\setminus E_m} |\chi_{\varepsilon_j}(u)-\chi(u,C^*)|\,du\leqslant 2(2\pi-\mathrm{mes}E_m)<\varepsilon/2.$$

对此 m 及 ε 有自然数 j_0, 对 $j>j_0$, 有

$$\int_{E_m} |\chi_{\varepsilon_j}(u)-\chi(u,C^*)|\,du<\varepsilon/2.$$

故对 $j>j_0$, 有 $\int_0^{2\pi} |\chi_{\varepsilon_j}(u)-\chi(u,C^*)|\,du<\varepsilon$. 此即表明

$$\lim_{j\to\infty}\|\chi_{\varepsilon_j}(u)-\chi(u,C^*)\|_L=0.$$

但 L 平均收敛含有 $\chi_{\varepsilon_j}(u)$ 对 $\chi(u,C^*)$ 的测度收敛, 得到矛盾. 命题 2 证完.

由命题 1,2 得

定理 1 对任意的 $K(t)\in L$ 定义的函数类 $\mathfrak{M}_X(X=C,L)$ 及满足 (1.5) 的 $K_n(t)$ 定义的 $L_n(f)$, 有

$$\mathscr{E}[\mathfrak{M}_L;L_n]_L\leqslant\mathscr{E}[\mathfrak{M}_C;L_n]_C=\frac{1}{\pi}E_1(K_n^*)_L.$$

为使上式左端有等号成立, 只须存在实数 C^*, 对任取的 $\varepsilon>0,\delta>0$, 有 $\varphi_\varepsilon\in H_L^0,\psi_\varepsilon\in H_M$, 使

$$\mathrm{mes}\,E\{u:|\chi_\varepsilon(u)-\chi(u,C^*)|\geqslant\delta,0\leqslant u\leqslant 2\pi\}<\varepsilon.$$

此时有

$$\mathscr{E}[\mathfrak{M}_L;L_n]_L=\mathscr{E}[\mathfrak{M}_C;L_n]_C=\frac{1}{\pi}\int_0^{2\pi}|K_n^*(u)-C^*|\,du.$$

反之, 如果(2.10), (2.11) 成立, 则上面的条件也是必要的.

推论 若存在实数 C^*, 对任取的 $\varepsilon>0$, 有 $\varphi_\varepsilon\in H_L^0$ 及 $\psi_\varepsilon\in H_M$, 使

$$\mathrm{mes}\,E\{u:\chi_\varepsilon(u)\neq\chi(u,C^*),0\leqslant u\leqslant 2\pi\}<\varepsilon,$$

则(2.10) 成立.

为应用方便起见, 下面给出能保证(2.10) 成立的一种充分条件.

定义 称 $K_n^*(t)$ 具有 N 性质($K_n^*\in N$), 若存在实数 C^*,β 及一自然数 p, 使

$$\sigma\chi(u,C^*)\cdot \operatorname{sgn}\sin(pu-\beta\pi)\geqslant 0\quad \text{a.e.},\tag{2.12}$$

$\sigma=\pm 1$.

定理 2 若 $K_n^*(t)\in N$，则

$$\mathscr{E}[\mathfrak{M}_L;L_n]_L=\mathscr{E}[\mathfrak{M}_C;L_n]_C=\frac{1}{\pi}E_1(K_n^*)_L$$
$$=\frac{1}{\pi}\left|\int_0^{2\pi}K_n^*(t)\operatorname{sgn}\sin(pt-\beta\pi)\mathrm{d}t\right|.\tag{2.13}$$

证明 不失一般性，不妨只考虑 $\beta=0$ 的情形．令 $x_k=\frac{k\pi}{p}$，$k=0,1,\cdots,2p-1$．以 x_k 为中心构作区间

$$\Delta_k=\left(x_k-\frac{\varepsilon}{4p},x_k+\frac{\varepsilon}{4p}\right),k=0,1,\cdots,2p-1.$$

令 ε 取得充分小，使 Δ_k 互不相交，构作函数

$$\varphi_\varepsilon(t)=\begin{cases}(-1)^k\dfrac{1}{\varepsilon},t\in\Delta \text{ 时},\\ 0,\qquad t\in[0,2\pi)\setminus\bigcup\limits_{k=0}^{2p-1}\Delta_k \text{ 时}.\end{cases}$$

$\varphi_\varepsilon(2\pi+t)=\varphi_\varepsilon(t)$．易见 $\varphi_\varepsilon(t)\in H_L^0$．再令 $\psi_\varepsilon(t)=\operatorname{sgn}\sin pt$，$\chi_\varepsilon(t)=\int_0^{2\pi}\varphi_\varepsilon(t+u)\operatorname{sgn}\sin pu\mathrm{d}u$，$\chi_\varepsilon(t)$ 是：

$$\chi_\varepsilon(t)=\begin{cases}(-1)^{k-1},t\in\left(x_k+\dfrac{\varepsilon}{4p},x_{k+1}-\dfrac{\varepsilon}{4p}\right),k=0,\cdots,2p-1.\\ \text{当 } t\in\Delta_k \text{ 时},\chi_\varepsilon(t)\text{ 是线性函数，在区间的端点保持连续}.\end{cases}$$

注意到

$$E\{t:\chi_\varepsilon(t)\neq\operatorname{sgn}\sin pt,0\leqslant t\leqslant 2\pi\}\subseteq\bigcup_{k=0}^{2p-1}\Delta_k,$$

而

$$\operatorname{mes}\left(\bigcup_{k=0}^{2p-1}\Delta_k\right)<2\varepsilon,$$

我们有

$$\operatorname{mes}E\{t:\chi_\varepsilon(t)\neq\operatorname{sgn}\sin pt,0\leqslant t\leqslant 2\pi\}<2\varepsilon.$$

由(2.6)得

$$\mathscr{E}[\mathfrak{M}_L;L_n]_L\geqslant\pi^{-1}\int_0^{2\pi}\{K_n^*(t)-C^*\}\chi_\varepsilon(t)\mathrm{d}t$$

$$= \frac{\sigma}{\pi}\int_0^{2\pi}\{K_n^*(t) - C^*\}\operatorname{sgn}\sin pt\,dt +$$

$$\frac{\sigma}{\pi}\int_0^{2\pi}\{K_n^*(t) - C^*\}\{\chi_\varepsilon(t) - \operatorname{sgn}\sin pt\}dt$$

$$= \frac{1}{\pi}\int_0^{2\pi} | K_n^*(t) - C^* | dt +$$

$$\frac{\sigma}{\pi}\int_0^{2\pi}\{K_n^*(t) - C^*\}\{\chi_\varepsilon(t) - \operatorname{sgn}\sin pt\}dt.$$

对任取的 $\varepsilon > 0$，可选取充分小的 $\delta > 0$，使对于 $\chi_\delta(t)$，有

$$\left|\frac{1}{\pi}\int_0^{2\pi}\{K_n^*(t) - C^*\}\{\chi_\delta(t) - \operatorname{sgn}\sin pt\}dt\right| < \varepsilon,$$

由此可得

$$\mathscr{E}[\mathfrak{M}_L; L_n]_L > \frac{1}{\pi}\int_0^{2\pi} | K_n^*(t) - C^* | dt - \varepsilon.$$

从而有

$$\mathscr{E}[\mathfrak{M}_L; L_n]_L = \frac{1}{\pi}\int_0^{2\pi} | K_n^*(t) - C^* | dt$$

$$= \frac{\sigma}{\pi}\int_0^{2\pi}\{K_n^*(t) - C^*\}\operatorname{sgn}\sin pt\,dt$$

$$= \frac{1}{\pi}\left|\int_0^{2\pi}K_n^*(t)\operatorname{sgn}\sin pt\,dt\right|.$$

定理 2 证完.

§3. 函数类 W_M^r, $\widetilde{W}_M^r$, W_L^r, $\widetilde{W}_L^r$ $(r = 1, 2, \cdots)$ 用弱正算子的逼近

置

$$K_r(t) = \sum_{k=0}^{\infty} k^{-r}\cos\left(kt + \frac{\pi r}{2}\right), \tag{3.1}$$

$$\widetilde{K}_r(t) = \sum_{k=1}^{\infty} k^{-r}\sin\left(kt + \frac{\pi r}{2}\right). \tag{3.2}$$

W_X^r, $\widetilde{W}_X^r(X = M, L)$ 的定义见 §1. 在这种情况下，作为 $K_n^*(t)$ 是

$$F_{nr}(t) = \sum_{k=1}^{\infty} \frac{1 - \lambda_k^{(n)}}{k^r}\cos\left(kt + \frac{\pi r}{2}\right), \tag{3.3}$$

$$\widetilde{F}_{nr}(t) = \sum_{k=1}^{\infty} \frac{1 - \lambda_k^{(n)}}{k^r}\sin\left(kt + \frac{\pi r}{2}\right). \tag{3.4}$$

置 $G_n(t)=\int_t^{\pi}K_n(u)\mathrm{d}u$,并设它满足条件

$$G_n(t)\geqslant 0,\quad 0\leqslant t\leqslant \pi. \tag{3.5}$$

我们称算子 $L_n(f)=\dfrac{1}{\pi}K_n * f$ 是弱正的,如果它的核 $K_n(t)$ 满足(3.5).

定理 3 若算子 $L_n(f)$ 是弱正的,则

$$\begin{aligned}\mathscr{E}[W_L^r;L_n]_L=\mathscr{E}[W_M^r;L_n]_C&=\frac{1}{\pi}E_1(F_{nr})_L\\&=\frac{1}{\pi}\left|\int_0^{2\pi}F_{nr}(t)\operatorname{sgn}\sin(t-\beta\pi)\mathrm{d}t\right|\\&=\frac{4}{\pi}\left|\sum_{\nu=0}^{\infty}\frac{(-1)^{\nu(r+1)}(1-\lambda_{2\nu+1}^{(n)})}{(2\nu+1)^{r+1}}\right|.\ n,r=1,2,\cdots\end{aligned} \tag{3.6}$$

此处 $\beta=0$,当 $r=1,3,5,\cdots;\beta=1/2,r=2,4,6,\cdots$ 时.

定理 3 的证明基于以下

命题 3 若 $G_n(t)\geqslant 0,0\leqslant t\leqslant\pi$,则 $F_{nr}(t)\in N(n,r=1,2,3,\cdots)$:

(1) $r=$ 奇数时,$\varepsilon_r\operatorname{sgn}F_{nr}(t)\cdot\operatorname{sgn}\sin t\geqslant 0$;

(2) $r=$ 偶数时,$\varepsilon_r\operatorname{sgn}\left\{F_{nr}(t)-F_{nr}\left(\dfrac{\pi}{2}\right)\right\}\operatorname{sgn}\cos t\geqslant 0.\ \varepsilon_r=\pm 1$.

证明 可仿[5].由于

$$K_n(t)\sim\frac{1}{2}+\sum_{k=1}^{\infty}\lambda_k^{(n)}\cos kt,$$

得知

$$\begin{aligned}G_n(t)=\int_t^{\pi}K_n(u)\mathrm{d}u&=\frac{\pi-t}{2}-\sum_{k=1}^{\infty}\frac{\lambda_k^{(n)}}{k}\sin kt\\&=\sum_{k=1}^{\infty}\frac{1-\lambda_k^{(n)}}{k}\sin kt=-F_{n1}(t).\end{aligned}$$

由条件(3.5)得

$$-\operatorname{sgn}F_{n1}(t)\operatorname{sgn}\sin t\geqslant 0\quad(n=1,2,\cdots). \tag{3.7}$$

注意到

$$\frac{\mathrm{d}}{\mathrm{d}t}F_{n,r+1}(t)=-F_{n,r}(t)\qquad(n,r=1,2,\cdots), \tag{3.8}$$

令 $r=1$,得

$$\operatorname{sgn}\left\{\frac{\mathrm{d}}{\mathrm{d}t}F_{n2}(t)\right\}\cdot\operatorname{sgn}\sin t\leqslant 0.$$

这说明 $F_{n2}(t)$ 在$[0,\pi]$单调下降,在$[\pi,2\pi]$单调上升. 再注意 $F_{n2}(t)$ 是偶函数,就得到

$$\operatorname{sgn}\left\{F_{n2}(t)-F_{n2}\left(\frac{\pi}{2}\right)\right\}\operatorname{sgn}\cos t\geqslant 0. \tag{3.9}$$

往下论证可以应用数学归纳法. 详细步骤略去. 命题 3 证完.

定理 3 包含了 В.П.Моторный[3] 的定理 2. 下面给出一条推论. 把 $K_n(t)$ 取为 Poisson 核

$$P_\rho(t)=\frac{1}{2}+\sum_{k=1}^{\infty}\rho^k\cos kt,0<\rho<1.$$

此时的 $L_n(f)$ 即 Poisson 积分

$$P_\rho(f;x)=\frac{1}{\pi}P_\rho * f.$$

作为 $K_n^*(t)$ 的是

$$F_{\rho r}(t)=\sum_{k=1}^{\infty}\frac{1-\rho^k}{k^r}\cos\left(kt+\frac{\pi r}{2}\right),$$

我们有

推论 $\mathscr{E}[W_L^r;P_\rho]_L$

$$\begin{aligned}&=\mathscr{E}[W_M^r;P_\rho]_C\\&=\frac{1}{\pi}E_1(F_{\rho r})_L=\frac{1}{\pi}\left|\int_0^{2\pi}F_{\rho r}(t)\operatorname{sgn}\sin(t-\beta\pi)\mathrm{d}t\right|\\&=\frac{4}{\pi}\left|\sum_{\nu=0}^{\infty}\frac{(-1)^{\nu(r+1)}(1-\rho^{2\nu+1})}{(2\nu+1)^{r+1}}\right|,r=1,2,3,\cdots\quad 0<\rho<1.\end{aligned}$$

$\beta=0,r=1,3,5,\cdots;\beta=1/2,r=2,4,6,\cdots$ 时.

定理 4 设算子 $L_n(f)$ 是弱正的,则

$$\begin{aligned}\mathscr{E}[\widetilde{W}_L^r;L_n]_L=\mathscr{E}[\widetilde{W}_M^r;L_n]_C&=\frac{1}{\pi}E_1(\widetilde{F}_{nr})_L\\&=\frac{1}{\pi}\left|\int_0^{2\pi}\widetilde{F}_{nr}(t)\operatorname{sgn}\sin(t-\beta\pi)\mathrm{d}t\right|\\&=\frac{4}{\pi}\left|\sum_{\nu=0}^{\infty}\frac{(-1)^{\nu r}(1-\lambda_{2\nu+1}^{(n)})}{(2\nu+1)^{r+1}}\right|,\\&\quad r=2,3,\cdots;n=1,2,\cdots.\end{aligned} \tag{3.10}$$

此处 $\beta=0,r=2,4,6,\cdots;\beta=1/2,r=3,5,\cdots$.

定理 4 的证明基于下列

命题 4 若 $G_n(t)\geqslant 0,0\leqslant t\leqslant\pi$,则 $\widetilde{F}_{nr}\in N(r=2,3,\cdots,n=1,$

2, …)：

(1) $r=2,4,6,\cdots$ 时，$\varepsilon_r \operatorname{sgn} \widetilde{F}_{nr}(t) \operatorname{sgn} \sin t \geqslant 0$；

(2) $r=3,5,7,\cdots$ 时，$\varepsilon_r \operatorname{sgn}\left\{\widetilde{F}_{nr}(t)-\widetilde{F}_{nr}\left(\frac{\pi}{2}\right)\right\} \operatorname{sgn} \cos t \geqslant 0$.

$\varepsilon_r=\pm 1$.

本命题可仿照命题 3 的证法. 由于

$$\frac{\mathrm{d}}{\mathrm{d}t}\widetilde{F}_{n,r+1}(t)=-\widetilde{F}_{nr}(t),$$

所以只要能证 $r=2$ 时命题的结论成立即可.

证明 注意到

$$G_n(t)=\sum_{k=1}^{\infty}\frac{1-\lambda_k^{(n)}}{k}\sin kt.$$

我们来考虑类 $\widetilde{W}_M^2$ 上用弱正算子 $L_n(f)$ 的一致逼近.

$$\begin{aligned}
L_n(f)-f(x) &= \frac{1}{\pi}\int_0^{\pi}\{f(x+u)+f(x-u)-2f(x)\}K_n(u)\mathrm{d}u \\
&= \frac{-1}{\pi}\{f(x+u)+f(x-u)-2f(x)\}G_n(u)\Big|_{u=0}^{u=\pi}+ \\
&\quad \frac{1}{\pi}\int_0^{\pi}\{f'(x+u)-f'(x-u)\}G_n(u)\mathrm{d}u \\
&= \frac{1}{\pi}\int_0^{\pi}\{f'(x+u)-f'(x-u)\}G_n(u)\mathrm{d}u \\
&= \frac{1}{\pi}\left\{\int_0^{\pi/2}+\int_{\pi/2}^{\pi}\right\}\{f'(x+u)-f'(x-u)\}G_n(u)\mathrm{d}u.
\end{aligned}$$

由于

$$\begin{aligned}
&\int_{\pi/2}^{\pi}\{f'(x+u)-f'(x-u)\}G_n(u)\mathrm{d}u \\
&= \int_0^{\pi/2}\{f'(x+\pi-u)-f'(x-\pi+u)\}G_n(\pi-u)\mathrm{d}u \\
&= \int_0^{\pi/2}\{f'(x+\pi-u)-f'(x+\pi+u)\}G_n(\pi-u)\mathrm{d}u.
\end{aligned}$$

从而有

$$\begin{aligned}
L_n(f;x)-f(x) = \frac{1}{\pi}\Big\{&\int_0^{\pi/2}\{f'(x+u)-f'(x-u)\}G_n(u)\mathrm{d}u+ \\
&\int_0^{\pi/2}\{f'(x+\pi-u)-f'(x+\pi+u)\}G_n(u)\mathrm{d}u\Big\}.
\end{aligned} \tag{3.11}$$

由于 $f\in\widetilde{W}_M^2$ 蕴含 $f'\in\widetilde{W}_M^1$，根据[6]，有

$$\sup_{f\in\widetilde{M}'}\max_x\left|f'(x+u)-f'(x-u)\right|\leqslant\sup_{\varphi\in\widetilde{M}'}\omega(\varphi,2u)_C$$

$$=\omega(\tilde{\varphi}_1,2u),0\leqslant u\leqslant\frac{\pi}{2}.$$ 此处

$$\tilde{\varphi}_1(t)=\frac{4}{\pi}\sum_{\nu=0}^{\infty}\frac{\sin(2\nu+1)t}{(2\nu+1)^2},$$

$$\omega(\tilde{\varphi}_1,2u)_C=\frac{8}{\pi}\sum_{\nu=0}^{\infty}\frac{\sin(2\nu+1)u}{(2\nu+1)^2},0\leqslant u\leqslant\pi/2.$$

同样有

$$\sup_{f\in\widetilde{M}'}\max_x\left|f'(x+\pi-u)-f'(x+\pi+u)\right|\leqslant\omega(\tilde{\varphi}_1;2u).$$

$$(0\leqslant u\leqslant\pi/2)$$

由(3.11)得

$$\begin{aligned}\mathscr{E}[\widetilde{W}_M^2;L_n]_C&\leqslant\frac{1}{\pi}\left\{\int_0^{\pi/2}\omega(\tilde{\varphi}_1,2u)_CG_n(u)\mathrm{d}u+\right.\\&\left.\int_0^{\pi/2}\omega(\tilde{\varphi}_1;2u)_CG_n(\pi-u)\mathrm{d}u\right\}\\&=\frac{1}{\pi}\int_0^{\pi}\frac{8}{\pi}\sum_{\nu=0}^{\infty}\frac{\sin(2\nu+1)u}{(2\nu+1)^2}\cdot\sum_{k=1}^{\infty}\frac{1-\lambda_k^{(n)}}{k}\sin ku\,\mathrm{d}u\\&=\frac{4}{\pi}\sum_{\nu=0}^{\infty}\frac{1-\lambda_{2\nu+1}^{(n)}}{(2\nu+1)^3}.\end{aligned}\tag{3.12}$$

另一方面，由(2.6)，依照定理2证明方法来做，即得

$$\begin{aligned}\mathscr{E}[\widetilde{W}_L^2;L_n]_L&\geqslant\frac{1}{\pi}\int_0^{2\pi}\widetilde{F}_{n2}(t)\cdot(-\operatorname{sgn}\sin t)\mathrm{d}t\\&=\frac{4}{\pi^2}\int_0^{2\pi}\sum_{k=1}^{\infty}\frac{1-\lambda_k^{(n)}}{k^2}\sin kt\sum_{\nu=0}^{\infty}\frac{\sin(2\nu+1)t}{2\nu+1}\mathrm{d}t\\&=\frac{4}{\pi}\sum_{\nu=0}^{\infty}\frac{1-\lambda_{2\nu+1}^{(n)}}{(2\nu+1)^3}.\end{aligned}\tag{3.13}$$

由(3.12)，(3.13)得

$$\begin{aligned}E_1(\widetilde{F}_{n2})_L&=4\sum_{\nu=0}^{\infty}\frac{1-\lambda_{2\nu+1}^{(n)}}{(2\nu+1)^3}=-\int_0^{2\pi}\widetilde{F}_{n2}(t)\operatorname{sgn}\sin t\,\mathrm{d}t\\&=\min_C\int_0^{2\pi}\left|\widetilde{F}_{n2}(t)-C\right|\mathrm{d}t=\int_0^{2\pi}\left|\widetilde{F}_{n2}(t)-C^*\right|\mathrm{d}t\\&=\int_0^{2\pi}\{\widetilde{F}_{n2}(t)-C^*\}\{-\operatorname{sgn}\sin t\}\mathrm{d}t.\end{aligned}$$

由于

$$\left|\widetilde{F}_{n2}(t)-C^*\right|\geqslant\{\widetilde{F}_{n2}(t)-C^*\}\{-\operatorname{sgn}\sin t\},$$

且

$$\left|\widetilde{F}_{n2}(t)-C^*\right|=\{\widetilde{F}_{n2}(t)-C^*\}\cdot\operatorname{sgn}\{\widetilde{F}_{n2}(t)-C^*\},$$

得

$$\{\widetilde{F}_{n2}(t)-C^*\}\{\operatorname{sgn}(\widetilde{F}_{n2}(t)-C^*)+\operatorname{sgn}\sin t\}\geqslant 0.$$

从而有

$$\int_0^{2\pi}\left|\widetilde{F}_{n2}(t)-C^*\right|\cdot\left|\operatorname{sgn}(\widetilde{F}_{n2}(t)-C^*)+\operatorname{sgn}\sin t\right|\mathrm{d}t=0.$$

这样一来,在点集

$$\mathscr{E}=\{t:\widetilde{F}_{n2}(t)-C^*\neq 0,0\leqslant t\leqslant 2\pi\}$$

上几乎处处有

$$-\operatorname{sgn}\{\widetilde{F}_{n2}(t)-C^*\}=\operatorname{sgn}\sin t$$

成立.从而得

$$-\operatorname{sgn}\{\widetilde{F}_{n2}(t)-C^*\}\operatorname{sgn}\sin t\geqslant 0\quad\text{a.e.}\tag{3.14}$$

由于 $\widetilde{F}_{n2}(t)$ 是奇函数,知必有 $C^*=0$,故得

$$-\operatorname{sgn}\widetilde{F}_{n2}(t)\cdot\operatorname{sgn}\sin t\geqslant 0.$$

命题4在 $r=2$ 时成立.证毕.

定理4包含了[3]的定理3.作为推论,我们取 f 的Poisson积分,注意这时有

$$\widetilde{F}_{\rho r}(t)=\sum_{k=1}^{\infty}\frac{1-\rho^k}{k^r}\sin\left(kt+\frac{\pi r}{2}\right).\tag{3.15}$$

可得

推论　当 $r=1,2,\cdots,0<\rho<1$ 时,有

$$\begin{aligned}\mathscr{E}[\widetilde{W}_L^r;P_\rho]_L=\mathscr{E}[\widetilde{W}_M^r;P_\rho]_C&=\frac{1}{\pi}E_1(\widetilde{F}_{\rho r})_L\\&=\frac{1}{\pi}\left|\int_0^{2\pi}\widetilde{F}_{\rho r}(t)\operatorname{sgn}\sin(t-\beta\pi)\mathrm{d}t\right|\\&=\frac{4}{\pi}\left|\sum_{\nu=0}^{\infty}\frac{(-1)^{\nu r}(1-\rho^{2\nu+1})}{(2\nu+1)^{r+1}}\right|.\end{aligned}\tag{3.16}$$

此处 $\beta=0,r=2,4,6,\cdots;\beta=1/2,r=1,3,5,\cdots$ 时.

证明　我们来验证 $\widetilde{F}_{\rho r}(t)\in N$:　(3.17)

(1) $r=1,3,5,\cdots$ 时,

$$\varepsilon_r \operatorname{sgn}\{\widetilde{F}_{\rho r}(t) - \widetilde{F}_{\rho r}(\pi/2)\} \operatorname{sgn} \cos t \geqslant 0;$$

(2) $r = 2, 4, 6, \cdots$ 时,

$$\varepsilon_r \operatorname{sgn} \widetilde{F}_{\rho r}(t) . \operatorname{sgn} \sin t \geqslant 0.$$

$\varepsilon_r = \pm 1$. $r \geqslant 2$ 的情形包含在命题 4 中, 故只须验证一下 $r = 1$ 的情形即可.

$$\widetilde{F}_{\rho 1}(t) = \sum_{k=1}^{\infty} \frac{\cos kt}{k} - \sum_{k=1}^{\infty} \frac{\rho^k}{k} \cos kt$$

$$= \ln \frac{1}{2\sin \frac{t}{2}} - \sum_{k=1}^{\infty} \frac{\rho^k}{k} \cos kt \qquad (0 < \rho < 1).$$

由
$$\frac{\mathrm{d}}{\mathrm{d}t} \widetilde{F}_{\rho 1}(t) = -\frac{1}{2} \cot \frac{t}{2} + \sum_{k=1}^{\infty} \rho^k \sin kt$$

$$= \rho \sin t \left\{ \frac{1}{(1-\rho)^2 + 4\rho \sin^2 \frac{t}{2}} - \frac{1}{4\rho \sin^2 \frac{t}{2}} \right\} < 0$$

$(0 < t \leqslant \pi)$, 得知 $\widetilde{F}_{\rho 1}(t)$ 在 $(0, \pi)$ 内是单调下降的. 它是偶函数. 故

$$\operatorname{sgn}\left\{ \widetilde{F}_{\rho 1}(t) - \widetilde{F}_{\rho 1}\left(\frac{\pi}{2}\right) \right\} = \operatorname{sgn} \cos t.$$

即 $r = 1$ 时, $\widetilde{F}_{\rho 1}(t) \in N$. 证毕.

我们还可给出保证 $\widetilde{F}_{nr}(t) \in N$ 的其他充分条件.

命题 5 设 $[0,1]$ 上的连续函数 $g(t)$ 满足 $g(1) = 0$, $t^{-1}\{1 - g(t)\} \downarrow (0 < t \leqslant 1)$, $\lambda_k^{(n)} = g\left(\frac{k}{n+1}\right)$. ($k \geqslant n+1$ 时, 规定 $\lambda_k^{(n)} = 0$). 则 $\widetilde{F}_{nr} \in N$:

(1) $r = 2, 4, 6, \cdots$ 时,

$$\varepsilon_r \operatorname{sgn} \widetilde{F}_{nr}(t) \cdot \operatorname{sgn} \sin t \geqslant 0;$$

(2) $r = 3, 5, \cdots$ 时,

$$\varepsilon_r \operatorname{sgn}\left\{ \widetilde{F}_{nr}(t) - \widetilde{F}_{nr}\left(\frac{\pi}{2}\right) \right\} \operatorname{sgn} \cos t \geqslant 0. \ \varepsilon_r = \pm 1.$$

证明 和命题 4 一样, 只需考虑 $r = 2$ 的情形.

$$-\widetilde{F}_{n2}(t) = \sum_{k=1}^{n} \frac{1 - g\left(\frac{k}{n+1}\right)}{k} \cdot \frac{\sin kt}{k} + \sum_{k=n+1}^{\infty} \frac{1}{k} \cdot \frac{\sin kt}{k}$$

$$= \sum_{k=1}^{n-1} \left\{ \frac{1 - g\left(\frac{k}{n+1}\right)}{k} - \frac{1 - g\left(\frac{k+1}{n+1}\right)}{k+1} \right\} \sum_{\mu=1}^{k} \frac{\sin \mu t}{\mu} +$$

$$\frac{1-g\left(\frac{n}{n+1}\right)}{n}\sum_{\mu=1}^{n}\frac{\sin\mu t}{\mu}+\sum_{k=\mu+1}^{\infty}\left(\frac{1}{k}-\frac{1}{k+1}\right)\sum_{\mu=1}^{k}\frac{\sin\mu t}{\mu}-$$

$$\frac{1}{n+1}\sum_{\mu=1}^{n}\frac{\sin\mu t}{\mu}$$

$$=\sum_{k=n+1}^{\infty}\left(\frac{1}{k}-\frac{1}{k+1}\right)\sum_{\mu=1}^{k}\frac{\sin\mu t}{\mu}+$$

$$\frac{1}{n+1}\sum_{k=1}^{n}\left\{\frac{1-g\left(\frac{k}{n+1}\right)}{\frac{k}{n+1}}-\frac{1-g\left(\frac{k+1}{n+1}\right)}{\frac{k+1}{n+1}}\right\}\sum_{\mu=1}^{k}\frac{\sin\mu t}{\mu}.$$

熟知(见[7],第106页)

$$\sum_{\mu=1}^{n}\frac{\sin\mu t}{\mu}>0,\text{当 }0<t<\pi\text{ 时},(n=1,2,\cdots).$$

由此得

$$\operatorname{sgn}\widetilde{F}_{n2}(t)=-\operatorname{sgn}\sin t.$$

命题证完.

定理5 若 $g(t)$ 如命题5中给出者,则对 $r=2,3,\cdots$,有

$$\mathscr{E}[\widetilde{W}_L^r,L_n]_L=\mathscr{E}[\widetilde{W}_M^r;L_n]_C=\frac{1}{\pi}E_1(\widetilde{F}_{nr})_L$$

$$=\frac{1}{\pi}\left|\int_0^{2\pi}\widetilde{F}_{nr}(t)\operatorname{sgn}\sin(t-\beta\pi)\mathrm{d}t\right|$$

$$=\frac{4}{\pi}\left|\sum_{\nu=0}^{\infty}\frac{(-1)^{\nu r}\left\{1-g\left(\frac{2\nu+1}{n+1}\right)\right\}}{(2\nu+1)^{r+1}}\right|.\tag{3.18}$$

此处 $g\left(\frac{k}{n+1}\right)=0,k\geqslant n+1$ 时;$\beta=0,r=2,4,6,\cdots$ 时;$\beta=1/2,r=3,5,7,\cdots$ 时.

推论 $g(t)=1-t^{\alpha}\quad 0<\alpha\leqslant1,0\leqslant t\leqslant1.\lambda_k^{(n)}=1-\left(\frac{k}{n+1}\right)^{\alpha}$.则

$$\mathscr{E}[\widetilde{W}_L^r,L_n]_L=\mathscr{E}[\widetilde{W}_M^r;L_n]_C$$

$$=\frac{1}{\pi}E_1(\widetilde{F}_{nr})_L$$

$$=\frac{4}{\pi}\left|\frac{1}{(n+1)^{\alpha}}\sum_{\nu=0}^{\left[\frac{n-1}{2}\right]}\frac{(-1)^{\nu r}}{(2\nu+1)^{r+1}}+\sum_{\nu=\left[\frac{n-1}{2}\right]+1}^{\infty}\frac{(-1)^{\nu r}}{(2\nu+1)^{r+1}}\right|.$$

$$(r = 2,3,\cdots, n = 1,2,\cdots). \tag{3.19}$$

§4. 类 $\mathfrak{M}_X^\omega$ $(X = C, L)$ 上的逼近

本节内 $\omega(t)$ 为上凸连续模. 同 §2(参看[3]), 有

$$\mathscr{E}[\mathfrak{M}_L^\omega; L_n]_L \leqslant \mathscr{E}[\mathfrak{M}_C^\omega; L_n]_C. \tag{4.1}$$

首先讨论(4.1) 内有等号成立的条件. 与(2.6) 一样, 有

$$\mathscr{E}[\mathfrak{M}_L^\omega; L_n]_L = \frac{1}{\pi}\sup_{\varphi\in H_L^\omega}\sup_{\psi\in H_M}\int_0^{2\pi}K_n^*(u)\left\{\int_0^{2\pi}\varphi(u+x)\psi(x)\mathrm{d}x\right\}\mathrm{d}u. \tag{4.2}$$

注意 $\chi(u) = \int_0^{2\pi}\varphi(u+x)\psi(x)\mathrm{d}x \in H_C^\omega$. 显然, 若以下两条件能满足:

(1) 有 $\tilde{\varphi} \in H_C^\omega$, 使

$$\mathscr{E}[\mathfrak{M}_C^\omega; L_n]_C = \frac{1}{\pi}\int_0^{2\pi}K_n^*(u)\tilde{\varphi}(u)\mathrm{d}u, \tag{4.3}$$

(2) 存在 $\varphi^*(u) \in H_L^\omega, \psi^*(u) \in H_M$, 使

$$\tilde{\varphi}(u) = \int_0^{2\pi}\varphi^*(u+x)\psi^*(x)\mathrm{d}x. \tag{4.4}$$

则(4.1) 内有等号成立.

从这一简单事实得

定理 6 若 $K_n^*(t)$ 为一奇函数, 且当 $0 \leqslant t \leqslant \pi$ 时, $K_n^*(t) \geqslant 0$ $(\leqslant 0)$, 则

$$\mathscr{E}[\mathfrak{M}_L^\omega, L_n]_L = \mathscr{E}[\mathfrak{M}_C^\omega; L_n]_C$$

$$= \frac{2}{\pi}\int_0^{\pi}K_n^*(u)\tilde{\omega}(u)\mathrm{d}u. \tag{4.5}$$

此处

$$\tilde{\omega}(u) = \begin{cases}\dfrac{1}{2}\omega(2u), & 0 \leqslant u \leqslant \dfrac{\pi}{2},\\[2mm] \dfrac{1}{2}\omega(2\pi - 2u), & \dfrac{\pi}{2} \leqslant u \leqslant \pi.\end{cases} \tag{4.6}$$

$$\tilde{\omega}(-u) = -\tilde{\omega}(u), \tilde{\omega}(u+2\pi) = \tilde{\omega}(u).$$

证明 $\tilde{\omega}(t) \in H_C^\omega$ 是熟知的(例如见[8], 第 188 页). В. И. Бердышев(见[3]) 证明:

函数

$$
\begin{cases}
\phi^*(u) = -\dfrac{1}{8}\dfrac{\mathrm{d}}{\mathrm{d}u}\omega(\pi - 2u), & 0 < u < \dfrac{\pi}{2},\\
\phi^*\left(\dfrac{\pi}{2} + u\right) = \phi^*\left(\dfrac{\pi}{2} - u\right), & 0 \leqslant u \leqslant \dfrac{\pi}{2},\\
\phi^*(-u) = -\phi^*(u),\\
\phi^*(u + 2\pi) = \phi^*(u)
\end{cases}
\tag{4.7}
$$

有 $\phi^*(u) \in H_L^{\omega}$. В. П. Моторный[3] 证明：

若取 $\psi^*(u) = \operatorname{sgn}\cos u \in H_M$，则

$$
\tilde{\omega}(t) = \int_0^{2\pi} \phi^*(t + u)\psi^*(u)\mathrm{d}u. \tag{4.8}
$$

从以上这些结果，为证定理 6，只要能证

$$
\mathscr{E}[\mathfrak{M}_C^{\omega}; L_n]_C = \frac{2}{\pi}\int_0^{\pi} K_n^*(t)\tilde{\omega}(t)\mathrm{d}t \tag{4.9}
$$

即可. 由

$$
f(0) - L_n(f;0) = \frac{1}{\pi}\int_0^{\pi}\{\varphi(t) - \varphi(-t)\}K_n^*(t)\mathrm{d}t,
$$

当 $0 \leqslant t \leqslant \pi/2$ 时，

$$
|\varphi(t) - \varphi(-t)| \leqslant \omega(2t).
$$

当 $\pi/2 \leqslant t \leqslant \pi$ 时，令 $t = \pi - t'$，有

$$
\begin{aligned}
|\varphi(t) - \varphi(-t)| &= |\varphi(\pi - t') - \varphi(-\pi + t')| \\
&= |\varphi(\pi - t') - \varphi(\pi + t')| \\
&\leqslant \omega(2t') = \omega(2\pi - 2t).
\end{aligned}
$$

由 $\tilde{\omega}(t)$ 的定义知

$$
|\varphi(t) - \varphi(-t)| \leqslant 2\tilde{\omega}(t) \quad \forall\, \varphi \in H_C^{\omega},
$$

但 $\tilde{\omega}(t) \in H_C^{\omega}$，故定理 6 得证.

定理 7 设 $L_n(f)$ 为弱正的，则当 $r = 1, 3, 5, \cdots$ 时，有

$$
\begin{aligned}
\mathscr{E}[W^r H_L^{\omega}; L_n]_L &= \mathscr{E}[W^r H_C^{\omega}; L_n]_C \\
&= \frac{2}{\pi}\left|\int_0^{\pi} F_{nr}(t)\tilde{\omega}(t)\mathrm{d}t\right|.
\end{aligned}
\tag{4.10}
$$

证明 因由命题 3 知 $\varepsilon_r \operatorname{sgn} F_{nr}(t) \cdot \operatorname{sgn}\sin t \geqslant 0$. ($r = 1, 3, 5, \cdots$). 故由定理 6 即得(4.10).

定理 8 若 $L_n(f)$ 为弱正的, $r=2,4,6,\cdots$, 则

$$\mathscr{E}[\widetilde{W}^r H_L^\omega; L_n]_L = \mathscr{E}[\widetilde{W}^r H_C^\omega; L_n]_C = \frac{2}{\pi}\left|\int_0^\pi \widetilde{F}_{nr}(t)\tilde{\omega}(t)\mathrm{d}t\right|. \quad (n=1,2,\cdots) \tag{4.11}$$

证明 由命题 4 知: $r=2,4,6,\cdots$ 时,

$$\varepsilon_r \operatorname{sgn}\widetilde{F}_{nr}(t)\cdot \operatorname{sgn}\sin t \geqslant 0.$$

$\varepsilon_r=\pm 1$. 故由定理 6 即得(4.11).

定理 7,8 包含了[3] 中的定理 4,5(又参看[9]).

定理 9 设 $g(t)$ 满足命题 5 的条件, 且

$K_n^\lambda(t)=\dfrac{1}{2}+\sum\limits_{k=1}^n \lambda_k^{(n)}\cos kt$, $\lambda_k^{(n)}=g\left(\dfrac{k}{n+1}\right)$. 则当 $r=2,4,6\cdots$ 时, 有

$$\mathscr{E}[\widetilde{W}^r H_L^\omega; L_n]_L = \mathscr{E}[\widetilde{W}^r H_C^\omega; L_n]_C = \frac{2}{\pi}\left|\int_0^\pi \widetilde{F}_{nr}(t)\tilde{\omega}(t)\mathrm{d}t\right| \quad (n=1,2,3,\cdots). \tag{4.12}$$

利用本文中给出的逼近度的解析表达式, 当算子 $L_n(f)$ 的核 $K_n^\lambda(t)$ 的傅立叶系数 $\lambda_k^{(n)}$ 具体给出时, 即可据以计算逼近度的渐近展开的主项.

参考文献

[1] Никольский С. М. Приближение функций тригонометрическими полиномами в среднем. Изв. А Н СССР, 1946, серuя матем. (10): 207 ~256

[2] Стечкин С. В., Теляковский С. А. О приближении дифференцируемых функций тригонометрическими полиномами в метрике *L*. *Труды Матем. цн-та А Н СССР*, 1967, (88): 20 ~ 29

[3] Моторный В. П. Приближение периодических функций тригонометрическими многочленами в среднем *Матем. заметки,* 1974, 16(1): 15 ~ 26

[4] Тиман А. Ф. *Теория Приближения функций действительного переменного.* М.: ФМ гиз, 1960, 183

[5] Никольский С. М. Приближение периодических функций тригонометрическими многочленами. Труды Матем цн-та АН СССР, 1945, (15): 30 ~ 31

[6] Моторный В. П. Об одном неравенстве для модулей гладкости периодической

функции с ограниченной производной. ДАН СССР, 1964, 154(1):45 ～ 47

[7] Зигмунд А. *Тригонометрические Ряды*. Т. 1, М.: Издат.《МИР》, 1965.106

[8] Корнейчук Н. П., *Экстремальные Задачи теории приближения*. М.: Издат.《НАУКА》, 1976.188

[9] Баскаков В. А. *Теория прибли. функций*. М.: Издат.《НАУКА》, 1975.28 ～ 32

二、B-核(广义 Bernoulli 核,CVD 核)宽度精确计算问题

II. Exact Evaluation of the Widths for B-kernel (Generalized Bernoulli Kernel, CVD Kernel)

Chin. Ann. of Math.
1981, 2(Eng. Issue)

一个解析的周期函数类的 L_1 宽度*

On n – width of a Class of Periodic Analytic Functions in L_1 Metric

1. Let $f(x)$ be a 2π-periodic real analytic function which has analytic continuation $f(x+\mathrm{i}y)=u(x,y)+\mathrm{i}v(x,y)$ in the region $-\infty<x<+\infty$, $-\delta<y<+\delta$ $(\delta>0)$, and the real part $u(x,y)$ satisfies the condition $\|u(\cdot,y)\|_{L_p}\leqslant 1(1\leqslant p\leqslant+\infty)$, $\forall y\in(-\delta,\delta)$, where

$$\|u(\cdot,y)\|_{L_p}=\begin{cases}\left\{\int_0^{2\pi}|u(x,y)|^p\mathrm{d}x\right\}^{1/p} & \text{for } 1\leqslant p<+\infty,\\ \operatorname*{ess\,sup}_x|u(x,y)| & \text{for } p=+\infty.\end{cases}$$

The class of all such functions is denoted by $H_\delta(L_p)$. It is well-known([1], p. 153) that for $1<p\leqslant+\infty$ we have

$$f(x)=(2\pi)^{-1}\int_0^{2\pi}H(x-t)g(t)\mathrm{d}t, \tag{1}$$

where $\|g\|_{L_p}\leqslant 1$ and for $p=1$

$$f(x)=(2\pi)^{-1}\int_0^{2\pi}H(x-t)\mathrm{d}\lambda(t), \tag{2}$$

where $\lambda(t)\in V[0,2\pi]$, $\int_0^{2\pi}|\mathrm{d}\lambda|\leqslant 1$,

$$H(t)=1+4\sum_{k=1}^{\infty}\frac{q^k}{1+q^{2k}}\cos kt\,(q=\mathrm{e}^{-\delta}) \tag{3}$$

---

* Manuscript received Apr. 24, 1980.

([2],p.207), and $V[0,2\pi]$ denotes the class of functions of bounded-variation on the interval $[0,2\pi]$. In the case $p=1$ we need to introduce the class

$$H_\delta^*(L_1)=\left\{f: f(x)=(2\pi)^{-1}\int_0^{2\pi}H(x-t)g(t)\mathrm{d}t,\ \|g\|_{L_1}\leqslant 1\right\}. \tag{4}$$

It is obvious that $H_\delta^*(L_1)\subset H_\delta(L_1)$ and the L_1-closure of $H_\delta^*(L_1)$ contains $H_\delta(L_1)$.

Let X_n be any n-dimensional subspace of L_p. For a given set of functions $\mathfrak{M}\subset L_p$ let us put

$$E(\mathfrak{M};X_n)_{L_p}=\sup_{f\in\mathfrak{M}}E(f;X_n)_{L_p}, \tag{5}$$

where $E(f;X_n)_{L_p}=\min_{\varphi\in X_n}\|f-\varphi\|_{L_p}$

denotes the best approximation of f when it is approximated by the subspace X_n in L_p metric. It is well-known that

$$d_n(\mathfrak{M})_{L_p}=\inf_{X_n}\sup_{f\in\mathfrak{M}}E(f;X_n)_{L_p} \tag{6}$$

is the n-width of $\mathfrak{M}$ in the sense of kolmogoroff in L_p metric, and

$$d'_n(\mathfrak{M})_{L_p}=\inf_{X_n}\inf_{\substack{A\\ A(\mathfrak{M})\subset X_n}}\sup_{f\in\mathfrak{M}}\|f-A(f)\|_{L_p} \tag{7}$$

is the n-dimensional linear width of $\mathfrak{M}$ in L_p metric, where A is any linear continuous mapping from $\mathfrak{M}$ to X_n. In what follows the $(2n-1)$-dimensional subspace of all trigonometric polynomials of order $\leqslant n-1$ will be denoted by T_n, and the best approximation of f by T_n in L_p will be written as

$$E_n(f)_{L_p}=E(f;T_n)_{L_p}=\min_{t_n\in T_n}\|f-t_n\|_{L_p}. \tag{8}$$

Ахиезер, Н. И.[2] has proved that

$$E_n(H_\delta(L_\infty))_{L_\infty}\overset{\text{def}}{=\!=}\sup_{f\in H_\delta(L_\infty)}E_n(f)_{L_\infty}=(2\pi)^{-1}E_n(H)_{L_1}$$

$$=\frac{4}{\pi}\sum_{k=1}^{\infty}\frac{(-1)^k}{(2k+1)\operatorname{ch}(2k+1)n\delta}. \tag{9}$$

From the result of [2] we know that

$$(\mathrm{i})\ f_{n,\delta}(x)=(2\pi)^{-1}\int_0^{2\pi}H(x-t)\operatorname{sgn}\cos nt\,\mathrm{d}t \tag{10}$$

is the function for which supremum in (9) is realized. Put

$$x_k = \frac{k\pi}{n}, \xi_k = \frac{\pi}{2n} + \frac{k\pi}{n}, k = 0, 1, \cdots, 2n - 1.$$

It is easily verified that

$$f_{n,\delta}(\xi_k) = 0, \quad f_{n,\delta}(x_k) = (-1)^k E_n(H)_{L_1} \tag{11}$$

and sgn $f_{n,\delta}(x) =$ sgn cos nx;

(ⅱ) there exists a trigonometric polynomial $T^*_{n-1}(t)$ of order $\leqslant n-1$ such that

$$\operatorname{sgn}\{H(t) - T^*_{n-1}(t)\} = \operatorname{sgn} \cos nt, \tag{12}$$

i.e., $H(t)$ satisfies Markov-Nikolskii's condition (see[3]). Therefore we have

$$E_n(H_\delta(L_1))_{L_1} \overset{\text{def}}{=} \sup_{f \in H_\delta(L_1)} E_n(f)_{L_1} = E_n(H^*_\delta(L_1))_{L_1}$$

$$= E_n(H_\delta(L_\infty))_{L_\infty} = (2\pi)^{-1} E_n(H)_{L_1} \tag{13}$$

and

$$E'_n(H_\delta(L_1))_{L_1} \overset{\text{def}}{=} \inf_{A_n} \sup_{f \in H_\delta(L_1)} \| f - A_n(f) \|_{L_1} = (2\pi)^{-1} E_n(H)_{L_1}, \tag{14}$$

where A_n is any linear continuous mapping from $H_\delta(L_1)$ to T_n.

2. Recently Forst, W. has proved in [4] (see also [5])

$$d_{2n-1}(H_\delta(L_\infty))_{L_\infty} = d_{2n}(H_\delta(L_\infty))_{L_\infty} = d'_{2n-1}(H_\delta(L_\infty))_{L_\infty}$$

$$= d'_{2n}(H_\delta(L_\infty))_{L_\infty} = (2\pi)^{-1} E_n(H)_{L_1}. \tag{15}$$

In this note we shall give an exact estimation of the $(2n-1)$-width number of $H_\delta(L_1)$ in L_1 metric. Our main results are the following two theorems.

Theorem 1

$$d_{2n-1}(H_\delta(L_1))_{L_1} = d_{2n-1}(H^*_\delta(L_1))_{L_1} = d'_{2n-1}(H_\delta(L_1))_{L_1}$$

$$= d'_{2n-1}(H^*_\delta(L_1))_{L_1}$$

$$= (2\pi)^{-1} E_n(H)_{L_1}, n = 1, 2, 3 \cdots. \tag{16}$$

Proof On account of $H^*_\delta(L_1) \subset H_\delta(L_1) \subset \overline{H^*_\delta(L_1)}$, it is easily proved that

$$d_{2n-1}(H_\delta(L_1))_{L_1} = d_{2n-1}(H^*_\delta(L_1))_{L_1},$$

$$d'_{2n-1}(H_\delta(L_1))_{L_1} = d'_{2n-1}(H_\delta^*(L_1))_{L_1}.$$

Putting $U_n^*(f;x) = (2\pi)^{-1}\int_0^{2\pi} T_{n-1}^*(x-t)\varphi(t)\mathrm{d}t$, $\|\varphi\|_{L_1} \leqslant 1$, we can derive

$$d_{2n-1}(H_\delta(L_1))_{L_1} \leqslant d'_{2n-1}(H_\delta(L_1))_{L_1} \leqslant (2\pi)^{-1}E_n(H)_{L_1}, \quad (17)$$

from the relations

$$d'_{2n-1}(H_\delta(L_1))_{L_1} \leqslant \sup_{f\in H_\sigma(L_1)} \|f - U_n^*(f)\|_{L_1} = (2\pi)^{-1}E_n(H)_{L_1}.$$

The validity of the last equality can be derived from a theorem of Nikolskii, S.M. (see [6], p. 91). So we need only to prove

$$d_{2n-1}(H_\delta(L_1))_{L_1} \geqslant (2\pi)^{-1}E_n(H)_{L_1}. \quad (18)$$

Let us define $2n$ functions $x_j(t)$ $(j = 0,1,\cdots,2n-1)$ such that

$$x_j(t) = \begin{cases} 1, & \xi_j < t \leqslant \xi_{j+1}, \\ 0, & t \in [0,2\pi] \setminus (\xi_j, \xi_{j+1}), j = 0,1,\cdots,2n-1 \end{cases}$$

$(\chi_j(t+2\pi) = \chi_j(t))$, and let X_{2n-1} be any $(2n-1)$-dimensional subspace of L_1. Suppose $X_{2n-1} = \mathrm{span}\{e_1(t),\cdots,e_{2n-1}(t)\}$. There exists a system of numbers $(\beta_0,\cdots,\beta_{2n-1}) \neq (0,\cdots,0)$ such that the function $h_\beta(t) = \sum_{j=0}^{2n-1}\beta_j\chi_j(t)$ satisfies $\int_0^{2\pi} h_\beta(t)\varphi(t)\mathrm{d}t = 0$, $\forall \varphi \in X_{2n-1}$, because the system of homogeneous linear equations $\int_0^{2\pi} e_i(t)h_\beta(t)\mathrm{d}t = \sum_{j=0}^{2n-1}\beta_j\int_{\xi_j}^{\xi_{j+1}} e_i(t)\mathrm{d}t = 0$ $(i = 1,\cdots,2n-1)$ has non-trival solutions. We may assume $\max_{0\leqslant j\leqslant 2n-1} |\beta_j| = 1$, and $\beta_\nu = (-1)^\nu$ for certain ν. For any $f \in H_\delta^*(L_1)$ we have ([6], pp. 25 ~ 26)

$$E(f;X_{2n-1})_{L_1} = \sup_{\substack{\|\varphi\|_{L_\infty}\leqslant 1 \\ \varphi\perp X_{2n-1}}} \int_0^{2\pi} f(t)\varphi(t)\mathrm{d}t \geqslant \left|\int_0^{2\pi} f(t)h_\beta(t)\mathrm{d}t\right|.$$

Therefore

$$\sup_{f\in H_\delta(L_1)} E(f;X_{2n-1})_{L_1} = \sup_{f\in H_\delta^*(L_1)} E(f;X_{2n-1})_{L_1}$$

$$= \sup_{f\in H_\delta^*(L_1)} \sup_{\substack{\|\varphi\|_{L_\infty}\leqslant 1 \\ \varphi\perp X_{2n-1}}} \int_0^{2\pi} f(t)\varphi(t)\mathrm{d}t$$

$$\geqslant \sup_{f\in H_\delta^*(L_1)}\left|\int_0^{2\pi} f(t)h_\beta(t)\mathrm{d}t\right|.$$

Taking account of

$$\int_0^{2\pi} f(t)h_\beta(t)\mathrm{d}t = \int_0^{2\pi}\left\{\frac{1}{2\pi}\int_0^{2\pi} H(t-u)g(u)\mathrm{d}u\right\}h_\beta(t)\mathrm{d}t$$

$$= \int_0^{2\pi} g(u)\left\{\frac{1}{2\pi}\int_0^{2\pi} H(u-t)h_\beta(t)\mathrm{d}t\right\}\mathrm{d}u \ (\|g\|_{L_1}\leqslant 1),$$

we obtain

$$\sup_{f\in H_\delta^*(L_1)}\left|\int_0^{2\pi} f(t)h_\beta(t)\mathrm{d}t\right| = \sup_{\|g\|_{L_1}\leqslant 1}\int_0^{2\pi} g(u)\left\{\frac{1}{2\pi}\int_0^{2\pi} H(u-t)h_\beta(t)\mathrm{d}t\right\}\mathrm{d}u$$

$$= \left\|\frac{1}{2\pi}\int_0^{2\pi} H(u-t)h_\beta(t)\mathrm{d}t\right\|_{L_\infty}$$

$$= \left\|\frac{1}{2\pi}\sum_{j=0}^{2n-1}\beta_j\int_{\xi_j}^{\xi_{j+1}} H(u-t)\mathrm{d}t\right\|_{L_\infty}$$

$$\geqslant \left\|\frac{1}{2\pi}\sum_{j=0}^{2n-1}(-1)^j\int_{\xi_j}^{\xi_{j+1}} H(u-t)\mathrm{d}t\right\|_{L_\infty}. \quad (19)$$

The last step of the inequalities in (19) can be demonstrated by Reductio ad absurdum. Suppose that

$$\left\|\sum_{j=0}^{2n-1}\beta_j\int_{\xi_j}^{\xi_{j+1}} H(u-t)\mathrm{d}t\right\|_{L_\infty} < \left\|\sum_{j=0}^{2n-1}(-1)^j\int_{\xi_j}^{\xi_{j+1}} H(u-t)\mathrm{d}t\right\|_{L_\infty}.$$

Then for the function

$$G(t) = \sum_{j=0}^{2n-1}[(-1)^j-\beta_j]\int_{\xi_j}^{\xi_{j+1}} H(t-u)\mathrm{d}u$$

we have $\qquad G(t) = \sum_{j\neq\nu}\alpha_j\int_{\xi_j}^{\xi_{j+1}} H(t-u)\mathrm{d}u,\ \alpha_j = (-1)^j-\beta_j,$

and sgn $G(x_k) = (-1)^k,\ k = 0,1,\cdots,2n$, which implies the existence of $2n$ distinct points $\eta_k\in[0,2\pi)(k = 1,\cdots,2n)$ such that $G(\eta_k) = 0$. But by lemma 2 in [4] the number of distinct zeros of $G(t)$ in $[0,2\pi)$ cannot exceed $2n-2$. Hence (19) is proved.

From (19) we obtain

$$d_{2n-1}(H_\delta^*(L_1))_{L_1} \geqslant \|f_{n,\delta}\|_{L_\infty}. \quad (20)$$

The proof of theorem 1 is completed.

3. Gelfand's width number of $H_\delta(L_1)$ in L_1 metric.

By definition[7] for $\mathfrak{M} \subset L_1$ and any n-dimensional subspace $X_n \subset L_\infty$

$$d^n(\mathfrak{M})_{L_1} \stackrel{\text{def}}{=} \inf_{X_n} \sup_{\substack{f\in\mathfrak{M} \\ f\perp X_n}} \| f \|_{L_1} \tag{21}$$

is called the n-width of $\mathfrak{M}$ in the sense of Gelfand in L_1 metric, where $f \perp X_n$ means $\int_0^{2\pi} f(t)\varphi(t)\mathrm{d}t = 0,\ \forall \varphi \in X_n$. For $\mathfrak{M} = H_\delta(L_1)$ we have

Theorem 2

$$d^{2n-1}(H_\delta(L_1))_{L_1} = \sup_{\substack{f\in H_\delta(L_1) \\ f\perp T_n}} \| f \|_{L_1} = (2\pi)^{-1} E_n(H)_{L_1}$$

$$= \| f_{n,\sigma} \|_{L_\infty},\ n = 1,2,\cdots \tag{22}$$

Proof For any system of $2n-1$ linearly independent functions $f_1,\cdots,f_{2n-1} \in L_\infty$ we can choose $\gamma = (\gamma_1,\cdots,\gamma_{2n}) \neq (0,\cdots,0)$ such that $\sum_{j=1}^{2n} |\gamma_j| = 1$ and the function $g(t) = (2\pi)^{-1}\sum_{j=1}^{2n}\gamma_j H(t-x_j)$ satisfies $\int_0^{2\pi} g(t)f_j(t)\mathrm{d}t = 0,\ j = 1,\cdots,2n-1$. Let us put $h_\delta(t) = \sum_{j=0}^{2n-1}\delta_i\chi_i(t)$, where $\lambda = (\delta_0,\cdots,\delta_{2n-1})$ is determined by the following condition

$$(2\pi)^{-1}\sum_{j=0}^{2n-1}\delta_i\int_{\xi_i}^{\xi_{i+1}} H(t-x_k)\mathrm{d}t = (\operatorname{sgn}\gamma_k)\| f_{n,\delta} \|_{L_\infty},\ k = 1,\cdots,2n. \tag{23}$$

We claim that there exists one and only one such system of numbers λ. Indeed, denoting

$$g_j(x) = (2\pi)^{-1}\int_{\xi_j}^{\xi_{j+1}} H(t-x)\mathrm{d}t \quad (j = 0,1,\cdots,2n-1), \tag{24}$$

we have

$$\det(g_j(x_k)) \neq 0. \tag{25}$$

In fact, for $\qquad f_{n,\delta}(x) = \sum_{j=0}^{2n-1}(-1)^j g_j(x)$

the relation

$$\sum_{j=0}^{2n-1}(-1)^j g_j(x_k) = (-1)^k \| f_{n,\delta} \|_{L_\infty} \quad (k = 1,\cdots,2n)$$

holds. If (25) is false, then the system of linear equations

$$\sum_{j=0}^{2n-1} g_j(x_k) z_j = (-1)^k \| f_{n,\delta} \|_{L_\infty} \quad (k = 1, \cdots, 2n)$$

will have infinitely many systems of solutions, so that for the solutions $(z_0, \cdots, z_{2n-1})$ the number $\max\limits_{0 \leqslant j \leqslant 2n-1} |z_j|$ may be as large as possible. On the other hand, for any system of solutions $(z_0, \cdots, z_{2n-1})$ the function

$$G_z(t) = \sum_{j=0}^{2n-1} z_j g_j(t)$$

satisfies $\qquad |G_z(x_k)| \leqslant \| f_{n,\delta} \|_{L_\infty}, k = 1, \cdots, 2n,$

so we conclude that $|z_j| \leqslant 1 (j = 0, 1, \cdots, 2n-1)$ from lemma 3 in [4]. Hence (25) holds, and the system of linear equations (23) has exactly one system of solutions $(\delta_0, \cdots, \delta_{2n-1})$, for which $\max\limits_{0 \leqslant j \leqslant 2n-1} |\delta_j| \leqslant 1$. Now let us notice that $g(t) \in H_\delta(L_1)$ and $g \perp X_{2n-1}$. So we have

$$\sup_{\substack{f \in H_\delta(L_1) \\ f \perp X_{2n-1}}} \| f \|_{L_1} \geqslant \| g \|_{L_1}.$$

From $\qquad \left| \int_0^{2\pi} g(t) h_\delta(t) \mathrm{d}t \right| \leqslant \| g \|_{L_1} \cdot \| h_\delta \|_{L_\infty} \leqslant \| g \|_{L_1},$

we obtain

$$\begin{aligned} \| g \|_{L_1} &\geqslant \left| \int_0^{2\pi} g(t) h_\delta(t) \mathrm{d}t \right| \\ &= \left| \frac{1}{2\pi} \int_0^{2\pi} \sum_{j=1}^{2n} \gamma_j H(t - x_j) \cdot \sum_{i=0}^{2n-1} \delta_i \chi_i(t) \mathrm{d}t \right| \\ &= \sum_{i=0}^{2n-1} \delta_i \sum_{j=1}^{2n} \gamma_j g_i(x_j) = \sum_{j=1}^{2n} \gamma_j \sum_{i=0}^{2n-1} \delta_i g_i(x_j) \\ &= \sum_{j=1}^{2n} |\gamma_j| \cdot \| f_{n,\delta} \|_{L_\infty} = \| f_{n,\delta} f \|_{L_\infty}. \end{aligned}$$

Therefore

$$d^{2n-1}(H_\delta(L_1))_{L_1} \geqslant \| f_{n,\delta} \|_{L_\infty}. \tag{26}$$

As to the upper bound of $d^{2n-1}(H_\delta(L_1))_{L_1}$, using a theorem of Nikolskii, S. M. ([6], p. 81) we have

$$\begin{aligned} d^{2n-1}(H_\delta(L_1))_{L_1} &\leqslant \sup_{\substack{f \in H_\delta(L_1) \\ f \perp T_n}} \| f \|_{L_1} = (2\pi)^{-1} E_n(H)_{L_1} \\ &= \| f_{n,\delta} \|_{L_\infty}. \end{aligned} \tag{27}$$

Theorem 2 is then proved.

References

[1] Тиман, А.Ф. *Теория приближения функций действительного переменного.* Москва: ФИЗ МАТГИЗ. 1960. 150 ~ 153

[2] Ахиезер, Н.И. *Лекции по теории аппроксимации.* Москва: ИЗДАТ 《НАУКА》. 1965. 207

[3] Дзядык, В.К. Исследования по современным проблемам конструктивной теории функций. Москва: Гос-издат. ФМ, 1961. 72

[4] Forst, W. Journal of Approximation theory, 1977, (19): 325 ~ 331

[5] Тихомиров, В.М.: УМН, 1960, 15: 3(93), 81 ~ 120

[6] Корнейчук, Н.П. *Экстремальные задачи теории приближения функчий.* Москва: Издат. 《НАУКА》. 1977

[7] Тихомиров, В.М.: Матем. сборник. 1969, 80(122): 290 ~ 304

SCIENTIA SINICA(Series A)
March, 1984, Vol. XXVII No. 3

一个广义样条函数类上的极值问题和有关的宽度问题

Some Extremal Problems on a Class of Perfect Splines

Abstract

Let $P(D) = \prod_{j=1}^{l}(D^2 - t_j^2)$ be a differential form, where $D = \frac{\mathrm{d}}{\mathrm{d}x}$, $l \geqslant 1$, $t_j \geqslant 0$, and $\Omega_p^{2l} = \{f(x): f^{(2l-1)}$ abs. continuous on $[0,1]$, $f^{(2k)}(0) = f^{(2k)}(1) = 0$ for $k = 0, \cdots, l-1$, $\| P(D)f \|_p \leqslant 1\}$. $d_n[\Omega_p^{2l}; L^s]$ is the Kolmogorov n-width of Ω_p^{2l} in $L^s[0,1]$. In this paper, we get the exact expressions of $d_n[\Omega_p^{2l}; L^s]$ for the two cases $p = \infty$, $1 \leqslant s \leqslant \infty$; $1 \leqslant p \leqslant \infty$, $s = 1$. As a consequence, we confirm a conjecture of A. Melkman and C. A. Micchelli.

Ⅰ Introduction

Let $q_r(D) = \prod_{j=1}^{l}(D^2 - t_j^2)$ be a linear differential form of order $r = 2l$, $(l \geqslant 1)$, where $D = \frac{\mathrm{d}}{\mathrm{d}x}$, and $t_1, \cdots, t_l \geqslant 0$. $\Omega^r_{p[0,1]}$ consists of $f(x)$ such that $f^{(2l-1)}(x)$ is absolutely continuous on $[0,1]$, $f^{(2k)}(0) = f^{(2k)}(1) = 0 (k = 0, \cdots, l-1)$ and $\| q_r(D)f \|_p \leqslant 1$, where $\| q_r(D)f \|_p = \left(\int_0^1 |q_r(D)f(x)|^p \mathrm{d}x\right)^{1/p}$ for $1 \leqslant p < \infty$, and $\| q_r(D)f \|_\infty = \operatorname*{ess\,sup}_{0\leqslant x\leqslant 1} |q_r(D)f(x)|$ for $p = \infty$. It is known that every

$f(x) \in \Omega^r_{p[0,1]}$ can be represented by

$$f(x) = \int_0^1 K(x,y)g(y)\mathrm{d}y, \tag{1.1}$$

where

$$K(x,y) = 2\sum_{k=1}^{\infty} \frac{\sin k\pi x \sin k\pi y}{q_r(\mathrm{i}k\pi)}, \quad \mathrm{i} = \sqrt{-1}, \tag{1.2}$$

is the Green function of the differential operator

$$q_r(D)f = 0, f^{(2k)}(0) = f^{(2k)}(1) = 0, k = 0, \cdots, l-1, \tag{1.3}$$

and $g(\cdot) = (q_r(D)f)(\cdot)$. It is known also that $(-1)^l K(x,y)$ is totally positive on $[0,1]\times[0,1]$. The Kolmogorov n-width of $\Omega^r_p[0,1]$ in $L^s[0,1]$ is defined by

$$d_n[\Omega^r_p; L^s] = \inf_{L_n} \sup_{f\in\Omega^r_p} \min_{\varphi\in L_n} \| f-\varphi \|_s, \tag{1.4}$$

where L_n is any n-dimensional linear subspace of $L^s[0,1]$. Let us put

$$d^n[\Omega^r_p; L^s] = \inf_{L^n} \sup_{f\in\Omega^r_p \cap L^n} \| f \|_s, \tag{1.5}$$

where L^n is any linear subspace of $L^s[0,1]$ of co-dimension n. Quantity(1.5)is known as the Gelfand n-width of $\Omega^r_p[0,1]$ in $L^s[0,1]$. Finally, the linear n-width of $\Omega^r_p[0,1]$ in $L^s[0,1]$ is defined as

$$d'_n[\Omega^r_p; L^s] = \inf_{L_n}\inf_{A_n}\sup_{f\in\Omega^r_p} \| f - A_n(f) \|_s, \tag{1.6}$$

where $A_n(f)$ is any continuous linear operator which maps $\Omega^r_p[0,1]$ into L_n. C. A. Micchelli and A. Pinkus[1] have found the exact estimation of $d_n[\Omega^r_p; L^s]$ for $p = s = \infty$ and $p = s = 2$. For $p = s = \infty$ we have[1]

Theorem A

$$d_n[\Omega^r_\infty; L^\infty] = \| P_n(x) \|_\infty, \tag{1.7}$$

$$P_n(x) = \int_0^1 K(x,y)\,\mathrm{sgn}\sin(n+1)\pi y\,\mathrm{d}y = \frac{4}{\pi}\sum_{k=0}^{\infty} \frac{\sin(2k+1)(n+1)\pi x}{(2k+1)q_r(i(2k+1)(n+1)\pi)}. \tag{1.8}$$

Let us put

$$\Lambda_n = \{\xi : \xi = (\xi_1, \cdots, \xi_n), 0 = \xi_0 < \xi_1 < \cdots < \xi_n < \xi_{n+1} = 1\}.$$

$$\Gamma_n = \{h_\xi(t) : h_\xi(t) = (-1)^j, \xi_j \leqslant t < \xi_{j+1}, j = 0, \cdots, m\}.$$

$$(\xi \in \Lambda_m, m \leqslant n).$$

$$P_\xi(x) = \int_0^1 K(x, y) h_\xi(y) \mathrm{d}y, \quad h_\xi \in \Gamma_n. \tag{1.9}$$

In fact, C. A. Micchelli and A. Pinkus[1] have proven

Proposition A $$\min_{h_\xi \in \Gamma_n} \| P_\xi(x) \|_\infty = \| P_n(x) \|_\infty. \tag{1.10}$$

For other values of p, s the exact estimation of the quantities (1.4), (1.5), (1.6) is still unknown. On the basis of the general results given by C. A. Micchell[2] for $p = \infty, 1 \leqslant s < \infty$ we may state the

Theorem B *For every $p \in [1, \infty)$ and every nature number n there exists $\xi^* = (\xi_1^*, \cdots, \xi_m^*) \in \Lambda_m (m \leqslant n)$ such that*

$$d_n[\Omega_\infty^r; L^p] = \| P_{\xi^*}(x) \|_p. \tag{1.11}$$

The proof of Theorem B is based on the foundation of

Proposition B *For every $p \in [1, \infty)$ and every positive integer n there exists a $\xi^* \in \Lambda_m (m \leqslant n)$ such that*

$$\min_{h_\xi \in \Gamma_n} \| P_\xi(x) \|_p = \| P_{\xi^*}(x) \|_p. \tag{1.12}$$

In [2] some characteristics of the extremal function $P_{\xi^*}(x)$ has been given, from which we can only derive the weaker conclusion $m \leqslant n$, where m is the number of points contained in ξ^*. Generally speaking, ξ^* depends on p and n, and it cannot be given concretely. In this paper we prove

Theorem 1 *Suppose $1 \leqslant p \leqslant \infty$, then*

$$\min_{h_\xi \in \Gamma_n} \| P_\xi(x) \|_p = \| P_n(x) \|_p. \tag{1.13}$$

This theorem shows that for kernel (1.2) the knots ξ^* of the extremal function $P_{\xi^*}(x)$, in fact, is independent of p: for each fixed n, $\xi^* = \left\{\frac{1}{n+1}, \cdots, \frac{n}{n+1}\right\}$. From Theorem 1 we obtain

Theorem 2 *Suppose $1 \leqslant s \leqslant \infty$, then*

$$(1)\ d_n[\Omega_\infty^r; L^s] = d^n[\Omega_{s'}^r; L^1] = \| P_n(x) \|_s. \tag{1.14}$$

$$(2)\ d_n'[\Omega_\infty^r; L^s] = \| P_n(x) \|_s. \tag{1.15}$$

(3) *span* $\left\{ K\left(x, \frac{1}{n+1}\right), \cdots, K\left(x, \frac{n}{n+1}\right) \right\}$ *is an extremal subspace*

for the Kolmogorov n-width and linear n-width of $\Omega^r_\infty[0,1]$ *in* $L^s[0,1]$.

The assertion (3) confirms a conjecture posed by A. Melkman and C. A. Micchelli in [3]. We have also

Theorem 3 *Suppose* $1 \leqslant p \leqslant \infty$, *then*

$$d_n[\Omega^r_p; L^1] = d^n[\Omega^r_\infty; L^{p'}] = \| P_n(x) \|_{p'}, \left(\frac{1}{p} + \frac{1}{p'} = 1 \right). \tag{1.16}$$

Besides these we obtain

Theorem 5 *For* $1 < p < \infty$, *we have*

$$d_n[\Omega^r_p; L^p] \geqslant \frac{(-1)^l}{q_r(i(n+1)\pi)}. \tag{1.17}$$

It is known from [1] *that the equality holds in* (1.17) *for* $p = 2$.

The proof of Theorem 1 is given in Sec. Ⅱ. The main idea consists in reducing Ω^r_∞ to a class of periodic functions of convolution type. In this case the kernel is no longer totally positive, but the method of approach of Zensykbaev[4] can be employed. The proofs of other theorems are given in Sec. Ⅲ ~ Sec. Ⅳ.

Ⅱ Proof of theorem 1

1. The case $p = \infty$

As it has been mentioned above, this case is contained in [1]. In fact, on account of [1] (cf. [1] p.p. 166 ~ 167, 170) for any $\xi \in \Lambda_n$ we have

$$\| P_n(x) \|_\infty \leqslant \max_{1 \leqslant j \leqslant n+1} \left| P_\xi\left(\frac{2j-1}{2(n+1)} \right) \right|.$$

Consequently

$$\| P_n(x) \|_\infty \leqslant \inf_{\xi \in \Lambda_n} \max_{1 \leqslant j \leqslant n+1} \left| P_\xi\left(\frac{2j-1}{2(n+1)} \right) \right| \leqslant \inf_{\xi \in \Lambda_n} \| P_n(x) \|_\infty.$$

Hence

$$\min_{\xi \in \Lambda_n} \| P_\xi(x) \|_\infty = \| P_n(x) \|_\infty. \tag{2.1}$$

Now we see that in the extremal problem (2.1) the domain of ξ may be assumed to be $\Lambda_m (m \leqslant n)$, for the value of $\| P_n(x) \|_\infty$ is strictly decreasing. The proof of the case $p = \infty$ is completed.

2. The case $1 < p < \infty$

We state the following proposition. Its validity can be derived from the theorems in [2].

Proposition C *For any* $p \in (1, \infty)$ *and any* n, *there exists* $\xi^* = (\xi_1^*, \cdots, \xi_m^*) \in \Lambda_m (m \leqslant n)$ *such that for* $P_{\xi^*}(x) = \int_0^1 K(x, y) h_{\xi^*}(y) \mathrm{d}y$ *the following facts hold*:

(1) $\min\limits_{h_\xi \in \Gamma_n} \| P_\xi(x) \|_p = \| P_{\xi^*}(x) \|_p$.

(2) $\int_0^1 | P_{\xi^*}(x) |^{p-1} \operatorname{sgn} P_{\xi^*}(x) K(x, \xi_j^*) \mathrm{d}x = 0, j = 1, \cdots, m$.

(3) $P_{\xi^*}(x)$ *in* $(0,1)$ *has exact* m *zeros* $\eta^* = (\eta_1^*, \cdots, \eta_m^*) \in \Lambda_m$. $P_{\xi^*}(x)$ *alternates between the points of* η^*, *and its rule of sign changes coincides with that of* $h_{\eta^*}(x)$, *i.e.*, $\operatorname{sgn} P_{\xi^*}(x) = h_{\xi^*}(x)$.

(4) *The function*

$$\int_0^1 K(x, y) | P_{\xi^*}(x) |^{p-1} \operatorname{sgn} P_{\xi^*}(x) \mathrm{d}x \tag{2.2}$$

has in $(0,1)$ *exact* m *zeros* $\xi^* = (\xi_1^*, \cdots, \xi_m^*)$ *and alternates between the points of* ξ^*, *and the rule of sign changes coincides with that of* $h_{\xi^*}(x)$, *i.e.*,

$$\operatorname{sgn} \int_0^1 K(x, y) | P_{\xi^*}(x) |^{p-1} \operatorname{sgn} P_{\xi^*}(x) \mathrm{d}x = h_{\xi^*}(y).$$

On the basis of Proposition C, we have only to prove: for every $p \in (1, \infty)$ and n, $m = n$ and $\xi^* = \left\{ \frac{j}{n+1} \right\}, j = 1, \cdots, n$. To this end we first give some preliminaries. For any $g \in L^\infty[0,1]$, let $\tilde{g}$ be the odd continuation with period 2. By introducing

$$K_r(x) = \sum_{k=1}^{\infty} \frac{\cos k\pi x}{q_r(ik\pi)}, \tag{2.3}$$

we can easily see that the function

$$\tilde{f}(x) = \int_{-1}^1 K_r(x - y) \tilde{g}(y) \mathrm{d}y \tag{2.4}$$

is the odd continuation of $f(x)$ with period 2, where

$$f(x) = \int_0^1 K(x, y) g(y) \mathrm{d}y.$$

Setting now

$$\widetilde{P}_{\xi^*}(x) = \int_{-1}^{1} K_r(x-y)\widetilde{h}_{\xi^*}(y)\mathrm{d}y \tag{2.5}$$

and

$$G(x) = \int_{-1}^{1} K_r(x-y)\,|\widetilde{P}_{\xi^*}(y)|^{p-1}\mathrm{sgn}\,\widetilde{P}_{\xi^*}(y)\mathrm{d}y, \tag{2.6}$$

we see that $G(x)$ is the odd continuation of (2.2) with period 2, so that we may write

$$\mathrm{sgn}\,G(x) = \widetilde{h}_{\xi} * (x). \tag{2.7}$$

Let us set

$$\widetilde{\Phi}(x) = \int_{-1}^{1} K_r(x-y)\widetilde{\varphi}(y)\mathrm{d}y,$$

where $\varphi \in L^{\infty}[0,1]$. It is obvious that $\widetilde{\Phi}^{(2l-1)}(x)$ is locally absolutely continuous, and

$$\widetilde{\Phi}^{(k)}(-1) = \widetilde{\Phi}^{(k)}(1), k = 0, \cdots, 2l-1. \tag{2.8}$$

By writing

$$q_{r-2}(x) = \prod_{j=1}^{l-1}(x^2 - t_j^2), \cdots, q_2(x) = (x^2 - t_1^2),$$

$$K_{r-2}(x) = \sum_{k=1}^{\infty}\frac{\cos k\pi x}{q_{r-2}(ik\pi)}, \cdots, K_2(x) = \sum_{k=1}^{\infty}\frac{\cos k\pi x}{q_2(ik\pi)},$$

we have the following identities

$$(D^2 - t_l^2)\widetilde{\Phi}(x) = \int_{-1}^{1} K_{r-2}(x-y)\widetilde{\varphi}(y)\mathrm{d}y,$$

$$(D^2 - t_l^2)(D^2 - t_{l-1}^2)\widetilde{\Phi}(x) = \int_{-1}^{1} K_{r-4}(x-y)\widetilde{\varphi}(y)\mathrm{d}y,$$

$$\cdots\cdots\cdots\cdots\cdots\cdots\cdots\cdots\cdots$$

$$q_r(D)\widetilde{\Phi}(x) = \widetilde{\varphi}(x).\ (\text{a.e.}). \tag{2.9}$$

Denoting

$$\widetilde{\Phi}_0(x) = \widetilde{\Phi}(x), \widetilde{\Phi}_1(x) = (D^2 - t_l^2)\widetilde{\Phi}_0(x), \cdots,$$

$$\widetilde{\Phi}_l(x) = (D^2 - t_1^2)\widetilde{\Phi}_{l-1}(x), \widetilde{\Phi}_l(x) = \widetilde{\varphi}(x), \tag{2.10}$$

we have a series of differential identities which can be verified immediately:

$$\widetilde{\Phi}_1(x) = \mathrm{e}^{t_l x}\frac{\mathrm{d}}{\mathrm{d}x}\left[\mathrm{e}^{-2t_l x}\left(\frac{\mathrm{d}}{\mathrm{d}x}\mathrm{e}^{t_l x}\widetilde{\Phi}_0(x)\right)\right],$$

$$\tilde{\Phi}_2(x) = e^{t_{l-1}x}\frac{d}{dx}\left[e^{-2t_{l-1}x}\left(\frac{d}{dx}e^{t_{l-1}x}\tilde{\Phi}_1(x)\right)\right],$$

……………………………

$$\tilde{\Phi}_l(x) = e^{t_1x}\frac{d}{dx}\left[e^{-2t_1x}\left(\frac{d}{dx}e^{t_1x}\tilde{\Phi}_{l-1}(x)\right)\right]. \quad (2.11)$$

(cf[5])

In the sequel we shall extend the set of points ξ^* on the real axis. For $k=-1\cdots,-(m+1)$ we define $\xi_k^*=-\xi_{|k|}^*$, for $|k|>m+1$ we define $\xi_k^*=\xi_{k'}^*$, where k' is an integer such that $|k'|\leqslant m+1$ and $k\equiv k'$ (mod $2m+2$). We write $\Delta_k=[\xi_{k-1}^*,\xi_k^*]$, $\Delta=\{\Delta_k\}(k=-m,\cdots,-1,0,1,\cdots,m+1)$, $|\Delta_k|=\xi_k^*-\xi_{k-1}^*$, $\delta=\min|\Delta_k|$, and

$$H_0(x) = G(x)+G(x+\delta). \quad (2.12)$$

Since $G(\xi_k^*)=0$ and $(-1)^kG(\xi_k^*+\delta)\geqslant 0$ for $k=0,\pm 1,\cdots,\pm(m+1)$, we have

$$(-1)^kH_0(\xi_k^*)\geqslant 0,\ k=0,\pm 1,\cdots,\pm(m+1), \quad (2.13)$$

and $H_0(\xi_k^*)=0$ if and only if $\xi_{k+1}^*-\xi_k^*=\delta$. Now we suppose that the quantities $|\Delta_k|$ are not equal to each other. From this hypothesis we deduce a contradiction.

2.2—1. First of all we consider the case when the function $H(x)$ is not identically zero on any interval Δ_k. From (2.13) we have

$$(-1)^k e^{t_l\xi_k^*}H_0(\xi_k^*)\geqslant 0,\ k=0,\pm 1,\cdots,\pm(m+1).$$

Thus on every Δ_k, there is a point α_k such that

$$\operatorname{sgn}\left[\frac{d}{dx}e^{t_l x}H_0(x)\right]_{x=\alpha_k} = (-1)^{k+1}, \quad (2.14)$$

$k=-m,\cdots,-1,0,1,\cdots,m+1$, $\alpha_{-m}<\cdots<\alpha_0<\cdots\alpha_{m+1}$, and $\alpha_k\in[-1,1]$. Noticing that $\operatorname{sgn}\left[\frac{d}{dx}e^{t_l x}H_0(x)\right]=\operatorname{sgn}[t_lH_0(x)+H'_0(x)]$ and that $t_lH_0(x)+H'_0(x)$ is a 2-periodic function, we may conclude that its number of sign changes in one period is even. So we have

$$v_c\left[\frac{d}{dx}e^{t_l x}H_0(x)\right]\geqslant 2(m+1),$$

where $v_c(f)$ denotes the number of sign changes in one period. Therefore in the interval $[-1,1]$ there must be a set of $2m+3$ points

$$\alpha_{-m-1}<\alpha_{-m}<\cdots<\alpha_{-1}<\alpha_0<\cdots<\alpha_{m+1}$$

(or $\alpha_{-m}<\cdots<\alpha_{-1}<\alpha_0<\cdots<\alpha_{m+1}<\alpha_{m+2}$) such that

$$\operatorname{sgn}\left[\frac{\mathrm{d}}{\mathrm{d}x}\mathrm{e}^{t_l x}H_0(x)\right]_{x=\alpha_k}=(-1)^{k+1},\ k=0,\ \pm 1,\cdots,\ \pm(m+1).$$

Repeating the same argument to the function $\mathrm{e}^{-2t_l x}\left[\frac{\mathrm{d}}{\mathrm{d}x}\mathrm{e}^{t_l x}H_0(x)\right]$, we confirm the existence of a set of $2m+3$ points in one period

$$\beta_{-m-1}<\cdots<\beta_{-1}<\beta_0<\beta_1<\cdots<\beta_{m+1}$$

such that

$$\operatorname{sgn}\left\{\mathrm{e}^{t_l x}\frac{\mathrm{d}}{\mathrm{d}x}\left[\mathrm{e}^{-2t_l x}\left(\frac{\mathrm{d}}{\mathrm{d}x}\mathrm{e}^{t_l x}H_0(x)\right)\right]\right\}_{x=\beta_k}=(-1)^{k+2},$$

so we have

$$v_c(H_1(x))\geqslant 2(m+1),$$

where $H_1(x)=(D^2-t_l^2)D_0(x)$. Writing $H_2(x)=(D^2-t_{l-1}^2)H_1(x)$, $\cdots$, $H_l(x)=(D^2-t_1^2)H_{l-1}(x)$ and repeating the same argument as before we obtain

$$v_c(H_j(x))\geqslant 2(m+1),\ j=1,\cdots,l. \tag{2.15}$$

Now we see that

$$\begin{aligned}&H_l(x)\\&=|\widetilde{P}_{\xi^*}(x)|^{p-1}\operatorname{sgn}\widetilde{P}_{\xi^*}(x)+|\widetilde{P}_{\xi^*}(x+\delta)|^{p-1}\operatorname{sgn}\widetilde{P}_{\xi^*}(x+\delta),\end{aligned}$$

and for any real numbers a and b

$$\operatorname{sgn}\{|a|^{p-1}\operatorname{sgn}a+|b|^{p-1}\operatorname{sgn}b\}=\operatorname{sgn}\{a+b\}.$$

Thus it follows that

$$\operatorname{sgn}(H_l(x))=\operatorname{sgn}\{\widetilde{P}_{\xi^*}(x)+\widetilde{P}_{\xi^*}(x+\delta)\},$$

and therefore

$$v_c[\widetilde{P}_{\xi}*(x)+\widetilde{P}_{\xi}*(x+\delta)]\geqslant 2(m+1). \tag{2.16}$$

Let us denote

$$\begin{aligned}\widetilde{P}_0(x)&\overset{\text{def}}{=\!=}\widetilde{P}_{\xi}*(x)+\widetilde{P}_{\xi}*(x+\delta)\\&=\int_{-1}^{1}K_r(x-y)[h_{\xi}*(y)+h_{\xi}*(y+\delta)]\mathrm{d}y,\end{aligned}$$

and $(D^2-t_l^2)\widetilde{P}_0=\widetilde{P}_1,\cdots,(D^2-t_1^2)\widetilde{P}_{l-1}=\widetilde{P}_l$, where

$$\widetilde{P}_l(x) \stackrel{\text{def}}{=} h_\xi * (x) + h_\xi * (x+\delta). \quad (2.17)$$

By the above argument, we can assert

$$v_c(\widetilde{P}_j) \geqslant 2(m+1), j = 1, \cdots, l. \quad (2.18)$$

However, on every interval Δ_k of length δ we have $\widetilde{P}_l(x) \equiv 0$. In one period, there are at least two such intervals because the points of ξ^* (extended) are symmetric with respect to zero. Thus we get

$$v_c(\widetilde{P}_l) \leqslant 2m.$$

This is a contradiction.

2.2—2. Assume that for some $\Delta_k \in \Delta$, $H_0(x) \equiv 0$. In what follows we say that such an interval Δ_k possesses property $\mathscr{P}_0$. In this case it follows from

$$H_0(\xi^*_{n-1}) = H_0(\xi^*_k) = 0$$

that $|\Delta_k| = |\Delta_{k+1}| = \delta$. By [6], an interval $D = [\xi^*_j, \xi^*_{j+q}] (q \geqslant 2)$ is called singular, if

(a) $|\Delta_{j+1}| = \cdots = |\Delta_{j+q}| = \delta$,

(b) one of $\Delta_{j+1}, \cdots, \Delta_{j+q}$ possesses property $\mathscr{P}_0$,

(c) $|\Delta_j|, |\Delta_{j+q+1}| > \delta$.

Δ_j and Δ_{j+q+1} do not possess property $\mathscr{P}_0$. It is obvious that there at least exists one singular interval under our assumption. We now point out that a singular interval has the following properties.

1° Different singular intervals can not be overlapping.

2° On every singular interval, $\widetilde{P}_l(x) \equiv 0$.

3° On every singular interval, $\widetilde{P}_{l-1}(x) \equiv 0$.

In fact, let D be a singular interval. Then there is an $\Delta_k \subset D$ which possesses $\mathscr{P}_0$. Thus we have $\widetilde{P}_{l-1}(x) \equiv 0$ on Δ_k. From $(D^2 - t_1^2)\widetilde{P}_{l-1}(x) = h_\xi * (x) + h_\xi * (x+\delta) = 0$ on D we have

(a) If $t_1 > 0$, then $\widetilde{P}_{l-1}(x) = c_1 e^{-t_1 x} + c_2 e^{t_1 x}$ on D. Now from the condition $\widetilde{P}_{l-1}(x) \equiv 0$ on D_k, $c_1 = c_2 = 0$ follows.

(b) If $t_1 = 0$, then $\widetilde{P}_{l-1}(x) = c_1 + c_2 x$ on D. From the condition $\widetilde{P}_{l-1}(x) \equiv 0$ on $D_k \subset D$, $c_1 = c_2 = 0$ also follows.

4° If $D = [\xi_j^*, \xi_{j+q}^*](q \geqslant 2)$ is singular, then $\mathrm{e}^{-2t_1 x}\left(\frac{\mathrm{d}}{\mathrm{d}x}\mathrm{e}^{t_1 x}\widetilde{P}_{l-1}(x)\right) \equiv 0$ on D, but on Δ_j and Δ_{j+q+1} it is continuous and essentially monotonic (not identifying constant).

In fact, from 3° we have $\mathrm{e}^{-2t_1 x}\left(\frac{\mathrm{d}}{\mathrm{d}x}\mathrm{e}^{t_1 x}\widetilde{P}_{l-1}(x)\right) \equiv 0$ on D. *On* Δ_j and Δ_{j+q+1}, $h_\xi * (x) + h_\xi * (x+\delta)$ keeps the sign constant and the function is not identical to zero. Thus from

$$\mathrm{e}^{t_1 x}\frac{\mathrm{d}}{\mathrm{d}x}\left[\mathrm{e}^{-2t_1 x}\left(\frac{\mathrm{d}}{\mathrm{d}x}\mathrm{e}^{t_1 x}\widetilde{P}_{l-1}(x)\right)\right] = h_\xi * (x) + h_\xi * (x+\delta),$$

we know that the function $\mathrm{e}^{-2t_1 x}\left(\frac{\mathrm{d}}{\mathrm{d}x}\mathrm{e}^{t_1 x}\widetilde{P}_{l-1}(x)\right)$ is continuous and essentially monotonic on Δ_j as well as on Δ_{j+q+1}.

5° Let D_1, D_2 be two different singular intervals in one period and $D_1 < D_2$ then between D_1 and D_2(D_2 and D_1^*, D_1^* is the periodic right-translation of D_1) there are at least two intervals contained in Δ.

In fact, if there were just one single $\Delta_j = [\xi_{j-1}^*, \xi_j^*]$ between D_1 and D_2, then from 5° the function $\mathrm{e}^{-2t_1 x}\left[\frac{\mathrm{d}}{\mathrm{d}x}\mathrm{e}^{t_1 x}\widetilde{P}_{l-1}(x)\right]$ would be continuous and essentially monotonic on Δ_j, but its values at the end points would be zero. This is impossible.

Assume that there is only one singular interval D in $[-1,1]$, $\Delta_k \supset D$ possesses property $\mathscr{P}_0$. We write $D = [\xi_j^*, \xi_{j+q}^*]$, $q \geqslant 2$. In this case on every interval $\Delta_k \in \Delta$ not contained in D, $H_0(x)$ is not identically zero. On $[\xi_{j+q-1}^*, \xi_{j+q}^*]$, $H_0(x)$ is also notidentically zero.

1° When $q =$ even number, on account of the fact that $H_0(x)$ is not identically zero on each interval of $\{\Delta_{-m}, \cdots, \Delta_0, \cdots, \Delta_{m+1}\} \setminus \{\Delta_{j+1}, \cdots, \Delta_{j+q-1}\}$, by repeating the same argument as in 2.2 ~ 1 we obtain a set of points $\{\tau_k\} \subset [-1,1]$, where $k \in \{-m, \cdots, -1, 0, 1, \cdots, m+1\} \setminus \{j+1, \cdots, j+q-1\}$, $\tau_k < \tau_{k'}$ for $k < k'$, and

$$\operatorname{sgn}\left(\frac{\mathrm{d}}{\mathrm{d}x}\mathrm{e}^{t_l x}H_0(x)\right)_{x=\tau_k} = (-1)^{k+1}.$$

Since q is even, there are $2(m+1) - q - 1$ sign changes, which are odd.

Therefore we may conclude

$$v_c\left[\frac{\mathrm{d}}{\mathrm{d}x}\mathrm{e}^{t_l x}H_0(x)\right]\geqslant 2(m+1)-q.$$

With the same argument, we obtain

$$v_c(H_j)\geqslant 2(m+1)-q,\quad j=1,\cdots,l,$$
$$v_c(\widetilde{P}_j)\geqslant 2(m+1)-q,\quad j=0,\cdots,l-1,$$

and also

$$v_c\left[\mathrm{e}^{-2t_1 x}\left(\frac{\mathrm{d}}{\mathrm{d}x}\mathrm{e}^{t_1 x}\widetilde{P}_{l-1}(x)\right)\right]\geqslant 2(m+1)-q.$$

On the other hand, since $\mathrm{e}^{-2t_1 x}\left(\frac{\mathrm{d}}{\mathrm{d}x}\mathrm{e}^{t_1 x}\widetilde{P}_{l-1}(x)\right)$ does not change its sign on $[\xi_{j-1}^*,\xi_{j+q+1}^*]$, but changes its sign at most once on each of the other intervals because it is monotonic on each of these, so that

$$v_c\left[\mathrm{e}^{-2t_1 x}\left(\frac{\mathrm{d}}{\mathrm{d}x}\mathrm{e}^{t_1 x}\widetilde{P}_{l-1}(x)\right)\right]\leqslant 2(m+1)-q-2.$$

This is a contradiction.

2° Assume that q is odd. By the same reasoning we can obtain, on the one hand,

$$v_c\left[\mathrm{e}^{-2t_1 x}\frac{\mathrm{d}}{\mathrm{d}x}(\mathrm{e}^{t_1 x}\widetilde{P}_{l-1}(x)\right]\geqslant 2(m+1)-q+1,$$

on the other hand

$$v_c\left[\mathrm{e}^{-2t_1 x}\frac{\mathrm{d}}{\mathrm{d}x}(\mathrm{e}^{t_1 x}\widetilde{P}_{l-1}(x))\right]\leqslant 2(m+1)-q-1.$$

This again is a contradiction.

Now we have to consider the remaining part of the case when in $[-1,1]$ there are $\mu(\mu\geqslant 2)$ singular intervals denoted by $D_1,D_2,\cdots,D_\mu$. Let us assume that D_k contains exactly q_k intervals from Δ and set $q=\sum_{k=1}^{\mu}q_k$. We assume also that among $\{q_k\}$ there are just μ' odd numbers. By the same reasoning as in the previous passage we may obtain, on the one hand,

$$v_c\left[\mathrm{e}^{-2t_1 x}\frac{\mathrm{d}}{\mathrm{d}x}(\mathrm{e}^{t_1 x}\widetilde{P}_{l-1}(x))\right]\geqslant 2(m+1)-q+\mu',$$

and on the other hand,

$$v_c\left[e^{-2t_1x}\frac{d}{dx}(e^{t_1x}\widetilde{P}_{l-1}(x))\right]\leqslant 2(m+1)-q-2\mu+\mu'.$$

This is also a contradiction.

To sum up, we have proven $|\Delta_k|=\dfrac{1}{m+1}(k=-m,\cdots,0,1,\cdots,m+1)$. It remains for us to prove $m=n$ in the following section.

2.2—3 Let us write

$$F_{j+1,r}(x)=\frac{4}{\pi}\sum_{k=0}^{\infty}\frac{\sin(2k+1)\pi x}{(2k+1)q_r(i(2k+1)(j+1)\pi)},\quad (2.19)$$
$$(j=0,1,2,\cdots).$$

By direct calculation we have

$$[(j+1)^2D^2-t_l^2]F_{j+1,r}(x)=F_{j+1,r-2}(x),$$

……………………………………………

$$[(j+1)^2D^2-t_1^2]F_{j+1,2}(x)=F_{j+1,0}(x),$$

$$F_{j+1,0}(x)=\frac{4}{\pi}\sum_{k=0}^{\infty}\frac{\sin(2k+1)\pi x}{2k+1}=\operatorname{sgn}\sin\pi x.\quad (2.20)$$

From this, we see that on $[0,1]$, $F_{j+1,2}(x)$ satisfies the differential equation

$$(j+1)^2y''-t_1^2y=1,\ y(0)=y(1)=0.$$

Assume that $t_1>0$, then

$$F_{j+1,2}(x)=\frac{-1}{(j+1)t_1}\left\{\sinh\frac{t_1(1-x)}{j+1}\int_0^x\sinh\frac{t_1y}{j+1}dy+\right.$$
$$\left.\sinh\frac{t_1x}{j+1}\int_x^1\sinh\frac{t_1(1-y)}{j+1}dy\right\}.$$

From this, we see that for any $x\in(0,1)$ the sequence $\{|F_{j+1,2}(x)|\}$ $(j=0,1,\cdots)$ is strictly decreasing. If $t_1=0$, then we simply have

$$F_{j+1,2}(x)=\frac{1}{2(j+1)^2}(x^2-x),\quad (0\leqslant x\leqslant 1).$$

For any $x\in(0,1)$ the sequence $\{|F_{j+1,2}(x)|\}$ is also strictly decreasing. We consider now $F_{j+1,4}(x)$. The second order differential operator

$$y''-\alpha^2y=0,\quad y(0)=y(1)=0,$$

has its Green function

$$K(x,y,\alpha)=\begin{cases}-\alpha^{-1}\sinh\alpha x\sinh\alpha(1-y), & x\leqslant y,\\ -\alpha^{-1}\sinh\alpha(1-x)\sinh\alpha y, & y\leqslant x,\end{cases}$$

when $\alpha > 0$. Let us first suppose $t_2 > 0$. By putting $\alpha = \dfrac{t_2}{j+1}$, we have

$$F_{j+1,4}(x) = \frac{-1}{(j+1)t_2}\left\{\sinh\frac{t_2(1-x)}{j+1}\int_0^x \sinh\frac{t_2 y}{j+1}F_{j+1,2}(y)\mathrm{d}y + \right.$$
$$\left.\sinh\frac{t_2 x}{j+1}\int_x^1 \sinh\frac{t_2(1-y)}{j+1}F_{j+1,2}(y)\mathrm{d}y\right\}.$$

From this expression, we see that for any $x \in (0,1)$ the sequence $\{|F_{j+1,4}(x)|\}$ is strictly decreasing. For $\alpha = \dfrac{t_2}{j+1} = 0$, the Green function of the differential operator

$$y'' = 0,\quad y(0) = y(1) = 0,$$

is

$$K(x,y,0) = \begin{cases} -x(1-y), & x \leqslant y, \\ -y(1-x), & y \leqslant x, \end{cases}$$

and in this case we have

$$F_{j+1,4}(x) = \frac{-1}{(j+1)^2}\left\{(1-x)\int_0^x yF_{j+1,2}(y)\mathrm{d}y + \right.$$
$$\left. x\int_x^1 (1-y)F_{j+1,2}(y)\mathrm{d}y\right\}.$$

From this it readily follows that for any $x \in (0,1)$ the sequence $\{|F_{j+1,4}(x)|\}$ is strictly decreasing. By induction, we conclude that for any $x \in (0,1)$ the sequence $\{|F_{j+1,r}(x)|\}$ is strictly decreasing. Therefore we have

$$\int_0^1 |F_{n+1,r}(x)|^p \mathrm{d}x < \int_0^1 |F_{m,r}(x)|^p \mathrm{d}x, \tag{2.21}$$

whenever $m \leqslant n$.

Now we can complete our proof. Obviously we have

$$\min_{h_\xi \in \Gamma_n} \| P_\xi(x) \|_p \leqslant \| P_n(x) \|_p. \tag{2.22}$$

It can easily be verified that

$$\| P_n \|_p^p = \int_0^1 |F_{n+1,r}(x)|^p \mathrm{d}x.$$

From this we readily obtain $m = n$. The proof for the case $1 < p < \infty$ is completed. Letting $p \to 1+$ we have, for all $p, 1 \leqslant p \leqslant \infty$, the validity of (1.13). Theorem 1 is proven.

Ⅲ Proofs of theorems 2 ~ 4

1. The Proof of Theorem 2

By the duality theorem (cf [7], [8]), we obtain

$$d_n[\Omega_\infty^r; L^s] = \mathrm{d}^n[\Omega_{s'}^r; L^1]. \tag{3.1}$$

It has been proven in [1] that for $h \in L^\infty[0,1]$ there exists linear functionals $\lambda_1(h), \cdots, \lambda_n(h)$ such that

$$\left| \int_0^1 K(x,y)h(y)\mathrm{d}y - \sum_{j=1}^n \lambda_j(h) K\left(x, \frac{j}{n+1}\right) \right| \leqslant | P_n(x) |$$

for any $x \in [0,1]$ and any $h(t) \in L^\infty[0,1]$ with $\| h \|_\infty \leqslant 1$. Thus we have

$$\begin{aligned} d_n[\Omega_\infty^r; L^s] &\leqslant d_n'[\Omega_\infty^r; L^s] \\ &\leqslant \sup_{\| h \|_\infty \leqslant 1} \Big\| \int_0^1 K(x,y)h(y)\mathrm{d}y - \\ &\qquad \sum_{j=1}^n \lambda_j(h) K\left(x, \frac{j}{n+1}\right) \Big\|_s \\ &\leqslant \| P_n(x) \|_s \end{aligned}$$

We remark that the first half of the proof of Theorem 3.2 in [2] can be completely applied to our particular case. Its method of approach is standard in approximation theory. Consequently, we have

$$d_n[\Omega_\infty^r; L^s] \geqslant \| P_n(x) \|_s .$$

Assertions (1) and (2) of Theorem 2 are proven. Writing now

$$X_n^o = \mathrm{span}\left\{ K\left(x, \frac{j}{n+1}\right) \right\}, \quad j = 1, \cdots, n,$$

$$E(\Omega_\infty^r; X_n^o)_s \stackrel{\mathrm{def}}{=\!=} \sup_{f \in \Omega_\infty^r} \min_{g \in X_n^o} \| f - g \|_s ,$$

then from

$$\begin{aligned} d_n[\Omega_\infty^r; L^s] &\leqslant E(\Omega_\infty^r; X_n^o)_s \\ &\leqslant \sup_{\| h \|_\infty \leqslant 1} \Big\| \int_0^1 K(x,y)h(y)\mathrm{d}y - \\ &\qquad \sum_{j=1}^n \lambda_j(h) K\left(x, \frac{j}{n+1}\right) \Big\|_s \\ &\leqslant \| P_n(x) \|_s , \end{aligned}$$

we obtain

$$E(\Omega_\infty^r; X_n^o)_s = d_n[\Omega_\infty^r; L^s] = d_n'[\Omega_\infty^r; L^s] = \| P_n(x) \|_s .$$

The assertion [3] is proven.

2. The Proof of Theorem 3

By the duality theorem

$$d_n[\Omega_p^r; L^1] = d^n[\Omega_\infty^r; L^{P'}] \tag{3.2}$$

$\left(\frac{1}{p} + \frac{1}{p'} = 1\right)$. In order to prove (1.16) we can apply Theorems 3.4 and 3.5 in [2]. For the case being considered, $K(x, y) = K(y, x)$ and so we have $K^T = K$ (where K^T is the conjugate operator of K, cf. [2]). By Theorem 3.5 in [2] we have

$$d_n[\Omega_p^r; L^1] \geqslant \| P_n(x) \|_{p'} .$$

On the other hand, for the kernel (1.2), Lemma 3.4 in [1] holds (cf [1], pp. 164 ~ 165):

$$K\begin{pmatrix} \frac{1}{n+1}, \cdots, \frac{n}{n+1} \\ \frac{1}{n+1}, \cdots, \frac{n}{n+1} \end{pmatrix} = \begin{vmatrix} K\left(\frac{1}{n+1}, \frac{1}{n+1}\right) \cdots K\left(\frac{1}{n+1}, \frac{n}{n+1}\right) \\ \vdots \qquad\qquad \vdots \\ K\left(\frac{n}{n+1}, \frac{1}{n+1}\right) \cdots K\left(\frac{n}{n+1}, \frac{n}{n+1}\right) \end{vmatrix} \neq 0.$$

As a result, the second half of the proof of Theorem 3.5 in [2] is entirely suitable to Ω_p^r. Theorem 3 is proven.

3. L^1-Approximation of $K(x, y)$

For arbitrary $u_i, v_i \in L[0,1] (i = 1, \cdots, n)$, we write

$$E_{11}(K) = \min_{u_i, v_i} \int_0^1 \int_0^1 \left| K(x, y) - \sum_{i=1}^n u_i(x) v_i(y) \right| dx dy. \tag{3.3}$$

In [2], an exact expression for (3.3) is given under the condition when $K(x, y)$ is any non-degenerate TP kernel. The function $K(x, y)$ is degenerate on the boundary of the square $[0,1] \times [0,1]$, nevertheless (cf [2], pp. 60 ~ 61) we still have

Theorem 4 $E_{11}(K) = \int_0^1 \int_0^1 | E(x, y) | dx dy,$ (3.4)

where

$$E(x,y)=K\begin{pmatrix} x & \frac{1}{n+1} & \cdots & \frac{n}{n+1} \\ y & \frac{1}{n+1} & \cdots & \frac{n}{n+1}\end{pmatrix}\cdot K\begin{pmatrix} \frac{1}{n+1} & \cdots & \frac{n}{n+1} \\ \frac{1}{n+1} & \cdots & \frac{n}{n+1}\end{pmatrix}^{-1},$$

and

$$K\begin{pmatrix} x & \frac{1}{n+1} & \cdots & \frac{n}{n+1} \\ y & \frac{1}{n+1} & \cdots & \frac{n}{n+1}\end{pmatrix}$$

$$= \begin{vmatrix} K(x,y) & K\left(x,\frac{1}{n+1}\right)\cdots K\left(x,\frac{n}{n+1}\right) \\ K\left(\frac{1}{n+1},y\right) & \\ \vdots & K\begin{pmatrix} \frac{1}{n+1} & \cdots & \frac{n}{n+1} \\ \frac{1}{n+1} & \cdots & \frac{n}{n+1}\end{pmatrix} \\ K\left(\frac{n}{n+1},y\right) & \end{vmatrix}.$$

Ⅳ Some lower bounds for $d_n[\Omega_p^r;\ L^p]$

In this section, we shall give some lower bounds for $d_n[\Omega_p^r; L^p], 1 < p < \infty$. The exact estimation of this quantity is known only for $p = 2$[1]. First, let us state one lemma of multipliers by Lu Shan-zhen.

Lemma (*Lu Shan-zhen*). *Suppose that the set of real numbers* $\Lambda = (\lambda_k : k = 0, \cdots, n, \lambda_0 = 0)$ *satisfies conditions*

$$\Delta\lambda_k = \lambda_k - \lambda_{k+1} \leqslant 0, \quad (0 \leqslant k \leqslant n-1),$$

$$\Delta^2\lambda_k = \lambda_k - 2\lambda_{k+1} + \lambda_{k+2} \geqslant 0, \quad (0 \leqslant k \leqslant n-2),$$

and

$$U_n(x, T_n) = \frac{a_0\lambda_0}{2} + \sum_{k=1}^{n} \lambda_k (a_k \cos kx + b_k \sin kx),$$

where $T_n(x) = \frac{a_0}{2} + \sum_{k=1}^{n}(a_k \cos kx + b_k \sin kx)$, *then*

$$\sup_{\|T_n\|_p \leqslant 1} \| U_k(x, T_n) \|_p = \lambda_n. \tag{4.1}$$

This lemma generalizes the well-known results of S. Bernstein and A. Zygmund (cf [9], [10]). On the basis of this, we can give the proof of theorem 5.

Setting $g(y)=\sum_{j=1}^{n+1}\alpha_j\sin j\pi y$, we can see

$$f(x)=\int_0^1 K(x,y)g(y)\mathrm{d}y$$

$$=\sum_{j=1}^{n+1}\frac{a_j}{q_r(\mathrm{i}j\pi)}\sin j\pi x,\quad \mathrm{i}=\sqrt{-1}.$$

If $\|g(y)\|_p\leqslant 1$, then by the above lemma we obtain

$$|q_r(\mathrm{i}(n+1)\pi)|^{-1}\left\|\sum_{k=1}^{n+1}q_r(\mathrm{i}k\pi)\alpha_k\sin k\pi x\right\|_p\leqslant\|g(y)\|_p. \tag{4.2}$$

Let us denote by

$$B_{n+1}^p=\left\{\sum_{k=1}^{n+1}\alpha_k\sin k\pi x:\left\|\sum_{k=1}^{n+1}\alpha_k\sin k\pi x\right\|_p\leqslant 1\right\}$$

the $(n+1)$-dimensional unit ball in $L^p[0,1]$. From (4.2) we have

$$|q_r(\mathrm{i}(n+1)\pi)|^{-1}B_{n+1}^p\subset\Omega_p^r.$$

Therefore

$$d_p[\Omega_p^r;L^p]\geqslant|q_r(\mathrm{i}(n+1)\pi)|^{-1}.$$

Theorem 5 is proved.

Remark. The exact estimation of $d_n[\Omega_p^r;L^p]$ is left unanswered. For sufficiently great n we have

$$d_n[\Omega_p^r;L^p]<\frac{4}{\pi}|q_r(\mathrm{i}(n+1)\pi)|^{-1}. \tag{4.3}$$

References

[1] Micchelli, C. A. & Pinkus, A. Trans. AMS, 1977, 234(1): 139 ~ 174

[2] ———, J. Appr. Th. 24, 1978. 51 ~ 77

[3] Melkman A. & Micchelli, C. A. Illinois J. Math. 1978, 22(4): 541 ~ 564

[4] Zensykbaev A. A. ДАН СССР. 1976, 227(2): 277 ~ 279

[5] Полиа Г. и Сеге. Г. Задачц ц теоремы цз анализа, II, M.: Госиздат. 1956, 259 ~ 260

[6] Маковоз Ю. И. Матем. заметки. 1979, 26(5): 805 ~ 812

[7] Triebel Hans. *Interpolation Theory*, *Function Spaces*, *Differential Operators*.

Amsterdam, North-Holland Publishing Company, 1978, 109

[8] 孙永生. 中国科学, 1982, 3:215 ～ 225

[9] Zygmund, A. PLMS, 1932, 34:435 ～ 446

[10] Бернштейн С. Н. *Собрание Сочинений, II, Москва, нздаТ.* АН СССР, 1954, 173 ～ 177

Approximation Theory and Its Applications
Oct., 1984, Vol. 1, No. 1

关于光滑函数类 $\Omega_p^{2r+\sigma}$ 上的单边逼近*

On One-sided Approximation of Class $\Omega_p^{2r+\sigma}$ of Smooth Functions

§ 1. Introduction

Let L_p be the class of 2π-periodic, measurable functions such that $\| f \|_p < +\infty$, and $L_p^k = \{f : f \in L_p, f^{(k-1)}$ abs. cont. and $f^{(k)} \in L_p\}$, where $p \geqslant 1$ is any real number and $k \geqslant 1$ any integer. For $f \in L_p$ and $\mathfrak{R} \subset L_p$ we put

$$E^+(f;\mathfrak{R})_p = \inf\{\| f - u \|_p : u \in \mathfrak{R}, u(t) \geqslant f(t) \text{ a.e.}\}, \tag{1.1}$$

$$E^-(f;\mathfrak{R})_p = \inf\{\| f - u \|_p : u \in \mathfrak{R}, u(t) \leqslant f(t) \text{ a.e.}\}, \tag{1.1'}$$

The quantity $E^+(f;\mathfrak{R})_p$ $(E^-(f;\mathfrak{R})_p)$ is called the best upper-approximation (lower-approximation) to f by the set $\mathfrak{R}$ in metric L_p. For a given set $\mathfrak{W} \subset L_p$ the quantities

$$E^+(\mathfrak{W};\mathfrak{R})_p = \sup\{E^+(f;\mathfrak{R})_p : f \in \mathfrak{W}\} \tag{1.2}$$

$$E^-(\mathfrak{W};\mathfrak{R})_p = \sup\{E^-(f;\mathfrak{R})_p : f \in \mathfrak{W}\} \tag{1.2'}$$

are called the best upper-approximation and lower-approximation to $\mathfrak{W}$ by $\mathfrak{R}$ (in metric L_p) respectively. The one-sided n-width of the set $\mathfrak{W}$ in L_p is

* Received February 20, 1984.
本文与黄达人合作.

defined by

$$\mathrm{d}_n^{\pm}[\mathfrak{W};L_p] = \inf_{M_n \subset L_p} \mathrm{E}^{\pm}(\mathfrak{W};M_n)_p, \tag{1.3}$$

where the infimum is taken over all n-dimensional subspaces M_n of L_p. Let

$q_{2r}(t) = \prod_{j=1}^{r}(t^2 - t_j^2)$, were $r \geqslant 0$ and $t_1, \cdots, t_r \geqslant 0$.

By setting $D = \dfrac{\mathrm{d}}{\mathrm{d}t}$ we introduce a differential form

$$Q_{2r+\sigma}(D) = D^{\sigma} \prod_{j=1}^{r}(D^2 - t_j^2 I) \tag{1.4}$$

where $\sigma \geqslant 0$ is an integer, and I the identity. By virtue of $Q_{2r+\sigma}(D)$ we define

$$N\,\Omega_p^{2r+\sigma} = \{f : f \in L_p^{2r+\sigma}, \| Q_{2r+\sigma}(D)f \|_p \leqslant N\} \tag{1.5}$$

$$N \cdot W_p^{\sigma} = \{f : f \in L_p^{\sigma}, \| D^{\sigma} f \|_p \leqslant N\} \tag{1.6}$$

where N is an arbitrary positive number. It is obvious that NW_p^{σ} can be obtianed from $N\,\Omega_p^{2r+\sigma}$ by setting $r = 0$. We are mainly concerned with the function class $\Omega_p^{2r+\sigma}(= 1.\,\Omega_p^{2r+\sigma})$. For $\sigma \geqslant 1$ It consists of all functions

$$f(x) = C + \pi^{-1}\int_0^{2\pi} K_{2r+\sigma}(x-t)h(t)\mathrm{d}t, \tag{1.7}$$

where $C \in \mathbf{R}$, $\| h \|_p \leqslant 1, \int_0^{2\pi} h(t)\mathrm{d}t = 0$, and the kernel is

$$K_{2r+\sigma}(x) = \sum_{\nu=1}^{\infty} \frac{\cos\left(\nu x - \dfrac{\sigma\pi}{2}\right)}{\nu^{\sigma} q_{2r}(\mathrm{i}\nu)}, \mathrm{i} = \sqrt{-1}. \tag{1.8}$$

If $r \geqslant 1, \sigma = 0$, we specially define

$$\Omega_p^{2r} = \{f : f = \pi^{-1}K_{2r} * h : \| h \|_p \leqslant 1, \int_0^{2\pi} h\ \mathrm{d}t = 0\} \tag{1.9}$$

For further necessity we introduce some auxiliary functions:

$$B_{2r+\sigma}(t;\lambda) = 2\sum_{\nu=1}^{\infty} \frac{\cos\left(\lambda\nu t - \dfrac{\sigma\pi}{2}\right)}{(\lambda\nu)^{\sigma} q_{2r}(\mathrm{i}\lambda\nu)}, \mathrm{i} = \sqrt{-1}, \tag{1.10}$$

$$\widetilde{B}_{2r+\sigma}(t;\lambda) = B_{2r+\sigma}(t\lambda^{-1};\lambda), \tag{1.11}$$

$$H_{2r+\sigma,+}(\lambda) = \sup_t B_{2r+\sigma}(t;\lambda), \tag{1.12}$$

$$H_{2r+\sigma,-}(\lambda) = \sup_t\{-B_{2r+\sigma}(t;\lambda)\}, \tag{1.13}$$

$$H_{2r+\sigma}(\lambda) = \frac{1}{2}\{H_{2r+\sigma,+}(\lambda) + H_{2r+\sigma,-}(\lambda)\}. \tag{1.14}$$

It is the purpose of this paper to consider the problem of computing the upper n-width $d_n^+[\Omega_p^{2r+\sigma}; L^s]$.

So far as we know, this problem remains unsolved even in the special case $r = 0$. Inspired by the recent work[1] of Korneichuk and others, we establish (in § 2) some comparison theorems of Kolmogorov type concerning the class $\Omega_\infty^{2r+\sigma,-}$, from which some sharp inequalities are derived. By means of these inequalities and other tools developed in[1] we are able to compute exactly $E^\pm(\Omega_p^{2r+\sigma,-}, T_{n-1})_s$ for $p = s = 1$ and $p = +\infty$, $s = 1$ in § 3, and $d_n^+[\Omega_1^{2r+\sigma}; L_1]$ in § 4. As a consequence we get the exact estimation of $d_n^+[W_1^\sigma; L_1]$.

§ 2. The comparison theorem concerning class $\Omega_\infty^{2r+\sigma,-}$

We first define the sub-operator of $Q_{2r+\sigma}(D)$ and its compl ementary-operator.

Definition 2.1 *Let*

$$Q_{2r+\sigma}(D) = D^\sigma \prod_{j=1}^{\infty}(D^2 - t_j^2 I),\ q_{2r}(t) = \prod_{j=1}^{\infty}(t^2 - t_j^2), \tag{2.1}$$

where the order of $t_1, \cdots, t_r$ in (2.1) is arbitrarily given. Then the operator given by

$$Q_{2l+\mu}(D) = D^\mu \prod_{j=1}^{l}(D^2 - t_j^2),\ q_{2l}(t) = \prod_{j=1}^{l}(t^2 - t_j^2) \tag{2.2}$$

$(0 \leqslant \mu \leqslant \sigma, 0 \leqslant l \leqslant r)$ *is called the sub-operator of $Q_{2r+\sigma}(D)$, and the operator*

$$\widetilde{Q}_{2l+\mu}(D) = D^{\sigma-\mu} \prod_{j=l+1}^{r}(D^2 - t_j^2) \tag{2.3}$$

is called the complementary-operator of $Q_{2l+\mu}(D)$ relative to $Q_{2r+\sigma}(D)$.

If $f \in L_p$, we denote

$$f_-(t) = \max\{0, -f(t)\},\ f_+(t) = \max\{0, f(t)\}. \tag{2.4}$$

For $\sigma \geqslant 1$, we define the class of functions

$$\Omega_p^{2r+\sigma,-} = \{f : f \in L_p^{2r+\sigma}, \| (Q_{2r+\sigma}(D)f)_- \|_p \leqslant 1.\} \quad (2.5)$$

And $\Omega_{p'}^{0-}$

$$\Omega_{p'}^{0-} = \{f : f \in L_p, \| f \|_p \leqslant 1, f(t) \leqslant 0\}. \quad (2.6)$$

We note that, if $Q_{2l+\mu}(D)$ is a sub-operator of $Q_{2r+\sigma}(D)$, then

$$\widetilde{Q}_{2l+\mu}(D) B_{2r+\sigma}(t;\lambda) = B_{2l+\mu}(t;\lambda), \quad (2.7)$$

where $B_{2l+\mu}(t;\lambda)$ is defined by (1.10) corresponding to $Q_{2l+\mu}(D)$.

The following lemma is obviously true.

Lemma 2.1 *$H_{2r+\sigma,\pm}(\lambda)$ and $H_{2r+\sigma}(\lambda)$ are continuous functions of $\lambda > 0$, moreover if $\sigma \geqslant 1$, when λ increases from 0^+ to $+\infty$, $H_{2r+\sigma,\pm}(\lambda)$ and $H_{2r+\sigma}(\lambda)$ strictly decrease from $+\infty$ to 0^+.*

From this lemma we know that, for given $\xi > 0$, the equation

$$H_{2r+\sigma}(\lambda(\xi)) = \xi \quad (2.8)$$

has unique solution $\lambda(\xi)$.

Lemma 2.2 *Let $Q_{2l+\mu}(D)$ be a sub-operator of $Q_{2r+\sigma}(D)$ $(1 \leqslant \mu \leqslant \sigma)$, K a positive integer and $\lambda > 0$ a fixed real number. If $f(t)$ is a periodic function with period $\frac{2\pi}{\lambda}K$, $f^{(2r+\sigma-1)}(t)$ is absolutely continuous and*

$$Q_{2r+\sigma}(D)\, f(t) \geqslant -1, \qquad \text{a.e.} \quad (2.9)$$

$$-H_{2r+\sigma}(\lambda) < f(t) < H_{2r+\sigma,+}(\lambda), \quad (2.10)$$

then

$$-H_{2l+\mu,-}(\lambda) \leqslant \widetilde{Q}_{2l+\mu}(D)\, f(t) \leqslant H_{2l+\mu,+}(\lambda). \quad (2.11)$$

Proof Suppose that the lemma is not true. Then without loss of generality we may assume

$$\max \widetilde{Q}_{2l+\mu}(D)\, f(t) > H_{2l+\mu,+}(\lambda). \quad (2.12)$$

We write

$$B_{2r+\sigma,h}(t;\lambda) = \frac{\lambda}{\pi h}\int_{t-\frac{\pi h}{2\lambda}}^{t+\frac{\pi h}{2\lambda}} B_{2r+\sigma}(t;\lambda)\,d\tau \quad (2.13)$$

$$B_{2l+\mu,h}(t;\lambda) = \widetilde{Q}_{2l+\mu}(D) B_{2r+\sigma,h}(t;\lambda) \quad (2.14)$$

$$B_{1.h}(t;\lambda) = Q_{2r+\sigma-1}(D) B_{2r+\sigma,h}(t;\lambda).$$

The functions

$B_{j,h}(t;\lambda)$ uniformly converge to $B_j(t;\lambda)$, $(j = 2r+\sigma, 2l+\mu)$ (2.15)

when $h \to 0^+$. Thus, if we set

$$M = \text{ess sup } Q_{2r+\sigma}(D)f(t) + 1, \tag{2.16}$$

we can choose $h > 0$ so small that

$$(1)\ h < \frac{1}{M}, \tag{2.17}$$

$$(2)\ \min_t B_{2r+\sigma,h}(t;\lambda) < f(t) < \max_t B_{2r+\sigma,h}(t;\lambda), \tag{2.18}$$

$$(3)\ \max_t \widetilde{Q}_{2l+\mu}(D)f(t) > \max B_{2l+\mu,h}(t;\lambda). \tag{2.19}$$

Choosing t_*, $\bar{t}$ and $\alpha > 1$ such that

$$\max_t \widetilde{Q}_{2l+\mu}(D)f(t) = \alpha \max_t B_{2l+\mu,h}(t;\lambda),$$

$$\max_t \widetilde{Q}_{2l+\mu}(D)f(t) = \widetilde{Q}_{2l+\mu}f(t_*),$$

$$\max_t B_{2l+\mu,h}(t;\lambda) = B_{2l+\mu,h}(t_* - \bar{t}, \lambda),$$

and considering the difference

$$\Delta(t) = B_{2r+\sigma,h}(t-\bar{t}) - \frac{1}{\alpha}f(t).$$

By (2.18) we know that $\Delta(t)$ has at least $2K$ distinct zeros in the interval $[t_* - \frac{\pi h}{\lambda}, t_* - \frac{\pi h}{\lambda} + \frac{2\pi}{\lambda}K)$. Because of $\Delta(t + \frac{2\pi}{\lambda}K) = \Delta(t)$, by using the generalized Rolle's theorem we may conclude that $\widetilde{Q}_{2l+\mu}(D)\Delta(t)$ has at least $2K$ distinct zeros in the same interval.

First we consider the case $2l+\mu > 1$. By the definition of t_* we have

$$\widetilde{Q}_{2l+\mu}(D)\Delta(t_*) = 0.$$

Since t_* is a local maximum for $\widetilde{Q}_{2l+\mu}(D)f(t)$ as well as for $B_{2l+\mu,h}(t - \bar{t};\lambda)$, we have

$$D\,\widetilde{Q}_{2l+\mu}(D)\Delta(t_*) = 0.$$

Hence for any $\xi \in \mathbf{R}$ we have

$$(D-\xi)\,\widetilde{Q}_{2l+\mu}(D)\Delta(t_*) = 0.$$

By applying the Rolle's theorem once more we see that $(D-\xi)\widetilde{Q}_{2l+\mu}(D)\Delta(t)$ has at least $2K+1$ distinct zeros in a $\frac{2\pi}{\lambda}K$ period. So that

$$Q_{2r+\sigma-1}(D)\Delta(t) = B_{1,h}(t-\bar{t};\lambda) - \frac{1}{\alpha}Q_{2r+\sigma-1}(D)f(t)$$

has at least $2K+1$ distinct zeros in $[t_* - \frac{h\pi}{\lambda}, t_* + \frac{2K-h}{\lambda}\pi)$. On the other hand, by the definition of h and M we know that

$$-1 < \frac{1}{\alpha} Q_{2r+\sigma}(D) f(t) < \frac{1}{h} - 1,$$

and that $B'_{1,h}(t-\bar{t};\lambda)$ is a step function which equals $\frac{2}{h}-1$ and -1 K times alternatively in $[t_* - \frac{\pi h}{\lambda}, t_* + \frac{2k-h}{\lambda}\pi)$. It follows that $Q_{2r+\sigma-1}(D)\Delta(t)$ has at most $2K$ zeros in a $\frac{2\pi}{\lambda}K$ period. This is a contradiction.

It remains to consider the case $2l+\mu=1$, i.e. $l=0, \mu=1$. Since $B_{1,h}(t-\bar{t})$ is a broken line, its slopes are $\frac{2}{h}-1$ and -1 in $[t_* - \frac{\pi h}{\lambda} + \frac{2\pi j}{\lambda}, t_* + \frac{2\pi j}{\lambda}]$, and $[t_* + \frac{2\pi j}{\lambda}, t_* - \frac{\pi h}{\lambda} + \frac{2\pi(j+1)}{\lambda}]$ $(j=0,\cdots,k-1)$ respectively. From the definition of $\Delta(t)$ and α we know that $Q_{2r+\sigma-1}(D)\Delta(t)$ has only one zero $t^{\cdot}_*$ in $[t_* - \frac{\pi h}{\lambda}, t_* - \frac{\pi h}{\lambda} + \frac{2\pi}{\lambda}]$ and $2K-2$ zeros in the other $2K-2$ monotonic intervals of $B_{1,h}(t-\bar{t})$. Hence (2.11) is true for the case $2l+\mu=1$. Thus we complete the proof.

By lemma 2.2 we obtain

Theorem 2.1 *Suppose that* $\sigma \geqslant 1, 2r+\sigma \geqslant 2, \widetilde{Q}_{2l+\mu}(D)$ *is a sub-operator of* $Q_{2r+\sigma}(D)$ $(1 \leqslant \mu \leqslant \sigma, 2l+\mu \leqslant 2r+\sigma-1)$. *If* $g \in \Omega_\infty^{2r+\sigma,-}$ *and for certain* $\lambda > 0$ g satisfies

$$-H_{2r+\sigma,-}(\lambda) \leqslant g(t) \leqslant H_{2r+\sigma,+}(\lambda) \tag{2.20}$$

Then

$$-H_{2l+\mu,-}(\lambda) \leqslant \widetilde{Q}_{2l+\mu}(D)\, g(t) \leqslant H_{2l+\mu,+}(\lambda). \tag{2.21}$$

Corollary 2.1 *Suppose that* $\sigma \geqslant 1, 2r+\sigma \geqslant 2, Q_{2l+\mu}(D)$ *is a sub-operator of* $Q_{2r+\sigma}(D)$ $(1 \leqslant \mu \leqslant \sigma, 2l+\mu \leqslant 2r+\sigma-1)$. *If* $g \in L_\infty^{2r+\sigma}$ *and* g *const*.

Then

$$\frac{E_1(\widetilde{Q}_{2l+\mu}(D)g)_\infty}{\|(Q_{2r+\sigma}(D)g)_-\|_\infty} \leqslant H_{2l+\mu}(\lambda), \tag{2.22}$$

where λ *is the solution of the following equation*

$$H_{2r+\sigma}(\lambda) = \frac{E_1(f)_\infty}{\|(Q_{2r+\sigma}(D)g)_-\|_\infty}. \tag{2.23}$$

The definition of f_- is given in (2.4), and $E_1(f)_p$ is the best L_p approximation to $f \in L_p$ by constants:

$$E_1(f)_p = \inf_{\xi \in \mathbf{R}} \| f - \xi \|_p. \tag{2.24}$$

With some modifications we can prove (cf. [1])

Theorem 2.2 *Let* $\sigma \geqslant 1, 2r + \sigma \geqslant 2$, *and* $g \in \Omega_\infty^{2r+\sigma,-}$ *satisfies* (2.20) *for certain* $\lambda > 0$. *Suppose that there exist* ξ *and* η *such that* $g(\xi) = B_{2r+\sigma}(\eta;\lambda)$.

(1) *If* $B'_{2r+\sigma}(\eta;\lambda) > 0$, *then* $g'(\xi) \leqslant B'_{2r+\sigma}(\eta;\lambda)$;

(2) *If* $B'_{2r+\sigma}(\eta;\lambda) < 0$, *then* $g'(\xi) \geqslant B'_{2r+\sigma}(\eta;\lambda)$.

Corollary 2.2 *Let* $\sigma \geqslant 1, 2r + \sigma \geqslant 2$, *and* $g \in \Omega_\infty^{2r+\sigma,-}$ *satisfies* (2.20) *for certain* $\lambda > 0$.

(1) *Assume that* $g(\xi_0) = B_{2r+\sigma}(\eta_0, \lambda)$ *and* (α, β) *is a monotonic interval of* $B_{2r+\sigma}(t;\lambda)$ *containing* η_0. *If* $B_{2r+\sigma}(t;\lambda)$ *increases in* (α, β), *then*

$$g(\xi_0 + u) \leqslant B_{2r+\sigma}(\eta_0 + u;\lambda), \qquad 0 \leqslant u \leqslant \beta - \eta_0,$$
$$g(\xi_0 - u) \geqslant B_{2r+\sigma}(\eta_0 - u;\lambda), \qquad 0 \leqslant u \leqslant \eta_0 - \alpha.$$

If $B_{2r+\sigma}(t;\lambda)$ *decreases in* (α, β), *then the two previous inequalities should be turned in contrary direction.*

(2) *If* $g(\xi_0) = B_{2r+\sigma}(\eta_0;\lambda), g(\xi_1) = B_{2r+\sigma}(\eta_1;\lambda)$, *and* η_0, η_1 *belong to the same monotonic interval of* $B_{2r+\sigma}(t;\lambda)$ *then*

$$|\xi_0 - \xi_1| \geqslant |\eta_0 - \eta_1|.$$

On the basis of Theorem 2.2 and the two corollaries by employing the rearrangement theory introduced in [1][2], we obtain another corollary.

Corollary 2.3 *Let* $2r + \sigma \geqslant 1, f \in \Omega_\infty^{2r+\sigma+1,-}$, *and for certain* $\lambda > 0$ f *satisfies*

$$E_1(f)_\infty \leqslant H_{2r+\sigma+1}(\lambda).$$

Then for any $p \in [1, \infty]$ *we have*

$$\| g' \|_p \leqslant \| \widetilde{B}_{2r+\sigma}(\cdot;\lambda) \|_p,$$

$$E_1(g')_p \leqslant E_1(\widetilde{B}_{2r+\sigma}(\cdot;\lambda))_p,$$

$$E_1^{\pm}(g')_p \leqslant E_1^{\pm}(\widetilde{B}_{2r+\sigma}(\cdot;\lambda))_p.$$

where

$$E_1^{+}(f)_p = \inf\{\|f-\xi\|_p : f(t)\leqslant \xi, \xi\in \mathbf{R}\}. \qquad (2.25)$$

Combining theorem 2.1 with Corollary 2.3 we obtain

Theorem 2.3 Suppose that $Q_{2l+\mu}(D)$ is a sub-operator of $Q_{2r+\sigma}(D)(2l+\mu\geqslant 1, 0\leqslant \mu\leqslant \sigma-1)$ and $\widetilde{Q}_{2l+\mu}(D)$ is the associated complementary-operator. If $g\in \Omega_{\infty}^{2r+\sigma,-}$ and for certain $\lambda>0$ g satisfies

$$E_1(g)_\infty \leqslant H_{2r+\sigma}(\lambda), \qquad (2.26)$$

then for any $p\in[1,\infty]$

$$\|\widetilde{Q}_{2l+\mu}(D)g\|_p \leqslant \|\widetilde{B}_{2l+\mu}(\cdot;\lambda)\|_p, \qquad (2.27)$$

$$E_1(\widetilde{Q}_{2l+\mu}(D)g)_p \leqslant E_1(\widetilde{B}_{2l+\mu}(\cdot;\lambda))_p, \qquad (2.28)$$

$$E_1^{\pm}(\widetilde{Q}_{2l+\mu}(D)g)_p \leqslant E_1^{\pm}(\widetilde{B}_{2l+\mu}(\cdot;\lambda))_p. \qquad (2.29)$$

§ 3. The best one-sided approximations to class $\Omega_p^{2r+\sigma}$ by trigonometric polynomials

We denote T_{n-1} the space of trigonometric polynomials of degr $n-1$. for $f\in L_p$ we define

$$E_n(f)_p = \inf\{\|f-u\|_p : u\in T_{n-1}\},$$

$$E_n^{+}(f)_p = \inf\{\|f-u\|_p : u\in T_{n-1}, \text{and } u(t)\geqslant f(t)\},$$

$$E_n^{-}(f)_p = \inf\{\|f-u\|_p : u\in T_{n-1}, \text{and } u(t)\leqslant f(t)\}.$$

The best one-sided approximations to a set $\mathfrak{W}\subset L_p$ by T_{n-1} are defined similarly.

Let the 2π-periodic function $K(t)$ be bounded and continuous in $(0, 2\pi)$ with

$$K(0) = \frac{1}{2}(K(0^+)+K(0^-)).$$

We denote

$$W_p^0(K) = \{f : f = c + \frac{1}{\pi}\int_0^{2\pi} K(t-\tau)h(\tau)\mathrm{d}\tau, h(t+2\pi) = h(t)$$

and $\int_0^{2\pi} h(t)\mathrm{d}t = 0\}$. (3.1)

From §1 we know that if $\sigma \geqslant 1$, then

$$W_p^0(K_{2r+\sigma}) = \Omega_p^{2r+\sigma}, \tag{3.2}$$

where the kernel $K_{2r+\sigma}(t)$ is defined by (1.8).

If we set

$$P_{2r+\sigma}(t;n) = \frac{1}{\pi}\int_0^{2\pi} K_{2r+\sigma}(t-\tau) \text{ sign sin } n\tau \mathrm{d}\tau, \tag{3.3}$$

then from [2][3]we have

Theorem 3.1 *Let* $2r+\sigma \geqslant 1, n = 1,2,\cdots$, *then*

$$E_n(K_{2r+\sigma})_1 = \pi \| P_{2r+\sigma}(\cdot;n) \|_\infty \tag{3.4}$$

$$E_n(\Omega_1^{2r+\sigma})_1 = E_n(\Omega_\infty^{2r+\sigma})_\infty = \| P_{2r+\sigma}(\cdot;n) \|_\infty \tag{3.5}$$

and for $p \in [1,\infty], \frac{1}{p}+\frac{1}{p'} = 1$.

$$E_n(\Omega_p^{2r+\sigma})_p \leqslant \frac{1}{\pi}E_n(K_{2r+\sigma})_1 = \| P_{2r+\sigma}(\cdot;n) \|_\infty \tag{3.6}$$

$$E_n(\Omega_p^{2r+\sigma})_1 = \| P_{2r+\sigma}(\cdot;n) \|_{p'}. \tag{3.7}$$

Definition 3.1 *If* $f(t)$ *is a* 2π-*periodic, differentiable function* $t_{n-1}(f;t) \in T_{n-1}$ *satisfies*

$t_{n-1}(f;t_k) = f(t_k), t'_{n-1}(f;t_k) = f'(t_k), k = 1,\cdots,n, (0 \leqslant t_1 < \cdots < t_n < 2\pi)$, *then we say that* $t_{n-1}(f;t)$ *interpolates* $f(t)$ *doubly at* $\{t_k\}_{k=1}^n$.

The following two lemmas can be found in [1].

Lemma 3.1 *Suppose* $f(t)$ *is a* 2π - *periodic and differentiable function, and*

$$\varphi_n(f;t) = \frac{1}{n}\sum_{k=1}^n f\left(t+\frac{2k\pi}{n}\right). \tag{3.8}$$

(1) *If* $\varphi_n(f;t)$ (*attains*) *its maximum and minimum at* α_n *and* β_n *respectively, then there is one and only one trigonometric polynomial* $t_{n-1}^*(f;t)(t_{n-1}^{**}(f;t))$, *interpolating* $f(t)$ *doubly at*

$$\left\{t_k^* = \alpha_n + \frac{2k\pi}{n}\right\}_{k=1}^n \left(\left\{t_k^{**} = \beta_n + \frac{2k\pi}{n}\right\}_{k=1}^n\right).$$

(2) *If the difference* $\delta^*(t) = t_{n-1}^*(f;t) - f(t)$ *does not change its*

sign in $(0,2\pi)$, *then*

$$E_n^+(f)_1 = E_1^+(\varphi_n(f))_1 = \| \max_u \varphi_n(f;u) - \varphi_n(f;\cdot) \|_1.$$

(3) *If the difference* $\delta^{**}(t) = f(t) - t_{n-1}^{**}(f;t)$ *does not change its sign, then*

$$E_n^-(f)_1 = E_1^-(\varphi_n(f))_1 = \| \varphi_n(f;\cdot) - \inf_u \varphi_n(f;u) \|_1.$$

Lemma 3.2 *For* $n = 1,2,\cdots$, *we have*

$$(1) E_n^+(W_1^0(K))_1 \leqslant \frac{1}{2\pi}(E_n^+(K)_1 + E_n^-(K)_1). \tag{3.9}$$

And furthermore, if

$$E_n^{\pm}(K)_1 = E_1^{\pm}(\varphi_n(K))_1, \tag{3.10}$$

then equality holds in (3.9).

$$(2) E_n^+(W_\infty^0(K))_1 \leqslant 2E_n(K)_1. \tag{3.11}$$

If there exists γ_0 *such that*

$$E_n(K)_1 = \left| \int_0^{2\pi} K(t) \operatorname{sign} \sin n(t-\gamma_0) \mathrm{d}t \right|,$$

then equality holds in (3.11).

Theorem 3.2 *Let* $\sigma \geqslant 1$, $K_{2r+\sigma}(t)$ *be given by* (1.8), $n = 1,2,\cdots$ *then*

$$(1) E_n^+(K_{2r+\sigma})_1 = \pi H_{2r+\sigma,+}(n), \tag{3.12}$$

$$E_n^-(K_{2r+\sigma})_1 = \pi H_{2r+\sigma,-}(n), \tag{3.13}$$

$$(2) E_n^+(\Omega_1^{2r+\sigma})_1 = H_{2r+\sigma}(n), \tag{3.14}$$

$$(3) E_n^+(\Omega^{2r+\sigma})_1 = 2\pi \| P_{2r+\sigma}(\cdot;n) \|_\infty. \tag{3.15}$$

Proof It is easy to verify

$$\varphi_n(\cos k.;t) = \begin{cases} \cos kt, & k = mn \\ 0, & k = mn. \end{cases} \tag{3.16}$$

Hence

$$\varphi_n(K_{2r+\sigma};t) = K_{2r+\sigma}(t;n), \tag{3.17}$$

where $K_{2r+\sigma}(t;n) = \dfrac{1}{2} B_{2r+\sigma}(t;n)$.

In case $2r+\sigma = 1$, from [1] we obtain (3.12) and (3.13).

If we additionally define $K'_2(0) = 0$ for the even function $K_2(t;n)$, then for $2r+\sigma \geqslant 2$, $K_{2r+\sigma}(t;n)$ is $\dfrac{2\pi}{n}$ - periodic and differentiable. Select

$\alpha_n, \beta_n \in \left[0, \frac{2\pi}{n}\right)$ such that $K_{2r+\sigma}(t; n)$ attains its maximum and minimum at α_n and β_n respectively. By lemma 3.1, there exists $t^*_{n-1}(K_{2r+\sigma}; t)$ $(t^{**}_{n-1}(K_{2r+\sigma}; t))$, which interpolates $K_{2r+\sigma}(t)$ doubly at

$$\left\{t_k^* = \alpha_n + \frac{2k\pi}{n}\right\}_{k=1}^n \quad \left(\left\{t_k^{**} = \beta_n + \frac{2k\pi}{n}\right\}_{k=1}^n\right).$$

By applying the generalized Rolle's theorem we see that

$$\delta^*(t) = t^*_{n-1}(K_{2r+\sigma}; t) - K_{2r+\sigma}(t)$$

$$\delta^{**}(t) = K_{2r+\sigma}(t) - t^{**}_{n-1}(K_{2r+\sigma}; t)$$

do not change sign. Hence

$$\begin{aligned} E_n^+(K_{2r+\sigma})_1 &= E_1^+(\varphi_n(K_{2r+\sigma}))_1 \\ &= \| \max_u \varphi_n(K_{2r+\sigma}; u) - \varphi_n(K_{2r+\sigma}; \cdot) \|_1 \\ &= \frac{1}{2}\int_0^{2\pi} [H_{2r+\sigma,+}(n) - B_{2r+\sigma}(t; n)]\mathrm{d}t = \pi H_{2r+\sigma,+}(n). \end{aligned}$$

Here we have used the following relations:

$$B_{2r+\sigma}(t; n) = 2K_{2r+\sigma}(t; n),$$

$$H_{2r+\sigma,+}(n) = \max_u B_{2r+\sigma}(u; n),$$

$$\int_0^{2\pi} B_{2r+\sigma}(t; n)\mathrm{d}t = 0.$$

From Lemma 3.2 we can derive (3.14). And by (3.3) we know that

$$E_n(K_{2r+\sigma})_1 = \pi \| P_{2r+\sigma}(\cdot; n) \|_\infty = \left\| \int_0^{2\pi} k(t-\tau)\,\mathrm{sign}\sin n\tau \mathrm{d}\tau \right\|_\infty$$

$$= \left| \int_0^{2\pi} K(t)\,\mathrm{sign}\sin n(t - r_n)\mathrm{d}t \right|.$$

Hence we obtain (3.15) from Lemma 3.2.

Analogously we can prove

Theorem 3.2′ *If* $\sigma \geqslant 1; n = 1, 2, \cdots$, *then*

$$E_n^-(\Omega_1^{2r+\sigma})_1 = H_{2r+\sigma}(n),$$

$$E_n^-(\Omega^{2r+\sigma})_1 = 2\pi \| P_{2r+\sigma}(\cdot; n) \|_\infty \cdot$$

Remark It is obvious that if $r \geqslant 1, \sigma = 0$, the results of theorem 3.2 and 3.2′ are still true for Ω_1^{2r} and Ω_∞^{2r} which are defined by (1.9)

§ 4. Exact estimation of $d_n^+[\Omega_1^{2r+\sigma}; L_1]$

In this section we compute the upper n-width of class $\Omega_1^{2r+\sigma}(\sigma \geqslant 2)$ in

L_1 metric (cf. (1.3)).

Lemma 4.1 *Suppose that* $Q_{2l+\mu}(D)$ *is a sub-operator of* $Q_{2r+\sigma}(D)$, $\widetilde{Q}_{2l+\mu}(D)$ *is the associated complementary-operator*, $p, q, s \in [1, \infty]$, $N > 0$ *then*

$$E^+(\Omega_p^{2l+\mu}; N\Omega_q^{2r+\sigma})_s = \sup_{f \in \Omega_{s'}^{2r+\sigma,-}} \{E_1(\widetilde{Q}_{2l+\mu}(D)f)_{p'} - NE_1(f)_{q'}\}, \tag{4.1}$$

where $\frac{1}{p} + \frac{1}{p'} = \frac{1}{q} + \frac{1}{q'} = \frac{1}{s} + \frac{1}{s'} = 1$.

Proof By the duality theorem of best one-sided approximation(cf. [1]) we have

$$E^+(\Omega_p^{2l+\mu}; N\Omega_q^{2r+\sigma})_s = \sup_{\varphi \in \Omega_p^{2l+\mu}} \Big\{ \sup_{\substack{f \in L_{s'} \\ \|(f)_-\|_{s'} \leqslant 1}} \int_0^{2\pi} f(t)\Psi(t)\mathrm{d}t - \sup_{u \in N\Omega_q^{2r+\sigma}} \int_0^{2\pi} u(t)f(t)\mathrm{d}t \Big\}. \tag{4.2}$$

Integrating by parts and using the duality theorem again we get the result.

Lemma 4.2 *If* $0 \leqslant \mu \leqslant \sigma - 1, 2l + \mu \geqslant 1$, $Q_{2l+\mu}(D)$ *is a sub-operator of* $Q_{2l+\sigma}(D)$, $\widetilde{Q}_{2l+\mu}(D)$ *is the associated complementary-operator, and* $f \in L_\infty^{2r+\sigma}$ *then*

$$\frac{E_1(\widetilde{Q}_{2l+\mu}(D)f)_p}{\|(Q_{2r+\sigma}(D)f)_-\|_\infty} \leqslant E_1(\widetilde{B}_{2l+\mu}(\cdot;\lambda))_p, \tag{4.3}$$

where λ *is the solution of*

$$H_{2r+\sigma}(\lambda) = \frac{E_1(f)_\infty}{\|(Q_{2r+\sigma}(D)f)_-\|_\infty}. \tag{4.4}$$

Proof Let $\xi_0 \in \mathbf{R}$ be the number such that

$$\|f(t) - \xi_0\|_\infty = \inf_{\xi \in \mathbf{R}} \|f - \xi\|_\infty = E_1(f)_\infty.$$

Since $Q_{2r+\sigma}(D)(f(t) - \xi) = Q_{2r+\sigma}(D)f(t)$, if we set

$$f_*(t) = \frac{f(t) - \xi_0}{\|(Q_{2r+\sigma}(D)f)_-\|_\infty},$$

and take λ such that satisfies (4.4), then $f_* \in \Omega_\infty^{2r+\sigma,-}$ and

$$E_1(f_*(t)) = H_{2r+\sigma}(\lambda).$$

Hence we derive (4.3) from (2.28) of Theorem 2.3.

Now for a semi-norm Ψ defined on L_1 we set

$$\Psi(\mathfrak{W}) = \sup\{\Psi(f); f \in \mathfrak{W}\}.$$

Concerning Ψ we have the following

Lemma 4.3 *Let $\Omega_1^{0,-}$ be given by (2.6), $\sigma \geqslant 2$, and $1 \leqslant \mu \leqslant \sigma - 1$. If ψ is an arbitrarily given semi-norm on L_1, then*

$$\frac{\psi(W_{p'}^{\mu})}{\psi(\Omega_1^{0,-})} \leqslant E_1(\widetilde{B}_\mu(\cdot;\lambda))_p, \tag{4.5}$$

where $p \in [1,\infty]$, $\frac{1}{p} + \frac{1}{p'} = 1$, and λ is the solution of

$$H_{2r+\sigma}(\lambda) = \frac{\psi(\Omega_1^{2r+\sigma})}{\psi(\Omega_1^{0,-})}. \tag{4.6}$$

Proot For arbitrarily given $f \in W_p^{\mu'}$ and $N > 0$, let $u_f^+(t)$ be the best upper approximant, of f by $N\Omega_1^{2r+\sigma}$ in the L_1 metric. Then

$$\begin{aligned}\psi(f) &\leqslant \psi(f - u_f^+) + \psi(u_f^+)\\ &\leqslant \| f - u_f^+ \|_1 \cdot \psi(\Omega_1^{0,-}) + \psi(N\Omega_1^{2r+\sigma})\\ &\leqslant \psi(\Omega_1^{0,-}) \cdot E^+(W_{p'}^{\mu}; N\Omega_1^{2r+\sigma}) + \psi(\Omega_1^{2r+\sigma}).\end{aligned}$$

For $\xi > 0$, we denote $\lambda(\xi)$ the solution of the euqation $H_{2r+\sigma}(\lambda(\xi)) = \xi$. By lemma 4.1 and 4.2 we obtain

$$\begin{aligned}\psi(f) &\leqslant \psi(\Omega_1^{0,-}) \sup_{g \in \Omega_\infty^{2r+\sigma,-}} \{E_1(Q_{2r+\sigma-\mu}(D)g)_p - NE_1(g)_\infty\} +\\ &\quad N\psi(\Omega_1^{2r+\sigma})\\ &\leqslant \psi(\Omega_1^{0,1}) \sup_{g \in \Omega_\infty^{2r+\sigma,-}} \left\{E_1\left(\widetilde{B}_\mu\left(t; \lambda\left(\frac{E_1(g)_\infty}{\| (Q_{2r+\sigma}(D)g)_- \|_\infty}\right)\right)\right)_p - \right.\\ &\quad NE_1(g)_\infty\} + N\psi(\Omega_1^{2r+\sigma})\\ &\leqslant \psi(\Omega_1^{0,-}) \sup_{\xi > 0}\{E_1\{\widetilde{B}_\mu(t;\lambda(\xi))\}_p - N\xi\} + N\psi(\Omega_1^{2r+\sigma}).\end{aligned} \tag{4.7}$$

Since

$$E_1(\widetilde{B}_\mu(t;\lambda))_p = \frac{1}{\lambda_\mu}, A_\mu$$

where $A_\mu > 0$ is a constant independent of λ, we can verify that when $\sigma \geqslant 2$

$$g(\lambda) = \frac{\frac{d}{d\lambda}\{E_1(\widetilde{B}_u(\cdot;\lambda))_p\}}{\frac{d}{d\lambda}H_{2r+\sigma}(\lambda)} \tag{4.8}$$

is a monotonic function. We note

$$\frac{d}{d\xi}\{E_1(\widetilde{B}_\mu(\cdot;\lambda(\xi)))_p\}=g(\lambda(\xi)).$$

If we take $\xi_0, \lambda_0=\lambda(\xi_0)$ such that

$$\xi_0=\frac{\psi(\Omega_1^{2r+\sigma})}{\psi(\Omega_1^{0,-})},\ H_{2r+\sigma}(\lambda(\xi_0))=\xi_0,$$

and $N_0=g(\lambda(\xi_0))$, then we have

$$\sup_{\xi>0}\{E_1(\widetilde{B}_\mu(t;\lambda(\xi)))_p-N\xi\}=E_1(\widetilde{B}_\mu(\cdot;\lambda(\xi_0)))P-N_0\xi_0.$$

4.5 follows immediately.

Now we can compute the upper n-width of class $\Omega_1^{2r+\sigma}$ in metric L_1. In [1] has been obtained the following

Lemma 4.4 *For $n=1,2,\cdots$*

$$d_{2n}^+(W_\infty^1;L_1)=2\pi\frac{\mathfrak{K}_1}{n}=\frac{\pi^2}{n},$$

where $\mathfrak{K}_1=\frac{\pi}{2}$ is the Favard constant.

Theorem 4.1 *For the operator $Q_{2r+\sigma}(D)$, $\sigma\geqslant 2$, $n=1,2,\cdots$*

$$d_{2n-1}^+[\Omega_1^{2r+\sigma};L_1]=d_{2n}^+[\Omega_1^{2r+\sigma};L_1]=H_{2r+\sigma}(n).\tag{4.9}$$

The trigonometric subspace T_{n-1} is optimal for the upper n-widths.

Proof We will first prove the following inequality:

$$d_{2n}^+[\Omega_1^{2r+\sigma};L_1]\geqslant H_{2r+\sigma}(n),\tag{4.10}$$

for any $2n$-dimension subspace $M_{2n}\subset L_1$, we consider the semi-norm ψ on L_1,

$$\psi(\mathrm{f})=E^+(f;M_{2n})_1.$$

Putting $\mu=1$, $p=1$ in lemma 4.3 we obtain

$$E^+(W_\infty^1;M_{2n})_1\leqslant E_1(\widetilde{B}(t;\lambda))_1\cdot E^+(\Omega_1^{0,-};M_{2n})_1,$$

where λ satisfies

$$H_{2r+\sigma}(\lambda)=\frac{E^+(\Omega_1^{2r+\sigma};M_{2n})_1}{E^+(\Omega_1^{0,-};M_{2n})_1}.$$

By lemma 4.4.

$$E^+(W_\infty^1;M_{2n})_1\geqslant\frac{\pi^2}{n}.\tag{4.11}$$

By the definition of $\Omega_1^{0,-}$ it is clear that

$$E^+(\Omega_1^{0,-};M_{2n}) \leqslant 1. \qquad (4.12)$$

On the other hand by direct calculation we can obtain

$$E_1(\widetilde{B}_1(t;\lambda))_1 = \frac{2\pi}{\lambda}\|P_1(\cdot;1)\|_\infty = \frac{\pi^2}{\lambda}. \qquad (4.13)$$

Combining lemma 2.1 with (4.11) – (4.13) we have

$$\lambda(E^+(\Omega_1^{2r+\sigma};M_{2n})_1) \leqslant n.$$

Hence

$$E^+(\Omega_1^{2r+\sigma};M_{2n})_1 \geqslant H_{2r+\sigma}(n).$$

Because $M_{2n} \subset L_1$ is arbitrarily given, we obtain (4.10). On the other hand, we have proved in theorem 3.2 that

$$E_n^+(\Omega_1^{2r+\sigma})_1 = H_{2r+\sigma}(n),$$

hence

$$d_{2n-1}^+(\Omega_1^{2r+\sigma};L_1) \leqslant H_{2r+\sigma}(n). \qquad (4.14)$$

Thus the theorem is proved.

Remark 1 By setting $r = 0$ in (4.9) we obtain the exact estimation of $d_n^+[W_1^\sigma;L_1]$.

2 Some 2n-dimensional subspaces of generalized periodic splines are also optimal for $d_{2n}^+[\Omega_1^{2r+\sigma};L_1]$. This fact has been established in our another paper[4].

References

[1] Н. П. Корнейчук, А. А. Лигун, В. Г. Доронин. *Аппроксимация с ограничениями.* Киев: Наукова, Думка, 1982. 219 ~ 235

[2] Н. П. Корнейчук. *Экстремальные Задачи приближения функций.* Москва: Наука, 1976, 252 ~ 286

[3] Huang Daren. On extremal subspace of $d_n[\Omega_p^r;L_1]$ (in Chinese) (to be published in 《Acta Mathematica Sinica》)

[4] Huang Daren, Sun Yongsheng. Upper-approximation of class $\Omega_p^{2r+\sigma}$ of smooth function (in Chinese) (to be published in 《Scientia Sinica》)

Approximation Theory and Its Applications
Mar., 1985, Vol. 1, No. 2

关于广义 Bernoulli 核的 n- 宽度*

On n-width of Generalized Bernoulli Kernel

§ 1. Introduction

Let $P(x) = x^r + a_1 x^{r-1} + \dots + a_r$ be a polynomial, zeros of which are all real, and $D = \frac{\mathrm{d}}{\mathrm{d}x}$. We denote $\widetilde{L}^r$ the class of 2π-periodic continuous functions such that $f^{(r-1)}$ are absolutely continuous on $[0, 2\pi]$ and $f^{(i)}(0) = f^{(i)}(2\pi)$, $\mathrm{i} = 1, \dots, \mathrm{r} - 1$. Put

$$\widetilde{\mathfrak{M}}_p^{(0)} = \left\{ f \in \widetilde{L}^r : \| P(D) f \|_p \leqslant 1, \int_0^{2\pi} P(D) f(x) \mathrm{d}x = 0 \right\}.$$

By [1], $f(x) \in \widetilde{\mathfrak{M}}_p^{(0)}$ provided that

$$f(x) = c_\sigma + \int_0^{2\pi} G(x-t) h(t) \mathrm{d}t, \qquad (1.1)$$

where $\| h \|_p \leqslant 1, \int_0^{2\pi} \cdot h(t) \mathrm{d}t = 0, c_\sigma = 0$. When $P(0) \neq 0$, and $c_\sigma \in \mathbf{R}$ is arbitrary when $P(0) = 0$. The kernel

$$G(x) = \frac{1}{2\pi_k} \sum_{k=-\infty}^{+\infty}{}' \frac{\mathrm{e}^{\mathrm{i}k, x}}{P(\mathrm{i}k)}, \qquad \mathrm{i} = \sqrt{-1}, \qquad (1.2)$$

where $\sum'$ means the term $k = 0$ is disappeared when $P(0) = 0$. Note that (cf. [1])

* Received February 21, 1984.
本文与黄达人合作.

$$D^{\sigma}P(D)G(x)\equiv 0,\quad 0<x<2\pi, \tag{1.3}$$

where $\sigma=0$ in the case $P(0)=0$, and $\sigma=1$ in the case $P(0)=0$. We also have

$$P(D)f(x)\xlongequal{\text{a.e.}}h(x). \tag{1.4}$$

when $p(x)=x^r$(1.2) gives $G(x)=\frac{1}{2\pi}\sum_{k=-\infty}^{+\infty}{}'\frac{e^{ik,x}}{(ik)^r}$. This is the Bernoulli kernel, the important role which plays in the approximation theory is well known. We call the function (1.2) generalized Bernoulli kernel. M. G. Krein[1] proved

Theorem A *For* $n=1,2,3\ldots$

$$\sup_{f\in\widetilde{\mathfrak{M}}_{\infty}^{(0)}}E_n(f)_c=\max_x\left|\int_0^{2\pi}G(x-t)\operatorname{sgn}\sin nt\,dt\right|$$
$$=\frac{2}{\pi}\max_x\left|\sum_{k=-\infty}^{+\infty}\frac{e^{i(2k+1)nx}}{(2k+1)P[(2k+1)ni]}\right| \tag{1.5}$$

where $E_n(f)_c\xlongequal{\text{df}}\min_{t_{n-1}}\|f-t_{n-1}\|_c$, t_{n-1} *is taken from the* $2n-1$ *dimensional trigonometric subspace* T_{n-1}, *and* $\|f\|_c\xlongequal{\text{df}}\max_X|f(x)|$.

Krein proved[1] for every n $G(x)$ satisfies A_n condition (viz., the MarkovNicolsky condition, cf, [2]). From this we know (1.5) is also valid for the L_1-approximation to $\widetilde{\mathfrak{M}}_1^{(0)}$ by the space T_{n-1}. In this paper we are mainly concerned with the exact estimation of the kolmogorov n-width, the Gelfand n-width and the linear n-width of the class $\widetilde{\mathfrak{M}}_p^{(0)}$ in metric L^s. Some special cases of this problem nave been considered. For example, the case $P(x)=x^r$ is well studied (cf. [3][4]). In [6] − [9] we have considered the case $P(x)=x^{\tau}\prod_{j=1}^{l}(x^2-t_j^2)(\tau\geqslant 0,t_1,\ldots,t_l>0)$, i.e., $P(D)$ is selfadjoint. Now we continue our research in [6] ~ [9]. Definitions and notations we refer to [7].

§ 2. One Minimum Norm Problem on a Class of Generalized Perfect Splines

Let $\xi=(\xi_1,\ldots,\xi_{2n})\in\mathbf{R}^{2n}$. We say that ξ is 2π-periodic, if $\xi_1<\xi_2$

$< \ldots < \xi_{2n} < \xi_1 + 2\pi$. We put

$$\Xi_{2n}^{(0)} = \{\xi \in \mathbf{R}^{2n} : 0 \leqslant \xi_1 < \ldots < \xi_{2n} < 2\pi\}.$$

For every $\xi \in \Xi_{2n}^{(0)}$ there is a 2π-periodic step function $h_\xi(x) = (-1)^{j-1}$ in $\xi_j < x < \xi_{j+1}, j = 1, 2, \ldots, 2n(\xi_{2n+1}) = 2\pi + \xi_1)$, and $h_\xi(\xi_j) = 0, j = 1, \ldots, 2n$. Writing

$$\Gamma_{2n} = \{h_\xi : \xi \in \Xi_{2m}^{(0)}, 1 \leqslant m \leqslant n\}, n = 1, 2, 3, \ldots,$$

$$\Pi_{2n} = \{f_\xi = c_\sigma + G * h_\xi : h_\xi \in \Gamma_{2n}\}$$

we consider the minimum norm problem

$$e_{2n}(\Pi; L^P) \stackrel{\text{df}}{=\!=} \min_{f_\xi \in \Pi_{2n}} \| f_\xi \|_p, \tag{2.1}$$

$1 \leqslant p \leqslant +\infty$. $n = 1, 2, 3, \ldots$. The minimum is attained on the set Π_{2n}. The following theorem gives the full information about the extremal functions.

Theorem 1 *For* $1 \leqslant p \leqslant +\infty$, $n = 1, 2, 3, \ldots$ *We have*

$$e_{2n}(\Pi; L^p) = \| \Phi_n \|_p, \tag{2.2}$$

where $\Phi_n(x) = \int_0^{2\pi} G(x - t) \operatorname{sgn} \sin nt \, dt$.

Proof First we consider the case $1 < p < +\infty$. Let an extremal function be $f_{\xi^*}(x) = c_\sigma^* + (G * h_{\xi^*})(x)$, $\xi^* = (\xi_1^*, \ldots, \xi_{2m}^*) \in \Xi_{2m}^{(0)}$, $m \leqslant n$ ($c_\sigma^* = 0$, for $\sigma = 0$). From [4] we know that

$$g(x) = \int_0^{2\pi} | f_{\xi^*}(\cdot) |^{p-1} \operatorname{sgn} f_{\xi^*}(\cdot) G(\cdot - x) d(\cdot) \tag{2.3}$$

satisfies $g(\xi_j^*) = 0, j = 1, \ldots, 2m$, and that the condition

$$\int_0^{2\pi} | f_{\xi^*}(x) |^{p-1} \operatorname{sgn} f_{\xi^*}(x) dx = 0 \tag{2.4}$$

holds when $\sigma = 1$. We claim that $\xi_1^*, \ldots, \xi_{2m}^*$ is the unique set of zeros of $g(x)$ in a period, and each ξ_j^* is simple. Suppose $g(x)$ in a period has more than $2m$ zeros (the multiplicity of each zero should be counted). If $P(x) = x^s \prod_{j=1}^{r-s} (x - \lambda_j)(\lambda_1, \ldots, \lambda_{r-s} \neq 0, s \geqslant 0)$, by considering

$$P^*(D) g(x) = (-1)^r | f_{\xi^*}(x)^{p-1} \operatorname{sgn} f_{\xi^*}(x),$$

where $P^*(D) = D^s(D + \lambda_1) \cdots (D + \lambda_{r-s})$ is the adjoint form of $P(D)$, and by applying the generalized Rolle theorem (for $D - \lambda_i$) or Rolle theorem (for D), then we may conclude that $P^*(D)g(x)$ has in $[0,2\pi)$ more than $2m$ zeros, and so does the function $f_{\xi^*}(x)$. Once more applying the Rolle theorem, in case $s = 0$ we know

$$g_1(x) = (D-\lambda_2)\cdots(D-\lambda_r)(G*h_{\xi^*})(x)$$

has more than $2m$ zeros in a period. But

$$(D-\lambda_1)g_1(x) = e^{-\lambda_1 x}\frac{d}{dx}[e^{\lambda_1 x}g_1(x)] = h_{\xi^*}(x),$$

so that $e^{\lambda_1 x}g_1(x)$ is strictly monotonic in every $[\xi_j^*, \xi_{j+1}^*]$, and therefore $g_1(x)$ has no more than $2m$ zeros in a period. This is a contradiction. In case $s > 0$, on account of

$$g_1(x) = D^{s-1}(D-\lambda_1)\cdots(D-\lambda_{r-s})f_{\xi^*}(x)$$

and $Dg_1(x) = h_{\xi^*}(x)$, we know that $g_1(x)$ is a linear segment in each $[\xi_j^*, \xi_{j+1}^*]$ with slope ± 1, so that $g_1(x)$ also has no more than $2m$ zeros in a period. This is also a contradiction. Now we put $\Delta_j = [\xi_j^*, \xi_{j+1}^*], j = 1, \cdots, 2m$, $|\Delta_j| = \xi_{j+1}^* - \xi_j^*$. Arguing similarly as in [3] [4] we may conclude $|\Delta_j| = \frac{\pi}{m}, j = 1, \cdots, 2m$, so that we obtain

$$e_{2n}(\Pi; L^p) = \| c_\sigma^* + \Phi_m(\cdot) \|_p.$$

For the case $\sigma > 0$ it is easy to prove $c_\sigma^* = 0$. In fact. (2.4) gives

$$\int_0^{2\pi} |\Phi_m(x) + c_\sigma^*|^{p-1}\operatorname{sgn}(\Phi_m(x) + c_\sigma^*)dx = 0,$$

which means c_σ^* is the best approximation number to Φ_m by $\mathbf{R}$ in L^p metric. This number is unique. On account of

$$\int_0^{2\pi} |\Phi_m(x)|^{p-1}\operatorname{sgn}\Phi_m(x)dx = 0$$

we obtain $c_\sigma^* = 0$.

It remains to prove m = n. Suppose m < n. Let $\xi^* \in (\xi_1, \xi_1 + 2\pi)$ be any point such that $\xi^* \neq \xi_j^*$ $(j = 1, \cdots, 2m)$, and $\varepsilon > 0$ be a number so small that $(\xi^* - \varepsilon, \xi^* + \varepsilon) \subset [\xi_j^*, \xi_{j+1}^*]$ for some j. Then $\xi(\varepsilon) \overset{\text{df}}{=} (\xi_1^*, \cdots, \xi_j^*, \xi^* - \varepsilon, \xi^* + \varepsilon, \xi_{j+1}^*, \cdots, \xi_{2m}^*) \in \Xi_{2m+2}$(i.e., 2π − periodic point set) and $2m + 2 \leqslant 2n$. So that $h_{\xi(\varepsilon)} \in \Gamma_{2n}$ · Write $\phi(\varepsilon) =$

$\|(G*h_{\xi(\varepsilon)})\|_p$. We have $\phi(0)=e_{2n}(\Pi;L^p)$ and $\lim\limits_{\varepsilon\to 0+}\varepsilon^{-1}[\phi(\varepsilon)-\phi(0)]=0$. This gives

$$\int_0^{2\pi}|\Phi_m(\cdot)|^{p-1}\operatorname{sgn}\Phi_m(\cdot)G(\cdot-\xi_j^*)d(\cdot)=0,$$

i.e., $g(x)=0$, which is impossible. So $m=n$ follows. We have proved $e_{2n}(\Pi;L^p)=\|\Phi_n\|_p$ for all $p\in(1,+\infty)$. By setting $p\to +1$ and $+\infty$ respectively we complete the proof.

It is worthy to point out

$$\min_{\substack{h_\xi\in\Gamma_{2n}\\ c_\sigma(h_\xi\perp 1)}}\|c_\sigma+(G*h_\xi)(\cdot)\|_p=\|\Phi_n\|_p \tag{2.5}$$

Arguing similarly as in [3] [4] we obtain: for $p\in[1,+\infty]$, $n=1,2,3,\cdots$,

$$d_{2n}[\widetilde{\mathfrak{M}}_\infty^{(0)};L^p]\geqslant\|\Phi_n\|_p, \tag{2.6}$$

$$d^{2n}[\widetilde{\mathfrak{M}}_\infty^{(0)};L^p]\geqslant\|\Phi_n\|_p. \tag{2.7}$$

In the special case $p=+\infty$ we have

Theorem 2 *If* $X=L_\infty$ or L^1, $n=1,2,3,\cdots$, *then we have*

$$\begin{aligned}&d_{2n-1}[\widetilde{\mathfrak{M}}_X^{(0)};X]=d_{2n}[\widetilde{\mathfrak{M}}_X^{(0)};X]=d'_{2n-1}[\widetilde{\mathfrak{M}}_X^{(0)};X]\\&=d'_{2n}[\widetilde{\mathfrak{M}}_X^{(0)};X]=d^{2n-1}[\widetilde{\mathfrak{M}}_X^{(0)};X]=d^{2n}[\widetilde{\mathfrak{M}}_X^{(0)};X]\\&=\|\Phi_n\|_c.\end{aligned} \tag{2.8}$$

In the following sections we will prove that equality holds in (2.6) and (2.7) for all p such that $1\leqslant p<+\infty$.

§ 3. Linear Interpolating Operators

Denote $S_{2n,r-1}$ a class of generalized 2π-periodic splines defined by $P(D)$

$$S_{2n,r-1}=\left\{f: f\in C^{r-2}, P(D)f(x)\equiv 0, x\in\left(\frac{k\pi}{n},\frac{k+1}{n}\pi\right), k=0,\cdots,2n-1.\right\}$$

It is easily seen that, in the case $p(0)\neq 0$

$$S_{2n,r-1}=\left\{f: f=\sum_{j=1}^{2n}c_jG\left(\cdot-\frac{j\pi}{n}\right), c_j\in\mathbf{R}\right\},$$

and in case $P(0)=0$

$$S_{2n,r-1} = \left\{\begin{array}{l} f: f = c_0 + \sum_{j=1}^{2n} c_j G\left(\cdot - \frac{j\pi}{n}\right) \\ c_j \in \mathbf{R}, \sum_{j=1}^{2n} c_j = 0 \end{array}\right\}.$$

From[1], $\Phi_n(x)$ has $2n$ simple zeros in $[0, 2\pi)$ $\beta, \beta + \frac{\pi}{n}, \cdots, \beta + \frac{2n-1}{n}\pi$.

Lemma 1 *For any* 2π*-periodic bounded function* f *there exists just one* $Q_{2n}(f) \in S_{2n,r-1}$ *such that*

$$Q_{2n}\left(f; \beta + \frac{j\pi}{n}\right) = f\left(\beta + \frac{j\pi}{n}\right), j = 1, \cdots, 2n. \tag{3.1}$$

Proof Let the zeros of $p(x)$ be $\lambda_1, \cdots, \lambda_r$. We construct the generalized polynomial $A_{r-1}(x, \lambda), A_r(x, \lambda)$ with step $\frac{\pi}{n}$ which is defined by $\lambda_1, \cdots, \lambda_r$ and $0, \lambda_1, \cdots, \lambda_r$ respectively (Concerning the generalized polynomials $A_{r-1}(x, \lambda), A_r(x, \lambda)$ we refer to [5], chapter 6). By the theorem of separation of zeros (cf. [5]) from $A_r(\beta, -1) = 0$ we get $A_{r-1}(\beta, -1) \neq 0$. Thus there does not exist nontrivial $s(x) \in S_{2n,r-1}$ such that $s\left(\beta + \frac{j\pi}{n}\right) = 0$ $(j = 0, \cdots, 2n-1)$, and the lemma follows.

Theorem 3 *For* $f \in \widetilde{\mathfrak{M}}_\infty^{(0)}$ *we have*

$$| f(x) - Q_{2n}(f; x) | \leqslant | \Phi_n(x) |, x \in [0, 2\pi]. \tag{3.2}$$

Proof (1) In case $P(0) = 0$ we may use the method of proof which is employed[2] in proving its theorem 10, 7, 2. We omit the details.

(2) In case $P(0) \neq 0$, for $f \in \widetilde{\mathfrak{M}}_\infty^{(0)}$ we have $f(x) = \int_0^{2\pi} G(x - t) h(t) \mathrm{d}t$, $\| h \|_\infty \leqslant 1$, $\int_0^{2\pi} h(t) \mathrm{d}t = 0$, and the remainder formula

$$f(x) - Q_{2n}(f; x) = \int_0^{2\pi} K(x, t) h(t) \mathrm{d}t,$$

where

$$K(x,t)=\frac{G\begin{pmatrix}x & \beta & \beta+\frac{\pi}{n} & \cdots & \beta+\frac{2n-1}{n}\pi\\ t & 0 & \frac{\pi}{n} & \cdots & \frac{2n-1}{n}\pi\end{pmatrix}}{G\begin{pmatrix}\beta & \beta+\frac{\pi}{n} & \cdots & \beta+\frac{2n-1}{n}\pi\\ 0 & \frac{\pi}{n} & \cdots & \frac{2n-1}{n}\pi\end{pmatrix}}$$

(Concerning this formula we refer to [3], p.219). For a fixed t $K(x,t)$ is the generalized periodic spline defined by the operator $p(D)$ with simple knots $\left\{t,\frac{k\pi}{n}\right\}$. On account of the properties of zeros and interpolation conditions of the spline $K(x,t)$ we may conclude that for any fixed $t\neq\frac{k\pi}{n}(k=1,\cdots,2n)$ the spline $K(x,t)$ has unique set of zeros $\left\{\beta+\frac{k\pi}{n}\right\}_{k=1}^{2n}$, and each of these is simple. Similarly, for a fixed x $K(x,t)$ is the generalized periodic spline defined by the adjoint operator $p^*(D)$ with simple knots $\left\{x,\beta+\frac{k\pi}{n}\right\}$. By the same reason as before for any fixed $x\neq\beta+\frac{k\pi}{n}(k=1,\cdots,2n)$ the spline $K(x,t)$ has unique set of simple zeros $\left\{\frac{k\pi}{n}\right\}_{k=1}^{2n}$. Specially taking $f=\Phi_n$ we obtain $K(x,t)=\operatorname{sgn}\Phi_n(x)\cdot\operatorname{sgn}\sin nt\,|\,K(x,t)\,|$, from which (3.2) follows.

As a consequence, from (3.2), combining (2.6), (2.7) we obtain (cf. [4]).

Theorem 4 *For every p such that $1\leqslant p\leqslant+\infty$, $n=1,2,3,\cdots$, we have*

(1) $$d_{2n}[\widetilde{\mathfrak{M}}_\infty^{(0)};L^p]=d'_{2n}[\widetilde{\mathfrak{M}}_\infty^{(0)};L^p]=\|\Phi_n\|_p. \tag{3.3}$$

$S_{2n,r-1}$ is optimal.

(2) $$d^{2n-1}[\widetilde{\mathfrak{M}}_\infty^{(0)};L^p]\geqslant d^{2n}[\widetilde{\mathfrak{M}}_\infty^{(0)};L^p]=\|\Phi_n\|_p. \tag{3.4}$$

In fact, on the left side of (3.4) equality holds. The proof of this fact depends on generalizing an inequality of Taikov. This will be given in the next section.

§ 4. Generalization of an Inequality of Taikov

One of the authors generalized (cf. [6]) an inequality of Taikov ([2], chap. 5, Th. 5.7.1, 5.7.2). By the same method used in [6] we can establish the Kolmogorov comparison theorem ([2], chap. 5) for the operator $P(D)$. And then by using the Kolmogorov comparison theorem we obtain

Theorem 5 *For any p such that $1 \leqslant p \leqslant +\infty$ and $n = 1, 2, 3, \cdots$ we have*

$$\sup_{\substack{f \in \widetilde{\mathfrak{M}}_{\infty}^{(0)} \\ f \perp T_{n-1}}} \| f \|_p = \| \Phi_n \|_p \tag{4.1}$$

As a consequence we have the

Corollary For any p such that $1 \leqslant p \leqslant +\infty$ and $n = 1, 2, 3, \cdots$

$$E_n(\widetilde{\mathfrak{M}}_p^{(0)})_L \overset{\mathrm{dt}}{=} \sup_{f \in \widetilde{\mathfrak{M}}_p^{(0)}} E_n(f)_L = \sup_{\substack{f \in \widetilde{\mathfrak{M}}_{\infty}^{(0)} \\ f \perp T_{n-1}}} \| f \|_{p'} = \| \Phi_n \|_{p'}. \tag{4.2}$$

$\left(\frac{1}{p} + \frac{1}{p'} = 1 \right)$

Write $\Phi_n^*(x) = \int_0^{2\pi} G(t - x) \operatorname{sgn} \sin nt \, dt$.

Theorem 6 *For any p such that $1 \leqslant p \leqslant +\infty$ and $n = 1, 2, 3, \cdots$*

(1) $d_{2n-1}[\widetilde{\mathfrak{M}}_p^{(0)}; L^1] = d_{2n}[\widetilde{\mathfrak{M}}_p^{(0)}]; L^1] = \| \Phi_n^* \|_{p'}$

$$(= \| \Phi_n \|_{p'}). \tag{4.3}$$

T_{n-1} is optimal for both $2n-1$ and $2n$-width. In addition to this, $S_{2n,r-1}^*$ is optimal for $2n$-width, where

$$S_{2n,r-1}^* = \left\{ f : f = \sum_{j=1}^{n} c_j G\left(\frac{j\pi}{n} - \cdot \right), c_j \in \mathbf{R} \right\}$$

in case $p(0) \neq 0$; and

$$S_{2n,r-1}^* = \left\{ f : f = c_0 + \sum_{j=1}^{n} c_j G\left(\frac{j\pi}{n} - \cdot \right), \sum_{j=1}^{n} = 0 \right\}.$$

in case $p(0) = 0$.

(2) $d^{2n-1}[\widetilde{\mathfrak{M}}_{\infty}^{(0)}; L^p] = \| \Phi_n \|_p$. $\qquad$ (4.4)

Proof The inequality

$$d_{2n}[\widetilde{\mathfrak{M}}_p^{(0)}; L^1] \geqslant \| \Phi_n^* \|_{p'}$$

can be obtained by the same method used in [4], and from (4.2) readily follows

$$d_{2n-1}[\widetilde{\mathfrak{M}}_p^{(0)}; L^1] \leqslant \| \Phi_n \|_{p'}.$$

It is worthy to note that $S_{2n,r-1}^*$ is the space of generalized 2π-periodic splines defined by the operator $P^*(D)$ adjoint to $P(D)$. For constructing the interpolating operator the knots of interpolation should be chosen at the 2n zeros of $\Phi_n^*(x)$ in a period.

For estimating $d^{2n-1}[\widetilde{\mathfrak{M}}_\infty^{(0)}; L^p]$ from below we have only to use (3.4). By using (4.1) we obtain (4.4). Theorem 6 is proven.

§ 5. Some Remarks

(1) For p such that $1 \leqslant p < +\infty$ we have only established

$$d_{2n-1}[\widetilde{\mathfrak{M}}_\infty^{(0)}; L^p] \geqslant d_{2n}[\widetilde{\mathfrak{M}}_\infty^{(0)}; L^p] \qquad (5.1)$$

We don't know whether the equality holds even in the simplest case $P(x) = x^r$. This problem is still left unsolved.

(2) Consider the class $\mathfrak{M}_p$ of (non-periodic) differentiable functions on $[0, 2\pi]$ such that $f^{(r-1)}$ are absolutely continuous and $\| P(D)f \|_p \leqslant 1$. We have

$$\mathfrak{M}_p = N_r(P) + \widetilde{\mathfrak{M}}_p,$$

where $N_r(P)$ is the null space of $P(D)$ and

$$\widetilde{\mathfrak{M}}_p = \{f : f = G * h, \| h \|_p \leqslant 1\}.$$

We may prove the following

Theorem 7 *For any p such that $1 \leqslant p \leqslant +\infty$, $N \to +\infty$*

$$d_N[\mathfrak{M}_\infty; L^p] \approx \| \Phi[\tfrac{N}{2}] \|_p, \qquad (5.2)$$

$$d_N[\mathfrak{M}_p; L^1] \approx \| \Phi[\tfrac{N}{2}] \|_{p'}, \qquad (5.3)$$

(3) Recently we obtained[8][9] a series of results concerning the one-sided n-width of the function class $\Omega_p^{2r+\sigma}$ which is defined by a self-adjoint operator $P(D) = D^s \prod_{j=1}^{l} (D^2 - t_j^2)$ $(s \geqslant 0, t_1, \cdots, t_l > 0)$. These results may be generalized without difficulty to the present case.

References

[1] М.Г.Крейн.К теории наилучшего приближения периодических функцнй.ДАН СССР,1938,Т.18,No.4 ~ 5:245 ~ 249

[2] Н.П.Корнейчук. *Extremal Problems in Approximation Theory* .М.: НАУКА, 1976.66

[3] A.Pinkus.On n-widths of periodic functions.J.D′Analyse Math.1979,V.(35):209 ~ 235

[4] Ю.И.Маковоз.Поперечники Соболевских классов и сплайны Наименьшее Уклоняющиеся от нуля.Мат.Зам.1979,Т.20,No.5:805 ~ 812

[5] A.Holland,B.Sahney. *The general problem of approximation and spline functions.* New York:McGraw - Hill.1979

[6] Huang Da-ren.On optimal subspace of $d_n[\Omega_p^r;L^1]$.Acta Math.Sinica (to appear)

[7] Sun Yong-sheng.Some extremal problems on a class of perfect splines.Scientia Sinica (Series A),1984,V.27,No.3:254 ~ 266

[8] Huang Da-ren, Sun Yong-sheng.On the upper-approximation of class $\Omega_p^{2r+\sigma}$ of smooth functions.Scientia Sinica,(to appear)

[9] Sun Yong-sheng, Huang Da-ren.On one-sided approximation of class $\Omega_p^{2r+\sigma}$ of smooth functions.ATA,1984,v.1,No.1:19 ~ 35

EAST JOURNAL ON APPROXIMATIONS
2000, Volume 6, Number 1, 53 ~ 63

带一个 B 核的周期卷积类的极子空间

Optimal Subspaces for a Periodic Convolution Class With a B-Kernel*

In this paper a new generalized version of the classical Landau-Kolmogorov and Stein inequalities is established on a convolution class of periodic functions with a B-kernel. On this basis some sets of optimal subspaces for the $2n$-dimensional Kolmogorov width of such function class are identified.

1. Preliminaries

Recently several papers have appeared which considered the problem of optimal subspaces for a periodic convolution class with a TP kernel ([11], [8], [9]). In this paper we continue our investigation in [9]. We consider the optimal subspaces for a periodic convolution class with a B-kernel in the sense of Pinkus.

1.1. Some preliminaries about a function satisfying B-condition

Definition 1 *Let $G(\cdot)$ be a real, 2π-periodic continuous function. We say that $G(\cdot)$ satisfies B-condition ($G \in B$), if for every choice of $0 \leqslant y_1 < \cdots < y_m < 2\pi$ ($m \geqslant 1$ any integer), the space*

$$W_m(G) := \left\{ b + \sum_{j=1}^{m} b_j G(\cdot - y_j) : b \in \mathbf{R}, \sum_{j=1}^{m} b_j = 0 \right\}, \qquad (1)$$

* Received September 20, 1999

is of dimension m, and $W_m(G)$ is a weak Haar space for odd m.

The concept of a function satisfying B-condition was introduced by Pinkus [6]. An important prototype of such a function is the well-known Bernoulli function

$$B_r(x):=\frac{1}{2\pi}\sum_{\substack{\nu=-\infty\\ \nu\neq 0}}^{\infty}\frac{e^{i\nu x}}{(i\nu)^r},\ r\in\mathbf{Z}_+. \tag{2}$$

The significance of this concept is illustrated by the fact that a series of fundamental results obtained in approximation theory for the Sobolev class $\widetilde{B}_p^r$ (which is a periodic convolution class with kernel $B_r(x)$) can be extended to a more general class of periodic functions defined by a kernel of convolution transformation that satisfies B-condition (see Pinkus [6]). The variation-diminishing property is very essential for the n-width problem. That is why we recall below two well-known assertions ([6, p. 129]) concerning this property.

Lemma 1 *Suppose that $G(\cdot)$ satisfies B-condition and $\phi(\cdot)$ is a nontrivial piecewise continuous, 2π-periodic function ($\phi\in\widetilde{H}$ (**T**)) satisfying $\int_0^{2\pi}\phi(t)\mathrm{d}t=0$ ($\phi\perp 1$). Denote* *

$$f(x)=a+\int_0^{2\pi}G(x-t)\phi(t)\mathrm{d}t,\quad a\in\mathbf{R}.$$

Then $S_c(f)\leqslant S_c(\phi)$, where $S_c(g)$ denotes the cyclic sign changes of the function g.

Pinkus pointed out in [6, p. 177] that if for each $m\in\mathbf{Z}_+$ $W_{2m-1}(G)$ is a Haar space of dimension $2m-1$, then Lemma 1 may be strengthened in the following way.

Denote by $\widetilde{Z}_c(g)$ the number of zeros of a function g (counted cyclically), where the zeros which are not sign changes are counted twice.

Lemma 2 *Let $G(\cdot)\in B$, and assume that for each $m\in\mathbf{Z}_+$ the class $W_{2m-1}(G)$ is a Haar space of dimension $2m-1$. Then for every*

* Here under the integration the factor $(2\pi)^{-1}$ is left out for simplification of symbols. It doesn't matter.

non-trivial $\phi \in \widetilde{H}(\mathbf{T}), \phi \perp 1$, *we have*

$$\widetilde{Z}_c(f) \leqslant S_c(\phi). \tag{3}$$

The condition on G in the last lemma will be denoted in the sequel as $G \in B^*$.

1.2. Smoothing

A function $G(\cdot)$ satisfying B-condition may be modified by a standard smoothing technique. For example, if we take the 2π-periodic kernel for heat conduction

$$\Phi_\sigma = \Phi(\sigma, t) := \frac{1}{\sqrt{2\pi}\sigma} \sum_{n=-\infty}^{\infty} \exp\left\{-\frac{1}{2\sigma^2}(t - 2n\pi)^2\right\}, \sigma > 0 \tag{4}$$

and convolute $G \in B$ with Φ_σ, then we get

$$G_\sigma := \Phi_\sigma * G, \quad \sigma > 0,$$

Lemma 3 ([6, p.62, 177]). *If* $G_\sigma = \Phi_\sigma * G, G \in B$, *then*

1) $G_\sigma \in B^* \cap C^\infty(\mathbf{T})$.

2) $\lim_{\sigma \to 0+} G_\sigma(x) = G(x)$

uniformly on $[0, 2\pi]$. *So, for any* $\phi \in L_\infty(\mathbf{T}), \phi \perp 1$, *it holds*

$$\| f_\sigma - f \|_{C(\mathbf{T})} \to 0, \sigma \to 0,$$

where $f = a + G * \phi, f_\sigma = a + G_\sigma * \phi$.

1.3. The generalized Euler spline associated with $G \in B$

The function

$$\Phi_n(G; x) := (G * \operatorname{sgn} \sin n(\cdot))(x), \quad n \in \mathbf{Z}_+, \tag{5}$$

is said to be the *generalized Euler spline* associated with G. In the following some known facts about $\Phi_n(G; \cdot)$ are listed. They are referred to [6].

1) $\Phi_n\left(G; x + \frac{\pi}{n}\right) = -\Phi_n(G; x)$, that is, $\Phi_n(G; x)$ is a $\frac{2\pi}{n}$-periodic function.

2) There is just one point $\gamma \in \left[0, \frac{\pi}{n}\right)$ such that $\Phi_n(G; \gamma) = 0$. Consequently $\Phi_n(G; x)$ has exactly 2n zeros in $[0, 2\pi)$, namely,

$$\gamma, \gamma + \frac{\pi}{n}, \cdots, \gamma + \frac{2n-1}{n}\pi.$$

3) In $\left[\gamma, \gamma + \frac{\pi}{n}\right)$ there is a point α such that

$$|\Phi_n(G;\alpha)| = \max_{\gamma \leqslant x \leqslant \gamma + \frac{\pi}{n}} |\Phi_n(G;x)| = \| \Phi_n(G;\cdot) \|_{C(\mathbf{T})}.$$

So there are 2n extremal points in $[0, 2\pi)$:

$$\alpha, \alpha + \frac{\pi}{n}, \cdots, \alpha + \frac{2n-1}{n}\pi.$$

4) $\Phi_n(G;x)$ is monotonic in $\left[\alpha + \frac{j-1}{n}\pi, \alpha + \frac{j\pi}{n}\right], j = 1, \cdots, 2n$.

5) For any given 2π-periodic continuous function g, there is a unique function $S_{2n}(x) = S_{2n}(g, x) \in W_{2n}^*(G)$ (sec Th. 3) such that

$$S_{2n}\left(g; \gamma + \frac{j\pi}{n}\right) = g\left(\gamma + \frac{j\pi}{n}\right), \quad j = 0, 1, \cdots, 2n-1.$$

Thus an interpolation operator $S_{2n}: C(\mathbf{T}) \to W_{2n}^*(G)$ is well-defined.

6) Let $G \in B$. Denote

$$\widetilde{B}_p(G) := \{f = a + G * \phi : a \in \mathbf{R}, \phi \in L_p(\mathbf{T}), \| \phi \|_{L_p} \leqslant 1, \phi \perp 1\}$$

where $1 \leqslant p \leqslant \infty$.

$\forall f \in \widetilde{B}_\infty(G)$ it holds

$$|f(x) - S_{2n}(f, x)| \leqslant |\Phi_n(G;x)|, \quad x \in [0, 2\pi). \tag{6}$$

Consequently

$$\sup_{f \in \widetilde{B}_\infty(G)} \| f - S_{2n}(f) \|_{C(\mathbf{T})} = \| \Phi_n(G;\cdot) \|_{C(\mathbf{T})}. \tag{7}$$

When $G(x) = B_r(x)$, the Bernoulli function, we denote $\widetilde{B}_p(G)$ by $\widetilde{B}_p^{(r)}$. We refer this classical case to Korneichuk's monograph [2].

2. Main results

2.1 Landau-Kolmogorov type and Stein type inequalities on a periodic convolution class with a *B*-kernel

The famous Landau-Kolmogorov as well as Stein inequality have many generalizations. (see [10, p. 454 ~ 459]). Here we give a new version of generalization.

Theorem 1 *Suppose* $G_1, G_2 \in B$, *and* $G := G_1 * G_2 \in B$. *If for some* $\phi \in L_\infty(\mathbf{T})$, $\| \phi \|_{L_\infty} \leqslant 1$, $\phi \perp 1$, $a \in \mathbf{R}$, $n \in \mathbf{Z}_+$, *and*

$$\| a + G * \phi \|_{C(\mathbf{T})} \leqslant \| \Phi_n(G;\cdot) \|_{C(\mathbf{T})},$$

then

$$\| G_2 * \phi \|_{C(\mathbf{T})} \leqslant \| \Phi_n(G_2;\cdot) \|_{C(\mathbf{T})}. \tag{8}$$

Proof In view of Lemma 3 it suffices to prove the theorem under the condition that $G_1, G_2 \in B^* \cap C^2(\mathbf{T})$. Suppose (8) is false. Then there exist a function $\phi \in L_\infty(\mathbf{T})$, $\| \phi \|_{L_\infty} \leqslant 1$, $\phi \perp 1$, and numbers $a \in \mathbf{R}$, $n \in \mathbf{Z}_+$ such that

$$\begin{aligned} &\| a + G * \phi \|_{C(\mathbf{T})} \leqslant \| \Phi_n(G;\cdot) \|_{C(\mathbf{T})}, \\ &\| G_2 * \phi \|_{C(\mathbf{T})} > \| \Phi_n(G_2;\cdot) \|_{C(\mathbf{T})}. \end{aligned} \tag{9}$$

Take $\beta > 1$ such that

$$\| \Phi_n(G_2;\cdot) \|_{C(\mathbf{T})} = \beta^{-1} \| G_2 * \phi \|_{C(\mathbf{T})}. \tag{10}$$

Choose points x_0, x_1 such that $\Phi_n(G_2; x - x_1)$ and $G_2 * \phi(x)$ attain their maximal values at one and the same point x_0. At the same time, we may assume that sign $(G_2 * \phi)(x_0) = \operatorname{sign} \Phi_n(G_2; x_0 - x_1) > 0$.

Set

$$\phi_n(t) = \operatorname{sign} \sin n(t - x_1) - \beta^{-1}\phi(t). \tag{11}$$

Clearly, $\phi_n \in L_\infty(\mathbf{T})$, $\phi_n \perp 1$, and sign $\phi_n(t) = \operatorname{sign} \sin n(t - x_1)$, i.e., $S_c(\phi_n) = 2n$. Denote $h_1(\cdot) := (G_2 * \phi_n)(\cdot)$, $h_2(\cdot) := (G_1 * h_1)(\cdot) - \beta^{-1}a$. We see that

$$\begin{aligned} h_2(\cdot) &= (G * \phi_n)(\cdot) - \beta^{-1}a \\ &= \Phi_n(G; \cdot - x_1) - \beta^{-1}[a + (G * \phi)(\cdot)]. \end{aligned} \tag{12}$$

On the basis of the relations

$$\| \Phi_n(G; \cdot - x_1) \|_{C(\mathbf{T})} = \| \Phi_n(G;\cdot) \|_{C(\mathbf{T})} > \beta^{-1} \| a + G * \phi \|_{C(\mathbf{T})}$$

and

$$\begin{aligned} \Phi_n\left(G; \alpha + \frac{j\pi}{n} \right) &= (-1)^j \Phi_n(G;\alpha) \\ &= (-1)^j \| \Phi_n(G;\cdot) \|_{C(\mathbf{T})}, (j = 0, 1, \cdots, 2n-1), \end{aligned}$$

we derive the estimate $S_c(h_2) \geqslant 2n$. It implies that $Z_c(h_2) \geqslant 2n$. Then, by Lemma 1,

$$S_c(h_2) = S_c(G_1 * h_1 - \beta^{-1}a) \leqslant S_c(h_1) \Rightarrow S_c(h_1) \geqslant 2n.$$

So that $Z_c(h_1) \geqslant 2n$. In addition, by assumption, $\Phi_n(G_2; x - x_1)$ and $(G_2 * \phi)(x)$ attain their maximal values at x_0, so from

$$h_1(\cdot) = \Phi_n(G_2; \cdot - x_1) - \beta^{-1}(G_2 * \phi)(\cdot)$$

and (10) we get $h_1(x_0) = 0$. Besides this, we also have $h'_1(x_0) = 0$. Thus x_0 is at least a double zero of $h_1(\cdot)$. To sum up, by Lemma 2 we get

$$2n + 1 \leqslant \tilde{Z}_c(h_1) \leqslant S_c(\phi_n) = 2n.$$

We arrive at a contradiction. The proof of Theorem 1 is completed.

Remark This is a generalized version of the famous Landau-Kolmogorov inequality. Analogous result for NCVD case is established in [9] (see also [5]).

Next we formulate the L_1-analog of Theorem 1.

Theorem 2 *Suppose* $G_1, G_2 \in B$ *and* $G = G_1 * G_2 \in B$. *If for some function* $\phi(\cdot) \in L_p(\mathbf{T})(1 \leqslant p < \infty)$, $\| \phi \|_{L_p} \leqslant 1$, $\phi \perp 1$, *and numbers* $a \in \mathbf{R}$ $n \in \mathbf{Z}_+$, *we have*

$$\| a + G * \phi \|_{L_p(\mathbf{T})} \leqslant \| \Phi_n(G; \cdot) \|_{C(\mathbf{T})},$$

then

$$\| G_2 * \phi \|_{L_p(\mathbf{T})} \leqslant \| \Phi_n(G_2; \cdot) \|_{C(\mathbf{T})}. \tag{13}$$

This is a generalized version of the famous Stein inequality. When $p = 1$, (13) is sharp. The proof of Theorem 2 may be done by using the technique of Stein [7]. Meanwhile, for the NCVD case, it was established already in [9]. The method of proof also applies to the present case, mutatis mutandis. So we omit the details.

2.2 Optimal subspaces for Kolmogorov 2n-width

Let the class of functions $\widetilde{B}_p(G)$ be defined as in Section 1, with a kernel $G \in B$. We shall describe in this section optimal subspaces for the Kolmogorov width

$$\mathrm{d}_{2n}(\widetilde{B}_p(G), L_p(\mathbf{T})), \quad p = 1, \infty.$$

Theorem 3 *Let G be a B-kernel*, $G_1 \in B$ *and* $G_2 = G_1 * G \in B$. *Then the Kolmogorov* $2n$*-width* $d_{2n}(\widetilde{B}_p(G), L_p(\mathbf{T}))$ *has an optimal subspace* $W^*_{2n}(G_2)$, *where* $p = 1, \infty$, *and*

$$W^*_{2n}(G_2) := \left\{ b + \sum_{j=1}^{2n} b_j G_2\left(\cdot - \frac{j\pi}{n} \right) : b, b_j \in \mathbf{R}, \sum_{j=1}^{2n} b_j = 0 \right\}.$$

Proof For each $f \in C(\mathbf{T})$, by a known duality theorem (see, for example, [2]), we have

$$E(f, W_{2n}^*(G_2), L_p(\mathbf{T}) := \min_{g \in W_{2n}^*(G_2)} \| f - g \|_{L_p(\mathbf{T})}$$

$$= \sup_{\substack{\| h \|_{L_{p'}} \leqslant 1, \\ h \perp W_{2n}^*(G_2)}} \left| \int_0^{2\pi} f(t) h(t) \mathrm{d}t \right|. \tag{14}$$

where $h \perp W_{2n}^*(G_2)$ means that

$$\int_0^{2\pi} h(t) \mathrm{d}t = 0,$$

$$\int_0^{2\pi} h(t) \left(G_2\left(t - \frac{j\pi}{n}\right) - G_2\left(t - \frac{2n-1}{n}\pi\right)\right) \mathrm{d}t = 0 \tag{15}$$

$$(j = 0, \cdots, 2n-2).$$

Set

$$H(x) := \int_0^{2\pi} G_2(t - x) h(t) \mathrm{d}t,$$

$$H^*(x) := -H(0) + \int_0^{2\pi} G_2(t - x) h(t) \mathrm{d}t.$$

Then (15) yields

$$H^*(0) = H^*\left(\frac{\pi}{n}\right) = \cdots = H^*\left(\frac{2n-1}{n}\pi\right) = 0. \tag{16}$$

Here we notice that

$$H^*(\cdot) \in \widetilde{B}_{p'}(G_2^*), \quad G_2^*(\cdot) := G_2(-\cdot).$$

Hence, by a result of Pinkus ([6, p.182, Th.4.11.]), for $p = 1$ we have

$$\| -H(0) + G_2^* * h \|_{C(\mathbf{T})} \leqslant \| \Phi_n(G_2^*; \cdot) \|_{C(\mathbf{T})}$$
$$= \| \Phi_n(G_2; \cdot) \|_{C(\mathbf{T})}. \tag{17}$$

When $p = \infty$ we have ([6, p.183, Th.4.15]).

$$\| -H(0) + G_2^* * h \|_{L_1(\mathbf{T})} \leqslant \| \Phi_n(G_2^*; \cdot) \|_{C(\mathbf{T})}$$
$$= \| \Phi_n(G_2; \cdot) \|_{C(\mathbf{T})}. \tag{18}$$

Case Ⅰ. $p = 1$. Consider the quantity

$$E(\widetilde{B}_1(G), W_{2n}^*(G_2), L_1(\mathbf{T})) := \sup_{f \in \widetilde{B}_1} E(f, W_{2n}^*(G_2), L_1(\mathbf{T}))$$

$$= \sup_{f \in \widetilde{B}_1} \sup_{\substack{\| h \|_{L_\infty} \leqslant 1, \\ h \perp W_{2n}^*(G_2)}} \left| \int_0^{2\pi} f(t) h(t) \mathrm{d}t \right| \text{(by(14))}$$

$$= \sup_{\substack{\|h\|_{L_\infty}\leqslant 1,\\ h\perp W_{2n}^*(G_2)}} \sup_{f\in \tilde{B}_2} \left| \int_0^{2\pi} f(t)h(t)\mathrm{d}t \right| \tag{19}$$

$$= \sup_{\substack{\|h\|_{L_\infty}\leqslant 1,\\ h\perp W_{2n}^*(G_2)}} \sup_{\substack{\|\phi\|_{L_1}\leqslant 1,\\ \phi\perp 1}} \left| \int_0^{2\pi} \left\{ \int_0^{2\pi} G(t-u)h(t)\mathrm{d}t \right\} \phi(u)\mathrm{d}u \right|$$

$$\leqslant \sup_{\substack{\|h\|_{L_\infty}\leqslant 1,\\ h\perp W_{2n}^*(G_2)}} \left\| \int_0^{2\pi} G^*(u-t)h(t)\mathrm{d}t \right\|_{C(\mathbf{T})}.$$

Applying (8) to (17) we obtain the bound

$$\| G^* * h \|_{C(\mathbf{T})} \leqslant \| \Phi_n(G^*;\cdot)\|_{C(\mathbf{T})} = \| \Phi_n(G;\cdot)\|_{C(\mathbf{T})}, \tag{20}$$

for each $h(\cdot)$ such that $\|h\|_{L_\infty} \leqslant 1, h \perp W_{2n}^*(G_2)$. So, from (19), (20), we get

$$E(\tilde{B}_1(G), W_{2n}^*(G_2), L_1(\mathbf{T})) \leqslant \| \Phi_n(G;\cdot)\|_{C(\mathbf{T})}. \tag{21}$$

But (21) yields

$$\mathrm{d}_{2n}(\tilde{B}_1(G), L_1(\mathbf{T})) = E(\tilde{B}_1(G), W_{2n}^*(G_2), L_1(\mathbf{T})),$$

which was to be shown.

Case Ⅱ. $p = \infty$. Consider

$$E(\tilde{B}_\infty(G), W_{2n}^*(G_2), C(\mathbf{T})) := \sup_{f\in \tilde{B}_\infty} \sup_{\substack{\|h\|_{L_1}\leqslant 1,\\ h\perp W_{2n}^*(G_2)}} \left| \int_0^{2\pi} f(t)h(t)\mathrm{d}t \right|$$

$$= \sup_{\substack{\|h\|_{L_1}\leqslant 1,\\ h\perp W_{2n}^*(G_2)}} \sup_{\substack{\|\phi\|_{L_\infty}\leqslant 1,\\ \phi\perp 1}} \left| \int_0^{2\pi} \left\{ \int_0^{2\pi} G(t-u)h(t)\mathrm{d}t \right\} \phi(u)\mathrm{d}u \right| \tag{22}$$

$$\leqslant \sup_{\substack{\|h\|_{L_1}\leqslant 1,\\ h\perp W_{2n}^*(G_2)}} \left\| \int_0^{2\pi} G^*(u-t)h(t)\mathrm{d}t \right\|_{L_1(\mathbf{T})}.$$

In this case, applying Stein type inequality (13) ($p = 1$) to (18), we get

$$\| G^* * h \|_{L_1(\mathbf{T})} \leqslant \| \Phi_n(G^*;\cdot)\|_{C(\mathbf{T})} = \| \Phi_n(G;\cdot)\|_{C(\mathbf{T})}, \tag{23}$$

for all $h(\cdot)$ such that $\|h\|_{L_1} \leqslant 1, h \perp W_{2n}^*(G_2)$. Consequently, from (22), (23) we derive

$$E(\tilde{B}_\infty(G), W_{2n}^*(G_2), C(\mathbf{T})) \leqslant \| \Phi_n(G;\cdot)\|_{C(\mathbf{T})}.$$

Which yields

$$d_{2n}(\widetilde{B}_{\infty}(G), C(\mathbf{T})) = E(\widetilde{B}_{\infty}(G), W_{2n}^{*}(G_2), C(\mathbf{T})).$$

The proof of Theorem 3 is completed.

In the following we formulate a series of corollaries which are of independent significance.

In Theorem 3 as G_1 we take the Bernoulli function $B_r(x)$, $r \in \mathbf{Z}_+$. (Strictly speaking, $B_r(x)$ satisfies B-condition only for $r \geqslant 2$. $B_1(x)$ satisfies all conditions of property B except that it has a jump discontinuity at $x = 0$. But we may include B_1 in the B-kernel class without any serious complications). Then $G_2 = B_r * G = G^{(r)}$, which is the r-fold periodic integral of G. Note that $G^{(r)} \in B$ for each $r \in \mathbf{Z}_+$ (see [6, p.132]). The following contains as a particular case a result due to Ligun [4].

Corollary 1 *When* $p = 1, \infty$ *the* $2n$*-dimensional Kolmogorov width* $d_{2n}(\widetilde{B}_p(G), L_p(\mathbf{T}))$ *has an infinite set of optimal subspaces*

$$\{W_{2n}^{*}(G^{(r)}) : r \in \mathbf{Z}_+\}.$$

As a special case of $\widetilde{B}_p(G)$ we take the Sobolev class $\widetilde{B}_p^{(r)}$ ($r \in \mathbf{Z}_+$, fixed). Since for any $G \in B$ we have $G * B_r = B_r * G = G^{(r)} \in B$, the following is an immediate consequence from Theorem 3.

Corollary 2 *When* $p = 1, \infty$, *the* $2n$*-dimensional Kolmogorov width* $d_{2n}(\widetilde{B}_p^{(r)}, L_p(\mathbf{T}))$ *has an infinite set of optimal subspaces, namely*

$$\{W_{2n}^{*}(G^{(r)}) : G \in B\}, (r \in \mathbf{Z}_+ \text{ fixed}).$$

The following lemma is due to Korneichuk ([3, p.235, Proposition 10.5.1(1)]).

Lemma 4 *Every optimal subspace* M_{2n} *of* $d_{2n}(\widetilde{B}_1^{(r)}, L_1(\mathbf{T}))$ *is optimal for* $d_{2n}(\widetilde{B}_p^{(r)}, L_1(\mathbf{T}))$, *where* $1 < p \leqslant +\infty$.

From Lemma 4 and Corollary 2 we obtain.

Corollary 3 *If* $1 \leqslant p \leqslant +\infty$, *then the* $2n$*-dimensional Kolmogorov width* $d_{2n}(\widetilde{B}_p^{(r)}, L_1(\mathbf{T}))$ *has an infinite set of optimal subspaces* $\{W_{2n}^{*}(G^{(r)}) : G \in B\}$, *where* $r \in \mathbf{Z}_+$, *fixed*.

The last corollary concerns the optimal subspaces of the class $W^r H^{\omega}$ that was deeply studied by Korneichuk [2]. According to Korneichuk, for a

$r \in \mathbf{Z}_+ \cup \{0\}$ and a convex-upward modulus of continuity $\omega(t)$, we say that a 2π-periodic continuous function f belongs to the class $W^r H^\omega$, provided $f^{(r)} \in C(\mathbf{T})$ and $\omega(f^{(r)}, t) \leqslant \omega(t), t \geqslant 0$. The following theorem is due to Korneichuk ([2, Chap. 10, Th. 10.7.4]).

Theorem K *Suppose that for some* $r \in \mathbf{Z}_+ \cup \{0\}$ *an N-dimensional subspace* $F_N \subset C(\mathbf{T})$ *is optimal for* $\mathrm{d}_N(\widetilde{B}_\infty^{r+1}, C(\mathbf{T}))$. *Then it is optimal also for* $\mathrm{d}_N(W^r H^\omega, C(\mathbf{T}))$.

From Theorem 4 and Corollary 2 we obtain.

Corollary 4 *Suppose* $r \in \mathbf{Z}_+ \cup \{0\}$, *and* $\omega(t)$ *is a convex-upward modulus of continuity. Then* $\mathrm{d}_{2n}(W^r H^\omega, C(\mathbf{T}))$ *has an infinite set of optimal subspaces* $\{W_{2n}^*(G^{(r+1)}) : G \in B\}$.

References

[1] Chen Di-rong. *n*-widths of some classes of periodic functions. Sci. China Ser. A. V. 1992. 35(1): 42 ~ 54

[2] N. P. Korneichuk. *Extremal Problems in Approximation Theory*. Moscow: Nauka, 1976. 240 ~ 245(in Russian)

[3] N. P. Korneichuk, A. A. Ligun and V. G. Doronin. *Approximation with Constraints*. Kiev: Naukova Dumka, 1982. 95 ~ 105(in Russian)

[4] A. Ligun. Exact inequalities for the upper bounds of seminorm on some classes of periodic functions. Mat. Zametki 1993. 13(5): 647 ~ 654 (in Russian)

[5] Nguen Thi Thieu Hoa. On an extremal problem for variation-diminishing transformations. *Vestnik Moskov. Univ. Ser. I Mat. Meh*. 1982, (5): 3 ~ 7 (in Russian)

[6] A. Pinkus. *N-widths in Approximation Theory*. New York: Springer-Verlag, 1985. 128 ~ 136

[7] E. M. Stein. Functions of exponential type. Ann. of Math. 1957, 65(3): 582 ~ 592

[8] Sun Yongsheng and Jiang Yanjie. Extremal problem for a convolution class of functions defined on the real line. East J. Approx. 1998, 4(4): 515 ~ 525

[9] Sun Yongsheng. Some remarks on the optimal subspaces of a convolution class with NCVD kernel. ATA. 1999, 15(4): 1 ~ 5

[10] Sun Yongsheng, *Approximation Theory of Functions*, Vol. 1, Beijing: Beijing Normal Univ. Press, 1989. 454 ~ 459 (in Chinese)

[11] V. M. Tikhomirov. On the optimal subspaces of classes of functions with cyclic variation-diminishing kernel. Vestnik Moskov. Univ. Ser. I. Mat. Meh. 1997, (4): 16 ~ 19 (in Russian)

三、Landau 不等式的扩充及其某些应用

III. Generalization of Landau's Inequality and its Applications

北京师范大学学报

1984，第四期

周期可微函数类上的某些极值定理*

Some Extremal Problems on the Class of Differentiable Functions

一、引言

在逼近论中，关于周期可微函数类的最佳逼近的极值问题的研究，已经达到十分精深和完善的地步. 长期以来，主要依靠苏联学派的努力，求出了函数类借助三角多项式子空间或多项式周期样条子空间的最佳逼近度在一系列重要情况下的精确值；计算了函数类在一些重要情况下宽度的精确值，并找出了极子空间(三角多项式子空间、多项式样条子空间等)；利用泛函分析的工具，以逼近论的对偶定理为基础，建立了逼近论中几个不同的极值问题之间的联系，从而找到了统一处理几类极值问题获致精确结果的有效途径. 苏联学派在这一方向的研究成果，还揭示了多项式样条类在解决逼近论的极值问题中的重要作用，例如，多项式 Euler 样条和 Bernoulli 样条在解决周期可微函数类的最佳逼近或宽度的精确估计问题时，起了非常关键的作用[1-3].

然而这一研究方向还有发展的余地. 近年来，由 I. J. Schoenberg, C. A. Micchelli 等人发展了的 $\mathscr{L}$ 样条理论有丰富的内容和广泛的应用. 而多项式样条无非是 $\mathscr{L}$ 样条的一个特例. $\mathscr{L}$ 样条在解决可微函数类的最佳逼近、宽度等极值问题中的作用问题，仍有待研究. 据我们了解，除 M. Г. Крейн1983 年的开创性工作[5] 以及近年来 В. Т. Шевалдин 的工作[6,13] 之

---

* 本文 1984 年 5 月 21 日发表.

外,这一方面的系统的工作尚待开展.我们近来在这一方面做了一些工作.本文目的是叙述一部分和广义的 Kolmogorov 比较定理有关的结果. Kolmogorov 于 1939 年在可微函数类内建立了导数的比较定理[1],它是在可微函数类上建立一系列精确不等式的基础.本文的中心内容是,对由任意常系数的常微分算子确定的可微函数类建立 Kolmogorov 比较定理,并由之推出一系列重要推论.

任给 n 次实系数多项式

$$P_n(\lambda)=\prod_{s=1}^{k}(\lambda_2-2\alpha_s\lambda+\alpha_s^2+\beta_s^2)\cdot\prod_{j=1}^{n-2k}(\lambda-\lambda_j)\qquad(1.1)$$

$n\geqslant 1,k\geqslant 0$.当 $k\geqslant 1$ 时 $\alpha_s,\beta_s\in\mathbf{R},\beta_s>0$,置 $\beta=\max\beta_s$.当 $n>2k$ 时,$\lambda_i\in\mathbf{R}$. $P_n(\lambda)$ 的零点是 $\alpha_s\pm \mathrm{i}\beta_s(k=1,\cdots,k),\lambda_1,\cdots,\lambda_{n-2k}$. $k=0$ 时 $P_n(\lambda)$ 仅有实零点.记 $D=\dfrac{\mathrm{d}}{\mathrm{d}x}$.

定义 1 给 $P_n(\lambda)$ 的零点任意排定一次序 $\alpha_1\pm\mathrm{i}\beta_1,\cdots,\alpha_k\pm\mathrm{i}\beta_k,\lambda_1,\cdots,\lambda_{n-2k}$.置 $P_0(D)=I$(恒等),$P_2(D)=D^2-2\alpha_1D+\alpha_1^2+\beta_1^2,\cdots,P_{2k}(D)=(D^2-2\alpha_kD+\alpha_k^2+\beta_k^2)\cdot P_{2k-2}(D),P_{2k+1}(D)=(D-\lambda_1)P_{2k}(D),\cdots,P_n(D)$.每一 $P_r(D)$ 称为 $P_n(D)$ 的子算子.如果 $n-r$ 次的多项式 $\hat{P}_r(\lambda)$ 满足 $\hat{P}_r(\lambda)\cdot P_r(\lambda)=P_n(\lambda)$,则称 $\hat{P}_r(D)$ 为算子 $P_r(D)$ 关于 $P_n(D)$ 的余算子.

定义 2 称 $f\in\mathscr{L}_p^{(n)}(P_n(D))$,若

(1)$f(x)\in C^{n-1}(\mathbf{R}),f(x)\in L^p(-\infty,\infty),(1\leqslant p\leqslant\infty)$.

(2)$f^{(n-1)}(x)$ 在 $\mathbf{R}$ 上分段连续可微,且在任何有限区间内 $f^{(n)}(x)$ 仅有的间断点是有限个第一类间断.

(3)$\|P_n(D)f\|_p<+\infty$.

以下记

$$A_p^{(n)}(P_n(D))=\{f\in\mathscr{L}_p^{(n)}(P_n(D)):\|P_n(D)f\|_p\leqslant 1\}.\qquad(1.2)$$

定义 3 称 $f\in L_p^{(n)}(P_n(D))$,若

(1)$f(x)\in C^{n-1}(\mathbf{R}),f\in L^p(-\infty,+\infty),1\leqslant p\leqslant+\infty$.

(2)$f^{(n-1)}(x)$ 在 $\mathbf{R}$ 上局部绝对连续.

(3)$\|P_n(D)f\|_p<+\infty$.

以下记

$$W_p^{(n)}(P_n(D)) = \{f \in L_p^{(n)}(P_n(D)) : \| P_n(D) f \|_p \leqslant 1\}. \quad (1.3)$$

明显有 $L_p^{(n)} \supset \mathscr{L}_p^{(n)}, W_p^{(n)} \supset A_p^{(n)}$. 两类间的关系可借助 Стеклов 函数沟通. $\mathscr{L}_p^{(n)}$类有较好的微分性质, 应用广义的 Rolle 定理可以先在 $\mathscr{L}_p^{(n)}$ 类上证明一些不等式, 然后借助 Стеклов 函数把结果拓广到 $L_p^{(n)}$ 类上. 我们先在下一节内叙述广义的 Rolle 定理.

二、广义的 Rolle 定理

熟知 Rolle 定理有很多推广的形式. 特别指出, Polya 把 Rolle 定理拓广到满足 W 条件的线性常微分算子上[4], 它在样条插值理论中有重要应用. 我们需要 Rolle 定理往任意常系数线性常微分算子 $P_n(D)$ 的拓广. 由于当 $P_n(\lambda)$ 有虚根时 $P_n(D)$ 不满足 W 条件, 所以 Polya 的结果包涵不了这种情形, 有必要另加讨论, 在这方面 М. Г. Крейн 已经给出了实质性的结果[5], 但没有证明. 我们在本节内给出 Крейн 的结果的一个完全初等的证明和一些补充.

引理 2.1 给定实系数多项式 $P(\lambda) = \lambda^2 + p\lambda + q$, $p^2 - 4q < 0$, 其二根为 $\alpha \pm i\beta$, $\beta > 0$. 若有 $f(x) \in C^2$ 及 $x_1 < x_2 < x_3$ 使

(1) $x_{j+1} - x_j < \frac{\pi}{2\beta}$, $(j = 1, 2)$.

(2) $f(x_j) = 0$, $j = 1, 2, 3$.

则存在 $\xi \in (x_1, x_3)$ 使 $P(D) f(\xi) = 0$.

证明 任取 $\varphi(x) \in C^2$ 有以下恒等式成立:

$$\begin{aligned} & e^{\alpha(x-x_0)} \sin \beta(x - x_0)(P(D)\varphi(x)) \\ & = D\left[\sin^2 \beta(x - x_0) D \frac{e^{\alpha(x-x_0)}}{\sin \beta(x - x_0)} \varphi(x)\right]. \end{aligned} \quad (2.1)$$

$x \neq x_0 + k\frac{\pi}{\beta}$, $k = 0, \pm 1, \pm 2, \cdots$. x_0 是任取的. 选择 τ_1 使 $\tau_1 < x_1 < x_3 < \tau_1 + \frac{\pi}{\beta}$, 置 $g(x) = \frac{e^{\alpha(x-\tau_1)}}{\sin \beta(x - \tau_1)} f(x)$, 则 $g(x_j) = 0$, $j = 1, 2, 3$. 由 Rolle 定理存在 $\xi_1 \in (x_1, x_2)$, $\xi_2 \in (x_2, x_3)$ 使 $g'(\xi_1) = g'(\xi_2) = 0$. 再置 $h(x) = \sin^2 \beta(x - \tau_1) g'(x)$. 对 $h(x)$ 在 (ξ_1, ξ_2) 内再用一次 Rolle 定理即得.

注 1 若在任一区间 $(a, b) \subset (x_1, x_3)$ 内 $P(D) f(x) \not\equiv 0$, 则 ξ 点可

以是 $P(D)f(x)$ 的变号(零)点.

注 2 条件 $f(x)\in C^2$ 可以减弱为 $f(x)\in C'$, f' 在 (x_1,x_3) 内分段连续可微, 而且 f'' 在 (x_1,x_3) 内仅有的间断点是有限个第一类间断. 对这样的 f, 若引理 2.1 的条件成立, 且注 1 的条件亦满足时, 则存在 $\xi\in(x_1,x_3)$ 为 $P(D)f(x)$ 的变号点.

注 3 若引理 2.1 的点 x_1,x_2,x_3 中有重合者(比如 $f(x_1)=0$, $f(x_2)=f'(x_2)=0$) 有同样结论成立.

引理 2.2 给定 $P(D)=D^2+pD+q$ 如前. 若有 $f(x)\in C^2$ 及 $x_1<x_2<x_3<x_4$ 满足(1) $x_{j+1}-x_j<\frac{\pi}{2\beta}(j=1,2,3)$. (2) $f(x_j)=0$, $j=1,2,3,4$. (3) 在任一区间 $(a,b)\subset(x_1,x_4)$ 内 $P(D)f(x)\not\equiv 0$. 则存在 $\xi_1,\xi_2\in(x_1,x_4)$ 使有

(1) $x_1<\xi_1<\xi_2<x_4$, $\xi_2-\xi_1<\frac{3\pi}{2\beta}$.

(2) $P(D)f(\xi_1)=P(D)f(\xi_2)=0$, 且 $P(D)f(x)$ 在 ξ_1,ξ_2 处交错变号.

证明 不失一般性, 我们认定 $f(x)$ 在 (x_1,x_2) 内取到正值.

(a) 如引理 2.1, 选择 τ_1 使 $\tau_1<x_1<x_3<\tau_1+\frac{\pi}{\beta}$, 置 $g_1(x)=\frac{e^{\alpha(x-\tau_1)}}{\sin\beta(x-\tau_1)}f(x)$, $h_1(x)=\sin^2\beta(x-\tau_1)g_1'(x)$. 则存在 $\eta_1\in(x_1,x_2)$, $g_1'(\eta_1)=0$, 在 η_1 左侧 $g_1'(x)\geqslant 0(\not\equiv 0)$. 在 η_1 右侧 $g_1'(x)\leqslant 0(\not\equiv 0)$, $h_1(x)$ 亦如此. 往下分两种情形讨论.

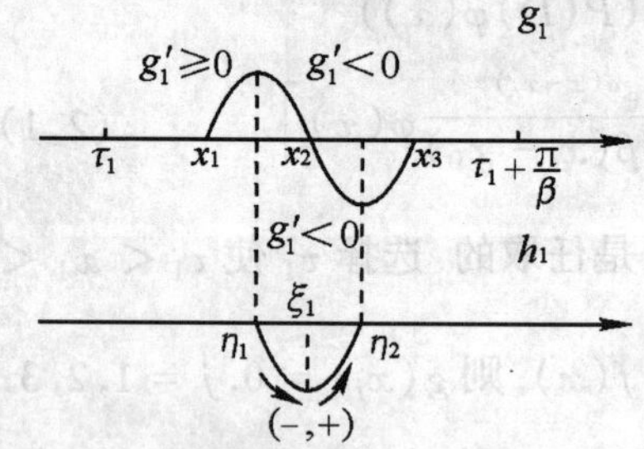

图 1

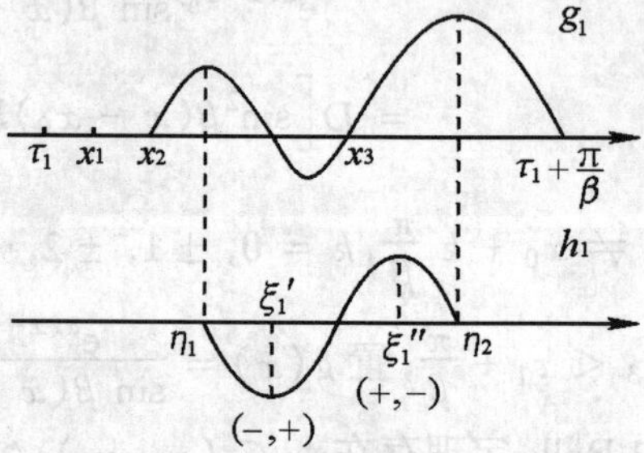

图 2

(1) $f(x)$ 在 (x_2, x_3) 内取到负值. 此时存在 $\eta_2 \in (x_2, x_3)$ 使 $g_1'(\eta_2) = 0$, 在 η_2 左侧(邻域) $g_1'(x) \leqslant 0 (\not\equiv 0)$, 在 η_2 右侧(邻域) $g_1'(x) \geqslant 0 (\not\equiv 0)$, $h_1(x)$ 亦如此. 由此可见 $h_1(\eta_1) = h_1(\eta_2) = 0$, $h_1(x)$ 在 η_1 的右侧(邻域) $\leqslant 0 (\not\equiv 0)$, 在 η_2 左侧(邻域) $\leqslant 0 (\not\equiv 0)$, 故有 $\xi_1 \in (\eta_1, \eta_2)$ 使 $h_1'(\xi_1) = 0$, 且 ξ_1 是 $P(D)f(x)$ 的 $(-, +)$ 变号点(见图 1).

(2) $f(x)$ 在 (x_2, x_3) 内无负值. 此时有 $\eta_2 \in (x_2, x_3)$, $g_1'(\eta_2) = 0$, 在 η_2 左侧(邻域) 有 $g_1'(x) \geqslant 0 (\not\equiv 0)$, 在 η_2 右侧(邻域) $g_1'(x) \leqslant 0 (\not\equiv 0)$, $h_1(x)$ 亦如此. 所以 $h_1(x)$ 在 (η_1, η_2) 内同时能有正值和负值, 且 $h_1(\eta_1) = h_1(\eta_2) = 0$. 这时有二点 $\xi_1', \xi_1'' \in (\eta_1, \eta_2)$ 满足 $h_1'(\xi_1') = h_1'(\xi_1'') = 0$, 而 ξ_1', ξ_1'' 各是 $P(D)f(x)$ 的 $(-, +)$, $(+, -)$ 变号点(见图 2).

(b) 转到 x_2, x_3, x_4. 选择 τ_2 使 $\tau_2 < x_2 < x_4 < \tau_2 + \frac{\pi}{\beta}$. 置 $g_2(x) = \frac{e^{\alpha(x-\tau_2)}}{\sin \beta(x - \tau_2)} f(x)$, $h_2(x) = \sin^2 \beta(x - \tau_2) g_2'(x)$. 在 (x_2, x_4) 内重复对 $f(x)$ 在 (x_1, x_3) 内做过的论证, 出现的情况不外乎以下几种:

(1) $f(x)$ 在 (x_2, x_3) 内取到负值, 在 (x_3, x_4) 内取到正值. 这属于(a)的(1), 只是差一符号. 故存在 $\xi_2 \in (x_2, x_4)$, $P(D)f(\xi_2) = 0$, ξ_2 是 $P(D)f(x)$ 的 $(+, -)$ 变号点. ξ_1, ξ_2 求得.

(2) $f(x)$ 在 (x_2, x_3) 内取到负值, 在 (x_3, x_4) 内取不到正值. 这属于(a) 的(2), 此时 $P(D)f(x)$ 在 (x_2, x_4) 内有二个反向变号(零) 点, 其中必有一个与 $P(D)f(x)$ 在 ξ_1 的变号方向相反, 把它取作 ξ_2 即得所求.

(3) $f(x)$ 在 (x_2, x_3) 内取不到负值, 在 (x_3, x_4) 内取到负值.

(4) $f(x)$ 在 (x_2, x_3) 内取不到负值, 在 (x_3, x_4) 内取不到负值.

最后二情形亦有同样结论成立. 引理 2.2 证完. 应用归纳推理便得.

引理 2.3 给定 $P(D) = D^2 + pD + q$ 如前. 若有 $f(x) \in C^2$ 及点 $x_1 < \cdots < x_n < x_{n+1} < x_{n+2}$ 满足

(1) $x_{j+1} - x_j < \frac{\pi}{2\beta}$, $(j = 1, \cdots, n+1)$.

(2) $f(x_j) = 0$, $j = 1, \cdots, n+2$.

(3) 在任一区间 $(a, b) \subset (x_1, x_{n+2})$ 内 $P(D)f(x) \not\equiv 0$. 则存在 n 个点 $\xi_1, \cdots, \xi_n$:

(4)$x_1 < \xi_1 < \cdots < \xi_n < x_{n+2}, \xi_{j+1} - \xi_j < \dfrac{3\pi}{2\beta}$.

(5)$P(D)f(\xi_j) = 0, j = 1, \cdots, n$.

(6)$P(D)f(x)$ 沿 $\xi_1, \xi_2, \cdots, \xi_n$ 交错变号.

附带指出:和引理2.1相仿,引理2.3也可给出三条注.这一点后面还要用到.

现在给出关于任意实系数的线性常微分算子 $P_n(D)$ 的广义 Rolle 定理.对 $n+1$ 个点 $x_1 < x_2 < \cdots < x_{n+1}$ 记 $\Delta_j = x_{j+1} - x_j$, $= \|\Delta\| \max\limits_{1\leqslant j\leqslant n} \Delta_j$.根据引理2.3及一阶微分算子 $D - \lambda_j (\lambda_j \in \mathbf{R})$ 拓广的 Rolle 定理便得

定理1 给定 $P_n(D)$.若有 $f(x) \in C^n$ 及 $x_1 < \cdots < x_{n+1}$ 满足

(1)$\|\Delta\| < \dfrac{\pi}{2 \cdot 3^{k-1} \beta}$($k \geqslant 1, k = 0$ 时此条件多余).

(2)$f(x_j) = 0, j = 1, \cdots, n+1$.

则存在 $\xi \in (x_1, x_{n+1})$ 使 $P_n(D)f(\xi) = 0$.若在任何区间$(a, b) \subset (x_1, x_{n+1})$ 内 $P_n(D)f(x) \not\equiv 0$,则 ξ 点可以是$P(D)f(x)$ 的变号点.

对此定理可以加上和引理2.1相仿的三条注.这一点应予注意.

三、标准函数

本节引入关于算子 $P_n(D)$ 的标准函数.首先给出广义 Bernoulli 函数.

定义4 给定一 n 次实系数多项式$P_n(\lambda)$,以下总规定 $P_n(\lambda) = 0$ 除 $\lambda = 0$ 可能是其根外,其他任何虚整点 $ik(k = \pm 1, \pm 2, \cdots)$ 均非此方程之根.由级数

$$G_n(x) = \frac{1}{2\pi} \sum_{\nu=-\infty}^{\infty}{}' \frac{e^{i\nu x}}{P_n(i\nu)} \tag{3.1}$$

定义的函数 $G_n(x)$ 称为广义 Bernoulli 函数,$\sum'$ 表示当$P_n(0) = 0$ 时和式中缺 $\nu = 0$ 的项.$G_n(x)$ 在$(0, 2\pi)$ 上的限制称为广义 Bernoulli 多项式.

关于 $G_n(x)$ 有以下简单命题成立.

引理3.1 $G_n(x)$ 有以下性质:

(1)$G_n(x) \in C^{n-2}(\mathbf{R})$.

(2) 在$(0,2\pi)$ 内 $G_n(x)$ n 阶连续可微, 区间端点是 $G_n^{(n)}(x)$ 的第一类间断点.

(3) 当 $P_n \neq 0$ 时有 $P_n(D)G_n(x) \equiv 0(0 < x < 2\pi)$; 而当 $P_n(0) = 0$ 时则有 $DP_n(D)G_n(x) \equiv 0(0 < x < 2\pi)$.

现在引入对研究 $G_n(x)$ 的三角多项式最佳平均逼近有重要作用的函数 $H_{n,\lambda}(x)$. 先设 $m \geqslant 1$ 是一正整数. 经计算给出

$$\sum_{j=0}^{2m-1}(-1)^j G_n\left(x+\frac{j\pi}{m}\right)=\frac{m}{\pi}\sum_{\nu=-\infty}^{+\infty}\frac{\mathrm{e}^{\mathrm{i}(2\nu+1)mx}}{P_n((2\nu+1)m\mathrm{i})}. \tag{3.2}$$

在(3.2) 的右边直接令 m 以 λ 代换($\lambda > \beta$), 给出函数

$$H_{n,\lambda}(x)=\frac{\lambda}{\pi}\sum_{\nu=-\infty}^{+\infty}\frac{\mathrm{e}^{\mathrm{i}(2\nu+1)\lambda x}}{P_n((2\nu+1)\lambda\mathrm{i})}. \tag{3.3}$$

由 $H_{n,\lambda}\left(x+\frac{\pi}{\lambda}\right)=-H_{n,\lambda}(x)$ 看出 $H_{n,\lambda}(x)$ 以$\frac{2\pi}{\lambda}$ 为周期. 下面引理给出它的一些简单性质.

引理 3.2 $H_{n,\lambda}(x)$ 有以下性质.

(1)$H_{n,\lambda}(x) \in C^{n-2}(\mathbf{R})$.

(2)$H_{n,\lambda}(x)$ 在$\left(0,\frac{\pi}{\lambda}\right)$ 内n 阶连续可微, 区间端点是 $H_{n,\lambda}^{(n)}(x)$ 的第一类间断点.

(3)$P_n(D)H_{n,\lambda}(x) \equiv 0$ 在$\left(0,\frac{\pi}{\lambda}\right)$ 内成立.

在 $G_n(x)$ 借助三角多项式的最佳平均逼近问题的精确解中给出解的极函数[5] 是

$$\int_0^{2\pi}G_n(x-t)\operatorname{sgn}\sin mt\,dt$$

$$=\frac{1}{\pi\mathrm{i}}\sum_{\nu=-\infty}^{+\infty}\frac{\mathrm{e}^{\mathrm{i}(2\nu+1)mx}}{(2\nu+1)P_n((2\nu+1)m\mathrm{i})},\mathrm{i}=\sqrt{-1}. \tag{3.4}$$

其中 m 是正整数. 在此表达式内以 $\lambda(\lambda > \beta)$ 替换 m. 给出

定义 5 称

$$\Phi_{n,\lambda}(x)=\frac{1}{\pi\mathrm{i}}\sum_{\nu=-\infty}^{+\infty}\frac{\mathrm{e}^{\mathrm{i}(2\nu+1)\lambda x}}{(2\nu+1)P_n((2\nu+1)\lambda\mathrm{i})} \tag{3.5}$$

为关于 $P_n(D)$ 的标准函数.

从定义直接看出

$$\Phi'_{n,\lambda}(x)=H_{n,\lambda}(x). \tag{3.6}$$

$$\Phi_{n,\lambda}\left(x+\frac{\pi}{\lambda}\right)=-\Phi_{n,\lambda}(x). \tag{3.7}$$

$$P_n(D)\Phi_{n,\lambda}(x)=\operatorname{sgn}\sin\lambda x, 0<x<\frac{\pi}{\lambda}. \tag{3.8}$$

我们需要 $\Phi_{n,\lambda}(x)$ 的零点、极值点和单调区间的信息. 当 $P_n(\lambda)=0$ 仅有实根时情况比较简单. 下面引理针对一般情形.

引理 3.3 设 $\lambda>2\beta$(当 $P_n(\lambda)=0$ 仅有实根时, $\beta=0$), 则

(1) 当 $n=1$ 时, $H_{n,\lambda}(x)$ 在 $\left[0,\frac{\pi}{\lambda}\right)$ 内恰有一变号点.

(2) 当 $n\geqslant 2$ 时, $H_{n,\lambda}(x)$ 在 $\left[0,\frac{\pi}{\lambda}\right)$ 内恰有一零点, 是单零点.

证明 (a) 首先考虑两特殊情形, $P_1(\lambda)=\lambda-\lambda_1(\lambda_1\in\mathbf{R})$, $P_2(\lambda)=\lambda^2-2\alpha\lambda+\alpha^2+\beta^2, 0<\beta<\frac{1}{2}$. 当 $n=1$ 时, 函数 $\sum\limits_{\nu=-\infty}^{+\infty}\frac{e^{i(2\nu+1)y}}{(2\nu+1)i-\lambda_1}$ 在 $[0,\pi)$ 内恰有一变号点(第一类间断), 这是因为

$$\sum_{\nu=-\infty}^{+\infty}\frac{e^{i(2\nu+1)y}}{(2\nu+1)i-\lambda_1}=\frac{1}{2}\left\{\sum_{\nu=-\infty}^{+\infty}{}'\frac{e^{i\nu y}}{\nu i-\lambda_1}-\sum_{\nu=-\infty}^{+\infty}{}'\frac{e^{i\nu(y+\pi)}}{\nu i-\lambda_1}\right\},$$

而当 $\lambda_1\neq 0$ 时

$$\sum_{\nu=-\infty}^{+\infty}\frac{e^{i\nu y}}{\nu i-\lambda_1}=\frac{-\pi}{\operatorname{sh}\lambda_1\pi}e^{\lambda_1(y-\pi)}, 0<y<2\pi,$$

$0, 2\pi$ 是其第一类间断点. 当 $\lambda_1=0$ 时

$$\sum_{\nu=-\infty}^{+\infty}{}'\frac{e^{i\nu y}}{\nu i}=\pi-y, 0<y<2\pi.$$

由此即知所断言者成立. 作变量代换 $y=\lambda x$, 即得引理 3.3 的(1) 之证.

转到 $n=2$, $P_2(\lambda)=\lambda^2-2\alpha\lambda+\alpha^2+\beta^2, 0<\beta<1/2$ 的情形. 注意此时存在 $a, b\in\mathbf{R}$ 使 $\sum\limits_{\nu=-\infty}^{+\infty}\frac{e^{i\nu y}}{P_2(i\nu)}=a e^{\alpha y}\cos(\beta y+b)$. 由于 $0<\beta<\frac{1}{2}$, 易见 $e^{\alpha y}\cos(\beta y+b)$ 在 $0\leqslant y\leqslant 2\pi$ 内部恰有一个极值点, 该点把 $[0,2\pi]$ 分成此函数的二个单调区间, 有相反单调方向. 由此便知

$$\sum_{\nu=-\infty}^{+\infty}\frac{e^{i(2\nu+1)y}}{P_2((2\nu+1)i)}=\frac{1}{2}\left\{\sum_{\nu=-\infty}^{+\infty}\frac{e^{i\nu y}}{P_2(i\nu)}-\sum_{\nu=-\infty}^{+\infty}\frac{e^{i\nu(y+\pi)}}{P_2(i\nu)}\right\}$$

在 $[0,\pi)$ 内恰有一单零点, 此外别无零点. 由此可见, 当 $n=2$, $P_2(x)=$

$x^2-2\beta x+\alpha^2+\beta^2,\lambda>2\beta(\beta>0)$ 时，注意到 $P_2((2\nu+1)\lambda i)=\lambda^2\left[(2\nu+1)^2i^2-2\left(\frac{\alpha}{\lambda}\right)(2\nu+1)i+\left(\frac{\alpha}{\lambda}\right)^2+\left(\frac{\beta}{\lambda}\right)^2\right],0<\frac{\beta}{\lambda}<\frac{1}{2}$ 便知 $\sum_{\nu=-\infty}^{+\infty}\frac{e^{i(2\nu+1)y}}{P_2((2\nu+1)\lambda i)}$ 在$[0,\pi)$ 内恰有一单零点. 作代换 $y=\lambda x$ 即得$n=2$ 时断语(2) 之证.

(b) 转到一般情形，证明可以在区间$\left[0,\frac{2\pi}{\lambda}\right)$ 上借助于广义 Rolle 定理用归谬法完成. 由 $H_{n,\lambda}\left(x+\frac{\pi}{\lambda}\right)=-H_{n,\lambda}(x),H_{n,\lambda}(x)$ 在$\left[0,\frac{2\pi}{\lambda}\right)$ 内至少有二个不同零点，其距离为$\frac{\pi}{\lambda}$. 假定 $n\geqslant 3,\lambda>2\beta$ 时 $H_{n,\lambda}(x)$ 在$\left[0,\frac{\pi}{\lambda}\right)$ 内零点个数多于(重零点要依其重次计算个数)，则 $H_{n,\lambda}(x)$ 在$\left[0,\frac{2\pi}{\lambda}\right)$ 内至少有 4 个零点，任二相邻零点距离$\leqslant\frac{\pi}{\lambda}$. 用 $P_n(D)$ 的真子算子 $P_2(D),\cdots,P_{2k}(D),P_{2k+1}(D),\cdots,P_{n-1}(D)$ 作用于 $H_{n,\lambda}(x)$，根据广义 Rolle 定理，并注意到 $H_{n,\lambda}(x)$ 以$\frac{2\pi}{\lambda}$ 为周期，最后推出结论与(a) 的结论相矛盾. 证完.

引理 3.4 设 $\lambda>2\beta,n\geqslant 2$ 则

(1) $\Phi_{n,\lambda}(x)$ 在$\left[0,\frac{\pi}{\lambda}\right)$ 内有唯一的极值点，在该点上 $\Phi'_{n,\lambda}(x)=0$.

(2) 若 x_0 是 $\Phi_{n,\lambda}(x)$ 的一极值点，则$\left[x_0,x_0+\frac{\pi}{\lambda}\right],\left[x_0+\frac{\pi}{\lambda},x_0+\frac{2\pi}{\lambda}\right]$ 是 $\Phi_{n,\lambda}(x)$ 的单调区间，$\Phi_{n,\lambda}(x)$ 在它们上面单调变化方向相反.

(3) $\Phi_{n,\lambda}(x)$ 在其单调区间内恰有一零点，它是单零点.

证明 根据引理 3.3 及(3.6) 即得.

四、Kolmogorov 比较定理

Kolmogorov 1939 年在可微函数类上建立了导数的比较定理[1]，该定理是在可微函数类上证明一系列精确不等式的基础. 本节把比较定理拓广到 $W_\infty^{(n)}(P_n(D))$ 类上，并由之导出一系列重要推论.

定理 2 （广义的 Kolmogorov 比较定理）

给定 $n\geqslant 1, P_n(D)$. 若对某 $\lambda > \Lambda = 4\cdot 3^{k-1}\beta$ 有 $f\in W_\infty^{(n)}(P_n(D))$ 满足

(1) $\| f\|_\infty \leqslant \|\Phi_{n,\lambda}\|_\infty$.

(2) 对 $\alpha, a, b\in\mathbf{R}$ 有 $f(\alpha) = \Phi_{n,\lambda}(a) = \Phi_{n,\lambda}(b)$, 此处 a, b 属于 $\Phi_{n,\lambda}(x)$ 的相邻的单调区间, 则

$$|f'(\alpha)|\leqslant \max\{|\Phi'_{n,\lambda}(a)|, |\Phi'_{n,\lambda}(b)|\}. \tag{4.1}$$

证明 注意到第一节末尾的说明, 只要证明定理对 $f\in A_\infty^{(n)}(P_n(D))$ 成立即可. 由于 $f(x)\in A_\infty^{(n)}(P_n(D))$ 含有 $f(x+t)\in A_\infty^{(n)}(P_n(D))$, 故不失一般性可以认为 $\alpha = a$ 或 b. 先考虑 $n\geqslant 2$ 的情形. 假设定理不成立, 则有 $f\in A_\infty^{(n)}(P_n(D))$ 对某 $\lambda > \Lambda$ 满足条件 $\| f\|_\infty \leqslant \|\Phi_{n,\lambda}\|_\infty, f(\alpha) = \Phi_{n,\lambda}(a) = \Phi_{n,\lambda}(b)$, 但(4.1) 不成立, 即 $|f'(\alpha)| > |\Phi'_{n,\lambda}(a)|, |f'(\alpha)| > |\Phi'_{n,\lambda}(b)|$. 由 $f, f', \Phi'_{n,\lambda}$ 的连续性, 有 $\rho > 1$ 及 a, b 的近旁的数 γ, γ' 使有

$$\frac{1}{\rho}f(\gamma) = \Phi_{n,\lambda}(\gamma) = \Phi_{n,\lambda}(\gamma'),$$

$$\frac{1}{\rho}|f'(\gamma)| > \max\{|\Phi'_{n,\lambda}(\gamma)|, |\Phi'_{n,\lambda}(\gamma')|\}.$$

我们要注意 $\alpha = a, \gamma, \gamma'$ 仍是取自 $\Phi_{n,\lambda}(x)$ 的一对相邻的单调区间内的数. 记 $\bar f(x) = \frac{1}{\rho}f(x)$, 则见 $\bar f\in A_\infty^{(n)}(P_n(D))$, $\|\bar f\|_\infty < \|\Phi_{n,\lambda}\|_\infty$, $\| P_n(D)\bar f\|_\infty < 1$. 只讨论一种典型情况就够了, 即

$$\bar f(\gamma) = \Phi_{n,\lambda}(\gamma)\geqslant 0, \bar f'(\gamma) > \Phi'_{n,\lambda}(\gamma) > 0. \tag{4.2}$$

(如果是 $\bar f'(\gamma) < 0$, 那么作简单的自变量平移变换挪到 γ' 点上去讨论就是了, 因 γ, γ' 在标准函数 $\Phi_{n,\lambda}(x)$ 的相邻单调区间内, 单调方向相反)

(1) 先假定 f 以 $M\cdot\frac{2\pi}{\lambda}$($M\geqslant 1$ 是整数) 为周期者. $M\cdot\frac{2\pi}{\lambda}$ 也是 $\Phi_{n,\lambda}(x)$ 的一个周期. 取包含 γ 点的长度是 $M\cdot\frac{2\pi}{\lambda}$ 的区间 Δ. 当 $\lambda > 2\beta$ 时, Δ 可划分为 $2M$ 个等长区间(两端处的区间必要时可以按周期 $M\cdot\frac{2\pi}{\lambda}$ 拼成一个整的小区间), 各为 $\Phi_{n,\lambda}(x)$ 的单调区间. 在每一单调区间上 $\Phi_{n,\lambda}(x)$ 在 $\pm\|\Phi_{n,\lambda}\|_\infty$ 间单调变动一次. 记包含 γ 点的单调区间为 Δ_ν,

由 $\bar{f}'(\gamma) > \Phi'_{n,\lambda}(\gamma) > 0$ 断定在区间 Δ_ν 上 $\bar{f}$ 的曲线穿过 $\Phi_{n,\lambda}(x)$ 的曲线至少三次，而在其余 $2M-1$ 个区间上，至少各穿过一次. 所以 $\Phi_{n,\lambda}(x)-\bar{f}(x)$ 在 Δ 内至少有 $2M+2$ 个零(变号) 点，任二相邻零点距离 $< \dfrac{2\pi}{\lambda}$. 若 $\lambda > \Lambda$，则对 $\Phi_{n,\lambda}(x)-\bar{f}(x)$ 连续实施微分算子序列 $P_2(D),\cdots,P_{2k}(D),\cdots,P_n(D)$ 的作用，则根据广义 Rolle 定理，知道函数 $P_n(D)[\Phi_{n,\lambda}(x)-\bar{f}(x)]$ 在 Δ 内至少有 $2M+2$ 个变号. 但 $P_n(D)[\Phi_{n,\lambda}(x)-\bar{f}(x)] = \operatorname{sgn}\sin\lambda x - \dfrac{1}{\rho}P_n(D)f(x)$，而 $\dfrac{1}{\rho}\|P_n(D)f\|_\infty < 1$，$P_n(D)[\Phi_{n,\lambda}(x)-\bar{f}(x)]$ 在 Δ 内变号数为 $2M$. 得到矛盾. 如果 $n=1$，这时 $P_1(D)=D-\lambda_1$，只要使用一次一阶微分算子的广义 Rolle 定理就行了. 往下转到一般情形.

(2) 设 $n\geqslant 2$，我们来证：若 $f\in A_\infty^{(n)}(P_n(D))$，则 $\|P_r(D)f\|_\infty < +\infty$，$r=2,\cdots,2k,2k+1,\cdots,n-1$.（$k=0$ 的情形在[7] 内已有证明）根据[6]，微分方程 $DP_n(D)y=0$ 的解空间有一组基 $\{\phi_0(x),\cdots,\phi_n(x)\}$ 在 $\left(0,\dfrac{\pi}{\beta}\right)$ 内是 ECT 组. 在 $\left(0,\dfrac{\pi}{\beta}\right)$ 内取 n 个点 $0<x_1<\cdots<x_n<\dfrac{\pi}{\beta}$，作一广义多项式

$$Q(x) = C\cdot D\begin{pmatrix}\phi_0 & \phi_1 & \cdots & \phi_n\\ x & x_1 & \cdots & x_n\end{pmatrix}$$

$$= C\begin{vmatrix}\phi_0(x) & \phi_1(x) & \cdots\cdots & \phi_n(x)\\ \phi_0(x_1) & \phi_1(x_1) & \cdots\cdots & \phi_n(x_1)\\ \cdots\cdots & \cdots\cdots & \cdots\cdots & \cdots\cdots\\ \phi_0(x_n) & \phi_1(x_n) & \cdots\cdots & \phi_n(x_n)\end{vmatrix}.$$

$C>0$ 是一常数，稍后确定之. $Q(x)$ 在 $\left(0,\dfrac{\pi}{\beta}\right)$ 内只有零点 x_j，且每一 x_j 是其变号点. 选取 C 足够大使有

(1) $\|P_n(D)Q\|_\infty > 1$. 由于 $DP_n(D)Q(x)\equiv 0$，知 $P_n(D)Q(x)$ 是一常数，只要 C 足够大这一条能满足.

(2) $\min\limits_{1\leqslant j\leqslant n-1}\left|Q\left(\dfrac{x_j+x_{j+1}}{2}\right)\right| \geqslant \|f\|_\infty$.

(3) 能找到 $c,d,0<c<x_1,x_n<d<\dfrac{\pi}{\beta}$ 使 $|Q(c)|=|Q(d)|=\|f\|_\infty$.

$Q(x)$ 在$[c,d]$上在 $\pm\|f\|_\infty$ 之间有 n 次振荡. 注意到 $P_2(D)=[D-(\alpha_1-\mathrm{i}\beta_1)][D-(\alpha_1+\mathrm{i}\beta_1)]$, 对任何函数 $z(x)\in C^1$ 有恒等式

$$|[D-(\alpha_1+i\beta_1)]z|=\mathrm{e}^{\alpha_1 x}\{[D\mathrm{e}^{-\alpha_1 x}z]^2+\beta_1^2(\mathrm{e}^{-\alpha_1 x}z)^2\}^{\frac{1}{2}}.$$

置 $m_1=\max\limits_{c\leqslant x\leqslant d}|[D-(\alpha_1+\mathrm{i}\beta_1)]Q(x)|$. 我们证

$$\|[D-(\alpha_1+\mathrm{i}\beta_1)]f\|_\infty\leqslant m_1. \tag{4.3}$$

假定(4.3) 不成立, 则存在 $x_0\in\mathbf{R}$ 使 $|[D-(\alpha_1+\mathrm{i}\beta_1)]f(x_0)|>m_1$. 在$[c,d]$内部可找到一点 x' 使$Q(x')=f(x_0)$. 置 $h(x)=f(x+x_0-x')-Q(x)$, 则 $h(x')=0$.

我们有

$$|[D-(\alpha_1+\mathrm{i}\beta_1)]f(x+x_0-x')\big|_{x=x'}>|[D-(\alpha_1+\mathrm{i}\beta_1)]Q(x)\big|_{x=x'}.$$

亦即

$$\{[D\mathrm{e}^{-\alpha_1 x}f(x+x_0-x')]^2+\beta_1^2(\mathrm{e}^{-\alpha_1 x}f(x+x_0-x'))^2\}^{\frac{1}{2}}_{x=x'}$$
$$>\{[D\mathrm{e}^{-\alpha_1 x}Q(x)]^2+\beta_1^2(\mathrm{e}^{-\alpha_1 x}Q(x))^2\}^{\frac{1}{2}}.$$

由 $Q(x')=f(x_0)$ 得

$$|D\mathrm{e}^{-\alpha_1 x}f(x+x_0-x')\big|_{x=x'}>|D\mathrm{e}^{-\alpha_1 x}Q(x)\big|_{x=x'}. \tag{4.4}$$

仿[7], x' 点的选择可使$f(x+x_0-x')$ 与 $Q(x)$ 的曲线在$[c,d]$内有 $n+1$个交点, 从而$h(x)$有$n+1$个零(变号)点. 如果$x_1,\cdots,x_n$ 选择时, 使其相邻点间距离充分小, 则可根据广义 Rolle 定理断定存在 $\xi\in(c,d)$ 使 $P_n(D)h(x)$ 在ξ 变号. 但这与 $\|P_n(D)Q\|_\infty>1$ 矛盾. (4.3) 得证.

令 $g(x)=[D-(\alpha_1+\mathrm{i}\beta_1)]f(x)$, 表示着 $P_n(D)=[D-(\alpha_1-\mathrm{i}\beta_1)]\cdot R(D)[D-(\alpha_1+\mathrm{i}\beta_1)]$. 由 $P_n(D)f=[D-(\alpha_1-\mathrm{i}\beta_1)]R(D)g$, 对 g 有 $\|g\|_\infty\leqslant m_1$, $\|[D-(\alpha_1-\mathrm{i}\beta_1)]R(D)g\|_\infty\leqslant 1$. 这表明 $g\in A_\infty^{(n-1)}([D-(\alpha_1-\mathrm{i}\beta_1)]R(D))$. 由于

$$\|[D-(\alpha_1-\mathrm{i}\beta_1)]R(D)g\|_\infty\leqslant 1\Leftrightarrow\|(D-\alpha_1)R(D)g\|_\infty\leqslant C_1$$

及 $\|R(D)g\|_\infty\leqslant C_2$ 对某 $C_1,C_2>0$. 往下不妨就 g 是实值函数的情形来证(否则由 $g=g_1+\mathrm{i}g_2$, 分别就 g_1,g_2 论证, 然后合并起来就是了). 由已得的结论 $\|[D-\alpha_1]R(D)g\|_\infty\leqslant 1$ 及 $\|g\|_\infty\leqslant m'_1$, 重复上面的论证便得 $\|(D-\alpha_1)g\|_\infty<+\infty$. 由此得 $\|[D-(\alpha_1-\mathrm{i}\beta_1)]g\|_\infty<+\infty$, 此事含有 $\|P_2(D)f\|_\infty<+\infty$. $r=2$ 证完. 往下借助数学归纳法

即可完成证明. 由此顺便得到一有用事实. 若 $f \in A_{\infty}^{(n)}(P_n(D))$, 则 $\| f^{(j)} \|_{\infty} < +\infty, j = 1, \cdots, n$. 下面转向正题.

(3) 仿照 Cavaretta[8], 作

$$g(x) = \begin{cases} 1, & -1 \leqslant x \leqslant 1. \\ (-1)^n (x-2)^n \sum_{j=0}^{n-1} \binom{n+j-1}{j} (x-1)^j, & 1 < x < 2. \\ (x+2)^n \sum_{j=1}^{n-1} \binom{n+j-1}{j} (x+1)^j, & -2 < x < 1. \\ 0, & |x| \geqslant 2. \end{cases}$$

可直接验证 $g(x) \in C^{n-1}(\mathbf{R})$. 又 $\| g^{(j)} \|_{\infty} < +\infty, j = 0, \cdots, n$, 因为 $g(x)$ 有紧支集. 今对 $k = 1, 2, 3, \cdots$ 置

$$F_k(x) = \bar{f}(x) g\left(\frac{\lambda x}{k\pi}\right), -2k\frac{\pi}{\lambda} \leqslant x \leqslant 2k\frac{\pi}{\lambda}$$

$$F_k\left(x + 4k\frac{\pi}{\lambda}\right) = F_k(x)$$

在(2) 内已证得 $\| D^j \bar{f} \|_{\infty} \leqslant A, j = 0, 1, \cdots, n$. 固定每一 x, 当 $k \to +\infty$ 时总有 $g\left(\frac{\lambda x}{k\pi}\right) \to 1$(对充分大的 k, 等于1), $D_x^j g\left(\frac{\lambda x}{k\pi}\right) \to 0$. 所以 $F_k(x) \to \bar{f}(x)$(k 充分大时相等), $P_n(D) \cdot F_k(x) \to P_n(D)\bar{f}(x)$. 由于 $\| \bar{f} \|_{\infty} < \| \Phi_{n,\lambda} \|_{\infty}$, 所以对充分大的 k 有 $\| F_k \| < \| \Phi_{n,\lambda} \|_{\infty}$ 以及 $\| P_n(D) F_k \|_{\infty} < 1$, 但由

$$F_k(\gamma) = \bar{f}(\gamma) = \Phi_{n,\lambda}(\gamma) = \Phi_{n,\lambda}(\gamma'),$$

$$| F'_k(\gamma) | \to | \bar{f}'(\gamma) | > | \Phi'_{n,\lambda}(\gamma) | \text{ 及 } | \Phi'_{n,\lambda}(\gamma') |$$

得到
$$| F'_k(\gamma) | > \max\{ | \Phi'_{n,\lambda}(\gamma) |, | \Phi'_{n,\lambda}(\gamma') | \}$$

当 k 充分大时成立, 这与(1) 中证得的结果矛盾. 定理 2 全证.

定理 2 可以稍微拓广为下面的形式.

定理 3 设对某 $\lambda > \Lambda$, 有 $f \in A_{\infty}^{(n)}(P_n(D))$ 满足

(1) $\| f \|_{\infty} \leqslant \| \Phi_{n,\lambda} \|_{\infty}$;

(2) 存在 $\alpha, a, b \in \mathbf{R}$ 使 $f(\alpha) = \Phi_{n,\lambda}(a) = \Phi_{n,\lambda}(b)$, a, b 属于 $\Phi_{n,\lambda}(x)$ 的相邻的单调区间.

则对任取的 $\gamma \in \mathbf{C}$ 有

$$|(D-\gamma)f(\alpha)| \leqslant \max\{|(D-\gamma)\Phi_{n,\lambda}(a)|, |(D-\gamma)\Phi_{n,\lambda}(b)|\}. \tag{4.5}$$

证明 仿定理 2 的证法，不妨只对 $\|f\|_\infty < \|\Phi_{n,\lambda}\|_\infty$ 及 $\|P_n(D)f\|_\infty < 1$ 的 $f \in A_\infty^{(n)}(P_n(D))$ 来证. 仍用反证法. 设定理 3 不成立，置 $f_1(x) = f(x+\alpha-a)$, $f_2(x) = f(x+\alpha-b)$, 则

$$|(D-\gamma)f(\alpha)| = |(D-\gamma)f_1(a)| > |(D-\gamma)\Phi_{n,\lambda}(a)|.$$

$$|(D-\gamma)f(\alpha)| = |(D-\gamma)f_2(b)| > |(D-\gamma)\Phi_{n,\lambda}(b)|.$$

记 $\alpha = \mathrm{Re}\gamma$，仿照定理 2 证明的(2) 便得

$$|De^{-\alpha x}f(x)\big|_{x=\alpha}| > |De^{-\alpha x}\Phi_{n,\lambda}(x)\big|_{x=a,b}|. \tag{4.6}$$

(此式与(4.4) 可对比) 往下基本重复前定理的证明步骤(先对 f 是周期情形证，然后利用 Cavaretta 的技巧过渡到一般) 即得. 证完.

有一种特殊情形值得一提. $P_n(D)$ 称为自共轭的，如果 $P_n(-D) = (-1)^n P_n(D)$，当 $P_n(D)$ 自共轭时，$A_\infty^{(n)}(P_n(D))$ 类的函数有对称性，意即 $f(x) \in A_\infty^{(n)}(P_n(D))$ 含有 $f(-x) \in A_\infty^{(n)}(P_n(D))$. 标准函数 $\Phi_{n,\lambda}(x)$ 的图象关于过极值点平行于纵坐标轴的直线对称. 这时比较定理有更为简洁的形式.

推论 1 若 $P_n(D)$ 自共轭，且对某 $\lambda > \Lambda$ 有 $f \in A_\infty^{(n)}(P_n(D))$ 满足定理 3 的条件，则对任取的 $\gamma \in \mathbf{C}$ 有

$$|(D-\gamma)f(\alpha)| \leqslant |(D-\gamma)\Phi_{n,\lambda}(a)|. \tag{4.7}$$

特别地，当 $\gamma = 0$ 时给出

$$|f'(\alpha)| \leqslant |\Phi'_{n,\lambda}(a)|. \tag{4.8}$$

证明 只需注意此时有 $\Phi'_{n,\lambda}(a) = -\Phi'_{n,\lambda}(b)$.

推论 2 在定理 2 或 3 的前提下，对任取的 $\gamma \in \mathbf{C}$ 有

$$\|(D-\gamma)f\|_\infty \leqslant \|(D-\gamma)\Phi_{n,\lambda}\|_\infty. \tag{4.9}$$

证明 假定 $\|(D-\gamma)f\|_\infty > \|(D-\gamma)\Phi_{n,\lambda}\|_\infty$. 取 $x_0 \in \mathbf{R}$，使 $|(D-\gamma)f(x_0)| = \|(D-\gamma)f\|_\infty$，取 $x', x'' \in \mathbf{R}$ 使 $f(x_0) = \Phi_{n,\lambda}(x') = \Phi_{n,\lambda}(x'')$，$x', x''$ 在 $\Phi_{n,\lambda}(x)$ 的相邻单调区间内. 置 $f_1(x) = f(x+x_0-x')$, $f_2(x) = f(x+x_0-x'')$，仿照定理 3 的证明可得

$$|De^{-\alpha x}f(x)\big|_{x=x_0}| > |De^{-\alpha x}\Phi_{n,\lambda}(x)\big|_{x=x',x''}|.$$

此处 $\alpha = \mathrm{Re}\gamma$. 从而导致矛盾. 证完.

推论3 设 $P_n(\lambda)=0$ 仅有实根. 若对某 $\lambda>0$ 有 $f\in A_{\infty}^{(n)}(P_n(D))$ 满足 $\|f\|_{\infty}\leqslant\|\Phi_{n,\lambda}\|_{\infty}$, 则对 $P_n(D)$ 的任一子算子 $P_r(D)$ 有

$$\|P_r(D)f\|_{\infty}\leqslant\|P_r(D)\Phi_{n,\lambda}\|_{\infty},(r=1,\cdots,n)\tag{4.10}$$

对 $\Phi_{n,\lambda}(x)$ 实现等号, 故不等式精确. 这便是 Sharma-Tzimbalario 的结果[7]. 当 $P_n(\lambda)=0$ 有共轭复根时, 对充分大的 λ 仍有类似于(4.10)的不等式成立. 但这时不等式不能直接从比较定理推出, 因为比较定理仅仅对一阶微分算子 $D-\gamma$ 得到了证明. 一般情况下不等式(4.10)的证明我们已在[9]中给出了. 此处不再重复.

推论4 设对某 $\lambda>\Lambda$ 有 $f\in A_{\infty}^{(n)}(P_n(D))$ 满足

(1) $\|f\|_{\infty}\leqslant\|\Phi_{n,\lambda}\|_{\infty}$.

(2) 存在 ξ_0 点使 $f(\xi_0)=\Phi_{n,\lambda}(\xi_0)$. $[c,d]$ 表示 $\Phi_{n,\lambda}(x)$ 的单调区间包含 ξ_0 者. 则

1) 若 $\Phi_{n,\lambda}(x)$ 在 $[c,d]$ 上升, 则

$$f(x)\leqslant\Phi_{n,\lambda}(x),x\in[\xi_0,d].\tag{4.11}$$

$$f(x)\geqslant\Phi_{n,\lambda}(x),x\in[c,\xi_0].\tag{4.12}$$

2) 若 $\Phi_{n,\lambda}(x)$ 在 $[c,d]$ 下降, 则(4.11), (4.12)的不等式反向.

推论5 设对某 $\lambda>\Lambda$ 有 $f\in A_{\infty}^{(n)}(P_n(D))$ 满足

(1) $\|f\|_{\infty}\leqslant\|\Phi_{n,\lambda}\|_{\infty}$.

(2) 有 $\xi_0,\xi_1,\eta_0,\eta_1\in\mathbf{R}$, 使 $f(\xi_0)=\Phi_{n,\lambda}(\eta_0)$, $f(\xi_1)=\Phi_{n,\lambda}(\eta_1)$, 且 η_0,η_1 属于 $\Phi_{n,\lambda}(x)$ 的一个单调区间, 则

$$|\xi_0-\xi_1|\geqslant|\eta_0-\eta_1|.\tag{4.13}$$

最后两条推论的证明与[1]中讨论过的情形很类似, 细节从略.

注记1 我们可以给出比定理2的形式更强的比较定理.

定理2′ 在定理2的前提下, 若 $\Phi'_{n,\lambda}(a)>0$, $\Phi'_{n,\lambda}(b)<0$, 则

$$\Phi'_{n,\lambda}(b)\leqslant f'(\alpha)\leqslant\Phi'_{n,\lambda}(a).\tag{4.14}$$

注意由(4.14)推出(4.1).

五、关于 $P_n(D)$ 的 Landau-Kolmogorov 型不等式

1913 年 Landau[10] 提出了建立可微函数的各阶导数的一致范数间的精确不等式的著名问题, 以及 Kolmogorov 对该问题的深刻研究[3] 是人们

所熟知的. Kolmogorov 的结果沿不同方向有许多拓广[7,11]. 本节给出一些这方面的新结果. 引入记号

$$\Gamma_{n,r}(\lambda) = \| P_r(D)\Phi_{n,\lambda} \|_\infty, r = 0,2,\cdots,2k,\cdots,n. \tag{5.1}$$

$P_r(D)$ 是 $P_n(D)$ 的子算子. 本节分成三段叙述.

1. ∞ − ∞ 型的 Landau-Kolmogorov 不等式

定理 4　给定 $P_n(D), n \geqslant 2$. 设 $F \in L_\infty^{(n)}(P_n(D)), \| P_n(D)F \|_\infty > 0$. 置 $g(x) = \| P_n(D)F \|_\infty^{-1} \cdot F(x)$, 若存在数 $\lambda(\| g \|_\infty) > \Lambda$ 使 $\| g \|_\infty \leqslant \Gamma_{n,0}(\lambda(\| g \|_\infty))$, 则有

$$\frac{\| P_r(D)F \|_\infty}{\| P_n(D)F \|_\infty} \leqslant \Gamma_{n,r}\left(\lambda\left(\frac{\| F \|_\infty}{\| P_n(D)F \|_\infty}\right)\right). \tag{5.2}$$

此处 $P_r(D)$ 是 $P_n(D)$ 的任一子算子.

证明　由 $g \in W_\infty^{(n)}(P_n(D))$ 而且对 $\lambda = \lambda(\| g \|_\infty)$ 有 $\| g \|_\infty \leqslant \| \Phi_{n,\lambda} \|_\infty = \Gamma_{n,0}(\lambda(\| g \|_\infty))$, 根据定理 2 的推论 3 即得(5.2). 证毕.

下面给出一特殊情形.

推论　设 $P_n(0) = 0, P_n(\lambda) = 0$ 仅有实根. 则对任取的 $F \in L_\infty^{(n)}(P_n(D)), \| P_n(D)F \|_\infty > 0$ 者, 必有 $\lambda(\| g \|_\infty)$ 使 $\| g \|_\infty \leqslant \Gamma_{n,0}(\lambda(\| g \|_\infty))$. 故此时有(5.2) 成立.

证明　此时, $\Gamma_{n,0}(\lambda)$ 作为 λ 的连续函数满足条件 $\lim\limits_{\lambda\to+\infty} \Gamma_{n,0}(\lambda) = 0, \lim\limits_{\lambda\to 0+} \Gamma_{n,0}(\lambda) = +\infty$. 故只需根据连续函数的中间值性质即可断定对 $\| g \|_\infty$ 有 $\lambda(\| g \|_\infty)$ 使 $\| g \|_\infty = \Gamma_{n,0}(\lambda(\| g \|_\infty))$ 成立. 所以有(5.2). 证完.

在某些特殊情况下, 比如 $P_n(D) = D^n$, 或 $P_n(D) = D^\sigma \prod\limits_{j=1}^{l}(D^2 - t_j^2)(t_1,\cdots,t_l > 0, \sigma \geqslant 1, \sigma + 2l = n)$ 时, $\lambda(\| g \|_\infty)$ 唯一. 特别当 $P_n(D) = D^n$ 时(5.2) 的右边有显式表示, 即 Kolmogorov 不等式.

例　(文献[7])

$P_2(D) = D^2 - \gamma^2, \gamma > 0$. 标准函数

$$\Phi_{2,\lambda}(x) = -\frac{1}{\gamma^2}\left[1 - \frac{\operatorname{ch}\left(x - \frac{\pi}{2\lambda}\right)\gamma}{\operatorname{ch}\frac{\pi\gamma}{2\lambda}}\right], 0 \leqslant x \leqslant \frac{\pi}{\lambda}.$$

$$\Phi_{2,\lambda}\left(x+\frac{\pi}{\lambda}\right)=-\Phi_{2,\lambda}(x).$$

$$\Gamma_{2,0}(\lambda)=\max_{0\leqslant x\leqslant\frac{\pi}{\lambda}}|\Phi_{3,\lambda}(x)|=\frac{1}{\gamma^2}\left(1-\frac{1}{\operatorname{ch}\frac{\gamma\pi}{2\lambda}}\right).$$

$$\Gamma_{2,1}(\lambda)=\max_{0\leqslant x\leqslant\frac{\pi}{\lambda}}|(D\pm\gamma)\Phi_{2,\lambda}(x)|$$

$$=\left\|-\frac{1}{\gamma}+\frac{2\mathrm{e}^{\gamma x}}{\gamma(1+\mathrm{e}^{\gamma\pi/\lambda})}\right\|_\infty=\frac{1}{\gamma}\operatorname{th}\frac{\gamma\pi}{2\lambda}.$$

$$\Gamma_{2,2}(\lambda)=1$$

注意 $\Gamma_{2,0}(\lambda)\downarrow 0+(\lambda\to+\infty)$, $\lim\limits_{\lambda\to 0+}\Gamma_{2,0}(\lambda)=\frac{1}{\gamma^2}$. 故任取 $f\in C^1$. 若 $\|f\|_\infty=M_0$. $\|(D^2-\gamma^2)f\|_\infty=M_2>0$, 则置 $g(x)=\frac{f(x)}{M_2}$, 当 $\|g\|_\infty=\frac{M_0}{M_2}<\gamma^{-2}$ 时方程 $\frac{M_0}{M_2}=\Gamma_{2,0}(\lambda)=\frac{1}{\gamma^2}\left(1-\operatorname{sech}\frac{\gamma\pi}{2\lambda}\right)$ 有唯一解 $\lambda\left(\frac{M_0}{M_2}\right)$. 对此有 $\|(D\pm\gamma)g\|_\infty\leqslant\Gamma_{2,1}\left(\lambda\left(\frac{M_0}{M_2}\right)\right)$. 计算 $\Gamma_{2,1}\left(\lambda\left(\frac{M_0}{M_2}\right)\right)$. 为此只要从 $\Gamma_{2,1}(\lambda)$ 的表达式及 $1-\operatorname{sech}\frac{\gamma\pi}{2\lambda}=\frac{M_0}{M_2}\gamma^2$ 消去 λ 即可:

$$\frac{1}{\gamma}\operatorname{th}\frac{\gamma\pi}{2\lambda}=\frac{1}{\gamma}\sqrt{1-\operatorname{sech}^2\frac{\gamma\pi}{2\lambda}}=\frac{1}{\gamma}\sqrt{1-\left(1-\gamma^2\frac{M_0}{M_2}\right)^2}$$

$$=\frac{1}{\gamma}\left(2\gamma^2\frac{M_0}{M_2}-\gamma^4\frac{M_0^2}{M_2^2}\right)^{\frac{1}{2}}.$$

所以
$$\|(D\pm\gamma)f\|_\infty\leqslant\sqrt{M_0(2M_2-\gamma^2M_0)}.\tag{5.3}$$

2. p－p 型 Landau-Kolmogorov 不等式

仿照[12], 应用卷积变换法可证

定理 5 若对某 $\lambda>\Lambda$ 有 $f\in W_p^{(n)}(P_n(D))$, $1\leqslant p<+\infty$ 满足 $\|f\|_p\leqslant\Gamma_{n,0}(\lambda)$, 则对 $P_n(D)$ 的任一子算子 $P_r(D)$ 有

$$\|P_r(D)f\|_p\leqslant\Gamma_{n,r}(\lambda).\tag{5.4}$$

当 $p=1$ 时不等式精确.

定理 5 的证明可完全仿照[12] 的方法做. 详细步骤从略.

现在考虑 $F\in L_p^{(n)}(P_n(D))$, 和定理 4 相仿, 成立着

定理6 给定 $P_n(D)$, $n\geqslant 2$. 设 $F\in L_p^{(n)}(P_n(D))$, $(1\leqslant p<+\infty)$,

$\|P_n(D)F\|_p > 0$. 置 $g(x) = \|P_n(D)F\|_p^{-1} \cdot F(x)$. 若存在数 $\lambda(\|g\|_p) > \Lambda$ 使 $\|g\|_p \leqslant \Gamma_{n,0}(\lambda\|g\|_p))$,则有

$$\frac{\|P_r(D)F\|_p}{\|P_n(D)F\|_p} \leqslant \Gamma_{n,r}\left(\lambda\left(\frac{\|F\|_p}{\|P_n(D)F\|_p}\right)\right). \tag{5.5}$$

$P_r(D)$ 是 $P_n(D)$ 的任一子算子.

推论 设 $P_n(0) = 0, P_n(\lambda) = 0$ 仅有实根, 则任取 $F \in L_p^{(n)}(P_n(D)), \|P_n(D)F\|_p > 0$ 时, 必有 $\lambda(\|g\|_p)$ 使 $\|g\|_p \leqslant \Gamma_{n,0}(\lambda(\|g\|_p))$. 故此时有(5.5) 成立. 但是只有 $p = 1$ 时不等式才精确.

例 仍考虑 $P_2(D) = D^2 - \gamma^2, \gamma > 0$. 任取 $F \in L_p^{(2)}$, 置 $M_0 = \|F\|_p, \|P_2(D)F\|_p = M_2(\neq 0)$, 则当$\dfrac{M_0}{M_2} < \gamma^{-2}$ 时有

$$\|(D \pm \gamma)F\|_p \leqslant \sqrt{M_0(2M_2 - \gamma^2 M_0)}. \tag{5.6}$$

3. ∞ - p 型 Landau-Kolmogorov 不等式

应用函数重排的工具可将[2]关于2π周期可微函数类 $\widetilde{W}^r_\infty$ 上的 p 范数的 Landau-Kolmogorov 不等式拓广到 $W_\infty^{(n)}(P_n(D))$ 的 2π 周期的子集上.

定义 6 称 $f \in \widetilde{\mathfrak{M}}_p^{(n)}(P_n(D))$, 若

$$f(x) = C_\sigma + \int_0^{2\pi} G_n(x-t)h(t)\mathrm{d}t. \tag{5.7}$$

此处 $\|h\|_p[0,2\pi] \leqslant 1, \int_0^{2\pi} h(t)\mathrm{d}t = 0, C_\sigma = 0$ 或为任意常数, 视 $P_n(0) \neq 0$ 或 $P_n(0) = 0$ 而定.

本节要用到周期函数的非增重排. 涉及重排的事实可参看[1,2]. 首先给出 $\widetilde{\mathfrak{M}}_\infty^{(n)}(P_n(D))$ 类的函数的微商重排的比较定理.

定理7 设 $P_n(0) = 0, n \geqslant 2$. 若对某 $\lambda > \Lambda$ 有 $f \in \widetilde{\mathfrak{M}}_\infty^{(n)}(P_n(D))$ 使 $\|f\|_\infty \leqslant \|\Phi_{n,\lambda}\|_\infty$, 则对每一 $x \in [0,2\pi]$ 有

$$\int_0^x p(f',u)\mathrm{d}u \leqslant \lambda \int_0^x p((\hat{\Phi}_{n,\lambda})',u)\mathrm{d}u. \tag{5.8}$$

此处 $p(g,u)$ 表示 g 在 2π 周期区间内的非增重排, 而

$$\hat{\Phi}_{n,\lambda}(x) \xlongequal{\mathrm{df}} \Phi_{n,\lambda}\left(\frac{x}{\lambda}\right). \tag{5.9}$$

不等式(5.8) 的证明可完全仿照[2] 的第 5 章对 $\widetilde{W}_M^{(r+1)}$ 类的同一定理的证法. 细节省略. 要指出的是, (5.8) 很容易推广到以 $P_n(D)$ 的子算子作用于 f 代替 f' 的情形.

推论 在定理 7 的条件下, 对 $P_n(D)$ 的任一子算子 $P_r(D)$, 倘 $P_r(0) = 0$, 便成立

$$\int_0^x p(P_r(D)f, u)\mathrm{d}u \leqslant \int_0^x p(P_r(\widehat{D)\Phi}_{n-1,\lambda}; u)\mathrm{d}u \tag{5.10}$$

$0 \leqslant x \leqslant 2\pi$. 此处 $\Phi_{n-1,\lambda}(x)$ 是关于 $P_{n-1}(D)$ 的标准函数, 而 $D \cdot P_{n-1}(D) = P_n(D)$.

证明 根据定理 2 的推论 3, 若记 $P_r(D) = DP_{r-1}(D)$, 则有 $\| P_{r-1}(D)f \|_\infty \leqslant \| P_{r-1}(D)\Phi_{n,\lambda} \|_\infty$. 注意到 $P_{r-1}(D)f \in \widetilde{\mathfrak{M}}_\infty^{(n-r+1)}(D\hat{P}_{r-1}(D))$, $\hat{P}_{r-1}(D)$ 是 $P_{r-1}(D)$ 关于 $P_{n-1}(D)$ 的余算子, $P_{r-1}(D)\Phi_{n,\lambda}(x)$ 是算子 $D\hat{P}_{r-1}(D)$ 的标准函数. 对 $P_{r-1}(D)f$ 应用 (5.8) 得

$$\int_0^x p((P_{r-1}(D)f)', u)\mathrm{d}u \leqslant \int_0^x \lambda p((P_{r-1}(\widehat{D)\Phi}_{n,\lambda})', u)\mathrm{d}u$$

但因 $\lambda((P_{r-1}(\widehat{D)\Phi}_{n-1,\lambda})') = (P_r(\widehat{D)\Phi}_{n-1,\lambda})(t)$. 这可以直接验证. 故得(5.10).

今利用(5.10) 及尚光明引理(见[2] 命题 5.4.8) 便得

定理 8 任给 $P_n(D)$, $n \geqslant 1$, 若 $f \in \widetilde{\mathfrak{M}}_\infty^{(n)}(P_n(D))$ 的周期积分 F 对某 $\lambda > \Lambda$ 满足 $\| F \|_\infty \leqslant \| \Phi_{n+1,\lambda} \|_\infty$, $\Phi_{n+1,\lambda}(x)$ 是 $DP_n(D) = P_{n+1}(D)$ 的标准函数, 则对 $P_n(D)$ 的任一子算子 $P_r(D)(r \geqslant 0)$ 有

$$\| P_r(D)f \|_p \leqslant \| P_r(\widehat{D)\Phi}_{n,\lambda} \|_p. (1 \leqslant p \leqslant \infty) \tag{5.11}$$

特别当 $r = 0$ 时有

$$\| f \|_p \leqslant \| \hat{\Phi}_{n,\lambda} \|_p. \tag{5.12}$$

(5.11) 略加变形可得 $\infty - p$ 型的 Landau-Kolmogorov 不等式:

设 $P_{n+1}(D)=DP_n(D), n\geqslant 1. P_{r+1}(D)=DP_r(D)$ 为 $P_{n+1}(D)$ 的子算子. 记

$$\Gamma_{n+1,r+1}(\lambda)_p=\|P_{r+1}(\widehat{D)\Phi}_{n+1,\lambda}\|_p \tag{5.13}$$

$(r=0$ 时, $P_1(D)=D)$.

定理 9　设有 2π 周期的 $F\in L_\infty^{(n+1)}(P_{n+1}(D)), \|P_{n+1}(D)F\|_\infty>0, g=\|P_{n+1}(D)F\|_\infty^{-1}\cdot F$, 若存在数 $\lambda(\|g\|_\infty)$ 使 $\|g\|_\infty\leqslant\Gamma_{n+1,0}(\lambda(\|g\|_\infty))$ 则有

$$\frac{P_{r+1}(D)F\|_p}{\|P_{n+1}(D)F\|_\infty}\leqslant\Gamma_{n+1,r+1}\left(\lambda\left(\frac{\|F\|_\infty}{\|P_{n+1}(D)F\|_\infty}\right)\right)_p, (1\leqslant p\leqslant\infty), r=0,1,\cdots,n. \tag{5.14}$$

注意 $p=\infty$ 的情形已包含于(5.2) 内. 当 $P_{n+1}(D)=D^{n+1}$ 时, (5.14) 可以写成显式(见[2] 第 6 章).

Тайков 不等式: 设 $f\in\widetilde{\mathfrak{M}}_\infty^{(n)}(P_n(D))$ 对某正整数 $m\geqslant 1$ 有 $f\perp T_{m-1}, T_{m-1}=\operatorname{span}\left\{1,\begin{matrix}\sin x\\ \cos x\end{matrix}\cdots\begin{matrix}\sin(m-1)x\\ ,\cos(m-1)x\end{matrix}\right\}$. 取 f 的周期积分 F 使满足 $\int_0^{2\pi}F(t)\mathrm{d}t=0$. 则 $F\in\widetilde{\mathfrak{M}}_\infty^{(n+1)}(DP_n(D))$. 而且也有 $F\perp T_{m-1}$. 由[5], 当 $m>\Lambda$ 时有 $G_{n+1}(x)\in A_m$(Markov 条件). 所以, 倘若 T_{m-1} 的 $m>\Lambda$, 就得到[1]

$$\|F\|_\infty\leqslant\|\Phi_{n+1,m}\|_\infty. \tag{5.15}$$

对 F 应用(5.12), 这里的 F' 是(5.12) 中的 f, 给出

$$\|F'\|_p\leqslant\|\Phi'_{n+1,m}\|_p, (1\leqslant p\leqslant+\infty). \tag{5.16}$$

(因此时 m 是整数, $\Phi_{n+1,m}(x)$ 仍有周期 2π). 由此得

定理 10　若 $m>\Lambda, 1\leqslant p\leqslant+\infty$, 则

$$\sup_{\substack{f\in\widetilde{\mathfrak{M}}_\infty^{(n)}\\ f\perp T_{m-1}}}\|f\|_p=\|\Phi_{n,m}\|_p. \tag{5.17}$$

证明　(5.16) 已经含有

$$\sup_{\substack{f\in\widetilde{\mathfrak{M}}_\infty^{(n)}\\ f\perp T_{m-1}}}\|f\|_p\leqslant\|\Phi'_{n+1,m}\|_p=\|\Phi_{n,m}\|_p$$

由于 $\Phi_{n,m}(x) \in \widetilde{\mathfrak{M}}_{\infty}^{(n)}(P_n(D))$ 而且与 T_{m-1} 正交，即得(5.17).

最后根据周期卷积类上三角多项式最佳逼近的对偶定理[1] 得

定理 11 对 $m > \Lambda, 1 \leqslant p \leqslant +\infty$ 有

$$E_m(\widetilde{\mathfrak{M}}_p^{(n)}(P_n(D)))_1 \overset{\text{df}}{=\!=} \sup_{f \in \widetilde{\mathfrak{M}}_p^{(n)}} E_m(f)_1$$

$$= \sup_{\substack{f \in \widetilde{\mathfrak{M}}_{\infty}^{(n)}(P_n^*(D)) \\ f \perp T_{m-1}}} \| f \|_{p'} = \| \Phi_{n,m}^* \|_{p'}. \tag{5.18}$$

此处 $\frac{1}{p} + \frac{1}{p'} = 1$，$P_n^*(D)$ 是 $P_n(D)$ 的共轭算子，$\Phi_{n,\lambda}^*(x)$ 是 $P_n^*(D)$ 的标准函数. 由于与 $P_n^*(D)$ 相对应的卷积类的核是 $G_n(x-t)$ 的转置 $G_n(t-x)$，那么 $P_n^*(D)$ 的标准函数是 $\Phi_{n,\lambda}(-x)$. 所以实际上有 $\| \Phi_{n,m}^* \|_{p'} = \| \Phi_{n,m} \|_{p'}$.

参考文献

[1] 考涅楚克. 逼近论的极值问题. 上海：上海科技出版社，1982

[2] Корнейчук, Лигун, Доронин. *Аппроксимация с ограничениями.* Киев: НАУКОВА ДУМКА, 1982

[3] Н. П. Корнейчук. Экстремальные свойства сплайнов. *Теория Прибп. Функц.* Москва: НАУКА, 1977. 237 ~ 248

[4] Беккенбах, Беллман. *Неравенства* . Москва:《МИР》, 1965. 201

[5] М. Г. Крейн. *ДАН СССР*, 1938, 18(4) ~ (5): 245 ~ 249

[6] В. Т. Шавалдин. Мамем. замемкц, 1981, 29(4): 603 ~ 622

[7] A. Sharma, J. Tzimbalario. Illinois JM, 20(1976), 443 ~ 455

[8] A. Cavarreta, AMM, 81(1974), 480 ~ 486

[9] 孙永生. 线性微分算子的 Landau-Kolmogorov 不等式. 科学通报，1985，30(7): 481 ~ 484

[10] E. Landau. Proc LMS, 13(1913), 2: 43 ~ 49

[11] 黄达人、王建忠，数学学报，26(1983)，No. 6 715 ~ 722

[12] E. M. Stein. Ann. Math. 1957, 65(3): 582 ~ 592

科学通报
1985, 第 7 期

线性微分算子的 Landau-Kolmogorov 型不等式*

Inequalities of Landau-Kolmogorov Type for Some Linear Differential Operators

§1.

Landau[1] 提出的关于可微函数的各阶导函数的一致范数间的精确不等式的著名问题, 以及 Kolmogorov[2] 对该问题的深刻研究是熟知的. 他们的工作有一系列沿不同方向的拓广(参看 Schoenberg[3], Cavaretta[4], Sharma 与 Tzimbalario[5], 黄达人与王建忠[6]). 本文给出任意常(实)系数线性常微分算子的 Landau-Kolmogorov 型不等式.

§2. 广义 Rolle 定理

Polya 给出了满足 W 条件的线性常微分算子的广义 Rolle 定理(见文献[7], 第四章, §16). 我们需要任意常系数的线性常微分算子的广义 Rolle 定理. 这样的常微分算子一般不满足 W 条件, 关于它的广义 Rolle 定理只能在一定限制条件下建立起来. 下面引理 1 包含了它的实质内容.

引理 1 (М.Г. Крейн[8]) 给定实系数二次式 $P(x) = x^2 + px + q$, $p^2 - 4q < 0$, 其两根为 $\alpha \pm i\beta$, $\beta > 0$. 若有 $f(x) \in C^2$ 及 $x_1 < \cdots < x_n < x_{n+1} < x_{n+2}$ 满足(1) $x_{j+1} - x_j < \dfrac{\pi}{2\beta}(j = 1, \cdots, n+1)$. (2) $f(x_j) = 0$, $j = 1, \cdots, n+2$. (3) 在任何区间 $(a, b) \subset (x_1, x_{n+2})$ 内 $P(D)f(x) \not\equiv$

---

* 本文 1984 年 3 月 7 日收到.

0,此处 $D=\frac{\mathrm{d}}{\mathrm{d}x}$.则存在 n 个 ξ 点:

(1) $x_1<\xi_1<\cdots<\xi_n<x_{n+2},\xi_{j+1}-\xi_j<\frac{2\pi}{2\beta}$.

(2) $P(D)f(\xi_j)=0,j=1,\cdots,n$.

(3) ξ_j 是 $P(D)f(x)$ 的变号点,变号方向交错改变.若 $f(x)\in C^1$, $f'(x)$ 在 (x_1,x_{n+2}) 内分段连续可微,且 $f''(x)$ 仅有的间断点是有限个第一类间断点,则存在 n 个 ξ 点成立着(1)、(3).

任给实系数多项式

$P_n(\lambda)=\prod_{s=1}^{k}(\lambda^2-2\alpha_s\lambda+\alpha_s^2+\beta_s^2)\prod_{j=1}^{n-2k}(\lambda-\lambda_j),k\geqslant 1,n\geqslant 2k,\alpha_s$, $\beta_s,\lambda_j\in\mathbf{R},\beta_s>0$.记 $\beta=\max\limits_{1\leqslant s\leqslant k}\beta_s$.对 $n+1$ 个点 $x_1<\cdots<x_{n+1}$ 置 $\Delta_j=x_{j+1}-x_j$, $\|\Delta\|=\max\limits_{1\leqslant j\leqslant n}\{\Delta_j\}$.

我们有

定理1 对于 $P_n(D)$, $f(x)\in C^n$ 及 $x_1<\cdots<x_{n+1}$,若(1) $\|\Delta\|<\frac{\pi}{2\cdot 3^{k-1}\beta}$,(2) $f(x_j)=0,j=1,\cdots,n+1$,(3)在任何区间 $(a,b)\subset(x_1,x_{n+1})$ 内 $P_n(D)f(x)\not\equiv 0$,则存在 $\xi\in(x_1,x_{n+1})$ 使 $P_n(D)f(\xi)=0$,且 $P_n(D)f(x)$ 在 ξ 点变号.若 $f(x)\in C^{n-1}$, $f^{(n-1)}$ 在 (x_1,x_{n+1}) 内分段连续可微,且 $f^{(n)}$ 在 (x_1,x_{n+1}) 内仅有的间断点是有限个第一类间断点,则存在 $\xi\in(x_1,x_{n+1})$ 使 $P_n(D)f(x)$ 在 ξ 变号.

§3. 算子 $P_n(D)$ 的子算子列及标准函数

定义1 多项式 $P_n(\lambda)$ 的零点排定一次序 $\alpha_1\pm\mathrm{i}\beta_1,\cdots,\alpha_k\pm i\beta_k$, $\lambda_1,\cdots,\lambda_{n-2k}$.置 $P_0(D)=1$(恒等), $P_2(D)=D^2-2\alpha_1D+(\alpha_1^2+\beta_1^2)$, $P_4(D)=(D^2-2\alpha_2D+\alpha_2^2+\beta_2^2)P_2(D),\cdots,P_{2k}(D)=(D^2-2\alpha_kD+\alpha_k^2+\beta_k^2)P_{2k-2}(D)$, $P_{2k+1}(D)=(D-\lambda_1)P_{2k}(D),\cdots,P_n(D)$ 是 $P_n(D)$ 的子算子列.

定义2 设 $\lambda>\beta$.置

$$\Phi_{n+1,\lambda}(x)=\frac{2}{\pi i}\sum_{\nu=-\infty}^{+\infty}\frac{\mathrm{e}^{i(2\nu+1)\lambda x}}{(2\nu+1)P_n((2\nu+1)\lambda i)}$$

为算子 $P_n(D)$ 的标准函数.

Шевалдин[9,10] 给出

引理 2 (1)$\Phi_{n+1,\lambda}\left(x+\frac{\pi}{\lambda}\right)=-\Phi_{n+1,\lambda}(x)$.

(2) 当 $\lambda>2\beta$ 时 $\Phi_{n+1,\lambda}(x)$ 在 $\left[0,\frac{\pi}{\lambda}\right)$ 内有唯一的零点,它是单的;有唯一的极值点.

(3)$P_n(D)\Phi_{n+1,\lambda}(x)=\operatorname{sgnsin}\lambda x$.

记 $\Gamma_{n,r}(\lambda)=\|P_r(D)\Phi_{n+1,\lambda}\|_\infty, r=0,2,\cdots,2k,2k+1,\cdots,n$. 注意 $\Gamma_{n,0}(\lambda)=\|\Phi_{n+1,\lambda}\|_\infty, \Gamma_{n,n}(\lambda)=1$.

定义 3 称 $f(x)\in A_\infty$, 如果

(1)$f(x)\in C^{n-1}(\mathbf{R}), f(x)=O(1)(|x|\to+\infty)$.

(2)$f^{(n-1)}$ 在 $\mathbf{R}$ 上分段连续可微,且在任一有界区间内 $f^{(n)}$ 仅有的间断点是有限个第一类间断点.

(3)$\|P_n(D)f\|_\infty\leqslant 1$.

§4. Landau-Kolmogorov 型不等式

定理 2 存在 $\Lambda\in\mathbf{R}$, 对每一 $\lambda>\Lambda$ 及 $f(x)\in A_\infty$, 若 $\|f\|_\infty\leqslant\Gamma_{n,0}(\lambda)$, 则 $\|P_r(D)f\|_\infty\leqslant\Gamma_{n,r}(\lambda), r=2,\cdots,2k,2k+1,\cdots,n-1$. 不等式精确:对标准函数 $\Phi_{n+1,\lambda}(x)(\in A_\infty)$ 等号成立.

证明 (1) 先设有以 $2N\frac{\pi}{\lambda}(N\geqslant 1$ 整数) 为周期的 $f(x)\in A_\infty$, 定理 2 对它不成立. 则 $\|f\|_\infty\leqslant\Gamma_{n,0}(\lambda)$, 但对某个 $\nu\in\{2,\cdots,2k,2k+1,\cdots,n-1\}$ 有 $\alpha>1$ 使 $\|P_\nu(D)f\|_\infty=\alpha\Gamma_{n,\nu}(\lambda), \lambda>0$ 充分大, 其具体下界后面明确. 我们由此推出矛盾. 选择 x_0 使 $|P_\nu(D)f(x_0)|=\|P_\nu(D)f\|_\infty$, 以及 x_1 使 $\alpha P_\nu(D)\Phi_{n+1,\lambda}(x_0-x_1)=P_\nu(D)f(x_0)$, 则见 $\Gamma_{n,\nu}(\lambda)=|P_\nu(D)\Phi_{n+1,\lambda}(x_0-x_1)|$. 置 $h(x)=\Phi_{n+1,\lambda}(x-x_1)-\frac{1}{\alpha}f(x)$. 我们有 $P_\nu(D)h(x_0)=0$, x_0 是 $P_\nu(D)f(x)$ 及 $P_\nu(D)\Phi_{n+1,\lambda}(x-x_1)$ 的局部极值点. 取一包含 x_0 的区间 $\left[c,c+\frac{2N\pi}{\lambda}\right)$, $\lambda>2\beta$, 则由引理 2, $\Phi_{n+1,\lambda}(x-x_1)$ 在 $\left[c,c+\frac{2N\pi}{\lambda}\right)$ 内在 $2N$ 个等距分布的点上符号交错地取到其极值, 从而在一个 $\frac{2N\pi}{\lambda}$ 周期区间内有 $2N$ 个变号(零)点, 任二相

邻零点距离 $<\frac{2\pi}{\lambda}$.

(a) $n=2k, k\geqslant 1$ 时. 对 $\nu<n$ 时有 $DP_\nu(D)f(x_0)=DP_\nu(D)\Phi_{n+1,\lambda}(x_0-x_1)=0$, 故 $DP_\nu(D)h(x_0)=0$. 若 λ 之大足使 $3^{k-1}\cdot\frac{2\pi}{\lambda}<\frac{\pi}{2\beta}$ 即 $\lambda>4\cdot 3^{k-1}\beta$, 则由引理 1(注意 $h(x)$ 的周期性), 知 $P_\nu(D)h(x)$ 在 $\frac{2N\pi}{\lambda}$ 周期区间内有 $2N$ 个交错变号(零)点. 再考虑到 x_0, 便可断定 $P_\nu(D)h(x)$ 在 $\frac{2N\pi}{\lambda}$ 周期内至少有 $2N+2$ 个零点. 继续应用引理 1 于 $P_{\nu+2}(D)h(x),\cdots,P_{2k}(D)h(x)$, 可以断定 $P_{2k}(D)h(x)$ 在 $\frac{2N\pi}{\lambda}$ 周期区间内变号个数大于 $2N$, 但这是不可能的, 因为 $P_n(D)h(x)=\operatorname{sgn}\sin\lambda(x-x_1)-\frac{1}{\alpha}P_n(D)f(x)$, $\frac{1}{\alpha}\|P_n(D)f(x)\|_\infty<1$. 得到矛盾.

(b) $n>2k, k\geqslant 2$ 时. 此时 $P_n(\lambda)=0$ 有实根 $\lambda_1,\cdots,\lambda_{n-2k}$. 若 $\nu<2k$, 重复前款的论证到 $P_{2k}(D)h(x)$ 之后, 继续应用(关于一阶算子 $D-\lambda_j$ 的) 广义 Rolle 定理对 $(D-\lambda_1)P_{2k}(D)h(x),\cdots,P_n(D)h(x)$ 进行论证, 同样导致矛盾. 若 $2k\leqslant\nu<n-1$, 直接对 $P_\nu(D)h(x)$ 应用文献[11]的第 6 章定理 6.8.1 的证法, 同样导出矛盾. 至于 $\nu=n-1$ 的情形稍微改变一下前面证法的细节亦推出矛盾.

(c) $n=3,\nu=2$ 的情形归结于(b) 的 $\nu=n-1$. 结果一样, 细节从略.

总之, 我们已经证得定理 2 对 A_∞ 内的周期函数成立.

(2) 一般情形. 先证: 若 $f(x)\in A_\infty$, 则对 $r=2,\cdots,2k,2k+1,\cdots,n-1,(n\geqslant 3)$, $\|P_r(D)f\|_\infty<+\infty$. 只需对 $r=2$ 来证就够了. 根据 Шевалдин[9], 方程 $DP_n(D)y=0$ 存在着在 $\left(0,\frac{\pi}{\beta}\right)$ 上是 ECT 组的基本解组 $\{\phi_0(x),\cdots,\phi_n(x)\}$. 在 $\left(0,\frac{\pi}{\beta}\right)$ 内取 n 个点 $0<x_1<\cdots<x_n<\frac{\pi}{\beta}$. 作一广义多项式

$$Q(x)=C\begin{vmatrix}\phi_0(x)\phi_1(x)\cdots\phi_n(x)\\ \phi_0(x_1)\phi_1(x_1)\cdots\phi_n(x_1)\\ \cdots \qquad \cdots\\ \phi_0(x_n)\phi_1(x_n)\cdots\phi_n(x_n)\end{vmatrix}.$$

$C>0$ 稍后确定. $Q(x)$ 在 $\left(0,\frac{\pi}{\beta}\right)$ 内除 x_j 外别无零点,且每一 x_j 是其变号点. 选择 C 足够大使

(1) $\| P_n(D)Q\|_\infty > 1$.

(2) $\min\limits_{1\leqslant i\leqslant n-1}\left|Q\left(\frac{x_i+x_{i+1}}{2}\right)\right| > \|f\|_\infty$.

(3) 存在 $c,d,0<c<x_1,x_n<d<\frac{\pi}{\beta},|Q(c)|=|Q(d)|=\|f\|_\infty$.

可以看出 $Q(x)$ 在 $[c,d]$ 上,在 $\|f\|_\infty$ 与 $-\|f\|_\infty$ 间振荡 n 次. 注意 $P_2(D)=[D-(\alpha_1-\mathrm{i}\beta_1)][D-(\alpha_1+\mathrm{i}\beta_1)]$ 以及恒等式 $|[D-(\alpha_1+\mathrm{i}\beta_1)]z|=\mathrm{e}^{\alpha_1 x}\{[D(\mathrm{e}^{-\alpha_1 x}z)]^2+\beta_1^2(\mathrm{e}^{-\alpha_1 x}z)^2\}^{\frac{1}{2}}$, 此处 $z(x)$ 是任一可微函数. 置

$$m_1=\max_{c\leqslant x\leqslant d}|[D-(\alpha_1+\mathrm{i}\beta_1)]Q(x)|.$$

我们证 $\|[D-(\alpha_1+\mathrm{i}\beta_1)]f\|_\infty\leqslant m_1$. 若此事不成立,则存在 $x_0\in\mathbf{R}$ 有 $|[D-(\alpha_1+\mathrm{i}\beta_1)]f(x_0)|>m_1$. 在 $[c,d]$ 内部可找到 x' 使 $Q(x')=f(x_0)$. 有

$$|[D-(\alpha_1+\mathrm{i}\beta_1)]f(x+x_0-x')|_{x'}>|[D-(\alpha_1+\mathrm{i}\beta_1)]Q(x)|_{x'}.$$

此即

$$\begin{aligned}&\{[D(\mathrm{e}^{-\alpha_1 x}f(x+x_0-x'))]^2+\beta_1^2(\mathrm{e}^{-\alpha_1 x}f(x+x_0-x'))^2\}_{x'}^{1/2}\\&\quad>\{[D(\mathrm{e}^{-\alpha_1 x}Q(x))]^2+\beta_1^2(\mathrm{e}^{-\alpha_1 x}Q(x))^2\}_{x'}^{1/2}.\end{aligned}$$

由于 $f(x_0)=Q(x')$,上式给出

$$|D(\mathrm{e}^{-\alpha_1 x}f(x+x_0-x'))|_{x'}>|D(\mathrm{e}^{-\alpha_1 x}Q(x))|_{x'}.$$

由此,仿文献[11],知 $f(x+x_0-x')$ 与 $Q(x)$ 的图象在 $[c,d]$ 内至少有 $n+1$ 个交点,即 $h(x)=f(x+x_0-x')-Q(x)$ 有 $n+1$ 个零点. 由定理1,存在 $\xi\in(c,d)$ 使 $P_n(D)h(x)$ 在 ξ 点变号,只要 $x_1,\cdots,x_n$ 在开始取时,其相邻点的最大间距适当小. 但 $\|P_n(D)Q\|_\infty>1$ 而 $P_n(D)Q(x)$ 是常数(因 $DP_n(D)Q(x)\equiv 0$),得到矛盾. 所以 $\|[D-(\alpha_1+\mathrm{i}\beta_1)]f\|_\infty\leqslant m_1$. 如果代替 f 我们取 $[D-(\alpha_1+\mathrm{i}\beta_1)]f$,代替 $P_n(D)$ 取

$$[D-(\alpha_1-\mathrm{i}\beta_1)]\prod_{s=2}^{k}(D^2-2\alpha_s D+\alpha_s^2+\beta_s^2)\cdot\prod_{j=1}^{n-2k}(\lambda-\lambda_j).$$

分开该算子的实部虚部分别论证,然后合并起来便得

$$\| [D - (\alpha_1 - i\beta_1)][D - (\alpha_1 + i\beta_1)]f \|_\infty < +\infty.$$

此即 $\| P_2(D)f \|_\infty < +\infty$. 借助数学归纳法便得 $\| P_r(D)f \|_\infty < +\infty$ 对 $r = 2, \cdots, 2k, 2k+1, \cdots, n-1$ 成立. 往下仿照 Cavaretta[4] 或文献 [11](第 263 ~ 264 页) 的处置方法即可完成定理 2 之证. 细节省略.

参考文献

[1] Landau, E. Proc. LMS, 1913, 13(2): 43 ~ 49

[2] Kolmogorov, A. Amer. MS Transactions, 1962, 1 ~ 2: 233 ~ 243

[3] Schoenberg, I. AMM, 1973, 80(2): 121 ~ 158

[4] Cavaretta, A. ibid., 1974, 81(5): 480 ~ 486

[5] Sharma, A. and Tzimbalario, J. Illinois JM, 1976, (20): 443 ~ 455

[6] 黄达人、王建忠. 数学学报. 1983, 26(6): 715 ~ 722

[7] Beckenbach, Bellman. *Inequalities*. New York: Springer-Verlag, 1961

[8] Крейн М. Докл. АН СССР, 1938, 18(4) ~ 18(5): 245 ~ 249

[9] Шевалдин В. Матем. Зам., 1981, 29(4): 603 ~ 622

[10] Шевалбин В. Матем. Зам., 1983, 33(5): 735 ~ 744

[11] Holland, A. and Sahney, B. *The General Problem of Approximation and Spline Functions*. New York: McGraw-Hill, 1979. 263 ~ 264

Approx. Theory & Its Appl.
Mar., 1990, 6:1

一个光滑函数类上微分算子的最优恢复

Optimal Recovery of Differential Operators on a Class of Smooth Functions*

Abstract

In this note, we study the optimal recovery of differential operators on a class of smooth functions and get three main theorems.

1. Introduction

Given $n \geqslant 2, P_n(\lambda) = \prod_{i=1}^{n} (\lambda - \lambda_j), \lambda_j \in \mathbf{R}, \mathrm{f} \in \mathrm{L}_\infty^{(n)}(\mathrm{P}_n(\mathrm{D}))$ if

(1) $f \in C^{n-1}(\mathbf{R}) \cap L_\infty(\mathbf{R})$,

(2) $f^{(n-1)}$ is a bsolutely continuous on any finite interval,

(3) $\| P_n(D) f \|_\infty < +\infty \left(D = \frac{\mathrm{d}}{\mathrm{d}x} \right)$.

Denote $K = W_\infty^{(n)}(P_n(D), \gamma) = \{f : f \in L_\infty^{(n)}(P_n(D)), \| P_n(D) f \|_\infty \leqslant \gamma\}$, where $\gamma > 0$ is any fixed number. We write $X_1 = L_\infty^{(n)}(P_n(D)), X_3 = \widetilde{C}(\mathbf{R}), \widetilde{C}(\mathbf{R})$ is the set of all bounded and uniformly continuous functions defined on $\mathbf{R}$, it is clear that $\widetilde{C}(\mathbf{R})$ is a linear space endowed with norm $\| f \| = \sup | f(x) |$, and that $L_\infty^{(n)}(P_n(D)) \subset \widetilde{C}(\mathbf{R})$. (Cf. [1]). We consider the optimal recovery problem for the functional Sf

* Received June 25, 1988.
Supported by National Natural Science Foundation of China.

$= f'(0)$ on K by using the information $If = f \in \tilde{C}(\mathbf{R})$, where the value of f is allowed to have a computation error within $\varepsilon > 0$. This means that instead of f we may take any $g \in \tilde{C}(\mathbf{R})$ such that $\| f - g \| \leqslant \varepsilon$. Write $S = \{f : f \in \tilde{C}(\mathbf{R}), \| f \| \leqslant 1\}$. Any mapping $\varphi : IK + \varepsilon S \to \mathbf{R}$ may serve as an approximating formula for $Sf = f'(0)$. Then the global error is

$$E_\varphi(\gamma, \varepsilon) = \sup\{ | f'(0) - \varphi(g) | : f \in K, \| f - g \| \leqslant \varepsilon, g \in \tilde{C}(\mathbf{R})\}.$$

The intrinsic error is

$$E(\gamma, \varepsilon) = \inf_\varphi E_\varphi(\gamma, \varepsilon) \tag{1}$$

We call some φ^0 optimal algorithm if $E_{\varphi^0}(\gamma, \varepsilon) = E(\gamma, \varepsilon)$. There are two main problems to be considered: it is required to obtain the exact expression for $E(\gamma, \varepsilon)$ and to construct optimal algorithms(formulas). The linear continuous mapping to be optimal is especially interesting. Several papers[3]-[4] deal with some special cases of this problem. In [3], C. A. Micchelli considered a special case, where the differential operator

$$P_n(D) = \prod_{j=1}^m (D^2 - \lambda_j^2), \ (n = 2m, \lambda_j \geqslant 0)$$

is formally self-adjoint. In this paper we consider this problem in its general case by a method quite different from that used in [3].

Write

$$E_0(\gamma, \varepsilon) = \sup\{ | f'(0) | : f \in K, \| f \| \leqslant \varepsilon\}, \tag{2}$$

$$\mathrm{d}(\gamma, \varepsilon) = \inf_{\alpha \in BV(\mathbf{R})} \{ \sup_{f \in K} | f'(0) - \int_{-\infty}^{+\infty} f(x) \mathrm{d}\alpha | + \varepsilon V_{\mathbf{R}}(\alpha)\}, \tag{3}$$

where $BV(\mathbf{R})$ is the calss of all functions of bounded-variation defined on $\mathbf{R}$. It coincides with the class of all linear continuous functienals on $\tilde{C}(\mathbf{R})$. $V_{\mathbf{R}}(\alpha)$ denotes the total variation of α on $\mathbf{R}$.

The following theorem is cited from Micchelli and Rivlin[4]

Theorem 1

$$E_0(\gamma, \varepsilon) = E(\gamma, \varepsilon) = \mathrm{d}(\gamma, \varepsilon). \tag{4}$$

There exists $\alpha^* \in BV(\mathbf{R})$ *such that*

$$\mathrm{d}(\gamma, \varepsilon) = \sup_{f \in K} \left\{ | f'(0) - \int_{-\infty}^{+\infty} f(x) \mathrm{d}\alpha^* | + \varepsilon V_{\mathbf{R}}(\alpha^*) \right\}.$$

If for some $f_0 \in K$, $\| f_0 \| \leqslant \varepsilon$ *holds* $E_0(\gamma, \varepsilon) = f'(0)$, *then*

(i) $$\varepsilon V_{\mathbf{R}}(\alpha^*) = \int_{-\infty}^{+\infty} f_0(x) \mathrm{d}\alpha^*, \tag{5}$$

(ii) $$\sup_{f \in K} \left| f'(0) - \int_{-\infty}^{+\infty} f(x) \mathrm{d}\alpha^* \right| = f_0'(0) - \int_{-\infty}^{+\infty} f_0'(x) \mathrm{d}\alpha^*. \tag{6}$$

α^* *identifies an optimal algorithm which is a linear continuous functional.*

2. Estimation of $E_0(\gamma, \varepsilon)$

We introduce the standard function (cf. [1])

$$\Phi_{n,\lambda}(x) = \frac{2}{\pi \mathrm{i}} \sum_{\nu=-\infty}^{+\infty} \frac{\mathrm{e}^{\mathrm{i}(2\nu+1)\lambda x}}{(2\nu+1) P_n[(2\nu+1)\lambda \mathrm{i}]}. \tag{7}$$

Let $r \geqslant 0$ be an integer, $P_r(\lambda)$ is a $r-$degree factor of $P_n(\lambda)$, $P_0(\lambda) = 1$. Denote

$$\Gamma_{n,r}(\lambda) = \| P_r(D) \Phi_{n,\lambda}(\cdot) \|_\infty, \; r = 0, 1, \cdots, n. \tag{8}$$

It is easy to verify

$$\Gamma_{n,0}(0+) = \begin{cases} +\infty & \text{if } P_n(0) = 0, \\ | P_n(0) |^{-1}, & \text{if } P_0 \neq 0 \end{cases} \tag{9}$$

and

$$\lim_{\lambda \to +\infty} \Gamma_{n,0}(\lambda) = 0. \tag{10}$$

Lemma 1 $\Gamma_{n,0}(\lambda)$ *is strictly decreasing and continuous on* $\mathbf{R}_+$.

Proof Put $P_n^*(\lambda) = \prod_{j=1}^{n} (\lambda + \lambda_1)$, and

$$Q_{2n}(D) = P_n^*(D) P_n(D) = D^\sigma \prod_{j=1}^{l} (D^2 - \lambda_j^2),$$

where $\lambda_1, \cdots, \lambda_l > 0$, $\lambda_{l+1} = \cdots = \lambda_n = 0$, $2n = 2l + \sigma$, $\sigma \geqslant 0$ is even. Then standard function relating to $Q_{2n}(D)$ may be given by

$$\Phi_{2n,\lambda}^*(x) = \frac{4(-1)^l}{\pi \lambda^l} \sum_{\nu=0}^{\infty} \frac{\cos\left[(2\nu+1)\lambda x - \frac{\sigma+1}{2}\pi\right]}{(2\nu+1)^{\sigma+1} \prod_{l=1}^{l} [(2\nu+1)^2 \lambda^2 + \lambda_j^2]} \tag{11}$$

Denote

$$\Gamma_{2n,0}^*(\lambda) = \| \Phi_{2n,\lambda}^*(\cdot) \| =$$

$$\frac{4}{\pi\lambda^{\sigma}}\sum_{k=0}^{\infty}\frac{(-1)^k}{(2k+1)^{\sigma+1}\prod_{l=1}^{l}[(2k+1)^2\lambda^2+\lambda_j^2]} \tag{12}$$

$\Gamma_{2n,0}^{*}(\lambda)$ is strictly decreasing, so that for any $\lambda_1, \lambda_2, 0<\lambda_1<\lambda_2$ we have $\|\Phi_{2n,\lambda_2}^{*}(\cdot)\| < \|\Phi_{2n,\lambda_1}^{*}(\cdot)\|$. By Landau − Kolmogorov inequality[1] we get

$$\|P_n^{*}(D)\Phi_{2n,\lambda_2}^{*}(\cdot)\| \leqslant \|P_n^{*}(D)\Phi_{2n,\lambda_1}^{*}(\cdot)\|,$$

i.e., $\Gamma_{n,0}(\lambda_2) \leqslant \Gamma_{n,0}(\lambda_1)$. We claim that $\Gamma_{n,0}(\lambda_2) < \Gamma_{n,0}(\lambda_1)$. Suppose $\Gamma_{n,0}(\lambda_2) = \Gamma_{n,0}(\lambda_1)$, then by a theorem due to Chen Han − lin[6] there exists a number c for which $\Phi_{n,\lambda_2}(x) = \Phi_{n,\lambda_1}(x+c)$. So we must have $\lambda_1 = \lambda_2$, which is a contradiction. Therefore $\Gamma_{n,0}(\lambda_2) < \Gamma_{n,0}(\lambda_1)$.

To prove the continuity of $\Gamma_{n,0}(\lambda)$ we may proceed as follows. Write $\Phi_n(x,\lambda) = \Phi_{n,\lambda}(x)$. This function is continuous at cach point $(x,\lambda) \in (-\infty,\infty)\times(0,\infty)$, and moreover, the equation

$$\frac{\partial}{\partial x}\Phi_n(x,\lambda) \equiv H_n(x,\lambda) = \frac{2\lambda}{\pi}\sum_{\nu=-\infty}^{+\infty}\frac{e^{i(2\nu+1)\lambda x}}{P_n[(2\nu+1)\lambda i]} = 0$$

for every $\lambda > 0$ has a single simple zero $x = x(\lambda), 0 \leqslant x(\lambda) < \frac{\pi}{\lambda}$, (cf. [1], p. 16, Lemma 3.3).

Let (x_0, λ_0) be a fixed point, where $x_0 = x(\lambda_0)$. Then $H_n(x_0,\lambda_0) = 0$ and $\frac{\partial}{\partial x}H_n(x_0,\lambda_0) \neq 0$. So by the implicit function theorem there exists $x = x(\lambda)$ satisfying

(1) $x(\lambda_0) = x_0$,

(2) $H_n(x(\lambda),\lambda) = 0$ in some neighbourhood (α,β) of λ_0 where $x(\lambda)$ is continuous.

We note that $x(\lambda)$ is an extremal point of $\Phi_{n,\lambda}(x)$, i.e.

$$\Gamma_{n,0}(\lambda) = \|\Phi_{n,\lambda}(\cdot)\| = |\Phi_n(x(\lambda),\lambda)|.$$

Therefore

$$\lim_{\lambda\to\lambda_0}\Gamma_{n,0}(\lambda) = \lim_{\lambda\to\lambda_0}|\Phi_n(x(\lambda),\lambda)| = |\Phi_n(x_0,\lambda_0)| = \Gamma_{n,0}(\lambda_0).$$

The proof is completed.

Lemma 2 *If* $P_n(0) = 0$, *or* $P_n(0) \neq 0$ *and* $\varepsilon' = \varepsilon\gamma^{-1} <$

$|P_n(0)|^{-1}$, *then*

$$E_0(\gamma,\varepsilon)=\gamma\|\Phi'_{n\lambda'}(\cdot)\|,\quad(\Gamma_{n,0}(\lambda')=\varepsilon'). \tag{13}$$

Proof In case that $P_n(0)=0$, or $P_n(0)\neq 0$ and $\varepsilon'<|P_n(0)|^{-1}$, by lemma 1 there exists an unique number $\lambda'>0$ such that $\varepsilon'=\Gamma_{n,0}(\lambda')$. Then we have

$$\begin{aligned}&E_0(\gamma,\varepsilon)\\&=\gamma\cdot\sup\{|f'(0)|:\|f\|\leqslant\varepsilon',\|P_n(D)f\|_\infty\leqslant 1\}\\&=\gamma\cdot\sup\{|f'(0)|:\|f\|\leqslant\Gamma_{n,0}(\lambda'),\|P_n(D)f\|_\infty\leqslant 1\}.\end{aligned}$$

On the basis of comparison theorem of Kolmogorov type for $W_\infty^{(n)}(P_n(D))$ ([1], P.17. Th.2) we have

$$\begin{aligned}\sup\{|f'(0)|:&\|f\|\leqslant\Gamma_{n,0}(\lambda'),\\&\|P_n(D)f\|_\infty\leqslant 1\}=\sup_{\{a,b\}}\sup\{|f'(0)|:f\in S_{a,b}(\lambda')\},\end{aligned}$$

where

$$\begin{aligned}S_{a,b}(\lambda')=\{f:&\|f\|\leqslant\Gamma_{n,0}(\lambda'),\|P_n(D)f\|_\infty\leqslant 1,\\&f(0)=\Phi_{n,\lambda'}(a)=\Phi_{n,\lambda'}(b)\}.\end{aligned}$$

On acount of (cf. [1])

$$\sup_{f\in S_{a,b}(\lambda')}|f'(0)|\leqslant\max\{|\Phi_{n,\lambda'}(a)|,|\Phi_{n,\lambda'}(b)|\},$$

so we have

$$\sup_{f\in S_{a,b}(\lambda')}|f'(0)|\leqslant\|\Phi_{n,\lambda'}(\cdot)\|_\infty.$$

Consequently

$$E_0(\gamma,\varepsilon)\leqslant\gamma\|\Phi'_{n,\lambda'}(\cdot)\|_\infty.$$

Now we choose a point x_0 such that $\Phi'_{n,\lambda'}(x_0)=\|\Phi'_{n,\lambda'}(\cdot)\|_\infty$. Then for the function $\Phi^*_{n,\lambda'}(x)=\Phi_{n,\lambda'}(x+x_0)$ it holds $\|\Phi^*_{n,\lambda'}(\cdot)\|_\infty=\Gamma_{n,0}(\lambda')$ and $\|P_n(D)\Phi^*_{n,\lambda'}(\cdot)\|_\infty\leqslant 1$, so that

$$E_0(\gamma,\varepsilon)\geqslant\gamma\|\Phi^{*\prime}_{n,\lambda'}(0)\|=\gamma\Phi'_{n,\lambda'}(x_0)=\gamma\|\Phi'_{n,\lambda'}(\cdot)\|_\infty.$$

Lemma 2 is proved.

The proof yields an extremal function $\gamma\Phi^*_{n,\lambda'}(x)$ attaining $E_0(\gamma,\varepsilon)$.

Lemma 3 *If* $P_n(0)\neq 0$ *and* $\varepsilon'=\gamma^{-1}\varepsilon\geqslant|P_n(0)|^{-1}$, *then*

$$E_0(\gamma,\varepsilon)=E_0(\gamma,\varepsilon_0)=\gamma\int_{-\infty}^{+\infty}|G'_n(\tau)|\,d\tau, \tag{14}$$

where

$$G_n(x) = (2\pi)^{-1}\int_{-\infty}^{+\infty} \frac{e^{ixu}}{P_n(iu)}du, \gamma^{-1}\varepsilon_0 = |P_n(D)|^{-1}. \quad (15)$$

Proof By Hirschman and Widder[6], any $f \in W_\infty^{(n)}(P_n(D))$ may be represented by

$$f(x) = \int_{-\infty}^{+\infty} G_n(x - t)\varphi(t)dt, \quad (16)$$

$\varphi(t) = P_n(D)f(t)$, a.e. Now by a theorem of S. Karlin ([7], Th. 13 – 1, (ii)), for $\varepsilon' = \gamma^{-1}\varepsilon \geqslant |P_n(0)|^{-1}$ we have

$$\sup\left\{|f'(0)| = \int_{-\infty}^{+\infty} |G'_n(\tau)|d\tau : \| f \| \leqslant \varepsilon', \| P_n(D)f \|_\infty \leqslant 1\right\}.$$

Therefore

$$E_0(\gamma, \varepsilon) = \gamma \cdot \sup\{|f'(0)| : \| f \| \leqslant \varepsilon',$$

$$\| P_n(D)f \|_\infty \leqslant 1\} = \gamma\int_{-\infty}^{+\infty} |G'_n(\tau)|d\tau.$$

An extremal function is given by

$$f_{0,\gamma}(x) = \gamma\int_{-\infty}^{+\infty} G_n(x - t)\varphi_0(t)dt, \varphi_0(t) = \operatorname{sgn} G'_n(-t). \quad (17)$$

It is worthy to point out that in this case $f_{0,\gamma}$ is extremal for all $\varepsilon \geqslant \varepsilon_0$. In fact, by (16), $f \in K = W_\infty^{(n)}(P_n(D), \gamma)$ we have

$$f(x) = \gamma\int_{-\infty}^{+\infty} G_n(x - t)\varphi(t)dt, \| \varphi \|_\infty \leqslant 1,$$

and

$$\| f \| \leqslant \gamma\int_{-\infty}^{+\infty} G_n(t)dt = \gamma \cdot |P_n(0)|^{-1} = \varepsilon_0.$$

Therefore for $\varepsilon \geqslant \varepsilon_0$ the class $\{f : \| f \| \leqslant \varepsilon, \| P_n(D)f \|_\infty \leqslant \gamma\} = W_\infty^{(n)}(P_n(D), \gamma)$ is independent of ε.

In what follows we characterize $f_{0,\gamma}$. By [6], (cf p.91, Th.5.1), $G_n(t) \geqslant 0$ and $G'_n(t)$ has just one change of sign.

Let t_0 be the point where $G'_n(t)$ changes sign. We have

$\operatorname{sgn} G'_n(t) = 1$, for $t < t_0$; 0, for $t = t_0$; and -1, for $t > t_0$.

And also

$$\int_{-\infty}^{+\infty} G_n(t)\mathrm{d}t = |P_n(0)|^{-1}, \quad \varphi_0(t) = \operatorname{sgn} G'_n(-t),$$

so that

$$\int_{-\infty}^{+\infty} G_n(x-t)\varphi_0(t)\mathrm{d}t = \int_{-\infty}^{+\infty} G_n(t)\varphi_0(x-t)\mathrm{d}t$$

$$= \int_{-\infty}^{+\infty} G_n(t)\operatorname{sgn} G'_n(t-x)\mathrm{d}t = \int_{-\infty}^{x+t_0} G_n(t)\mathrm{d}t - \int_{x+t_0}^{+\infty} G_n(t)\mathrm{d}t.$$

From this we derive that $f_{0,\gamma}(x) \uparrow$ and

$$\lim_{x\to+\infty} f_{0,\gamma}(x) = +\gamma|P_n(0)|^{-1}, \lim_{x\to-\infty} f_{0,\gamma}(x) = -\gamma|P_n(0)|^{-1}.$$

Consequently $\| f_{0,\gamma} \| = \varepsilon_0$. For every $\varepsilon \geqslant \varepsilon_0, E_0(\gamma,\varepsilon) = E_0(\gamma,\varepsilon_0) = f'_{0,\gamma}(0)$, which is independent of ε.

To sum up, we have proved

Theorem 2

(1) *If* $P_n(0) = 0$ *or* $P_n(0) \neq 0$ *and* $\gamma^{-1}\varepsilon = \varepsilon' < |P_n(0)|^{-1}$, *Then*

$$E_0(\gamma,\varepsilon) = E(\gamma,\varepsilon) = \mathrm{d}(\gamma,\varepsilon) = \gamma \| \Phi'_{n,\lambda'}(\cdot) \|_{\infty},$$

$\gamma\Phi_{n,\lambda'}(x+x_0)$ *is an extremal function attaining* $E_0(\gamma,\varepsilon)$.

(2) *If* $P_n(0) \neq 0$ *but* $\gamma^{-1}\varepsilon \geqslant |P_n(0)|^{-1}$, *then*

$$E_0(\gamma,\varepsilon) = E(\gamma,\varepsilon) = \mathrm{d}(\gamma,\varepsilon) = \gamma\int_{-\infty}^{+\infty} |G_n(\tau)|\mathrm{d}\tau.$$

$f_{0,\gamma}(x)$ *is an extremal function attaining* $E_0(\gamma,\varepsilon)$. *It is independent of* ε.

3. Linear optimal algorithm

(1) Case I. $P_0(0) = 0$ or $P_n(0) \neq 0$ and $\gamma^{-1}\varepsilon < |P_n(0)|^{-1}$. By Theorem 2, $\gamma\Phi_{n,\lambda'}(x+x_0)$ is an extremal function which attains $E_0(\gamma,\varepsilon)$. By Theorem 1 there exists $\alpha^* \in BV(\mathbf{R})$ such that

(i) $$\gamma^{-1}\varepsilon V_{\mathbf{R}}(\alpha^*) = \int_{-\infty}^{+\infty} \Phi_{n,\lambda'}(x+x_0)\mathrm{d}\alpha^*. \tag{18}$$

(ii) $\forall\ f \in K$

$$\gamma^{-1}\left| f'(0) - \int_{-\infty}^{+\infty} f(x)\mathrm{d}\alpha^* \right| \leqslant \Phi'_{n,\lambda'}(x_0) - \int_{-\infty}^{+\infty} \Phi_{n,\lambda'}(x+x_0)\mathrm{d}\alpha^*. \tag{19}$$

Condition (i) is equivalent to

$$\int_{-\infty}^{+\infty}\Phi_{n,\lambda'}(x+x_0)\mathrm{d}\alpha^* = \|\Phi'_{n,\lambda'}\|\int_{-\infty}^{+\infty}|\mathrm{d}\alpha^*|. \tag{20}$$

Denote the set of all extremal points of $\Phi_{n,\lambda'}(x+x_0)$ by $\left\{\alpha_0+\frac{k\pi}{\lambda'}\right\}_{k=-\infty}^{+\infty}$, where we convince $\Phi_{n,\lambda'}(\alpha_0+x_0) = \|\Phi_{n,\lambda'}(\cdot)\|$. Then

$$\Phi_{n,\lambda'}\left(x_0+\alpha_0+\frac{k\pi}{\lambda'}\right) = (-1)^k\Phi_{n,\lambda'}(x_0+\alpha_0).$$

Write $\Delta_k = \left[\alpha_0+x_0+\frac{k\pi}{\lambda'}, \alpha_0+x_0+\frac{k+1}{\lambda'}\pi\right]$, $(k = 0,1,2,\cdots)$, in which $\Phi_{n,\lambda'}(x)$ is strictly monotone. Let $\Delta = [\alpha,\beta]\subseteq\Delta_k$. From (20) it is easily seen that

$$\left|\int_\alpha^\beta\Phi_{n,\lambda'}(x+x_0)\mathrm{d}\alpha^*\right| = \|\Phi_{n,\lambda'}\|\cdot\bigvee_\alpha^\beta(\alpha^*). \tag{21}$$

If $\bigvee_\alpha^\beta(\alpha^*) > 0$, then we must have

$$\left|\int_\alpha^\beta\Phi_{n,\lambda'}(x+x_0)\mathrm{d}\alpha^*\right| \leqslant \max_{\alpha\leqslant x\leqslant\beta}|\Phi_{n,\lambda'}(x+x_0)|\cdot\bigvee_\alpha^\beta(\alpha^*) < \|\Phi_{n,\lambda'}\|\bigvee_\alpha^\beta(\alpha^*),$$

which contradicts with (21). So we have $\bigvee_\alpha^\beta(\alpha^*) = 0$, which entails $\alpha^*(x) = a_k$, $\forall x\in\left(\alpha_0+x_0+\frac{k\pi}{\lambda'}, \alpha_0+x_0+\frac{(k+1)\pi}{\lambda'}\right)$, $(k = 0,1,2\cdots)$ where a_k are some constants. Moreover, from (20) we also derive

$$(-1)^k(a_k-a_{k-1})\geqslant 0,\quad k = 0,1,2,\cdots \tag{22}$$

Write $c_k = a_k - a_{k-1}$. We get

$$\bigvee_{\mathbf{R}}(\alpha^*) = \sum|c_k| < +\infty, \tag{23}$$

$$\int_{-\infty}^{+\infty}\Phi_{n,\lambda'}(x+x_0)\mathrm{d}\alpha^* = \gamma^{-1}\varepsilon\sum|c_k|. \tag{24}$$

We claim that the linear functional

$$S(g) = \sum_{k=-\infty}^{+\infty}c_k g\left(\alpha_0+x_0+\frac{k\pi}{\lambda'}\right), g\in\widetilde{C}(\mathbf{R}) \tag{25}$$

is an optimal algorithm. In fact, writing $S(g) = \int_{-\infty}^{+\infty}g(x)\mathrm{d}\alpha^*$, by (19) we have

$\sup\{\gamma^{-1}|f'(0)-S(f)|: \|f\|\leqslant\varepsilon';$

$\| P_n(D)f \|_\infty \leqslant 1\} = \| \Phi'_{n,\lambda'}(\cdot) \|_\infty - S(\Phi_{n,\lambda'})$. Thus for any $g \in \widetilde{C}(\mathbf{R})$, $\| f - g \| \leqslant \varepsilon$,

$$| f'(0) - S(g) | \leqslant | f'(0) - S(f) | + \| f - g \| \cdot \| V_{\mathbf{R}}(\alpha^*) \|$$
$$\leqslant \gamma \| \Phi'_{n,\lambda'} \|_\infty - \varepsilon \sum | c_k | + \varepsilon V_{\mathbf{R}}(\alpha^*) = E_0(\gamma, \varepsilon).$$

(2) Case Ⅱ. $P_n(0) \neq 0$ and $\gamma^{-1}\varepsilon \geqslant | P_n(0) |^{-1}$. By Theorem 1 and Theorem 2. (ii), there exists $\alpha^* \in BV(\mathbf{R})$ such that

(i) $\| f_{0,\gamma} \| \cdot V_{\mathbf{R}}(\alpha^*) = \int_{-\infty}^{+\infty} f_{0,\gamma}(x) \mathrm{d}\alpha^*$. (26)

(ii) $\forall f \in K$

$$| f'(0) - \int_{-\infty}^{+\infty} f(x) \mathrm{d}\alpha^* | \leqslant f'_{0,\gamma}(0) - \int_{-\infty}^{+\infty} f_{0,\gamma}(x) \mathrm{d}\alpha^*. \tag{27}$$

$f_{0,\gamma}(x)$ is strict monotone, so that $\bigvee_{\alpha}^{\beta}(\alpha^*) = 0$ for every finite interval $[\alpha, \beta]$. Thus α^* yields the zero functional $s(g) = 0$ which is optimal in this case. This fact may also be verified directly. From (16) $\forall\ f \in W_\infty^{(n)}(P_n(D), \gamma) = K$, $f = G_n * \varphi$, $\| \varphi \|_\infty \leqslant \gamma$. We have

$$\| f \|_\infty \leqslant \gamma \int_{-\infty}^{+\infty} | G_n(x - t) | \, \mathrm{d}t = \gamma \, | P_n(0) |^{-1} = \varepsilon_0.$$

So that for $S(g) = 0$

$$\sup\{ | f'(0) - S(g) | : f \in K ; \| f - s \| \leqslant \varepsilon \}$$
$$= \sup\{ | f'(0) | : f \in K \} = E_0(\gamma, \varepsilon_0) = f_{0,\gamma}'(0).$$

i.e., $s(g)$ is optimal.

To sum up, we have proved

Theorem 3

(1) *If $P_n(0) = 0$ or $P_n(0) \neq 0$ and $\gamma^{-1}\varepsilon < | P_n(0) |^{-1}$, then the functional $s(g)$ defined by (25) yields an optimal algorithm.*

(2) *If $P_n(0) \neq 0$, $\gamma^{-1}\varepsilon \geqslant | P_n(0) |^{-1}$, then the zero functional $s(g) = 0$ yields an optimal algorithm.*

Remark. C. A. Micchelli[3] for a self-adjoint operator $P_{2m}(D)$ constructed the linear optimal approximating expression (algorithm) of $s(f) = f'(0)$ with remainder formulae

$$f'(0) - \sum c_\nu f\left(\nu - \frac{1}{2}\right) = -\int_{-\infty}^{+\infty} H\left(x + \frac{1}{2}\right)(P_{2m}(D)f(x))\mathrm{d}x. \tag{28}$$

He used the $\mathscr{L}$-spline technique to construct $H(x)$, analyzed its changes of sign, and on this basis obtained the exact estimation for $E_0(\gamma, \varepsilon)$ (in case $P_n(0) = 0$ or $P_n(0) \neq 0$ and $\gamma^{-1}\varepsilon < | P_n(0) |^{-1}$).

Our method of approach is quite different from that in [3]. We obtained the exact estimation for $E_0(\gamma, \varepsilon)$ by using the comparison theorem of Kolmogorov on the function class $W_\infty^{(n)}(P_n(D))$. This makes it unnecessary to construct the remainder formula (28). But (28) has its independent meaning. The meaningful problem to construct the remainder formula (28) for **arbitrary** linear differential operator $P_n(D) = \prod_{j=1}^{n}(D - \lambda_j)$ is still left open.

References

[1] Sun Yongsheng. Some Extremal Problems on a Class of Differentiable Functions. J. Beijing Nor-mal Univ., Series of Nat. Sci. 1984, (4): 11 ~ 20

[2] Shoenberg, I. J. The Elementary Cases of Landau Problem of Inequalities Between Derivatives. AMM, 1973, (80): 121 ~ 158

[3] Micchelli, C. A. On an Optimal Method for the Numerical Differentiation of Smooth Functions. Jour. Approx. Theory. 1976, 18(3): 189 ~ 204

[4] Micchelli, C. A. & Rivlin, T. J. A Survey of Optimal Recovery In: Micchelli edited *Optimal Estimation in Approximation Theory*. New York: Plenum Press(1977), 1 ~ 68

[5] Chen Hanlin & Hu Yingsheng. On the Uniqueness of the Extremal Functions of Landau Type Problem for the Differential operator $L_{n+1}(D) = \prod_{j=0}^{n}(D - t_j)$, Research Memorandum. Institute of Math., 1985, Acad. Sinica No. 9

[6] Hirchman, I. Widder, H. *The Convolution Transform*. N. Y.: Princeton Math. Series, 1955

[7] Karlin, S. Oscillatory Perfect Splines and Related Extremal Problems. *Studies in Spline functions and Approximation Theory*. New York: Acad. Press, 1976. 371

Apporn. Theory & its Appl.
May, 1992, 8:1

一个线性微分算子的 Hardy-Littlewood-Polya 不等式及有关优化问题*

Hardy-Littlewood-Polya Inequality for a Linear Differential Operator and Some Related Optimal Problems

Abstract

In this paper a generalized version of the classical Hardy-Littlewood-Polya inequality is given. Furthermore, the Stechkin's problem for a linear differential operator is solved in $L_2(\mathbf{R})$, and the optimal recovery problem for such differential operator is considered.

1. Introduction

Given $n \geqslant 2, P_n(\lambda) = \prod_{j=1}^{n}(\lambda - \lambda_j), \lambda_j \in \mathbf{R}, i = 1, \cdots, n$. Let $P_r(\lambda)$ be a factor of $P_n(\lambda)$ of order $r, 0 \leqslant r \leqslant n (P_0(\lambda) = 1)$. Define a class of smooth functions as follows.

$L_p^n = \{f \in L_p(\mathbf{R}) \mid f^{(n-1)}$ is absolutely continuous on any finite interval and $f^{(n)} \in L_p(\mathbf{R})\}$,

$$W_p(P_n(D)) = \{f \in L_p^n \mid \| P_n(D) f \|_p \leqslant 1\},$$

* Received Dec, 21, 1990, Revised Mar. 28, 1991.
Supported by the National Fund of Natural Sciences.

* 此文与陈迪荣合作.

Where $p \in [1, \infty]$, $\|\cdot\|_p$ is the usual $L_p(\mathbf{R})$-norm, $D = \frac{d}{dx}$. In this paper we mainly concern about $p = 2$ and denote $\|\cdot\|_2$ by $\|\cdot\|$ for simplicity.

Among the $L_p(\mathbf{R})$-norm of $f \in L_p^n$ and its derivatives $f^{(k)}, f^{(n)} (1 \leqslant k \leqslant n-1)$ holds inequality which is known as the Landau-kolmogorov inequality (for $p = 2$, the Hardy-Littlewood-Polya inequality). When $p = 1, 2$ and $+\infty$ the exact constants in these inequalities are known. Being closely related to many optimal problems in approximation theory, (see [1], [2]) the Landau-Kolmogorov inequality has been studied and extended in various directions (see [5], [6]). In this note we give a generalized version of the Hardy-Littlewood-Polya inequality (H-L-P inequality), and in connection with it we consider several related optimal problems.

(1) *The optimal smoothing problem* for the class of functions $W_2(P_r(D))$. It is to be considered the best $L_2(\mathbf{R})$-approximation of $W_2(P_r(D))$by $N \cdot W_2(P_n(D))$, where $N > 0$ is a fixed number, and

$$N \cdot W_2(P_n(D)) = \{Nf \mid f \in W_2(P_n(D))\}.$$

Set

$$F(N) := \sup_{f \in W_2(P_r(D))} \left\{ \inf_{g \in N \cdot W_2(P_n(D))} \| f - g \| \right\},$$

which is the best deviation to be calculated.

(2) *The optimal linear smoothing problem* for $W_2(P_r(D))$. Let $N > 0$ be a fixed number, and $\mathscr{M}(N)$ be the set of all linear continuous mappings S from $L_2^r \to L_2^n$ such that $\| P_n(D)(Sf) \| \leqslant N \| P_r(D) f \|$ for all $f \in L_2^r$. Set

$$G(N) := \inf_{S \in \mathscr{M}(N)} \left\{ \sup_{f \in W_2(P_r(D))} \| f - S(f) \| \right\},$$

which is the best linear deviation to be calculated.

(3) *The Stechkin problem*. It is to be considered the optimal approximation for the differential operator $P_r(D)$ by the set of bounded linear operators T from $L_2 \to L_2$ such that the operator-norm of T is bounded by N. Set

$$E(N):=\inf_{\|T\|\leqslant N}\left\{\sup_{f\in W_2(P_n(D))}\|P_r(D)f-Tf\|\right\}.$$

It is required to get the exact expression of $E(N)$ and to construct, if possible, optimal linear operators, i.e., to identify T_0, $\|T_0\|\leqslant N$ such that

$$E(N)=\sup_{f\in W_2(P_n(D))}\|P_r(D)f-T_0f\|.$$

(4) *The optimal recovery problem* for $P_r(D)$ on the class of functions $W_2(P_n(D))$ in $L_2(\mathbf{R})$. A detailed formulation of this problem will be given in §4.

A special case $P_n(\lambda)=\lambda^n$ of all these problems has been thoroughly investigated by Soviet mathematicians (see[1 ~ 4]).

2. A generalized version of H-L-P inequality

Theorem 2.1. *Denote* $\varphi_r(t)=|P_r(\mathrm{i}t)|$, $(1\leqslant r\leqslant n)$, $\mathrm{i}=\sqrt{-1}$. *Then for any* $f\in L_2^n$, $f\neq 0$ *we have*

$$\|P_r(D)f\|\leqslant\|f\|\,\varphi_r\circ\varphi_n^{-1}\left(\frac{\|P_n(D)f\|}{\|f\|}\right).\tag{2.1}$$

The inequality is sharp in the sense that, $\forall\,\varepsilon>0$, $\exists\, f\in L_2^n$, $f\neq 0$ *such that*

$$\|P_r(D)f\|>\|f\|\,\varphi_r\circ\varphi_n^{-1}\left(\frac{\|P_n(D)f\|}{\|f\|}\right)-\varepsilon,\tag{2.2}$$

($P_r(t)$ *fixed*).

Proof Put $\psi_j(t)=\varphi_j^2(t)$, $j=r,n$, and $\hat{\psi}_r=\psi_n/\psi_r$ is even on $\mathbf{R}$. Its inverse function ψ_j^{-1} is in $C^\infty(0,\infty)$. Write

$$\tilde{f}(t)=\frac{1}{\sqrt{2\pi}}\int_{\mathbf{R}}f(x)\mathrm{e}^{-\mathrm{i}tx}\mathrm{d}x.$$

By Plancherel's Theorem,

$$\|P_j(D)f\|^2=\int_{\mathbf{R}}\psi_j(t)\,|\,\tilde{f}(t)\,|^2\mathrm{d}t=\int_0^\infty\psi_j(t)\mathrm{d}\mu(t),$$

where $\mathrm{d}\mu=2\,|\,\tilde{f}(t)\,|^2\mathrm{d}t$. The function $\psi(t)=\psi_n\circ\psi_r^{-1}(t)$ is convex on its domain. Actually, notice that $\hat{\psi}_r$, ψ_r, $\hat{\psi}'_r/\psi'_r$ are strictly increasing for t

$\geqslant 0$, from

$$\psi''(t) = \frac{1}{\psi'_r(x)} \cdot \frac{\mathrm{d}}{\mathrm{d}x}\left(\frac{\psi'_n(x)}{\psi'_r(x)}\right)$$

$$= \frac{1}{\psi'_r(x)} \cdot \frac{\mathrm{d}}{\mathrm{d}x}\{\hat{\psi}_r(x) + \hat{\psi}'_r(x)\psi_r(x)/\psi'_r(x)\}, x = \psi_r^{-1}(t),$$

we see that $\psi(t)$ is strictly convex down-ward for $t \geqslant \psi_r(0)$. Thus by Jensen's inequality for $f \neq 0$ we have

$$\psi\left(\frac{\| P_r(D)f \|^2}{\| f \|^2}\right) \leqslant \frac{\int_0^\infty \psi \circ \psi_r \ (t)\mathrm{d}\mu}{\int_0^\infty \mathrm{d}\mu} = \frac{\| P_n(D)f \|^2}{\| f \|^2}.$$

From the relation between ψ_j and φ_j, (2.1) follows.

For any $\lambda > 0$ and $\delta \in (0, \lambda)$ there is an $f_\delta \in L_2^n$ for which $\| P_n(D)f_\delta \| = 1$ and supp $\tilde{f}_\delta \subseteq [\lambda - \delta, \lambda + \delta]$. For fixed λ and $\delta \to 0+$ we have

$$\| f_\delta \| \to [\varphi_n(\lambda)]^{-1}, \ \| P_r(D)f_\delta \| \to [\hat{\varphi}_r(\lambda)]^{-1}, \qquad (2.3)$$

where $\hat{\varphi}_r = \varphi_n/\varphi_r$. By taking $f = f_\delta$ in (2.1) and letting $\delta \to 0+$ we get the same limit on both sides of (2.1). So (2.2) follows. Theorem 1.1 is proven.

Corollary. *Given $f \in W_2(P_n(D))$. If for some $\lambda > 0$ it holds*

$$\| f \| \leqslant [\varphi_n(\lambda)]^{-1}, \qquad (2.4)$$

then for any sub-operator $P_r(D)$ we have

$$\| P_r(D)f \| \leqslant [\hat{\varphi}_r(\lambda)]^{-1}. \qquad (2.5)$$

(2.5) is sharp: $\forall \lambda > 0, \varepsilon > 0$ and for any $P_r(D)$, there is a $f \in W_2(P_n(D))$ such that (2.4) and $\| P_r(D)f \| > [\hat{\varphi}_r(\lambda)]^{-1} - \varepsilon$ hold.

Proof From(2.4) there is a $\lambda' \geqslant \lambda$ such that $\| f \| = [\varphi_n(\lambda')]^{-1}$. Write

$\lambda'' = \varphi_n^{-1}(\| P_n(D)f \| / \| f \|)$. Then

$$\varphi_n(\lambda'') = \| P_n(D)f \| \cdot \varphi_n(\lambda') \leqslant \varphi_n(\lambda').$$

So that $\lambda'' \leqslant \lambda'$. By Theorem 2.1

$$\| P_r(D)f \| \leqslant \frac{\varphi_r(\lambda'')}{\varphi_n(\lambda')} \leqslant [\hat{\varphi}_r(\lambda')]^{-1} \leqslant [\hat{\varphi}_r(\lambda)]^{-1}.$$

(2.5) is proved. The sharpness of (2.5) is deduced from Theorem 2.1.

Denote $H_n(\lambda) = [\varphi_n(\lambda)]^{-1}, \hat{H}_r(\lambda) = [\hat{\varphi}(\lambda)]^{-1}$,

$$\Omega(t): = \hat{H}_r \circ H_n^{-1}(t), \lambda = H_n^{-1}(t).$$

Then(2.5) may be reformulated as follows.

If for some $\lambda > 0$ and $f \in L_2(P_n(D)): = \{f \in L_2^n \mid \| P_n(D)f \| < \infty\}$, $\| f \| \leqslant H_n(\lambda)$, then

$$\frac{\| P_r(D)f \|}{\| P_n(D)f \|} \leqslant \Omega\left(\frac{\| f \|}{\| P_n(D)f \|}\right), \tag{2.6}$$

which is sharp. (2.6) plays very important role in solving the related problems (1) and (2).

3. Solution of problems (1)—(3)

Lemma 3.1 *$\Omega(u)$ is given in §2. Put*

$$t_0 = \begin{cases} |t_1 \cdots t_n|^{-1}, & \text{if } t_1 \cdots t_n \neq 0, \\ +\infty, & \textit{otherwise}. \end{cases}$$

Let $P_r(\lambda)$ be a factor of $P_n(\lambda)$ such that $P_r(0) = 0$ when $P_n(0) = 0$. Then

(1) *$\Omega(t) \in C^1[0, t_0)$ and is strictly increasing.*

(2) *$\Omega'(0+) = +\infty$.*

(3) *$t^{-1}\Omega(t)$ is strictly decreasing on $[0, t_0)$, and furthermore, when $t_0 = +\infty$, we have $t^{-1}\Omega(t) \downarrow 0 (t \to +\infty)$.*

The proof is elementary. We omit the details. In [7] we introduced a class of modulous of continuity which is called A-modulous. Lemma 3.1 means that $\Omega(t) = \hat{H}_r \circ H_n^{-1}(t)$ is A-modulous. By use of quite similar method we can prove the analogy of Equivalence theorems on the class of smooth functions $L_p(P_n(D))$(concerning Equivalence theorem we refer to [7]). Then by the Equivalence theorem and the sharpness of (2.6) we have

Theorem 3.1 *Given $N > 0$. If $P_r(0) = 0$ provided that $P_n(0) = 0$, then*

$$F(N) = \sup_{t>0}\{\Omega(t) - Nt\}. \tag{3.1}$$

Now we are going to prove

Theorem 3.2 *Given* $N > 0$. *If* $P_r(0) = 0$ *provided that* $P_n(0) = 0$, *then*

$$E(N) = F(N). \tag{3.2}$$

Proof Here we follow the method of proof in [4]. Let $\tilde{f}$ be the Fourier transform of $f \in L_2(\mathbf{R})$. $\forall f \in L_2^n$.

$$(P_r(D)f)(x) = \frac{1}{\sqrt{2\pi}}\int_{\mathbf{R}} \tilde{f}(\lambda)\varphi_r(\lambda)e^{i\lambda x}\operatorname{sgn} P_r(i\lambda)d\lambda,$$

where

$$\operatorname{sgn} P_r(i\lambda) = \begin{cases} P_r(i\lambda)/\mid P_r(i\lambda)\mid, & P_r(i\lambda) \neq 0, \\ 0, & \text{otherwise.} \end{cases}$$

Define $g(\lambda)$:

$$g(\lambda) = \begin{cases} \varphi_r(\lambda), & \text{when } \varphi_r(\lambda) \leqslant N, \\ N, & \text{when } \varphi_r(\lambda) > N. \end{cases}$$

We construct a linear operator $T(L_2(\mathbf{R}) \to L_2(\mathbf{R}))$ in the following way:

For $f \in L_2(\mathbf{R})$, we define

$$(Tf)(x) := \frac{1}{\sqrt{2\pi}}\int_{\mathbf{R}} g(\lambda)\tilde{f}(\lambda)e^{i\lambda x}\operatorname{sgn} P_r(i\lambda)d\lambda. \tag{3.3}$$

Obviously, we have $\| T \| = \| g(\lambda) \|_\infty = N$, and for any $f \in W_2(P_n(D))$

$$\begin{aligned} \| P_r(D)f - Tf \| &= \frac{1}{\sqrt{2\pi}} \left\| \int_{\mathbf{R}} \frac{\varphi_r(\lambda) - g(\lambda)}{\varphi_n(\lambda)} \cdot \varphi_n(\lambda) \cdot \operatorname{sgn} P_r(i\lambda)e^{i\lambda x}d\lambda \right\| \\ &\leqslant \sup_{\lambda \in \mathbf{R}} \frac{\varphi_r(\lambda) - g(\lambda)}{\varphi_n(\lambda)} \| P_n(D)f \| \leqslant \sup_{\lambda \in \mathbf{R}} \frac{\varphi_r(\lambda) - N}{\varphi_n(\lambda)} \\ &= \sup_{t \in [0, t_0)} \{\Omega(t) - Nt\}, t = [\varphi_n(\lambda)]^{-1}. \end{aligned}$$

Thus we have $E(N) \leqslant F(N)$. Suppose for some $N > 0$ and $\varepsilon > 0$, $E(N) \leqslant F(N) - \varepsilon$. From the properties of $\Omega(t)$ it follows that there exists some $t' \in (0, t_0)$ such that

$$E(N) \leqslant \Omega(t') - Nt' - \varepsilon.$$

So that for every $f \in W_2(P_n(D))$

$$\begin{aligned} \| P_r(D)f \| &\leqslant \Omega(t') - Nt' - \varepsilon + N \| f \| \\ &= [\hat{\varphi}_r(\lambda')]^{-1} - N[\varphi_n(\lambda')]^{-1} + N \| f \| - \varepsilon, \end{aligned} \tag{3.4}$$

where $t' = [\varphi_n(\lambda')]^{-1}$. But from the proof of Theorem 2.1 it follows that there is a $f_\delta \in W_2(P_n(D))$ such that

$$\| f_\delta \| \to [\varphi_n(\lambda')]^{-1},\ \| P_r(D) f_\delta \| \to [\hat{\varphi}_r(\lambda')]^{-1}, (\delta \to 0+).$$

By taking $f = f_\delta$ in (3.4) and letting $\delta \to 0+$ we obtain a relation which contradicts with (3.4). Therefore $E(N) = F(N)$.

Similar to [1] we may prove

Theorem 3.3 *Given* $N > 0$. *If* $P_r(0) = 0$ *provided that* $P_n(0) = 0$, *then*

$$G(N) = E(N).$$

4. Optimal Recovery of $P_r(D)$.

Let $\gamma > 0$ be a fixed number, and $\mathscr{K} \subset L_2^n$ be defined by

$$\mathscr{K} = \{ f \mid f \in L_2^n,\ \| P_n(D) f \| \leqslant \gamma \}.$$

Denote $\mathscr{X} = L_2(\mathbf{R})$. Consider the optimal recovery of the differential operator $S = P_r(D)$ on $\mathscr{K}$ in $L_2(\mathbf{R})$ by using the information. If $f \in \gamma$ with an error bound $\varepsilon > 0$. Denote the unit ball of γ by γ^0. Any mapping

$$\varnothing : I(\mathscr{K}) + \varepsilon\gamma^0 \to \gamma$$

is an algorithm (approximation formula) for S. The global error is

$$E_\varphi(\gamma, \varepsilon) = \sup\{ \| Sf - \varnothing(g) \| : f \in \mathscr{K},\ \| f - g \| < \varepsilon \},$$

and the intrinsic error is defined as

$$E(\gamma, \varepsilon) = \inf_\varphi E_\varphi(\gamma, \varepsilon).$$

Some algorithm φ_0 is said to be optimal, provided that $E_{\varphi_0}(\gamma, \varepsilon) = E(\gamma, \varepsilon)$ holds. It is required to get the exact expression for $E(\gamma, \varepsilon)$ and to construct the optimal algorithm. A linear continuous operator to be optimal (if it exists) is especially interesting. The special case $P_n(D) = D^n$ of this problem was considered by Melkman and Micchelli[9]. In this section we solve this problem in its general case by using a method quite different from that used in [9]. Put

$$E_0(\gamma, \varepsilon) = \sup\{ \| P_r(D) f \| ; f \in \mathscr{K},\ \| f \| \leqslant \varepsilon \}.$$

Then we have

$$E_0(\gamma, \varepsilon) \leqslant E(\gamma, \varepsilon).$$

Theorem 4.1 *Let $P_r(D)$ be a sub-operator of $P_n(D)$. Then*

(1) *If $P_n(0) = P_r(0) = 0$, or $P_n(0) = 0$ and $\varepsilon' := \varepsilon\gamma^{-1} < [\varphi_n(0)]^{-1}$, then*

$$E(\gamma, \varepsilon) = E_0(\gamma, \varepsilon) = \gamma[\hat{\varphi}_r(\lambda')]^{-1},$$

where $\lambda' = \varphi_n^{-1}(\gamma\varepsilon^{-1})$.

(2) *If $\varepsilon' = \varepsilon\gamma^{-1} \geqslant [\varphi_n(0)]^{-1}$, then*

$$E(\gamma, \varepsilon) = E_0(\gamma, \varepsilon) = \gamma[\hat{\varphi}_r(0)]^{-1}.$$

Proof (1) Let T be the operator given in Theorem 3.2. Then for $N > 0$ we have

$$\| P_r(D)f - Tg \| \leqslant \| P_r(D)f - Tf \| + \| T \| \varepsilon \leqslant \gamma \sup_{t>0}\{\Omega(t) - Nt\} + N\varepsilon. \tag{4.1}$$

From the property of $\Omega(t)$ (see [7]) there is a number $N_0 > 0$ such that

$$\gamma \sup_{t\in[0,t_0)} \{\Omega(t) - N_0 t\} = \gamma\left\{\Omega\left(\frac{\varepsilon}{\gamma}\right) - N_0\frac{\varepsilon}{\gamma}\right\}. \tag{4.2}$$

So that

$$E(\gamma, \varepsilon) \leqslant \sup\{ \| P_r(D)f - Tg \| : f \in K, \| f - g \| < \varepsilon\}$$
$$\leqslant \gamma\Omega\left(\frac{\varepsilon}{\gamma}\right) - N_0\varepsilon + N_0\varepsilon = \gamma[\hat{\varphi}_r(\lambda')]^{-1}.$$

Applying the method of proof of Theorem 1 we obtain

$$E_0(\gamma, \varepsilon) = \gamma[\hat{\varphi}_r(\lambda')]^{-1}.$$

(1) is proved.

(2) Suppose $P_n(0) \neq 0$, then every $f \in \mathscr{K}$ is represented by

$$f(x) = \int_{\mathbf{R}} G(x - t)\varphi(t)\mathrm{d}t,$$

where $\varphi(t) = P_n(D)f(t)$ a.e., and

$$G(x) = \frac{1}{2\pi}\int_{\mathbf{R}} \frac{\mathrm{e}^{\mathrm{i}xu}}{P_n(\mathrm{i}u)}\mathrm{d}u.$$

Thus we have $(P_r(D)G)(x) = (2\pi)^{-1}\int_{\mathbf{R}} [\hat{P}_r(\mathrm{i}u)]^{-1}\mathrm{e}^{\mathrm{i}xu}\mathrm{d}u$, where $\hat{P}_r(\mathrm{i}u) = \dfrac{P_n(\mathrm{i}u)}{P_r(\mathrm{i}u)}, 0 \leqslant r \leqslant n - 1$. So that

$$(P_r(D)f)(x) = \frac{1}{2\pi}\int_{\mathbf{R}}\left\{\int_{\mathbf{R}}\frac{e^{i(x-t)u}du}{\hat{P}_r(iu)}\right\}\varphi(t)dt = \frac{1}{\sqrt{2\pi}}\int_{\mathbf{R}}\frac{\tilde{\varphi}(u)}{\hat{P}_r(iu)}e^{iux}du.$$

From this it is readily derived $\| P_r(D)f \| \leqslant \gamma[\hat{\varphi}_r(0)]^{-1}$. Take $T = 0$, then we obtain

$$E(\gamma,\varepsilon) \leqslant \sup\{\| P_r(D)f - Tg \| : f \in \mathscr{K}, \| f - g \| < \varepsilon\}$$

$$= \sup_{f\in K}\| P_r(D)f \| \leqslant \gamma[\hat{\varphi}_r(0)]^{-1}.$$

The equality $E_0(\gamma,\varepsilon) = \gamma[\hat{\varphi}_r(0)]^{-1}$ may be proved analogously as before. We omit the details. (2) is proven.

From Theorem 4.1 we get

Theorem 4.2

(1) *If* $P_r(0) = P_n(0)$, *or* $P_n(0) \neq 0$ *and* $\varepsilon' = \varepsilon\gamma^{-1} < [\varphi_n(0)]^{-1}$, *Then the operator T defined in* (3.3) *is optimal, where the parameter N is defined by the relation* (4.2).

(2) *If* $P_n(0) \neq 0$, *and* $\varepsilon\gamma^{-1} \geqslant [\varphi_n(0)]^{-1}$, *then* $T = 0$ *is optimal.*

References

[1] Arestov, V. V. On Some Extremal Problems for Differentiable Functions of One Variable, Proc. Steklov Inst. Math., 1975, 138: 1 ~ 29

[2] Korneichuk, N. P., Ligun, A. A., Doronin, V. D. *Approximation with Constraints*. Kiev Naukova Dumka, 1982

[3] Stechkin. S. B. Best Approximation of Linear Operators. Math. Notes (Mat. Zametki, In Russian), 1967, 1: 91 ~ 99

[4] Subbotin. Ju. N., Taikov, L. V. Best Approximation of a Differential Operator in L_2. Mat. Zametk, 1968, 3: 157 ~ 164

[5] Sun Yongsheng. Landau-Kolmogorv Type Inequality for an Ordinary Linear Differential Operator. Bulletin of Sciences (Kexue tongbao, In Chinese), 1985, 7: 481 ~ 484

[6] Sharma, A., Tzimbalario, J. Landau Type Inequality for Some Linear Differential Operators. Illinois JM, 1976, 20: 433 ~ 455

[7] Sun Yongsheng. *Approximation Theory* Vol. 1. Beijing: Beijing Normal Univ. Press, 1989. 503 (in Chinese)

[8] Sun Yongsheng, Fang Gensun. *Approximation Theory* Vol. 2. Beijing: Beijing Normal Univ, Press, 1990 (in Chinese)

[9] Melkman. A., Micchelli, C. A. Optimal Estimation of Linear Operators in Hilbert Spaces from Inaccurate Data. SIAM J. numer. anal., 1979, 16(1): 87 ~ 105

[10] Hirschman. I., Widder. D. *The Convolution Transform*. N. Y.: Princeton Press, 1955

[11] Sun Yongsheng. Optimal Recovery of a Differential Operator on a Class of Smooth Functions. JATA, 1990, 6(1): 23 ~ 31

EAST JOURNAL ON APPROXIMATIONS
1998, Volume 4, Number 4: 515 ~ 525

定义在实直线上函数的卷积类的极值问题*

Extremal Problem for a Convolution Class of Functions Defined on the Real Line

In this paper a generalized version of the classical Hardy-Littlewood-Pólya inequality is given for a convolution class with a TP kernel. Furthermore, the problem of optimal subspaces on the class is considered.

1. Introduction

In a recent paper [14] Tikhomirov proved that for the convolution class of 2π-periodic functions

$$W_{\infty}^{k}(\mathbf{T}) := \left\{ x(\cdot) \,\middle|\, x(\cdot) = K * u(\cdot),\ \| u(\cdot) \|_{L_{\infty}(\cdot)} \leqslant 1 \right\},$$

with a cyclic variation diminishing kernel $K(\cdot)$, the $2n$-dimensional Kolmogorov width $d_{2n}(W_{\infty}^{k}(\mathbf{T}), C(\mathbf{T}))$ has an infinite sequence of optimal subspaces as follows:

$$S_{2n}^{K_m} := \operatorname{span}\left\{ K_m\left(\cdot - \frac{j\pi}{n} \right), j = 0, \cdots, 2n-1 \right\}, m \in \mathbf{Z}_+,$$

where $K_1 = K$, $K_m = K * \cdots * K$, $m \geqslant 2$. The purpose of this paper is to show that some analogous results may be obtained for the case where the convolution class of functions is defined on the real line with a TP kernel.

---

* The authors are supported by the National Natural Science Found of China (No 19671012) and by the Doctoral Program Foundation of State Educational Commission of China.

* 本文与蒋艳杰合作.

To formulate and prove our results, we must first introduce some definitions and preliminary results.

Let $G(t)$ be a function defined on the real line $\mathbf{R}$. It is said to be *totally positive* (TP), if $\forall\, n \geqslant 1, t_1 < \cdots < t_n, \tau_1 < \cdots < \tau_n$, it holds $\det(G(t_i - \tau_j)) \geqslant 0$. In addition, if $\int_{\mathbf{R}} G(t)\mathrm{d}t = 1$, then $G(t)$ is called a *Pólya frequency density* (we write $G \in PF$ for simplicity). For a complete information about the characteristics of the PF density we refer the reader to [6], We confine ourselves to recall the following:

Definition 1. *We say that an entire function $\psi(z)$ belongs to the Laguerre – Pólya class ($\psi \in L - P$), if*

$$\psi(z) = \mathrm{e}^{-cz^2+bz} \prod_{k=1}^{\theta} \left(1 - \frac{z}{a_k}\right) \mathrm{e}^{z/a_k}, \tag{1.1}$$

where $c \geqslant 0, b, a_k \in \mathbf{R}, 0 < c + \sum_{k=1}^{\theta} a_k^{-2} < \infty, \theta = n \in \mathbf{Z}_+$, or $\theta = \infty$ (we write $\deg \psi = \theta$).

The $L - P$ class essentially characterizes the set of TP functions on the real line. Actually, the following is well-known (see, for example, [6]).

Theorem A. *If $G \in PF$, then the reciprocal $\psi(z)$ of the Laplace transform of G belongs to $L - P$. Moreover, we have*

$$G(x) = \frac{1}{2\pi} \int_{-\infty}^{\infty} \frac{\mathrm{e}^{iux}}{\psi(iu)} \mathrm{d}u. \tag{1.2}$$

Conversely, for every $\psi(z) \in L - P$, the integral (1.2) defines a function $G \in PF$.

Throughout this paper we assume that $G(t) \in PF$. Let $V(\mathbf{R})$ be the set of all functions $\mu(\cdot)$ of bounded variation on $\mathbf{R}$. For $1 \leqslant p \leqslant \infty$, denote

$$W_p(G) := \{f : f = G * h, h \in L_p(\mathbf{R}), \| h \|_p \leqslant 1\},$$

$$W_V(G) := \{f : f = G *' \mu, \mu \in V(\mathbf{R}), \operatorname{var} \mu(\cdot) \leqslant 1\},$$

where $\| \cdot \|_p$ is the usual $L_p(\mathbf{R})$-norm, $(G * h)(x) = \int_{\mathbf{R}} G(x - t) h(t)\mathrm{d}t$, and $(G *' \mu)(x) = \int_{\mathbf{R}} G(x - t)\mathrm{d}\mu(t)$.

In the recent years, some extremal problems of approximation theory, such as the average Kolmogorov width, the optimal recovery, the Landau-Kolmogorov inequality, etc., have been studied on $W_p(G)$ in the space $L_p(\mathbf{R}), 1 \leqslant p \leqslant \infty$ (see [12], [7], [1] and [2]). In this paper we give firstly a generalized version of the classical Hardy-Littlewood-Pólya inequality (Section 2). Secondly, we consider the optimal subspaces for $W_p(G), p = 1, 2, \infty$ (Section 3). An essential step in this research is to extend the Schoenberg-Micchelli's cardinal $\mathscr{L}$-spline interpolation theory to the more general case where the cardinal $\mathscr{L}$-spline space is associated with an infinite order differential operator $\psi(D)$, $\psi(z)(\in L-P)$ being the reciprocal of Laplace transform of G, $\deg \psi := \theta = \infty$. This extension was initiated by Sun[12], and developed by Li[7] and Chen[1]. Some fundamental facts are needed for our further consideration, so that we cite them here in a series of lemmas.

Lemma 1[1] *Let $G \in PF$. Denote*

$$\widetilde{G}(x) = \sum_{k=-\infty}^{\infty} (-1)^k G(x-k).$$

Then $\widetilde{G}$ has exactly one zero y_0 in $[0,1)$. For any $x_0 \in [0,1) \setminus \{y_0\}$, there exists a function

$$L(x) = L(G, x) := \sum_{k \in \mathbf{Z}} c_k G(x-k),$$

such that:

(1) $L(x_0 + k) = \delta_{0,k}, k \in \mathbf{Z}$ (*δ_{ik} is the Kronecker symbol*).

(2) *There exists a constant $\beta_0 > 0$ such that for every $\beta \in (0, \beta_0)$ holds the estimate*

$$|L(x)| \leqslant A(\beta) e^{-\beta|x|}, x \in \mathbf{R},$$

$A = A(\beta)$ is some constant depending on β.

(3) $G(x) = \sum_{k \in \mathbf{Z}} G(x_0 + k) L(x-k)$.

Lemma 2[2] *Let $G \in PF$ and $\Phi(x) = \Phi(G, x) := (G * \operatorname{sgn} \sin \pi(\cdot))(x)$. Then $\Phi(x)$ has exactly one zero x_0 in $[0,1)$. Moreover, $\Phi'(x_0) = 2\widetilde{G}(x_0) \neq 0$.*

$\Phi(x)$ is the Euler spline associated with $\psi(D)$, where $\psi(z)(\in L -$

P) is the reciprocal of the Laplace transform of the kernel G.

Lemma 3[10] *Let $G \in PF$. Denote*

$$\widetilde{\widetilde{G}} = \sum_{k \in \mathbf{Z}} G(x - 2k).$$

Then $\widetilde{\widetilde{G}}$ is NCVD (non-degenerate cyclic variation diminishing).

Let $K(t)$ is NCVD, $K(t+2) = K(t)$. Consider the following isoperimetric problem:

$$\| K * h \|_{q[0,2]} \to \sup, h \in D_n, \quad (1.3)$$

where $n \in \mathbf{Z}_+$, $p, q \in (1, \infty)$, and

$$D_n = \{h(t) : \| h \|_{p[0,2]} \leqslant 1, h(t + n^{-1}) = -h(t), h(t)\operatorname{sgn} \sin n\pi t \geqslant 0\}.$$

The following theorem is due to Chen [3].

Theorem B *Let $p, q \in (1, \infty)$. Denote*

$$\lambda_n(p, q, K) = \sup_{h \in D_n} \| K * h \|_{q[0,2]}.$$

Then $\forall_n \in \mathbf{Z}_+$, $\lambda_n(p, q, K) > 0$. There exists a continuous function $h_n(x) \in D_n$ such that:

(1) $\| K * h_n \|_{q[0,2]} = \lambda_n(p, q, K)$, $\| h_n \|_{p[0,2]} = 1$.

(2) $$\int_0^2 K(x-y) \left| (K * h_n)(x) \right|^{q-1} \operatorname{sgn}(K * h_n)(x) \mathrm{d}x = \lambda_n^q \left| h_n(y) \right|^{p-1} \operatorname{sgn} h_n(y), \ \forall y \in [0,2].$$

(3) *For every $n \in \mathbf{Z}_+$ there exists an $\varepsilon_n \in \{1, -1\}$ and some $\alpha_n \in \left[0, \frac{1}{n}\right)$ for which*

$$\varepsilon_n (K * h_n)(x) \operatorname{sgn} \sin n\pi(x - \alpha_n) \geqslant 0.$$

In what follows, for any $G \in PF$, we denote $\lambda_n(p, q, \widetilde{\widetilde{G}})$ by $\lambda_n(p, q, G)$. When $p = q = \infty$, we define $\lambda_n(\infty, \infty, G) = \| \Phi(n^{-1}G(n^{-1}t), \cdot) \|_\infty$. The number $\lambda_1(p, p, G)(1 \leqslant p \leqslant \infty)$ is essentially connected with the average 1-Kolmogorov width of $W_p(G)$ in $L_p(\mathbf{R})$, and also with the optimal error bounds of functions taken from $W_p(G)$ and approximated by its cardinal $\mathscr{L}$ − spline interpolating functions $s(f, x) = \sum_{j \in \mathbf{Z}} f(x_o + j) \cdot L(x - j)$ belonging to

$$S_p(G) := \overline{\operatorname{span}\{G(\cdot - j), j \in \mathbf{Z}\}} \tag{1.4}$$

where $\overline{A}$ denotes the closure of A in $L_p(\mathbf{R})$.

Theorem C [1][4]. *Suppose that $G \in PF, 1 \leqslant p \leqslant \infty$. Then the average 1-Kolmogorov width of $W_p(G)$ in $L_p(\mathbf{R})$* is

$$\begin{aligned}\overline{d}_1(W_p(G), L_p(\mathbf{R})) &= E(W_p(G), S_p(G), L_p(\mathbf{R})) \\ &= \sup_{f \in W_p(G)} \| f - s(f) \|_p = \lambda_1(p, p, G),\end{aligned} \tag{1.5}$$

where

$$E(W_p(G), S_p(G), L_p(\mathbf{R})) := \sup_{f \in W_p} \inf_{g \in S_p} \| f - g \|_p,$$

$$\lambda_1(1,1,G) = \lambda_1(\infty, \infty, G) = \| \Phi(G, \cdot) \|_\infty,$$

$$\lambda_1(2,2,G) = | \psi(\mathrm{i}\pi) |^{-1}.$$

Theorem D [8]. *Let $G \in PF$. Then*

$$\sup_{f \in W_V(G)} \| f - s(f) \|_1 = \lambda_1(\infty, \infty, G).$$

Denote by $\overline{\dim} S_p(G)$ the average dimension of $S_p(G)$ in $L_p(\mathbf{R})$.

Lemma 4 [1]. *For any $1 \leqslant p \leqslant \infty$ we have $\overline{\dim}\, S_p(G) = 1$.*

The above lemma shows that $S_p(G)$ is an optimal subspace for $\overline{d}_1(W_p(G), L_p(\mathbf{R}))$. For the concept of average dimension of a subspace in $L_p(\mathbf{R})$ we refer to [15].

2. A generalized version of H-L-P inequality and L-K inequality

Lemma 5 [12]. *Given $G \in PF$, there exists a sequence of real polynomials $\{P_n(t)\}_{n=1}^{\infty}$ satisfying the following conditions:*

(1) $P_n(0) = 1$,

(2) *$P_n(t)$ has only real zeros,*

(3) $\lim\limits_{n \to \infty} P_n(z) = \psi(z)$ *uniformly on $| z | \leqslant A$, where A is an arbitrary constant.*

Theorem 2.1 *Let $\Lambda_k \in PF$,*

$$(\psi_k(s))^{-1} = \int_{-\infty}^{\infty} e^{-st}\Lambda_k(t)dt, \quad k = 1,2,$$
$$\Lambda = \Lambda_1 * \Lambda_2,$$
$$\psi(s) = \psi_1(s)\psi_2(s),$$

and

$$f(x) = \int_{-\infty}^{\infty} \Lambda(x-t)h(t)dt, \ \| h \|_2 \leqslant 1.$$

If for some $\lambda > 0$, $\| f \|_2 \leqslant \frac{1}{|\psi(i\lambda)|}$, *then*

$$\| \Lambda_2 * h \|_2 \leqslant \frac{1}{|\psi_2(i\lambda)|}.$$

Proof

(i) First we show that if $\| f \|_2 < \frac{1}{|\psi(i\lambda)|}$, then

$$\| \Lambda_2 * h \|_2 \leqslant \frac{1}{|\psi_2(i\lambda)|}. \tag{2.1}$$

Indeed, otherwise there exists some $f = \Lambda * h$ with $\| h \|_2 \leqslant 1$ and $\| f \|_2 < \frac{1}{|\psi(i\lambda)|}$ such that

$$\| \Lambda_2 * h \|_2 > \frac{1}{|\psi_2(i\lambda)|}. \tag{2.2}$$

By Lemma 5, there exist two sequences of real polynomials $\{P_{n,k}(s)\}_{n=1}^{\infty}$ such that $P_{n,k}(0) = 1$, and for any $A > 0$, $\lim_{n\to\infty} P_{n,k}(s) = \psi_k(s)$ uniformly on the circle $|s| \leqslant A$, $k = 1,2$, respectively. Put

$$\Lambda_{k,n}(x) = \frac{1}{2\pi}\int_{-\infty}^{\infty} \frac{e^{iux}}{P_{n,k}(iu)}du \quad k = 1,2, \quad n = 1,2,\cdots$$

By [12] we have $\lim_{n\to\infty} \| \Lambda_{k,n}(x) - \Lambda_k(x) \|_p = 0$, $p = 1,2,\infty$, $k = 1,2$. Set $\Lambda_n = \Lambda_{1,n} * \Lambda_{2,n}$, $P_n = P_{n,1}P_{n,2}$. Thus, for sufficiently large n, by (2.2) we have

$$\| \Lambda_{2,n} * h \|_2 > \frac{1}{|P_{n,2}(i\lambda)|}. \tag{2.3}$$

This is due to the relations

$$\lim_{n\to\infty} \| \Lambda_{2,n} * h \|_2 = \| \Lambda_2 * h \|_2, \quad \lim_{n\to\infty} \frac{1}{|P_{n,2}(i\lambda)|} = \frac{1}{|\psi_2(i\lambda)|}.$$

Further, for sufficiently large n, we have

$$\| \Lambda_n * h \|_2 \geqslant \frac{1}{|P_n(\mathrm{i}\lambda)|}. \quad (2.4)$$

Indeed, otherwise, there exists a subsequence $\{n_k\}$ of natural numbers such that

$$\| \Lambda_{n_k} * h \|_2 < \frac{1}{|P_{n_k}(\mathrm{i}\lambda)|}. \quad (2.5)$$

Since the kernel $\Lambda_{n_k} = \Lambda_{1,n_k} * \Lambda_{2,n_k}$ corresponds to the linear differential operator $P_{n_k}(D) = P_{n_k,1}(D)P_{n_k,2}(D)$ which is of finite order, we may use the generalized H-L-P inequality [5]. Accordingly, we have

$$\| \Lambda_{2,n_k} * h \|_2 \leqslant \frac{1}{|P_{n_k,2}(\mathrm{i}\lambda)|}. \quad (2.6)$$

This is in contradiction with (2.3), thus (2.4) holds. Letting $n \to \infty$ in (2.4), we get

$$\| \Lambda * h \|_2 \geqslant \frac{1}{|\psi(\mathrm{i}\lambda)|}.$$

This is in contradiction with the assumption $\| f \|_2 < \frac{1}{|\psi(\mathrm{i}\lambda)|}$, so (2.1) holds.

(ii) Secondly, we show that if $\| f \|_2 = \frac{1}{|\psi(\mathrm{i}\lambda)|}$, then $\| \Lambda_2 * h \|_2 \leqslant \frac{1}{|\psi_2(\mathrm{i}\lambda)|}$. For $0 < \alpha < 1$, set $f_\alpha = \Lambda * (\alpha h)$. Then $\| f_\alpha \|_2 < \frac{1}{|\psi(\mathrm{i}\lambda)|}$. By (i) we have

$$\| \Lambda_2 * (\alpha h) \|_2 \leqslant \frac{1}{|\psi_2(\mathrm{i}\lambda)|}. \quad (2.7)$$

Letting $\alpha \to 1$ in (2.7), we get

$$\| \Lambda_2 * h \|_2 \leqslant \frac{1}{|\psi_2(\mathrm{i}\lambda)|}.$$

The proof of Theorem 2.1 is completed.

By an argument completely similar to that in the proof of Theorem 2.1, we may prove the following generalized L − K inequality.

Theorem 2.2 *Let* $\Lambda_k \in PF, \lambda > 0, \Phi(\Lambda_k, \lambda_x) := (\Lambda_k * \operatorname{sgn} \sin \lambda\pi(\cdot))(x), k = 1, 2, \Lambda = \Lambda_1 * \Lambda_2,$ *and* $f(x) = \Lambda * h = \Lambda_1 * \Lambda_2 * h,$

$\| h \|_\infty \leqslant 1$. *If* $\| f \|_\infty \leqslant \| \Phi(\Lambda, \lambda\ \cdot) \|_\infty$, *then* $\| \Lambda_2 * h \|_\infty \leqslant \| \Phi(\Lambda_2, \lambda\ \cdot) \|_\infty$.

To prove Theorem 2.2, we need only to replace $\| \cdot \|_2$ and $| \psi(i\lambda) |^{-1}$ by $\| \cdot \|_\infty$ and $\| \Phi(\Lambda, \lambda\ \cdot) \|_\infty$, respectively, and instead of the generalized H − L − P inequality [5] to use the generalized L − K inequality of Sharma-Tzimbalario [11] or Sun (see[13, Th.6.2 − 3]). We omit the details.

Remark. Theorem 2.2 has been proved in [9] by a complicated method different from that we just sketched.

3. Optimal subspaces for $W_p(G)$

Let $G \in PF$, $m \in \mathbf{Z}_+$, $G_1 = G$, $G_m = G * \cdots * G$, $m \geqslant 2$. It is easy to see that $G_m \in PF$. Now we show our next result.

Theorem 3.1 *Let* $G \in PF$, $m \in \mathbf{Z}_+$. *Then*

(a) $\bar{d}_1(W_1(G), L_1(\mathbf{R})) = E(W_1(G), S_1(G_m), L_1(\mathbf{R})) = \| \Phi(G, \cdot) \|_\infty$;

(b) $\bar{d}_1(W_\infty(G), L_\infty(\mathbf{R})) = E(W_\infty(G), S_\infty(G_m), L_\infty(\mathbf{R})) = \| \Phi(G, \cdot) \|_\infty$;

(c) $\bar{d}_1(W_2(G), L_2(\mathbf{R})) = E(W_2(G), S_2(G_m), L_2(\mathbf{R})) = | \psi(i\pi) |^{-1}$.

By changing the scale we can get

Theorem 3.2 *Let* $G \in PF$, $n \in \mathbf{R}^+$, $m \in \mathbf{Z}_+$. *Then*

$$\bar{d}_n(W_p(G), L_p(\mathbf{R})) = E(W_p(G), S_{p,n}(G_m), L_p(\mathbf{R})) = \| \Phi(n^{-1}G(n^{-1}t), \cdot) \|_\infty,$$

$$p = 1, \infty,$$

$$\bar{d}_n(W_2(G), L_2(\mathbf{R})) = E(W_2(G), S_{2,n}(G_m), L_2(\mathbf{R})) = | \psi(in\pi) |^{-1},$$

where

$$S_{k,n}(G_m) := \overline{\mathrm{span}\{ G_m(\cdot - jn^{-1}) : j \in \mathbf{Z} \}} = \{ \phi(nt) : \phi \in S_k(G_m(n^{-1} \cdot)) \},$$

$$\overline{\dim} S_{k,n}(G_m) = n, \quad k = 1, 2, \infty.$$

Let X be a real normed space, X^* the dual of X, and $< x^*, x >$ the value of a functional $x^* \in X_*$ at an element $x \in X$. If L is a linear subspace of X, then

$$L^{\perp}:=\{x^{*}\in X^{*}\mid \langle x^{*},x\rangle=0\ \forall x\in L\}$$

denotes the annihilator of L.

Proof of Theorem 3.1 By Lemma 4 and Theorem C, we need only to prove that

$$E(W_1(G),S_1(G_m),L_1(\mathbf{R}))\leqslant\|\Phi(G,\cdot)\|_{\infty},\tag{3.1}$$

$$E(W_\infty(G),S_\infty(G_m),L_\infty(\mathbf{R}))\leqslant\|\Phi(G,\cdot)\|_{\infty},\tag{3.2}$$

$$E(W_2(G),S_2(G_m),L_2(\mathbf{R}))\leqslant|\psi(\mathrm{i}\pi)|^{-1}.\tag{3.3}$$

First we prove (3.1). By Duality Theorem [13], we have

$$E(W_1(G),S_1(G_m),L_1(\mathbf{R}))=\sup_{x(\cdot)\in W_1(G)}\ \sup_{\substack{\|y(\cdot)\|_\infty\leqslant 1\\ y(\cdot)\in(S_1(G_m))^{\perp}}}\int_{\mathbf{R}}x(t)y(t)\mathrm{d}t.$$

Suppose $x(\cdot)\in W_1(G)$, $x=G*h$, $\|h\|_1\leqslant 1$, $y(\cdot)\in(S_1(G_m))^{\perp}$, and $\|y(\cdot)\|_\infty\leqslant 1$. Then $f(t)=(G_m*y(-\cdot))(t)\in W_\infty(G_m)$, and $f(k)=0,k\in\mathbf{Z}$.

By Theorem C, $\|f\|_\infty\leqslant\|\Phi(G_m,\cdot)\|_\infty$, i.e.,

$$\|f\|_\infty=\|(G_{m-1}*G*y(-t))(\cdot)\|_\infty\leqslant\|\Phi(G_m,\cdot)\|_\infty.\tag{3.4}$$

By Theorem 2.2 and (3.4) we have

$$\|(G*y(-t))(\cdot)\|_\infty\leqslant\|\Phi(G,\cdot)\|_\infty.\tag{3.5}$$

This yields

$$\begin{aligned}\int_{\mathbf{R}}x(t)y(t)\mathrm{d}t&=\int_{\mathbf{R}}\int_{\mathbf{R}}G(t-\tau)h(\tau)\mathrm{d}\tau y(t)\mathrm{d}t\\&=\int_{\mathbf{R}}h(\tau)\mathrm{d}\tau\int_{\mathbf{R}}G(t-\tau)y(t)\mathrm{d}t\\&\leqslant\left\|\int_{\mathbf{R}}G(t-\tau)y(t)\mathrm{d}t\right\|_\infty\leqslant\|\Phi(G,\cdot)\|_\infty\end{aligned}$$

and (3.1) is proved.

Now we prove (3.2). Note that $W_\infty(G)\subset C_0$. It is easy to see that

$$E(W_\infty(G),S_\infty(G_m),L_\infty(\mathbf{R}))\leqslant E(W_\infty(G),S_\infty(G_m)\cap C_0,C_0)\tag{3.6}$$

where $C_0:=C_0(\mathbf{R})$ is the set of all continuous functions $x(\cdot)$ on $\mathbf{R}$ such that $x(t)\to 0$ as $t\to\infty$, with $\|x(\cdot)\|_{C_0}:=\max\{|x(t)|\mid t\in\mathbf{R}\}$. By

Duality Theorem [13], we have

$$
\begin{aligned}
&E(W_\infty(G), S_\infty(G_m)\cap C_0, C_0)\\
&\leqslant \sup_{x(\cdot)\in W_\infty(G)} \sup_{\substack{\mathrm{var}y(\cdot)\leqslant 1\\ dy\in(S_\infty(G_m)\cap C_0)^\perp}} \int_{\mathbf{R}} x(t)\mathrm{d}y(t). \qquad (3.7)
\end{aligned}
$$

Let $x(\cdot)\in W_\infty(G)$, $x=G*u$, $\|u\|_\infty\leqslant 1$, $dy\in(S_\infty(G_m)\cap C_0)^\perp$, and var $y(\cdot)\leqslant 1$. Then we have $f(t):=(G_m*'y(-\cdot))(t)\in W_V(G_m)$, and $f(k)=0$, $k\in\mathbf{Z}$. By Theorem C,

$$\|f\|_1\leqslant\|\Phi(G_m,\cdot)\|_\infty.$$

Set $z(t)=(G_m*g)(t)$. Let $\Lambda^{-i}z:=G_{m-i}*g$, $1\leqslant i\leqslant m$. Define

$$\phi(t):=\int_{\mathbf{R}} f(t-\tau)\mathrm{sgn}\Lambda^{-(m-1)}f(-\tau)\mathrm{d}\tau. \qquad (3.8)$$

Then we have

$$
\begin{aligned}
\Lambda^{-(m-1)}\phi(t) &= (G*'y(-\cdot)*\mathrm{sgn}\Lambda^{-(m-1)}f(-\cdot))(t)\\
&=\int_{\mathbf{R}}\Lambda^{-(m-1)}f(t-\tau)\mathrm{sgn}\Lambda^{-(m-1)}f(-\tau)\mathrm{d}\tau. \qquad (3.9)
\end{aligned}
$$

Hence, by (3.8),

$$\|\phi(\cdot)\|_\infty\leqslant\|f\|_1\leqslant\|\Phi(G_m,\cdot)\|_\infty. \qquad (3.10)$$

Moreover,

$$\|\Lambda^{-(m-1)}\phi(t)\|_\infty \geqslant |\Lambda^{-(m-1)}\phi(0)| = \int_{\mathbf{R}}|\Lambda^{-(m-1)}f(\tau)|\mathrm{d}\tau = \|\Lambda^{-(m-1)}f\|_1. \qquad (3.11)$$

But in view of (3.8),

$$\|\Lambda^{-m}\phi(\cdot)\|_\infty\leqslant \mathrm{var}y(\cdot)\leqslant 1. \qquad (3.12)$$

By Theorem 2.2 and (3.10) − (3.12),

$$\|\Lambda^{-(m-1)}f\|_1\leqslant\|\Lambda^{-(m-1)}\phi(t)\|_\infty\leqslant\|\Phi(G,\cdot)\|_\infty. \qquad (3.13)$$

Hence, we have

$$
\begin{aligned}
\int_{\mathbf{R}}x(t)\mathrm{d}y(t) &\leqslant \left|\int_{\mathbf{R}}\int_{\mathbf{R}}G(t-\tau)u(\tau)\mathrm{d}\tau\mathrm{d}y(t)\right|\\
&\leqslant \int_{\mathbf{R}}\left|\int_{\mathbf{R}}G(t-\tau)\mathrm{d}y(t)\right|\mathrm{d}\tau = \|\Lambda^{-(m-1)}f\|_1
\end{aligned}
$$

$$\leqslant \| \Phi(G, \cdot) \|_\infty .$$

The last inequality is justified by (3.13).

Now (3.2) follows from (3.6) and (3.7). By Theorem 2.1, similarly to the arguments in the proof of (3.1), we get (3.3). The proof of Theorem 3.1 is completed.

References

[1] Chen Dirong. On the cardinal spline interpolation corresponding to the infinite order differential operators. Acta Math. Sinica(N.S.), 1994, (3):315 ~ 324

[2] Chen Dirong. The comparison theorems of Kolmogorov type on a convolution class of functions. Northeast. Math. J., 1994, 10(1):47 ~ 52

[3] Chen Dirong. On the n-widths of some classes of periodic functions. Sci. China Ser. A, 1992, 35(1):42 ~ 54

[4] Chen Dirong. Optimal recovery and average n-K widths for some function classes in $L_2(\mathbf{R})$. ATA., 1991, 7(3):45 ~ 62

[5] Chen Dirong AND Sun Yongsheng. Hardy-Littlewood-Pólya inequality for a linear differential operator and some related optimal problems. ATA., 1992, 8(1):50 ~ 58

[6] I. I. Hirschman and D. V. Widder. *The Convolution Transform*. N. Y.: Princeton Press, 1955. 37

[7] Li Chun. Cardinal spline interpolation with PF density. preprint

[8] Li Xiaoshen. Optimal recovery for a class of convolutions of functions. Northeast. Math. J., 1993, 9(3):377 ~ 387

[9] Nguen Thi Thieu Hoa. On extremal problem for variation-diminishing transformations. Vestnik Moskov. Univ. Ser. I Mat. Meh., 1982, (5): 3 ~ 7 (in Russian)

[10] A. Pinkus. *N-width in Approximation Theory*. Berlin: Springer-Verlag, 1985

[11] A. Sharma and J. Tzimbalario. Landau type inequalities for some linear differential operators. Illinois J. Math., 1976, (20):433 ~ 455

[12] Sun Yongsheng. Optimal interpolation for convolution function class. Northeast. Math, J., 1990, (1):9 ~ 25

[13] Sun Yongsheng. *Approximation Theory* I. Beijing: Beijing Normal Univ. Press,

1989 437. (in Chinese) 437

[14] V.M. Tikhomirov. On the optimal subspaces of classes of functions with cyclic variation-diminishing kernel. Vestnik Moskov. Univ. Ser. I Mat. Meh., 1997, (4): 16 ~ 19 (in Russian)

[15] V.M. Tikhomirov. On the approximation characteristics of smooth functions of several variables. pp. 183 ~ 188; In: S.L. Sobolev, ed., Theory of Cubature Formulas and Numerical Mathematics, Proc. Conf., Novosibirsk, 1978, Novosibirsk, Nauka, 1980 (in Russian)

East Journal On Approximations
2001, Volume 7, Number 3: 375 ~ 380

逼近论中 Hardy-Littlewood-Polya 不等式的广义版本及相关优化问题*

A Generalized Version of Hardy-Littlewood-Polya Inequality and Some Related Optimal Problems in Approximation Theory

1. The problem

Let $G(\cdot)$, $K(\cdot)$ be real 2π-periodic continuous functions. We say that $G(\cdot)$ satisfies Property B (in brief, $G \in B$), if for every choice of $0 \leqslant y_1 < \cdots < y_m < 2\pi (m \geqslant 1)$,

$$X_m(G) := \left\{ b + \sum_{j=1}^{m} b_j G(\cdot - y_j) \mid b \in \mathbf{R}, \sum_{j=1}^{m} b_j = 0 \right\},$$

is of dimension m, and $X_m(G)$ is a weak Haar space on T for odd m. Recall that the m-dimensional space $Z_m = \text{span}\{u_1, \cdots, u_m\}$ is called a *week Haar space* on T if

$$U\begin{pmatrix} 1, \cdots, m \\ x_1, \cdots, x_m \end{pmatrix} := \det(u_i(x_j))_{i,j=1}^{m}$$

is of one sign (i.e., non-negative or non-positive) for all choice of $0 \leqslant x_1 < \cdots < x_m < 2\pi$.

Also, we say that $K(\cdot)$ is *non-degenerate cyclic variation diminishing* (in brief, $K \in \text{NCVD}$), if for every choice of $0 \leqslant y_1 < \cdots < y_m < 2\pi (m \geqslant 1)$,

---

* *Received May* 18, 2001.

* 本文与汪和平合作.

$$Y_m(K) := \text{span}\{K(\cdot - y_1), \cdots, K(\cdot - y_m)\},$$

is of dimension m, and $Y_m(K)$ is a weak Haar space for odd m.

The concept of Property NCVD and Property B have been introduced by Pinkus([5], pp. 126 ~ 129). For a given $p, 1 \leqslant p \leqslant \infty$, we denote

$$\widetilde{K}_p(K) := \{K * h \mid \|h\|_p \leqslant 1\},$$

$$\widetilde{B}_p(G) = \{a + G * h \mid a \in \mathbf{R}, h \perp 1, \|h\|_p \leqslant 1\},$$

where

$$(K * h)(x) = \int_0^{2\pi} K(x - t) h(t) \mathrm{d}t,$$

$\|h\|_p$ is the usual $L_p(T)$-norm, and $h \perp 1$ means $\int_0^{2\pi} h(t)\mathrm{d}t = 0$. A series of results have been published which concern the set of optimal subspaces for the $2n$—Kolmogorov widths $\mathrm{d}_{2n}(\widetilde{K}_p(K), L_p(\mathrm{T}))$, $\mathrm{d}_{2n}(\widetilde{B}_p(G), L_p(\mathrm{T}))$, $p = 1, \infty$ (see [6 ~ 8]). Among other things, it has been revealed in [7] that the above results rely heavily upon a new generalized version of the famous Landau-Kolmogorov inequality and the Stein inequality. Actually, we have established the following two theorems (see [7]).

Theorem A *Suppose that $G_1, G_2 \in B$, and $G := G_1 * G_2 \in B$. If for $\varphi \in L_\infty(T)$, $\|\varphi\|_{L\infty} \leqslant 1$, $\varphi \perp 1$, $a \in \mathbf{R}$, $n \in \mathbf{Z}_+$, we have*

$$\|a + G * \varphi\|_{C(\mathrm{T})} \leqslant \|\Phi_n(G; \cdot)\|_{C(\mathrm{T})}, \tag{1}$$

then it holds

$$\|G_2 * \varphi\|_{C(\mathrm{T})} \leqslant \|\Phi_n(G_2; \cdot)\|_{C(\mathrm{T})}$$

where $\Phi_n(G; x) := (G * \mathrm{sgn}\ \sin\ n(\cdot))(x)$ is the generalized Euler spline associated with G. The inequality is sharp in the sense that for $\forall\, \varepsilon > 0$ there exists a 2π-periodic function φ satisfying $\|\varphi\|_{L\infty} \leqslant 1$, $\varphi \perp 1$, and a real number a such that (1) is fulfilled while

$$\|G_2 * \varphi\|_{C(\mathrm{T})} > \|\Phi_n(G_2; \cdot)\|_{C(\mathrm{T})} - \varepsilon.$$

Theorem B *Suppose that $G_1, G_2 \in B$, $G := G_1 * G_2 \in B$ and $1 \leqslant p < \infty$. If for a certain function $\varphi \in L_p(\mathbf{T})$ such that $\|\varphi\|_{L_p(\mathrm{T})} \leqslant 1$, $\varphi \perp 1$, and numbers $a \in \mathbf{R}$, $n \in \mathbf{Z}_+$, we have*

$$\| a + G * \varphi \|_{L_p(\mathbf{T})} \leqslant \| \Phi_n(G;\cdot) \|_{C(\mathbf{T})}, \tag{2}$$

then it holds

$$\| G_2 * \varphi \|_{L_p(\mathbf{T})} \leqslant \| \Phi_n(G_2;\cdot) \|_{C(\mathbf{T})}.$$

Besides, for $p = 1$ this inequality is sharp in the sense that for $\forall \varepsilon > 0$ there exists a 2π-periodic function φ satisfying $\| \varphi \|_{L_1(\mathbf{T})} \leqslant 1, \varphi \perp 1$, and a real number a such that (2) is fulfiled (for $p = 1$) while

$$\| G_2 * \varphi \|_{L_1(\mathbf{T})} > \| \Phi_n(G_2;\cdot) \|_{C(\mathbf{T})} - \varepsilon.$$

Similar results hold in the case $K \in$ NCVD(see [6]).

In connection with these results we raise the following open problem.

Open problem Prove or disprove the following.

Conjecture A *Suppose that $G_1, G_2 \in B$ and $G = G_1 * G_2 \in B$. If for a certain function $\varphi \in L_2(\mathrm{T})$ such that $\| \varphi \|_{L_2} \leqslant 1, \varphi \perp 1$, and numbers $a \in \mathbf{R}$, $n \in \mathbf{Z}_+$, we have*

$$\| a + G * \varphi \|_{L_2}(\mathbf{T}) \leqslant | a_n^{(1)} \cdot a_n^{(2)} |,$$

then it holds

$$\| G_2 * \varphi \|_{L_2(\mathbf{T})} \leqslant | a_n^{(2)} |$$

where

$$a_n^{(j)} := \frac{1}{2\pi} \int_0^{2\pi} G_j(t) \mathrm{e}^{-int} \mathrm{d}t \, (j = 1, 2)$$

is the Fourier coefficient of $G_j(\cdot)$ (which is a real function).

Analogous proposition may be formulated in the case $K \in$ NCVD.

Recall that according to the famous Hardy-Littlewood-Pólyaiinequality, for every

$$\begin{aligned} f \in L_2^r &:= \{ f \in L_2(\mathrm{T}) \mid \| f^{(r)} \|_{L_2(\mathbf{T})} \leqslant 1 \} \\ &= \{ f \in L_2(\mathrm{T}) \mid f = a + B_r * f^{(r)}, \| f^{(r)} \|_{L_2(\mathbf{T})} \leqslant 1 \} \\ &= \tilde{B}_2(B_r), \end{aligned}$$

$f \neq$ const, $r, k \in \mathbf{Z}_+$, $k \in (0, r)$, we have

$$\left(\frac{\| f^{(r-k)} \|_2}{\| f^{(r)} \|_2} \right)^{1/k} \leqslant \left(\frac{\| f \|_2}{\| f^{(r)} \|_2} \right)^{1/r}.$$

Thus the inequality formulated above is equivalent to the

Hardy-Little-wood-Pólya inequality in the classical case, when $G_1(x) = B_s(x), G_2(x) = B_k(x)$ and $G(x): = G_1 * G_2(x) = B_{s+k}(x)$, where

$$B_r(x): = \frac{1}{2\pi}\sum_{\substack{v=-\infty \\ v\neq 0}}^{\infty}\frac{e^{ivx}}{(iv)^r}$$

is the Bernoulli function, $s, k \in \mathbf{Z}_+$.

If the conjecture is true, then a series of new results about the optimal subspaces for the $2n$-Kolmogorov width $d_{2n}(\widetilde{B}_2(G), L_2(\mathrm{T}))$ could be derived readily, for example, the following corollaries.

Corollary 1 *Let G be a B-kernel, $G_1 \in B$, and $G_2: = G_1 * G \in B$. Then the $2n$-width $d_{2n}(\widetilde{B}_2(G), L_2(\mathrm{T}))$ has an optimal subspace*

$$W_{2n}(G_2): = \left\{b + \sum_{j=1}^{2n} b_j G_2\left(\cdot - \frac{j\pi}{2n}\right) \middle| b \in \mathbf{R}, \sum_{j=1}^{2n} b_j = 0\right\}.$$

In Corollary 1, as G_1 we take the Bernoulli function $B_r(x), r \in \mathbf{Z}_+$. Strictly speaking, $B_r(x)$ satisfies B-condition for $r \geqslant 2$, $B_1(x)$ satisfies all conditions of property B except the continuity requirement(it has a jump discontinuity at $x = 0$). But we may include B_1 in the B-kernel class without any serious complications. Then $G_2 = B_r * G = G^{(r)}$, which is the r-fold periodic integral of G. Note that $G^{(r)} \in B, r \in \mathbf{Z}_+$ (see [5], p. 132). Consequently, we derive the following statement from Corollary 1.

Corollary 2 *Let G be a B-kernel. Then the $2n$-width $d_{2n}(\widetilde{B}_2(G), L_2(\mathrm{T}))$ has an infinite set of optimal subspaces $\{W_{2n}(G^{(r)}) \mid r \in \mathbf{Z}_+\}$.*

As a special case of $\widetilde{B}_2(G)$ we take the Sobolev class $\widetilde{B}_2{}^{(r)}: = \widetilde{B}_2(B_r)$ (with a fixed $r \in \mathbf{Z}_+$). Since for any $G \in B$ we have $G * B_r = B_r * G = G^{(r)} \in B$, the following is true.

Corollary 3 *The $2n$-width $d_{2n}(\widetilde{B}_2{}^{(r)}, L_2(\mathrm{T}))$ has an infinite set of optimal subspaces $\{W_{2n}(G^{(r)}) \mid G \in B\}$.*

2. Partial results

Given $r \geqslant 2$, $P_r(\lambda) = \Pi_{j=1}^r(\lambda - \lambda_j), \lambda_j \in \mathbf{R}$, the generalized Bernoulli function associated with the differential operator $P_r(D)\left(D = \frac{\mathrm{d}}{\mathrm{d}x}\right)$ is defined by

$$G^{(P_r)}(t):=\frac{1}{2\pi}\sum_{v=-\infty}^{\infty}{}'\frac{e^{ivt}}{P_r(iv)},\qquad i=\sqrt{-1}.$$

where the symbol $\sum'$ means that when $P_r(0)=0$ the corresponding term in the sum is absent. It is known that $G^{(P_r)}\in B$ if $P_r(0)=0$, and $G^{(P_r)}\in$ NCVD, if $P_r(0)\neq 0$. By means of the class of generalized Bernoulli functions Chakhiev(see[1])defined a class RA_c as follows.

Definition. *We say that a* 2π*-periodic continuous function* $\Psi(\cdot)$ *belongs to* RA_c, *if there exists a sequence of polynomials* $\{P_{r_j}(\lambda)\}_{j=1}^{+\infty}$ *such that*

$$P_{r_j}(\lambda)=\prod_{s=1}^{r_j}(\lambda-\lambda_s^{r_j}),\ \lambda_s^{(r_j)}\in\mathbf{R},\text{ and }\lim_{j\to\infty}\|\Psi(\cdot)-G^{(P_{r_j})}(\cdot)\|_{C(\mathbf{T})}=0.$$

It is easily seen that $\Psi(\cdot)\in B$, if $\Psi\perp 1$, and $\Psi(\cdot)\in$ NCVD otherwise. So that we get a subest RA_c of B $\cup$ NCVD. In this section, we give an affirmative answer to the open problem on the subclass RA_c.

Theorem 1 *Given* $n\geqslant 2$, $P_n(\lambda)=\Pi_{j=1}^n(\lambda-\lambda_j)$, $\lambda\in\mathbf{R}$. *Let* $P_r(\lambda)$ *be a factor of* $P_n(\lambda)$ *of degree* r, $0\leqslant r\leqslant n$. *Then, for any* 2π*-periodic differentiable function* f *such that* $\|P_n(D)f\|_{L_2}<\infty$, *it holds*

$$\|P_r(D)f\|_{L_2}\leqslant\|f\|_{L_2}\cdot\varphi_r\circ\varphi_n^{-1}\left(\frac{\|P_n(D)f\|_{L_2}}{\|f\|_{L_2}}\right)$$

where

$$\varphi_r(t)=|P_r(it)|,\ (1\leqslant r\leqslant n),\ i=\sqrt{-1}.$$

The inequality is sharp in the sense that for $\forall\varepsilon>0$ *there exists a* 2π*-periodic function* f *such that* $f\neq 0$, $\|P_n(D)f\|_{L_2}<+\infty$, *and*

$$\|P_r(D)f\|_{L_2}>\|f\|_{L_2}\cdot\varphi_r\circ\varphi_n^{-1}\left(\frac{\|P_n(D)f\|_{L_2}}{\|f\|_{L_2}}\right)-\varepsilon.$$

The proof of Theorem 1 follows the same line as in [2]. We omit the details. It is easily seen that in the special case when $P_n(D)=D^n$ this is the classical Hardy-Littlewood-Polya inequality. As a consequence of Theorem 1 we have

Theorem 2 *Suppose that* $G_1, G_2 \in RA_c, G = G_1 * G_2$. *If* $\int_0^{2\pi} G(t)\mathrm{d}t = 0$, *then* $G \in B$. *Furthermore, if for* $\varphi \in L_2(\mathrm{T})$, $\| \varphi \|_{L_2} \leqslant 1, \varphi \perp 1, a \in \mathbf{R}$, *we have*

$$\| a + G * \varphi \|_{L_2} \leqslant | a_n^{(1)} \cdot a_n^{(2)} |,$$

then it holds

$$\| G_2 * \varphi \|_{L_2} \leqslant | a_n^{(2)} |.$$

The formulation for the case $\int_0^{2\pi} G(t)\mathrm{d}t \neq 0$ is analogous.

The proof of Theorem 2 can be reduced to the case when G_1, G_2 are both generalized Bernoulli functions. For such case the proof follows the same line as in [2] (see [2], p. 53, Corollary).

Remark The class B ∪ NCVD is essentially wider than its subset RA_c. Actually, from [3] we know that there exists a function $\Phi(\cdot) \in$ NCVD such that $\Phi(\cdot) \overline{\in} RA_c$. So, it seems that one needs a new method in order to solve the problem.

References

[1] M.A. Chakhiev. Linear differential operator with real spectrals and optimal quadrature formulas. Izv. Akad. Nauk SSSR Ser. Mat., 1984, (48): 1078 ~ 1109 (in Russian)

[2] Chen Dirong and Sun Yongsheng. Hardy-Littlewood-Polya inequality for a linear differential operator and some related optimal problems. ATA., 1992, 8(1): 50 ~ 58

[3] J.C. Mairhuber, I.J. Schoenberg and R.E. Williamson. On variation-diminishing transformations of the circle. Rend. Circ. Mat. Palermo, 1979, 8(2): 241 ~ 270

[4] Nguen Thi Thien Hoa. An extremal problem for variation-diminishing transformations. Vestnik Moskov. Univ. Ser. I Mat. Meh., 1982, (5): 3 ~ 7 (in Russian)

[5] A. Pinkus. *n-Widths in Approximation Theory*. New York: Springer-Verlag, 1985. 126 ~ 129

[6] Sun Yongsheng. Some remarks on the optimal subspaces of a convolution class with NCVD kernel. ATA., 1999, 15(4): 1 ~ 5

[7] Sun Yongsheng. Optimal subspaces for a periodic convolution class with a B-kernel. East J. Approx, 2000, 6(1): 53 ~ 63

[8] V. M. Tikhomirov. On the optimal subspaces of functions with cyclic variation-diminishing kernel. Vestnik Moskov. Univ. Ser. I Mat. Meh., 1997, (4): 16 ~ 19(in Russian)

四、全实轴上光滑函数类的逼近及其最优恢复

IV. Approximation and Best Recovery of the Smooth Function Classes on the Real Line

Approximation Theory and Its Applications
Dec., 1986 Vol. 2, No. 4

关于一个可微函数类的最优插值(Ⅰ)*

On Optimal Interpolation for a Differentiable Function Class (Ⅰ)

§ 1. Introduction

Given and integer $n \geqslant 1$. Let $W_\infty^{(n)}(\mathbf{R})$ be a class of differentiable functions defined on $\mathbf{R}$, where for each $f \in W_\infty^{(n)}(\mathbf{R}) f^{(n-1)}$ is absolutely continuous on any finite interval and $\| f^{(n)} \|_{\infty(\mathbf{R})} \leqslant 1$. We denote $k = W_\infty^{(n)}(\mathbf{R}) \cap L^\infty(\mathbf{R})$. For any integer $k \geqslant 1$ we choose in $[-k, k]$ $2k$ points arbitrarily, ie.,

$$\Xi_{2k} =: \{-k \leqslant \xi_{-k+1} < \xi_{-k+2} < \cdots < \xi_k < k\},$$

and we denote $\widetilde{\Xi}_{2k}$ the continuation of Ξ_{2k} with period $2k$ on the whole real line. The set of all such denumerable sets (for all $k = 1, 2, 3, \cdots$) is denoted by Θ. For any $f \in K$ let

$$I_\Xi(f) = \{\cdots, f(\xi_{-j}), \cdots, f(\xi_0), \cdots f(\xi_j), \cdots\} \tag{1}$$

where

$$\Xi = \{\cdots < \xi_{-j} < \cdots < \xi_0 < \cdots < \xi_j < \cdots\} \in \Theta$$

$I_\Xi(f)$ is called the information operator. Our problem is to recover optimally the value of $f \in K$ at a given point $\tau \in \mathbf{R}$ by using the limited information $I_\Xi(f)$ about f, where $\Xi \in \Theta$ is a denumerable set for sampling. Assuming

---

* Projects Supported by the Science Fund of the Chinese Academy of Sciences.

error-free information we have the intrinsic error

$$E(K,\Xi,\tau)=:\inf_{A}\sup_{f\in K}|f(\tau)-AI_{\Xi}(f)|,\tag{2}$$

in which A is any mapping (algorithm) from $I_{\Xi}K$ into $\mathbf{R}$. According to [1], if we set

$$e(K,\Xi,\tau)=:\sup_{\substack{f\in K\\ I_{\Xi}(f)=0}}|f(\tau)|,\tag{3}$$

then

$$E(K,\Xi,\tau)\geqslant e(K,\Xi,\tau).\tag{4}$$

In this note we put $E(K,\Xi)=\|E(K,\Xi,\cdot)\|_{\infty}$ and consider the following extremal problem.

Problem Denote

$$E(K)=:\inf_{\Xi\in\Theta}E(K,\Xi).\tag{5}$$

It is required to obtain the exact expression for $E(K)$ and to determine a denumerable set $\Xi_0\in\Theta$ for which the infimum in (5) is achieved.

In what follows we will give a solution to this problem.

§ 2. Estimation for E(K) from above

Denote

$$\Xi_0=\{\cdots,-j,\cdots,0,\cdots,j,\cdots\}.\tag{6}$$

We have

$$E(K)\leqslant\|E(K,\Xi_0,\cdot)\|_{\infty}\leqslant \sup_{f\in K}\|f(\tau)-A_0I_{\Xi_0}(f;\tau)\|_{\infty},\tag{7}$$

where A_0 is defined by

$$A_0I_{\Xi_0}(f;\tau)=\sum_{j=-\infty}^{\infty}f_jL(\tau-j+\alpha),\tag{8}$$

$f_j=f(j)$, $L(x)$ is the fundamental $\mathscr{L}$ - spline associated with the differential operator D^n. From [2] we know

(1) $L(\alpha+j)=\delta_{0j}$, (9)

where $\alpha=0$ for even n, and $\alpha=\frac{1}{2}$ for odd n.

(2) There exist constants $A,B>0$ such that $|L(x)|\leqslant Ae^{-B|x|}$

for any $x \in \mathbf{R}$. By [7] [8] we have

$$\sup_{f \in K} \| f(\tau) - A_0 I_{\Xi_0}(f;\tau) \|_\infty = \| E_n(\cdot) \|_\infty, \tag{10}$$

where $E_n(x)$ is the Euler spline: $E_n(x+1) = -E_n(x)$, $|E_n^{(n)}(x)| = 1$. So we obtain

$$E(K) \leqslant \| E_n(\cdot) \|_\infty. \tag{11}$$

§ 3. Estimation for E(K) from below

From (4) we get

$$E(K) \geqslant \inf_{\Xi \in \Theta} \| \mathrm{e}(K, \Xi, \cdot) \|_\infty. \tag{12}$$

3-1. Periodic Case

For every positive integer k we put

$$\widetilde{K}_{2k} = \{ f : f \in K, f(x+2k) = f(x) \}.$$

Now for any $\widetilde{\Xi}_{2k} \in \Theta$ we consider the following problem:

$$\| e(\widetilde{K}_{2k}, \widetilde{\Xi}_{2k}, \cdot) \|_\infty = \sup_{\substack{f \in \widetilde{K}_{2k} \\ I_{\widetilde{\Xi}_{2k}}(f) = 0}} \| f(\tau) \|_\infty. \tag{13}$$

As the 2π-periodic case which had been studied by Velikin[3] we may apply the antipodal theorem of Borsuk[4] and have

Lemma 1 *For any given $\widetilde{\Xi}_{2k}$ there exists $f(x) = f(\widetilde{\Xi}_{2k}; x) \in \widetilde{K}_{2k}$ such that*

(1) $I_{\widetilde{\Xi}_{2k}}(f) = 0$.

(2) $|f^{(n)}(\widetilde{\Xi}_{2k}; x)| = 1$, and $f^{(n)}$ *in* $[-k, k)$ *has exactly* $2k$ *points of sign change. In other words, f is a 2k-periodic perfect spline with 2k knots in a period.*

Obviously

$$\sup_{\substack{f \in \widetilde{K}_{2k} \\ I_{\widetilde{\Xi}_{2k}}(f) = 0}} \| f(\cdot) \|_\infty \geqslant \| f(\widetilde{\Xi}_{2k}; \cdot) \|_\infty. \tag{14}$$

Let $\widetilde{\Pi}_{2k}$ be the subset of $\widetilde{K}_{2k}$ which consists of all perfect splines f_η with free knots such that the number of knots in a period is no more than $2k$. Thus for every $f_\eta \in \widetilde{\Pi}_{2k}$ we have $|f_\eta^{(n)}(x)| = 1$, $V_C(f_\eta^{(n)}) \leqslant 2k$. The minimum sup-norm problem on $\widetilde{\Pi}_{2k}$ is quite similar to the one which had been considered for the 2π-periodic case. We have

Lemma 2 *For any k*

$$\min_{f_\eta \in \widetilde{\Pi}_{2k}} \| f_\eta(\cdot) \|_\infty = \| E_n(\cdot) \|_\infty. \tag{15}$$

The proof of (15) is analogous to that given by Pinkus[5] or by Makovoz[6] for 2π-periodic case. We omit the details. From (14) (15) we derive

$$\sup_{\substack{f\in \widetilde{K}_{2k} \\ I_{\widetilde{\Xi}_{2k}}(f)=0}} \| f(\cdot) \|_\infty \geqslant \| E_n(\cdot) \|_\infty. \tag{16}$$

To consider (13) for a special $\widetilde{\Xi}_{2k} = \Xi_0$, we have

Lemma 3 For any k

$$\sup_{\substack{f\in \widetilde{K}_{2k} \\ I_{\widetilde{\Xi}_{2k}}(f)=0}} \| f(\cdot) \|_\infty = \| E_n(\cdot) \|_\infty. \tag{17}$$

(17) May be proved by the method of approach of [6]for 2π-periodic case. We omit the details also. On account of (16) (17) we get

Lemma 4 *For any k*

$$\min_{\widetilde{\Xi}_{2k}} \| e(\widetilde{K}_{2k}, \widetilde{\Xi}_{2k}, \cdot) \|_\infty = \| e(\widetilde{K}_{2k}, \Xi_0, \cdot) \|_\infty = \| E_n(\cdot) \|_\infty. \tag{18}$$

3-2. General Case

The general case may be treated readily. For any $\Xi \in \Theta$, by definition, we have $\Xi = \widetilde{\Xi}_{2k}$ for some k, and

$$\| e(K, \Xi, \cdot) \|_\infty = \sup_{\substack{f\in K \\ I_\Xi(f)=0}} \| f(\cdot) \|_\infty \geqslant \sup_{\substack{f\in \widetilde{K}_{2k} \\ I_{\widetilde{\Xi}_{2k}}(f)=0}} \| f(\cdot) \|_\infty$$

$$\geqslant \| E_n(\cdot) \|_\infty, \text{ by (16)}.$$

Therefore

$$\inf_{\Xi\in\Theta} \| e(K, \Xi, \cdot) \|_\infty \geqslant \| E_n(\cdot) \|_\infty,$$

and by (12) we obtain $E(K) \geqslant \| E_n(\cdot) \|_\infty$. Combining with (11) we get

Theorem

$$E(K) = \| E_n(\cdot) \|_\infty. \tag{19}$$

$A_0 I \Xi_0(f)$ *is an optimal interpolating operator.*

4. Remarks

(1)The above result may be extended to the differentiable function class defined by the self-adjoint operator

$$P_n(D) = D^\sigma \prod_{j=1}^{l} (D^2 - t_j^2), \sigma \geqslant 0, n = \sigma + 2l, t_j > 0.$$

(2) We don't know whether the theorem is true for non-periodic denumerable sets of sampling, if these sets are reasonably chosen. To be precise, we get

$$\Xi = \{\cdots < \xi_{-j} < \cdots < \xi_0 < \cdots < \xi_j < \cdots\} \tag{20}$$

where ξ_j may be chosen arbitrarily in $[j-1, j)$, $j = 0, \pm 1, \pm 2, \cdots$, and denote Θ^* the set of all such Ξ. We conjecture that in this case our theorem still holds true. There is no difficulty in obtaining estimation from above.

References

[1] Micchelli, C. A., Rivlin, T. J. A Survey of optimal recovery. In: C. Micchel and others edited: *Optimal Estimation in Approximation Theory*. N. Y.: Plenum Press, 1977. 1 ~ 68

[2] Micchelli, C. A., Cardinal $\mathscr{L}$-splines. In: S. Karlin edited: *Studies in spline functions and approximation theory*. N. Y.: Acad. Press, 1976. 203 ~ 250

[3] Великин, В. Л. Матем. Заметки, 1977, т. 22 No 5: 663 ~ 670

[4] Borsuk, K. Fund. Math., 1933, (20): 177 ~ 191

[5] Pinkus, A. J. D'analyse Math., 1979, (35): 209 ~ 235

[6] Makovoz, Y. I. Матем, Заметки. 1979, T. 20, No. 5, 805 ~ 812

[7] Schoenberg, I. J. On the remainders and convergence of cardinal spline interpolation for almost periodic functions. In *Studies in spline functions and approximation theory*. N. Y.: Acad. Press, 1976

[8] d'Boor, Carl, Schoenberg, I. J. Cardinal interpolation and spline function VIII. Lecture Notes in Math., N. Y.: Springer-Verlag, 501, 1976

KEXUE TONGBAO
August, 1987, Vol · 32, No · 16

可微函数类的最优恢复(二重取样)

Optimal Interpolation on Some Class of Differentiable Functions* (Double Sampling)

Ⅰ. Introduction

Given a positive integer $n \geqslant 2$. Let $W_{\infty}^{(n)}(\mathbf{R})$ be the set of differentiable functions defined on $\mathbf{R}$ where for each $f \in W_{\infty}^{(n)}(\mathbf{R})$ $f^{(n-1)}$ is absolutely continuous on any interval bounded on $\mathbf{R}$ and $\| f^{(n)} \|_{\infty} \overset{\mathrm{df}}{=} \operatorname{ess}\sup_{x \in \mathbf{R}} | f(x) | \leqslant 1$. Denote $K = W_{\infty}^{(n)}(\mathbf{R}) \cap L^{\infty}(\mathbf{R})$. Let Ξ be the set of all denumerable sets $\tilde{\xi} = \{\xi_j\}_{j=-\infty}^{+\infty}$ such that for every $\tilde{\xi}$ there is some α, $2j \leqslant \xi_j - \alpha < 2(j+1)$ for $j = 0, \pm 1, \pm 2, \cdots$, $\tilde{\Xi}_{2k}$ be the subset of Ξ where each $\tilde{\xi} \in \tilde{\Xi}_{2k}$ is $2k$-periodic, viz., $\xi_{j+2k} = \xi_j$ for $j = 0, \pm 1, \pm 2, \cdots$. Denote $\tilde{\Xi} = \bigcup_{k=1}^{\infty} \tilde{\Xi}_{2k}$, and $I_{\tilde{\xi}}(f) = \{(f(\xi_j))_{j=-\infty}^{+\infty}, (f'(\xi_j))_{j=-\infty}^{+\infty}\}$. Fix a $\tau \in \mathbf{R}$ and consider the problem of recovering optimally the value of $Uf = f(\tau)$ on the set K by using the information $I_{\tilde{\xi}}(f)$ about $f \in K$, where $\tilde{\xi} \in \tilde{\Xi}$ is a given denumerable set of sampling. Assuming the information to be error-free, following [1] we define the intrinsic error.

$$E(K, \tilde{\xi}, \tau) \overset{\mathrm{df}}{=} \inf_{A} \sup_{f \in K} | f(\tau) - AI_{\tilde{\xi}}(f) |, \tag{1}$$

where A is any mapping (algorithm) from $I_{\tilde{\xi}}(K)$ into $\mathbf{R}$. Denote as in [2]

---

* Project supported by the Science Fund of Academia Sinica.

$$e(K, \tilde{\xi}, \tau) \stackrel{\text{df}}{=} \sup_{\substack{f \in k \\ I_{\tilde{\xi}}(f)=0}} |f(\tau)|. \tag{2}$$

According to [1] we have

$$E(K, \tilde{\xi}, \tau) \geqslant e(K, \tilde{\xi}, \tau). \tag{3}$$

Now we put

$$E(K, \tilde{\xi}) \xlongequal{\text{d}f} \| E(K, \tilde{\xi}, \cdot) \|_{\infty}, \tag{4}$$

$$E(K) \xlongequal{\text{d}f} \inf_{\tilde{\xi} \in \tilde{\Xi}} E(K, \tilde{\xi}). \tag{5}$$

In this note our objective is to obtain the exact expression of $E(K)$ and an optimal set of sampling points $\tilde{\xi}_0 \in \tilde{\Xi}$, viz., a $\tilde{\xi}_0 \in \tilde{\Xi}$ for which $E(K) = E(K, \tilde{\xi}_0)$.

Ⅱ. Estimation of e (K) From Below

First of all, we note that

$$E(K) \geqslant \inf_{\tilde{\xi} \in \tilde{\Xi}} \| e(K, \tilde{\xi}, \cdot) \|_{\infty}.$$

(1) To begin with, for an arbitrary integer $k \in \mathbf{N}$ we put

$\tilde{K}_{2k} = \{f : f \in K, f(x + 2k) = f(x)\}$.

On this class of $2k$ − periodic functions we consider the following extremal problem:

$$\| e(\tilde{K}_{2k}, \tilde{\xi}_k, \cdot) \|_{\infty} = \sup_{\substack{f \in \tilde{K}_{2k} \\ I_{\tilde{\xi}_k}(f)=0}} \| f(\cdot) \|_{\infty}, \tag{6}$$

where $\tilde{\xi}_k \in \tilde{\Xi}_k$. We have

Lemma 1 *For any $\tilde{\xi}_k \in \tilde{\Xi}_k$, there exists a function $f(x) = f(\tilde{\xi}_k; x) \in \tilde{K}_{2k}$ such that*

(ⅰ) $I_{\tilde{\xi}_k}(f) = 0$;

(ⅱ) $f(x) \geqslant 0$;

(ⅲ) $| f^{(n)}(\tilde{\xi}_k; x) | = 1$.

and $f^{(n)}(\tilde{\xi}_k; x)$ has exactly $2k$ points of sign changes in $[-k, k)$.

It is worthy to note that $f(\tilde{\xi}_k; x)$ is a $2k$ − periodic perfect spline of degree n with $2k$ knots in a period. The proof of Lemma 1 is analogous to that of the 2π – periodic case (cf. [3], or[4]). Obviously we have

$$\sup_{\substack{f\in \widetilde{K}_{2k}\\ I_{\tilde{\xi}_k}(f)=0}} \| f(\cdot)\|_\infty \geqslant \| f(\tilde{\xi}_k;\cdot)\|_\infty. \tag{7}$$

Let $\widetilde{\Pi}_{2k}$ be the subset of $\widetilde{K}_{2k}$ in which every $f_\eta \in \widetilde{\Pi}_{2k}$ satisfies that

(ⅰ) f_η is a perfect spline function of degree n,

(ⅱ) $f_\eta \geqslant 0$(or $\leqslant 0$),

(ⅲ) f_η has $2m$ free knots in $[-k,k]$ and $m\leqslant k$.

Lemma 2 $\forall_k \in \mathbf{N}$

$$\min_{f_\eta\in \widetilde{\Pi}_{2k}} \| f_\eta(\cdot)\|_\infty = \| E_n(x) + \| E_n(\cdot)\|_\infty \|_\infty.$$

$E_n(x)$ is the Euler spline of degree n.

This lemma may be proved by the method in [4], where the 2π – periodic case was considered. From Lemmas 1 and 2 we conclude that for every $\tilde{\xi}_k \in \widetilde{\Xi}_{2k}$ there is a $f(\tilde{\xi}_k;x)\in \widetilde{\Pi}_{2k}$ such that

$$\| f(\tilde{\xi}_k;\cdot)\|_\infty \geqslant \| E_n(x) + \| E_n(\cdot)\|_\infty \|_\infty. \tag{9}$$

So we have

Lemma 3 $\forall\ \tilde{\xi}_k$

$$\| e(\widetilde{K}_{2k},\tilde{\xi}_k,\cdot)\|_\infty \geqslant \| E_n(x) + \| E_n(\cdot)\|_\infty \|_\infty. \tag{10}$$

The next lemma extends a result of Velikin[5] for the 2π-periodic case.

Lemma 4 $\forall k\in \mathbf{N}$

$$\sup_{\substack{f\in \widetilde{K}_{2k}\\ I_{\tilde{\xi}_0}(f)=0}} \| f(\cdot)\|_\infty = \| E_n(x) + \| E_n(\cdot)\|_\infty \|_\infty, \tag{11}$$

where $\tilde{\xi}_0 = \{2j\}_{j=-\infty}^{+\infty}\left(\textit{or}\ \left\{2j+\frac{1}{2}\right\}_{j=-\infty}^{+\infty}\right).$

From the above four lemmas we derive $\left(\widetilde{K} \overset{\text{df}}{=} \bigcup_{n=1}^{\infty} \widetilde{K}_{2k}\right)$

$$\begin{aligned} E(\widetilde{K}) &\geqslant \min_{\xi_k} \| e(\widetilde{K}_{2k},\tilde{\xi}_k,\cdot)\|_\infty = \| e(\widetilde{K}_{2k},\tilde{\xi}_0,\cdot)\|_\infty \\ &= \| E_n(x) + \| E_n(\cdot)\|_\infty \|_\infty. \end{aligned} \tag{12}$$

(2)We turn to the general case. For every $\tilde{\xi}\in \widetilde{\Xi}$ there exists some integer k such that $\tilde{\xi}\in \widetilde{\Xi}_{2k}$. So we get

$$\| e(K,\xi,\cdot)\|_\infty \overset{\text{df}}{=} \sup_{\substack{f\in K\\ I_{\tilde{\xi}}(f)=0}} \| (\cdot)\|_\infty \geqslant \sup_{\substack{f\in \widetilde{K}_{2k}\\ I_{\tilde{\xi}_k}(f)=0}} \| f(\cdot)\|_\infty \geqslant$$

$$\| E_n(x) + \| E_n(\cdot) \|_\infty \|_\infty .$$

Consequently

$$E(K) \geqslant \| E_n(x) + \| E_n(\cdot) \|_\infty \|_\infty . \tag{13}$$

Ⅲ. Estimation of $E(K)$ From Above

For a given $\tau \in \mathbf{R}$ there is a $k \in \mathbf{N}$ such that $|\tau| < k$. Take $\tilde{\xi}_k \in \tilde{\Xi}_{2k}$ arbitrarily, and $\alpha \in (0,1)$. Following the method of Cavaretta[6] for $f \in K$ we set

$$F_k(f_h; x) \overset{\text{df}}{=} \alpha f(x+h) g\left(\frac{x}{k}\right), \quad |x| \leqslant 2k.$$

$$F_k(f_h; x+4k) = F_k(f_h; x) \quad x \in \mathbf{R},$$

where h is a parameter, $f_h(\cdot) = f(\cdot + h)$, and $g(x)$ is the function defined by Cavaretta in [6]. Denote $M_n(f) \overset{\text{df}}{=} \| g^{(n)} \|_\infty$. In [6] it has been proved that

$$M_n(F_k(f_h)) \underset{k \to +\infty}{\longrightarrow} \alpha M_n(f_h) = \alpha \| f^{(n)} \|_\infty < 1.$$

Thus for every $f \in K$ there exists a $k \in \mathbf{N}$ sufficiently large such that

$$\| F_k^{(n)}(f_h) \|_\infty \leqslant 1$$

for all $h \in \mathbf{R}$. So we have $F_k(f_h; x) \in \tilde{K}_{4k}$. If $I_{\tilde{\xi}_k}(f) = 0$, then from $F'_k(x) = \alpha f'(x) \times g\left(\frac{x}{k}\right) + \frac{\alpha}{k} g'\left(\frac{x}{k}\right) f(x)$ we also have $I_{\tilde{\xi}_k}(F_k(f)) = 0$, and similarly $I_{\tilde{\xi}_{k-h}}(F_k(f_h)) = 0$, where $\xi_k - h$ is the $-h$ translation of the point set $\tilde{\xi}_k$. According to Smolyak[7] for $\tilde{K}_{4k}$ and $I_{\tilde{\xi}_k}(f)$ there is a linear functional $L: (I_{\tilde{\xi}_k}(\tilde{K}_{4k}) \to \mathbf{R})$ such that

$$\inf_A \sup_{\phi \in \tilde{K}_{4k}} | \phi(\tau) - AI\tilde{\xi}_k(\phi) |$$

$$= \sup_{\phi \in \tilde{K}_{4k}} | \phi(\tau) - LI_{\tilde{\xi}_k}(\phi) | = \sup_{\substack{\phi \in \tilde{K}_{4k} \\ I_{\tilde{\xi}_k}(\phi) = 0}} | \phi(\tau) |,$$

where A is any functional from $I_{\tilde{\xi}_k}(\tilde{K}_{4k})$ into $\mathbf{R}$. Now we define a mapping A_0(algorithm) from $I_{\tilde{\xi}}(K)$ into $\mathbf{R}$ as follows. $\forall f \in K$ for $\{f_h\}(h \in \mathbf{R})$ we first determine an integer k which is the least one of all k such that

$$\| F_k^{(n)}(f_h) \|_\infty < 1, \quad |\tau| < k,$$

where, as before, $F_k(f_h, \tau) = \alpha f(\tau + h)$, because $g\left(\frac{\tau}{k}\right) = 1$. Then we define an algorithm A_0:

$$A_0 I_{\tilde{\xi}}(f_h) = \alpha^{-1} L I_{\tilde{\xi}_{k+h}}(F_k(f)). \tag{14}$$

Then

$$\alpha \mid f_h(\tau) - A_0 I_{\tilde{\xi}}(f_h) \mid = \mid F_k(f_h, \tau) - L I_{\tilde{\xi}_k}(F_k(f_h)) \mid \leqslant \sup_{\substack{g \in \tilde{K}_{4k} \\ I_{\tilde{\xi}_k}(g) = 0}} \mid g(\tau) \mid \leqslant \sup_{\substack{g \in \tilde{K}_{4k} \\ I_{\tilde{\xi}_k}(g) = 0}} \| g \|_\infty .$$

The last relation holds for all $h \in \mathbf{R}$ so that we have

$$\alpha \| f(\tau) - A_0 I_{\tilde{\xi}}(f) \|_\infty \leqslant \sup_{\substack{\phi \in \tilde{K} k \\ I_{\tilde{\xi}_k}(\phi) = 0}} \| \phi \|_\infty . \tag{15}$$

Denote $G(\tilde{\xi}) = \alpha \sup_{f \in K} \| f(\tau) - A_0 I_{\tilde{\xi}}(f) \|_\infty$. We know that there exist some $\tilde{\xi}$ for which $G(\tilde{\xi}) < +\infty$. Let $\{\eta_N\}$ be a sequence of numbers $\eta_N \downarrow 0$, $\eta_N > 0$. For every N there is an $f_N \in K$ such that

$$\| f_N(\tau) - A_0 I_{\tilde{\xi}}(f_N) \|_\infty > G(\tilde{\xi}) - \eta_N .$$

Then from

$$\begin{aligned} \| \alpha E(K, \tilde{\xi}, \cdot) \|_\infty &\leqslant \alpha \sup_{f \in K} \| f(\tau) - A_0 I_{\tilde{\xi}}(f) \|_\infty < \\ &\| f_N(\tau) - A_0 I_{\tilde{\xi}}(f_N) \|_\infty + \eta_N < \\ &\sup_{\substack{\phi \in \tilde{K}_{4k_N N} \\ I_{\tilde{\xi}_k} N(\phi) = 0}} \| \phi \|_\infty + \eta_N \end{aligned}$$

$(N = 1, 2, \cdots)$, where $k_N \in \mathbf{N}$ is the integer corresponding to f_N. Therefore

$$\begin{aligned} \alpha \inf_{\tilde{\xi}} \| E(K, \tilde{\xi}, \cdot) \|_\infty &\leqslant \inf_{\tilde{\xi}_{k_N}} \sup_{\substack{\phi \in \tilde{K}_{4k_N} \\ I_{\tilde{\xi}_{k_N}}(\phi) = 0}} \| \phi \|_\infty + \eta_N \\ &= \| E_n(x) + \| E_n(\cdot) \|_\infty \|_\infty + \eta_N . \end{aligned}$$

(cf. [12]). By letting $N \to +\infty$ and $\alpha \to 1$ we obtain

$$E(K) \leqslant \| E_n(x) + \| E_n(\cdot) \|_\infty \|_\infty . \tag{16}$$

To sum up we have proved

Theorem. *Let $n \geqslant 2$, Ξ, $I_{\tilde{\xi}}(f)$, K be given as in Section* Ⅰ. *Then we have*

$$E(K) = \| E_n(x) + \| E_n(\cdot) \|_\infty \|_\infty . \tag{17}$$

The corresponding 2π-periodic case of this problem was solved by Velikin in [5].

References

[1] Micchelli, C. A. & Rivlin, T. J. A survey of optimal recovery. In C. Miccbelli ed.: *Optimal Estimation in Approximation Theory*. New York: Plenum Press, 1977. 1 ~ 68

[2] Sun Yongsheng. On optimal interpolation for a differentiable function class (Ⅰ). ATA, 1986, 2(4): 49 ~ 54

[3] Моторный, В. П. Изв. АН СССР, Сер. Матем., 1974, 38: 583 ~ 614

[4] Лигун, А. А. Изв. ВУЗ, Матем., 1984, 5: 32 ~ 38

[5] Великин, В. Л. Матем. Заметки., 1977, 22(5): 663 ~ 670

[6] Cavaretta, Jr. A. S. AMM, 1974, 5: 480 ~ 486

[7] Марчук, А. Г. и Осипенко, К. Ю. Матем. Зам., 1975, 17(3): 359 ~ 368

可微函数类的最优恢复(多重取样)*

Optimal Interpolation On Some Class Of Differentiable Functions(Multi-Sampling)

This paper is devoted to studying the problem of recovering optimally the value of $Uf = f(t)$ by using the r-fold information in a fixed denumerable set of samping. For $r \geqslant 1$, we determine the optimal set of sampling points and find an optimal algorithm.

1. Introduction

Given positive integers $n, r, n \geqslant r$, we say $f \in S_n$, if on any finite interval $f^{(n-1)}$ is absolutely continuous and $f^{(n)}$ has a finite set of points of first kind discountinuity. We set

$$W_\infty^{(n)}(\mathbf{R}) = \{f \in S_n; \| f^{(n)} \|_\infty \leqslant 1\},$$

$$K := \{f \in W_\infty^{(n)}(\mathbf{R}); f(x + 2kr) = f(x), \text{for some } k\},$$

$$F(\mathbf{R}) = \{f; \text{there exists a constant } \beta > 0, \text{ such that } | f(x) | = 0(| x |)^\beta), \text{for } | x | \to \infty\}$$

Let $\tilde{\theta}_k$ be the set of sequences $\tilde{\xi}$, where

$$\tilde{\xi} = \{\xi_j\}_{j=-\infty}^{\infty}, \nu r \leqslant \xi_\nu - \alpha < r(\nu + 1)$$

for some α and $\xi_{\nu+2rk} = \xi_\nu + 2rk, \nu \in \mathbf{Z}$ for some k.

$$\tilde{\theta} = \bigcup_{k=-\infty}^{\infty} \tilde{\theta}_k.$$

* 此文与黄达人合作.

We denote

$$I(\tilde{\xi};f) = \{(f^{(i)}(\xi_j))_{j=-\infty}^{+\infty}\}_{i=0}^{r-1}$$

For a fixed $t \in \mathbf{R}$ we consider the problem of recovering optimally the value of $Uf = f(t)$ by using the information $I(\tilde{\xi};f)$, where $\tilde{\xi} \in \tilde{\theta}$ is a fixed denumerable set of sampling. Assuming error-free inforamtion we have the intrinsic error

$$E(K,\tilde{\xi};t):= \inf_{A}\sup_{f\in K} | Uf - A(I(\tilde{\xi};f)) |,$$

where A is any mapping (algorithm) from $I(\tilde{\xi};K)$ into $\mathbf{R}$.

According to [3], if we set

$$e(K,\tilde{\xi};t):= \sup\{| f(t) |, f \in K, I(\tilde{\xi};f) = 0\},$$

then we have

$$E(K,\tilde{\xi};t) \geqslant e(K,\tilde{\xi};t).$$

As in [3], we put

$$E(K,\tilde{\xi}):= \| E(K,\tilde{\xi};\cdot) \|_{\infty}$$

$$E(K):= \inf\{E(K,\tilde{\xi});\tilde{\xi} \in \tilde{\theta}\}$$

We consider the extremal problem as follows.

Problem A. *It is required to obtain the exact expression for* $E(K)$ *and determine the optimal set of points of sampling* $\xi_0 \in \tilde{\theta}$, *namely for which*

$$E(K) = E(K,\tilde{\xi}_0).$$

And if possible, we are also concerned with the construction of the optimal algorithm.

As far as we know, this problem has been solved only for $r = 1$.[4] For $r = 2$, the exact expression for $E(K)$ is obtained, but no information is given about the optimal algorithm.[5] In this paper we completely solve Problem A for $r \geqslant 1$.

2. Generalized Euler spline En, $r(x)$

To begin with, we must give a generalization of Euler spline function, which is crutial in constructing the Hermite cardinal spline interpolating operator. For fixed $n \geqslant r \geqslant 1$ and $k \in \mathbf{N}$ we put

$$\widetilde{K}_{2kr} = \{f; f \in K, f(x+2kr) = f(x)\}.$$

Lemma 2.1[2] *For every $\tilde{\xi} \in \tilde{\theta}$, there exists an $f(x) = f(\tilde{\xi}; x) \in \widetilde{K}_{2kr}$, such that*

i) $I(\tilde{\xi}; f) = 0$.

ii) $|f^{(n)}(\tilde{\xi}; x)| = 1$.

iii) $V_C(f^{(n)}; [-kr, kr)) = 2kr$, *where V_C is the number of sign changes in one period.*

Definition 2.1 *In* Lemma 2.1 *if we choose $\tilde{\xi}^0 = \{\xi_\nu^0 = r\nu\}$, and assume $f^{(n)}(\tilde{\xi}^0; 0^+) = 1$, then there is exact one perfect spline satisfying the condition, which we call the generalized Euler spline and denote it $E_{n,r}(x)$.*

Now for fixed n, r we take a subset of $\widetilde{K}_{2kr}$,

$K_{2k}: = \{f; f \in \widetilde{K}_{2kr}, |f^{(n)}(x)| = 1. V_C(f^{(n)}; [-kr, kr)) \leqslant 2kr\}$,

which consists of periodic perfect splines of degree n.

Lemma 2.2[2] *Given $n \geqslant r \geqslant 1$, then*

$\min\{\|f\|_\infty; f \in K_{2k}, I(\tilde{\xi}; f) = 0$, *for some* $\tilde{\xi} \in \tilde{\theta}\} = \|E_{nr}(\cdot)\|_\infty$.

Then follows the lower bound of $E(K)$.

Theorem 2.1 *For given $n \geqslant r \geqslant 1$,*

$$E(K) \geqslant \|E_{nr}(\cdot)\|_\infty.$$

3. Upper estimation of $E(K)$

Given $n \geqslant r$, let $\{\eta_j\}_{-\infty}^{\infty}$ be the set of all knots of $E_{nr}(x)$, and let S_{nr} be the set of $n-1$ degree polynomial splines with simple knots $\{\eta_j\}$. We consider the following interpolatin problem:

For given $f \in F(\mathbf{R}) \cap S_n$, it is required to find

(3.1) $H(x) = H_{nr}(f; x) \in S_{nr} \cap F(\mathbf{R})$ such that

$$H^{(j)}(\nu r) = f^{(j)}(\nu r), j = 0, 1, \cdots, r-1, \nu \in \mathbf{Z}.$$

We denote $\xi_j = rj$, $(j \in \mathbf{Z})$. and let S_{nr}^* be the set of $n-1$ degree polynomial splines with the multiplicity r knots $\{\xi_j\}$, and we also consider the dual interplation problem:

(3.2)For given $f \in F(\mathbf{R})$, it is required to find $T(x) = T_{n,r}(f;x)$ such that

$$T(\eta_\nu) = f(\eta_\nu), \nu \in \mathbf{Z}.$$

Theorem 3.1 *Interpolation problem* (3.1) *and* (3.2) *are both uniquely solvable.*

If $f^{(n)} \in L^\infty(\mathbf{R})$, *we have the formulas*:

$$f(x) - H(f;x) = \frac{1}{(n-1)!}\int_{-\infty}^{\infty} K_{nr}(x;t) f^{(n)}(t)\mathrm{d}t.$$

$$f(t) - T(f;t) = \frac{1}{(n-1)!}\int_{-\infty}^{\infty} K_{nr}^{*}(t;x) f^{(n)}(x)\mathrm{d}x.$$

Lemma 3.1 *For given* $n \geqslant r \geqslant 1$.

$$K_{nr}(x;t) = (-1)^n K_{nr}^{*}(t;x).$$

By using Lemma 3.1, we can analyze the sign structure of $K_{nr}(x;t)$ and draw the following conclusion.

Theorem 3.2 *Given* $n \geqslant r \geqslant 1$, *then*

$$\sup_{f W_\infty^{(n)}(\mathbf{R})} \| f(\cdot) - H(f_j;\cdot) \|_\infty = \| E_{nr}(\cdot) \|_\infty,$$

$$E(K) \leqslant \sup\{ \| f \|_\infty; f \in \mathbf{K}. I(\tilde{\xi}^0; f) = 0\} = \| E_{n,r}(\cdot) \|_\infty.$$

Combining Theorem 2.1 With Theorem 3.2 we obtain our main result.

Theorem 3.3 *For given* $n \geqslant r \geqslant 1$, *the equality*

$$E(K) = \| E_{n,r}(\cdot) \|_\infty$$

holds. And the Hermite cardinal spline interpolating operator is an optimal algorithm.

References

[1] de Boor, C. and Schoenberg, I.J. Cardinal interpolation and spline functions VIII, in "Lecture Notes in Math. 501", 1976

[2] Huang, D. and Bojanov, B.D. Periodic monosplines and perfect splines of least norm, to appear

[3] Miccelli, C.A. and Rivlinm, T.J. A survey of optimal recovery. In "*Optimal Estimation in Approximation Theory*" (eds. C.A. Micchelli and T.J. Rivlin). N.Y.: Plenum Press, 1977. pp. 1 ~ 54

[4] Sun, Yongshen. Optimal interpolation on some classes of differentiable functions(Ⅰ), to appear

[5] Sun Yongshen. Optimal interpolation on some classes of differentiable functions (Ⅱ), to appear

Northeastern Math.
J. 1990, 6(1)

R 上的一个卷积函数类上的最优插值*

Optimal Interpolation on a Convolution Class of Functions on **R**

Abstract

In this paper we consider an optimal interpolation problem on some classes of functions, defined on **R**, of convolution type, where the kernel is an even *PF* density. The exact expression for optimal error, an optimal set of sampling points, and an optimal linear algorithm are obtained. Moreover, a cardinal $\mathscr{L}$-spline function class relating to an infinite-order differential operator is introduced.

Key Words Convolution Class; *PF* Desity; Optimal Interpolation; Optimal Sets of Sampling Point

§ 1. Formulation of Problem

Let $G(t)$ be a continuous function defined on **R**. It is said to be a *PF* density, if it satisfies

$$\int_{-\infty}^{+\infty} G(t)\mathrm{d}t = 1,$$

and is totally positive on **R**, i.e., $\forall\, n \geqslant 1, t_1 < \cdots < t_n, \tau_1 < \cdots < \tau_n$, it holds that

---

* Received Mar. 23, 1988.
Supported by the National Natural Science Foundation of China.

$$\det(G(t_i - \tau_j))_{i,j=1}^{n} \geqslant 0 \tag{1.1}$$

(For simplicity we write $G \in PF$). Suppose there is an even function $G(t) \in PF$. Consider the function class

$$S_G^\infty = \{f : f = G * \varphi, \varphi \in L^\infty(\mathbf{R}), \|\varphi\|_\infty \leqslant 1\} \tag{1.2}$$

which is a convolution class with kernel G. If $G(t) \in C^\infty(\mathbf{R})$, then $S_G^\infty \subset C^\infty(\mathbf{R})$. In this paper we consider the following optimal recovery problem on S_G^∞.

For every $k \in \mathbf{Z}_+$ we choose in $[-k, k)$ $2k$ points $-k \leqslant \xi_{-k+1} < \cdots < \xi_k < k$ arbitrarily, and denote by $\tilde{\xi}_{2k}$ the $2k$-periodic continuation of the set $\{\xi_{-k+1}, \cdots, \xi_k\}$ on the real line. The set of all such $\tilde{\xi}_{2k}$ is denoted by Ξ_{2k}. Then we set $\Theta = \bigcup_{k=1}^{\infty} \Xi_{2k}$. For every $f \in S_G^\infty$ and a fixed $\xi \in \Theta$ the sequence

$$I_\xi(f) = \{f(\xi_j)\}_{j=-\infty}^{+\infty} \in l^\infty \tag{1.3}$$

is said to be the sample of f on ξ, and the mapping $I_\xi : S_G^\infty \to l^\infty$ defined by (1.3) is said to be an information operator. Fix a point $\tau \in \mathbf{R}$, $\tau \neq \xi_j (\forall j \in \mathbf{Z})$. Any mapping $A : I_\xi(S_G^\infty) \to \mathbf{R}$ is called an algorithm, which is actually a functional defined on the image of S_G^∞ under I_ξ. If we take $A(I_\xi(f))$ as an approximative representation of $f(\tau)$, then the global error is

$$\sup\{|f(\tau) - A(I_\xi(f))| : f \in {}_G^\infty\}. \tag{1.4}$$

According to Micchelli and Rivlin[1], we have the intrinsic error

$$E(S_G^\infty, \xi, \tau) = \inf_A \sup_{f \in S_G^\infty} |f(\tau) - A(I_\xi(f))|, \tag{1.5}$$

and

$$E(S_G^\infty, \xi, \tau) \geqslant \mathrm{e}(S_G^\infty, \xi, \tau), \tag{1.6}$$

where

$$\mathrm{e}(S_G^\infty, \xi, \tau) = \sup\{|f(\tau)| : f \in S_G^\infty, I_\xi(f) = 0\}. \tag{1.7}$$

The quantity defined in (1.5) depends on τ, and is very difficult to calculate. Now we set

$$E(S_G^\infty, \xi) = \sup_\tau E(S_G^\infty, \xi, \tau), \tag{1.8}$$

and consider

$$E(S_G^\infty) = \inf\{E(S_G^\infty, \xi) : \forall\, \xi \in \Theta\}. \tag{1.9}$$

We pose the following two problems:

(1) *To seek for the exact expression for* $E(S_G^\infty)$.

(2) *To determine an optimal information operator as well as an optimal algorithm realizing* (1.9).

This is the so-called *optimal interpolation problem* on S_G^∞ over Θ in the matric L^∞. For various classes of functions defined on some compact sets the optimal interpolation problem has been considered by many authors (ξf. [2] ~ [7]). The author in [5] posed and considered an optimal interpolation problem concerning a class of differentiable functions defined on the whole real axis. A fundamental result was obtained in [5,6], and some further results are given in a recent paper [7]. This paper is a continuation of [5]. Our basic method is to approximate the kernel $G(t)$ by generalized Bernoulli kernels relating to ordinary differential operators with real characteristic roots. As far as we know, Karlin[8] first employed this method to get a Landau-Kolmogorov type inequality for functions in S_G^∞. Using a similar idea Chahkiev[9] and Fang[10] solved some extremal problems on some classes of periodic functions of convolution type.

§ 2. Kernels of Finite Order

Let

$$P_n(\lambda) = \lambda^n + \sum_{j=1}^{n} a_j \lambda^{n-j}$$

be a real polynomial such that $P_n(0) = 1$, and suppose $P_n(\lambda) = 0$ has only real roots. In [11] the generalized 2π-periodic Bernoulli function corresponding to the differential form $P_n(D)$ was introduced as follows:

$$G_n(x) = (2\pi)^{-1} \sum_{k=-\infty}^{+\infty} \frac{e^{ikz}}{P_n(ik)}. \tag{2.1}$$

Here a generalized (non-periodic) Bernoulli functon is defined by the Fourier integral:

$$G_n(x) = (2\pi)^{-1}\int_{-\infty}^{+\infty} \frac{e^{iuz}}{P_n(iu)}du. \tag{2.2}$$

We say $f \in W_\infty^{(n)}(P_n(D))$, if it satisfies

(1) $f \in C^{n-1}(\mathbf{R}) \cap L^\infty(\mathbf{R})$,

(2) $f^{(n-1)}$ is absolutely continuous on every finite interval on $\mathbf{R}$,

(3) $\operatorname{ess\,sup}_{x\in\mathbf{R}} |P_n(D)f(x)| \leqslant 1$.

Concerning $W_\infty^{(n)}(P_n(D))$ the following representation theorem is well-known (see [12]).

Theorem A *$f \in W_\infty^{(n)}(P_n(D)) \Leftrightarrow$ exists a $\varphi \in L^\infty(\mathbf{R})$, $\|\varphi\|_\infty \leqslant 1$, such that $f = G * \varphi$.*

Denote

$$E_n(x) = (G_n * \operatorname{sgn}\sin\pi(\cdot))(x)$$
$$= \frac{1}{\pi i}\sum_{v=-\infty}^{+\infty} \frac{e^{i(2v+1)\pi x}}{(2v+1)P_n(i(2v+1)\pi)}. \tag{2.3}$$

Consider the optimal interpolation problem on $W_\infty^{(n)}(P_n(D))$. A special case $P_n(D) = D^n$ (This is different from that mentioned in §1, because the equation $\lambda^n = 0$ has a zero $\lambda = 0$ of order n) was considered in [5]. We pointed out in [5] that the fundamental result obtained there may be extended to a more general case, when the class of differentiable functions is defined by a self-adjoint operator

$$P_n(D) = D^\sigma \prod_{j=1}^{l}(D^2 - t_j^2) \quad (\sigma \geqslant 0,\ n = \sigma + 2l,\ t_j > 0).$$

But the self-adjoint condition may be removed easily. Then we may formulate

Theorem 1

$$E(W_\infty^{(n)}(P_n(D))) = \|E_n(\cdot)\|_\infty. \tag{2.4}$$

On the basis of the cardinal $\mathscr{L}$-spline interpolation theory due to Micchelli we obtain the solution for the problem of determining optimal information operator and optimal algorithm.

Theorem B *There exist a set of points $\{\nu + \alpha\}_{\nu\in\mathbf{Z}}$ and a function $L_n(x)$, such that*

(1) $L_n(x) \in C^{n-2}(\mathbf{R})$.

(2) $P_n(D) L_n(x) = 0, \nu < x < \nu + 1, \forall \nu \in \mathbf{Z}$.

(3) $L_n(\alpha + \nu) = \delta_{0\nu}, \forall \nu \in \mathbf{Z}$.

(4) *There are constants* $A, B > 0$ *for which*

$$|L_n(x)| \leqslant A e^{-B|x|}, \forall x \in \mathbf{R}.$$

(5) *The cardinal interpolating series*

$$S_n(f, x) = \sum_{\nu=-\infty}^{+\infty} f(\alpha + \nu) L_n(x - \nu)$$

yields

$$\sup_{f \in W_\infty^{(n)}} \| f - S_n(f) \|_\infty = \| E_n(\cdot) \|_\infty. \tag{2.5}$$

It is known that (cf. [13]) $\alpha = \frac{1}{4}[1 + (-1)^{n-1}]$. Theorem B may be found in [13 ~ 15]. Theorem B reveals that $\{\nu + \alpha\}_{\nu \in \mathbf{Z}}$ is an optimal set of sampling points, and $S_n\{f, \cdot\}$ is an optimal (linear) algorithm.

§ 3. Approximation of a *PF* Density by a Series of Kernels of Finite Order

First we introduce

Definition[12] *An integral function* $\Psi(s)$ *is said to belong to* $\mathscr{L}$-$\mathscr{P}$ *class* ($\Psi \in \mathscr{L}$-$\mathscr{P}$), *if*

$$\Psi(s) = e^{-\gamma s^2 + \delta s} \prod_\nu \left(1 - \frac{s}{a_\nu}\right) e^{s/a_\nu}, \tag{3.1}$$

where $\gamma \geqslant 0$, δ, $a_\nu \in \mathbf{R}$, *and* $0 < \gamma + \sum_\nu a_\nu^{-2} < +\infty$.

We have (cf. [12])

Theorem C *Suppose* $G(t) \in PF$. *Then there is a* $\Psi(s) \in \mathscr{L}$-$\mathscr{P}$, *such that*

$$(\Psi(s))^{-1} = \int_{-\infty}^{+\infty} e^{-st} G(t) dt \tag{3.2}$$

holds in some vertical stripped region containing $s = 0$. *In this case*

$$G(x) = \frac{1}{2\pi i} \int_{-i\infty}^{+i\infty} \frac{e^{xs}}{\Psi(s)} ds. \tag{3.3}$$

There is a sequence of polynomials $\{P_n(x)\}_{n\geqslant 2}$, *with the properties that* $P_n(x)$ *is of order* n, $P_n(0)=1$, $P_n(\lambda)=0$ *has only real roots, and moreover, for every* $A>0$ *it holds that*

$$\lim_{n\to\infty} P_n(s) = \Psi(s)$$

uniformly on the circle $|s|\leqslant A$.

Notice that

$$G(x) = \frac{1}{2\pi}\int_{-\infty}^{+\infty}\frac{e^{ixt}}{\Psi(it)}dt, \tag{3.4}$$

and $\Psi(s)$ will be even for an even function $G(x)$. In this special case, the set $\{a_\nu\}$ is symmetric with respect to the origin, and

$$\Psi(s) = e^{-\gamma s^2}\prod_\nu\left(1-\frac{s^2}{a_\nu^2}\right). \tag{3.5}$$

Besides these, every polynomial $P_n(x)$ in Theorem C may be even, and then each polynomial $P_n(x)$ is of order $2n$.

In what follows we always assume that the set $\{a_\nu\}$ is infinite, and each polynomial in the sequence $\{P_n(x)\}_{n\geqslant 2}$ is even and of order $2n$.

Lemma 1

$$\lim_{n\to\infty}\| G_n(\cdot) - G(\cdot)\|_\infty = 0. \tag{3.6}$$

Proof According to Hirschman and Widder (see [12], P.52, Theorem 5.3) for every positive integer N we have

$$|\Psi(iu)|^{-1}\leqslant C|u|^{-N}, \tag{3.7}$$

where $C>0$ is a constant. We assert that

$$|P_n(iu)|^{-1}\leqslant C|u|^{-N}, \tag{3.8}$$

where the constant C may be taken to be independent of n. In fact, if we denote the zeros of $P_n(x)$ by $\pm a_j(n), j=1,\cdots,n$, then from

$$|P_n(iu)| = \prod_{j=1}^{n}\left(1+\frac{u^2}{a_j^2(n)}\right),$$

we get for $n>N$

$$|P_n(iu)|\leqslant a_1^2(n)\cdots a_\gamma^2(n)\cdot|u|^{-N},$$

where $\gamma=[N/2]+1$. Suppose the sequence of zeros of $\Psi(s)$ satisfies $|a_1|\leqslant|a_2|\leqslant\cdots$, and also $|a_1(n)|\leqslant|a_2(n)|\leqslant\cdots\leqslant|a_n(n)|$.

Then by [12] (cf. [12], P.44)

$$\lim_{n\to\infty} |a_j(n)| = |a_j|, \quad j = 1, \cdots, v.$$

So there must be a constant C independent of n, for which (3.8) holds. We write

$$G_n(x) - G(x) = (2\pi)^{-1}\int_{-\infty}^{+\infty}\left\{\frac{1}{P_n(iu)} - \frac{1}{\Psi(iu)}\right\}e^{iux}du$$

$$= (2\pi)^{-1}\left\{\int_{|u|\leqslant T} + \int_{|u|>T}\right\}, \quad T > 0.$$

Taking $N = 3$, by (3.7), (3.8) we see that for any $\varepsilon > 0$, there exists $T > 0$ such that

$$\left|\int_{|u|>T}\right| < \pi\varepsilon.$$

When $|u| \leqslant T$, there is $n_0 \in \mathbf{Z}_+$ such that for every $n > n_0$

$$\left|\frac{1}{P_n(iu)} - \frac{1}{\Psi(iu)}\right| < \frac{\varepsilon}{8T\pi}.$$

Consequently,

$$\left|(2\pi)^{-1}\int_{-T}^{T}\right| < \frac{\varepsilon}{2} \quad \text{for } n > n_0.$$

Lemma 1 is proved.

Now we are going to prove

$$\lim_{u\to\infty} \| G_u(\cdot) - G(\cdot)\|_1 = 0. \tag{3.9}$$

Lemma 2 *For any fixed positive integer $\nu \geqslant 2$ there exists a constant $M > 0$ independent of n such that*

$$|x^\nu G_n(x)| \leqslant M, \quad \forall\, x \in \mathbf{R}, n \in \mathbf{Z}_+. \tag{3.10}$$

Proof

$$P_n(s) = \prod_{j=1}^{n}\left(1 - \frac{s^2}{a_j^2(n)}\right).$$

A direct calculation yields

$$\left|\left(\frac{1}{P_n(iu)}\right)'\right| \leqslant \frac{2|u|}{|P_n(iu)|}\cdot\sum_{j=1}^{n} a_j^{-2}(n).$$

By [12] (pp.43 ~ 44, Theorem 3.3) the sequence $\left\{\sum_{j=1}^{u} a_j^{-2}(n)\right\}_{n=1}^{+\infty}$ is bounded, so we have

$$\left|\left(\frac{1}{P_n(iu)}\right)'\right| \leqslant \frac{C_1^{(1)} \mid u \mid}{\mid P_n(iu) \mid}, \tag{3.11}$$

where $C_1^{(1)}$ is some constant. Continuing this process, we arrive at the following conclusion: for every $\nu \in \mathbf{Z}_+$

$$\left|\left(\frac{1}{P_n(iu)}\right)^{(\nu)}\right| \leqslant \frac{C_v^{(\nu)} \mid u \mid^v + \cdots + C_0^{(v)}}{\mid P_n(iu) \mid}, \tag{3.12}$$

where $C_0^{(\nu)}, \cdots, C_v^{(\nu)}$ are some constants which depend only on ν and $\Psi(s)$. Thus we have for $n > v$

$$\lim_{|u| \to \infty}\left(\frac{1}{P_n(iu)}\right)^{(k)} = 0, \quad k = 1, \cdots, \nu - 1. \tag{3.13}$$

Moreover, for $n > v + 1$ we have

$$\left(\frac{1}{P_n(iu)}\right)^{(\nu)} \in L(\mathbf{R}). \tag{3.14}$$

Hence,

$$G_n(x) = \left(\frac{i}{x}\right)^{\nu}(2\pi)^{-1}\int_{-\infty}^{+\infty}\left(\frac{1}{P_n(iu)}\right)^{(\nu)} \mathrm{e}^{ixu}\mathrm{d}u, \quad x \neq 0,$$

and

$$\mid x^{\nu}G_n(x) \mid \leqslant (2\pi)^{-1}\int_{\infty}^{+\infty}\left|\left(\frac{1}{P_n(iu)}\right)^{(\nu)}\right| \mathrm{d}u. \tag{3.15}$$

The quantity standing on the right hand side of (3.15) is bounded. This fact can be verified easilly on the basis of the following relation:

$$\lim_{n \to \infty}\left(\frac{1}{P_n(iu)}\right)^{(\nu)} = \left(\frac{1}{\Psi(iu)}\right)^{(\nu)}$$

uniformly on every finite interval. We omit the details. Lemma 2 is proved.

Corollary

$$\| G_n - G \|_1 \to 0 \quad (n \to \infty).$$

Lemma 3 *Let*

$$P_n(D) = \prod_{j=1}^{n}(D - t_j), \quad t_j \in \mathbf{R}.$$

For any $\xi \in \Xi_{2k}$ and any $n \in \mathbf{Z}_+$

$$\sup\{ \| f \|_{\infty} : f \in W_{\infty}^{(n)}(P_n(D)), I_{\xi}(f) = 0\} \geqslant \| E_n(\cdot) \|_{\infty}.$$

There exists a $2k$-periodic function $f_n \in W_{\infty}^{(n)}(P_n(D))$ such that $I_{\xi}(f_n) = 0$, and $\| f_n \|_{\infty} \geqslant \| E_n(\cdot) \|_{\infty}$.

In [5] this lemma was proved for the special case $P_n(D) = D^n$. The general case may be treated similarly.

Theorem 2

$$E(S_G^{\infty}) = \| E(\cdot) \|_{\infty}, \tag{3.16}$$

where $E(x) = (G * \operatorname{sgn} \sin \pi(\cdot))(x)$.

Proof For any $f \in S_G^{\infty}$, we have $f = G * \varphi$ for some $\varphi \in L^{\infty}(\mathbf{R})$, $\| \varphi \|_{\infty} \leqslant 1$. Put $f_n = G_n * \varphi$. Denote by $S_n(f_n, \cdot)$ the cardinal $\mathscr{L}$-spline interpolation operator defined by the differential operator $P_n(D)$. We can write

$$S_n(f_n, \cdot) = \sum_{\nu=-\infty}^{+\infty} f_n(\nu) L_n(\cdot - \nu),$$

because $P_n(D)$ is a self-adjoint differential operator of order $2n$, so that we may take $a = 0$. Then from that

$$\begin{aligned} \| f - S_n(f_n) \|_{\infty} &\leqslant \| f - f_n \|_{\infty} + \| f_n - S_n(f_n) \|_{\infty} \\ &= \| G * \varphi - G_n * \varphi \|_{\infty} + \| f_n - S_n(f_n) \|_{\infty} \\ &\leqslant \| G - G_n \|_1 + \| E_n(\cdot) \|_{\infty}, \end{aligned}$$

and

$$\lim_{n \to \infty} \| E_n(\cdot) - E(\cdot) \|_{\infty} = 0, \tag{3.17}$$

(which may be easily proved) we get

$$\| E_n(\cdot) \|_{\infty} \to \| E(\cdot) \|_{\infty}, \quad n \to \infty.$$

By the corollary of Lemma 2, for any $\delta > 0$ there is some $n_0 \in \mathbf{Z}_+$, such that for every $n > n_0$

$$\| G - G_n \|_1 < \delta, \ \| E_n(\cdot) \|_{\infty} < \| E(\cdot) \|_{\infty} + \delta.$$

So from that $\| f - S_n(f_n) \|_{\infty} < \| E(\cdot) \|_{\infty} + 2\delta$ we derive $E(S_G^{\infty}) \leqslant \| E(\cdot) \|_{\infty} + 2\delta$. By taking limit $\delta \to 0^+$ we obtain

$$E(S_G^{\infty}) \leqslant \| E(\cdot) \|_{\infty}.$$

Now we are going to prove $E(S_G^{\infty}) \geqslant \| E(\cdot) \|_{\infty}$. We need to make use of the following relation, which is readily derived from (1.6):

$$E(S_G^{\infty}) \geqslant \inf_{\xi \in \Theta} \sup \{ \| f \|_{\infty} : f \in S_G^{\infty}, I_{\xi}(f) = 0 \}.$$

Fix a $\xi \in \Xi_{2k}$. By Lemma 3 for any $\delta > 0$ and $n \in \mathbf{Z}_+$ there exists a $2k$-periodic function $f_n \in W_{\infty}^{(n)}(P_n(D))$ such that $I_{\xi}(f_n) = 0$, and

$$\| f_n \|_\infty \geqslant \| E_n(\cdot) \|_\infty - \delta.$$

Notice that $f_n = G_n * \varphi_n$, $\| \varphi_n \|_\infty \leqslant 1$. By a theorem of functional analysis it is known that every bounded sequence in $L^\infty(\mathbf{R})$ is $^*w$-compact, so that there is a subsequence $\{\varphi_{n_j}\} \subset \{\varphi_n\}$, such that $\varphi_{n_j} \xrightarrow{*w} \varphi_0$, where φ_0 is some function in $L^\infty(\mathbf{R})$, $\| \varphi_0 \|_\infty \leqslant 1$. Put $f_0 = G * \varphi_0$. Then we must have $I_\xi(f_0) = 0$. Actually, if we write $I_\xi(f_0) = \{f_0(\xi_j)\}_{j \in z}$, where

$$f_0(\xi_j) = \int_{-\infty}^{+\infty} G(\xi_j - t)\varphi_0(t)\mathrm{d}t, j \in \mathbf{Z},$$

and denote $f_{n_j}^* = G * \varphi_{n_j}$, then for every $x \in \mathbf{R}$

$$\lim f_{n_j}^*(x) = f_0(x).$$

But

$$\| f_{n_j}^* - f_{n_j} \|_\infty \leqslant \| G - G_{n_j} \|_1 \to 0 \quad (j \to \infty),$$

so that for every $x \in \mathbf{R}$ we also have

$$\lim f_{n_j}(x) = f_0(x).$$

Thus, from $I_\xi(f_{n_j}) = 0$ we derive $I_\xi(f_0) = 0$. f_0 is $2k$-periodic. $\{f_{n_j}\}$ is uniformly bounded on $\mathbf{R}$ and equi-continuous. In fact, we first point out that from Lemmas 1 and Lemmas 2 for any positive integer $N \geqslant 2$ there is a constant $M > 0$ independent of n such that

$$|G_n(x)| \leqslant M(1 + |x|^N)^{-1}. \tag{3.18}$$

Then we get

$$\| f_{n_j} \|_\infty \leqslant \int_{+\infty}^{+\infty} |G_{n_j}(x)| \mathrm{d}x \leqslant M \int_{-\infty}^{+\infty} \frac{\mathrm{d}x}{1 + |x|^N} \leqslant M_1.$$

Thus, the uniform boundedness is derived. From that

$$\begin{aligned}
&\| f_{n_j}(\cdot + h) - f_{n_j}(\cdot) \|_\infty \\
\leqslant\ & \| f_{n_j}(\cdot + h) - f_{n_j}^*(\cdot + h) \|_\infty + \| f_{n_j}^*(\cdot + h) - f_{n_j}^*(\cdot) \|_\infty + \\
& \| f_{n_j}^*(\cdot) - f_{n_j}(\cdot) \|_\infty \\
\leqslant\ & 2 \| G - G_{n_j} \|_1 + \| G(\cdot - h) - G(\cdot) \|_1
\end{aligned}$$

and

$$\| G - G_{n_j} \|_1 \to 0 \qquad (j \to \infty),$$

$$\| G(\cdot - h) - G(\cdot) \|_1 \to 0 (h \to 0),$$

we get the equi-continuity of $\{f_{n_j}\}$. Notice that all functions f_{n_j} have the same $2k$-period. So by the Arzela theorem there must exist a subsequence of $\{f_{n_j}\}$ which converges uniformly on $\mathbf{R}$. For simplicity we denote such subsequence by $\{f_{n_j}\}_{j \in \mathbf{Z}}$. Then

$$\lim_{j \to \infty} \| f_{n_j} - f_0 \|_\infty = 0.$$

Consequently, for any $\delta > 0$ there is a sufficiently large j_0 such that for every $j > j_0$

$$\| f_0 \|_\infty \geqslant \| f_{n_j} \|_\infty - \delta \geqslant \| E_{n_j}(\cdot) \|_\infty - \delta,$$

and hence

$$\sup\{ \| f \|_\infty : f \in S_G^\infty, I_\xi(f) = 0 \} \geqslant \| E_{n_j}(\cdot) \|_\infty - \delta.$$

By taking limit $(j \to +\infty, \delta \to 0^+)$ we get

$$\sup\{ \| f \|_\infty : f \in S_G^\infty, I_\xi(f) = 0 \} \geqslant \| E(\cdot) \|_\infty,$$

which gives

$$\inf_{\xi \in \Theta} \sup\{ \| f \|_\infty : f \in S_G^\infty, I_\xi(f) = 0 \} \geqslant \| E(\cdot) \|_\infty.$$

Therefore, $E(S_G^\infty) \geqslant \| E(\cdot) \|_\infty$. Theorem 2 is proved.

§ 4. Optimal Interpolating Operator

Micchelli[13] thoroughly studied the cardinal $\mathscr{L}$-spline interpolation problem on the class $W_\infty^{(n)}(P_n(D))$. By [13] the fundamental function for cardinal interpolation may be represented by

$$L_n(x) = (2\pi)^{-1} \int_{-\pi}^{\pi} E_n(x, e^{iu}) du, \tag{4.1}$$

where

$$E(x, e^{iu}) = F_n(x, u) / F(\alpha, u),$$

and

$$F_n(x, u) = \sum_{k=-\infty}^{+\infty} \frac{e^{i(2k\pi + u)x}}{P_n(i(2k\pi + u))}. \tag{4.2}$$

In the present case $\alpha = 0$, $L_n(x)$ has the following integral form:

$$L_n(x) = (2\pi)^{-1} \int_{-\infty}^{+\infty} \frac{e^{iux}}{P_n(iu) F_n(0, u)} du. \tag{4.3}$$

All the above formulas may be found in [13] (pp. 223 ~ 224). Denote

$$F(x,u)=\sum_{k=-\infty}^{+\infty}\frac{e^{i(2k\pi+u)x}}{\Psi(i(2k\pi+u))},\tag{4.4}$$

$$E(x,e^{iu})=F(x,u)/F(\alpha,u),\tag{4.5}$$

$$L(x)=(2\pi)^{-1}\int_{-\pi}^{\pi}E(x,e^{iu})du.\tag{4.6}$$

When $\alpha=0$ the function $L(x)$ has the following Fourier integral form:

$$L(x)=(2\pi)^{-1}\int_{-\infty}^{+\infty}\frac{e^{iux}}{\Psi(iu)F(0,u)}du.\tag{4.7}$$

In the following theorem we list some basic properties of $L(x)$.

Theorem 3

(1) $L(x)\in C^{\infty}(\mathbf{R})$;

(2) $L(\nu)=\delta_{0\nu},\ \nu\in\mathbf{Z}$;

(3) $\|L_n(\cdot)-L(\cdot)\|_{\infty}\to 0\ (n\to\infty)$;

(4) $(F(0,2\pi u))^{-1}$ *is a 1-periodic function, whose Fourier series converges absolutely. Denote its Fourier coefficients by* $\{C_\nu\}_{\nu\in\mathbf{Z}}$. *Then we have*

$$L(x)=\sum_{\nu=-\infty}^{+\infty}G_\nu G(x+\nu);\tag{4.8}$$

(5) *There exist constants* $A,B>0$ *such that*

$$|L(x)|\leqslant Ae^{-B|x|},\ x\in\mathbf{R}.\tag{4.9}$$

Proof (1) By (3.7), if we take $\tau=2k\pi+u$, $|u|\leqslant\pi$, then for any fixed positive integer N we have

$$|\Psi(i(2k\pi+u))|^{-1}=O(|2k\pi+u|^{-N-2})\quad(|k|\geqslant 1).$$

So $F(x,u)\in C^N(\mathbf{R})$ with respect to the variable x. On account of the fact that N can be arbitrarily large, and the function

$$\frac{\partial^N}{\partial x^N}F(x,u)=\sum_{k=-\infty}^{+\infty}\frac{i^N(2k\pi+u)^Ne^{i(2k\pi+u)x}}{\Psi(i(2k\pi+u))}$$

is continuous with respect to the variable u, we conclude that the function

$$L^{(N)}(x)=(2\pi)^{-1}\int_{-\pi}^{\pi}\frac{\partial^N}{\partial x^N}E(x,e^{iu})du\tag{4.10}$$

is continuous. Thus we have proved that $L(x)\in C^{\infty}(\mathbf{R})$.

(2) It is easy to verify that $F(\nu,u)=e^{i\nu u}F(0,u)$, so that

$$L(\nu) = (2\pi)^{-1}\int_{-\pi}^{\pi} e^{i\nu u} du = 0, \quad \nu \neq 0$$

and $L(0) = 1$.

(3) First we have

$$|L_n(x) - L(x)| \leqslant$$

$$(2\pi)^{-1}\int_{-\pi}^{\pi} |E_n(x, e^{iu}) - E(x, e^{iu})| du \leqslant$$

$$(2\pi)^{-1}\int_{-\pi}^{\pi} \left| \frac{F_n(x,u) - F(x,u)}{F_n(0,u)} \right| du +$$

$$(2\pi)^{-1}\int_{-\pi}^{\pi} \frac{|F(x,u)| \cdot |F(0,u) - F_n(0,u)|}{|F_n(0,u)| \cdot |F(0,u)|} du$$

$$= \text{I}(x) + \text{II}(x).$$

In order to estimating $\text{I}(x)$, we notice that

$$F_n(x,u) - F(x,u)$$

$$= e^{iux}\left\{ \sum_{k=-\infty}^{+\infty} \left[\frac{1}{P_n(i(2k\pi + u))} - \frac{1}{\Psi(i(2k\pi + u))} \right] e^{i2k\pi x} \right\}$$

$$= e^{iux}\left\{ \sum\nolimits_{|k| \leqslant N} + \sum\nolimits_{|k| > N} \right\}.$$

The integer N will be chosen later. For an arbitrarily fixed N the relation

$$\lim_{n \to \infty} P_n(s) \doteq \Psi(s)$$

holds uniformly on the circle $|s| \leqslant (2N+1)\pi$. Hence for any $\varepsilon > 0$, there exists $n_0 \in \mathbf{Z}_+$ such that for every $n > n_0$ and u, $|u| \leqslant \pi$

$$\left| \sum\nolimits_{|k| \leqslant N} \right| < \varepsilon/2.$$

Moreover, from (3.7) we have

$$\sum_{|k| > N} \frac{1}{|\Psi(i(2k\pi + u))|} \leqslant |a_1|^2 \sum_{|k| > N} \frac{1}{(2k\pi + u)^2} = O(N^{-1}).$$

On account of

$$P_n(s) = \prod_{j=1}^{n} \left(1 - \frac{s^2}{a_j^2(n)} \right),$$

where $|a_1(n)| \leqslant \cdots \leqslant |a_n(n)|$, we see that

$$|P_n(i(2k\pi + u))|^{-1} \leqslant a_1^2(n)(2k\pi + u)^{-2}.$$

Then from

$$\lim_{n \to \infty} a_1(n) = a_1$$

where a_1 is a real zero of $\Psi(s)$, we obtain

$$\sum_{|k|>N} \frac{1}{|P_n(\mathrm{i}(2k\pi+u))|} \leqslant Ca_1^2 \sum_{|k|>N} \frac{1}{(2k\pi+u)^2} = O(N^{-1}),$$

where C is a constant. To sum up we have derived

$$\left(\sum_{|k|\leqslant N} + \sum_{|k|>N}\right)\mathrm{e}^{\mathrm{i}ux} = O(N^{-1}).$$

Thus we have proved that for any $\varepsilon > 0$, there exists $n_0 \in \mathbf{Z}_+$ such that for every $n > n_0$, arbitrary $x \in \mathbf{R}$ and $u: -\pi \leqslant u \leqslant \pi$, the relation

$$|F_n(x,u) - F(x,u)| < \varepsilon$$

holds.

Now let us notice that

$$\lim_{n\to\infty} F_n(0,u) = F(0,u)$$

uniformly holds for $|u| \leqslant \pi$ and $F(0,u) \neq 0$. By the continuity of $F(0,u)$ there must be a positive number C' such that

$$|F(0,u)| \geqslant C' > 0, \text{for } |u| \leqslant \pi.$$

So for sufficiently great n we have

$$|F_n(0,u)| \geqslant C'/2 > 0, \text{for } |u| \leqslant \pi.$$

From this fact we derive Ⅰ$(x) < 2\varepsilon/C'$. Similarly we can get an analogous result for Ⅱ(x). Thus (3) is proved.

(4) First we notice that $F(0,2\pi u)$ is a differentiable function of period 1, and $F(0,2\pi u) \neq 0$. Hence the Fourier series of $(F(0,2\pi u))^{-1}$ converges absolutely. Now we are going to prove $(\Psi(\mathrm{i}u))^{-1}\mathrm{e}^{\mathrm{i}ux} \in L(\mathbf{R})$ (x is fixed), and that it is of bounded-variation on $\mathbf{R}$. Actually, from $(\Psi(\mathrm{i}u))^{-1} \in C(\mathbf{R})$ and

$$(\Psi(\mathrm{i}u))^{-1} = O(|u|^{-3}) \quad (u \to \infty)$$

we obtain $(\Psi(\mathrm{i}u))^{-1}\mathrm{e}^{\mathrm{i}ux} \in L(\mathbf{R})$. To prove $(\Psi(\mathrm{i}u))^{-1}\mathrm{e}^{\mathrm{i}ux} \in BV(\mathbf{R})$ we have only to verify

$$\frac{\mathrm{d}}{\mathrm{d}u}[(\Psi(\mathrm{i}u))^{-1}\mathrm{e}^{\mathrm{i}ux}] \in L(\mathbf{R}).$$

Direct computation yields

$$\frac{\mathrm{d}}{\mathrm{d}u}[(\Psi(\mathrm{i}u))^{-1}\mathrm{e}^{\mathrm{i}ux}] = \mathrm{i}x(\Psi(\mathrm{i}u))^{-1}\mathrm{e}^{\mathrm{i}xu} - \mathrm{i}\mathrm{e}^{\mathrm{i}xu}\cdot(\Psi(\mathrm{i}u))^{-1}[(\Psi(\mathrm{i}u))^{-1}\Psi'(\mathrm{i}u)],$$

$$(\Psi(\mathrm{i}u))^{-1}\Psi'(\mathrm{i}u) = -\mathrm{i}2\gamma u + 2\mathrm{i}u\sum_{k=1}^{\infty}\frac{1}{u^2+a_k^2}.$$

Thus we obtain

$$\left|\frac{\Psi'(\mathrm{i}u)}{\Psi(\mathrm{i}u)}\right| \leqslant 2\gamma\,|\,u\,| + 2\,|\,u\,|\sum_{k=1}^{\infty}\frac{1}{a_k^2} = 2\gamma'\,|\,u\,|,$$

where

$$\gamma' = \gamma + \sum a_k^{-2} < +\infty.$$

Therefore

$$\left|\frac{\mathrm{d}}{\mathrm{d}u}\left(\frac{\mathrm{e}^{\mathrm{i}xu}}{\Psi(\mathrm{i}u)}\right)\right| \leqslant C\gamma'\,|\,u\,|\cdot|\,\Psi(\mathrm{i}u)\,|^{-1} \in L(\mathbf{R}).$$

According to a theorem due to Hardy and Young (see [16], P. 141)

$$L(x) = (2\pi)^{-1}\int_{-\infty}^{+\infty}\frac{\mathrm{e}^{\mathrm{i}ux}\,\mathrm{d}u}{\Psi(\mathrm{i}u)F(0,u)} = \sum_{\nu=-\infty}^{\infty}C_\nu\int_{-\infty}^{+\infty}\varphi(u)\mathrm{e}^{\mathrm{i}2\nu\pi u}\mathrm{d}u,$$

Where the Fourier expansion of $(F(0,2\pi u))^{-1}$ has been used:

$$(F(0,2\pi u))^{-1} = \sum_{\nu=-\infty}^{\infty}C_\nu\mathrm{e}^{-\mathrm{i}2\nu\pi x}, \sum|C_\nu| < +\infty,$$

and

$$\varphi(u) = \Psi(\mathrm{i}u)^{-1}\mathrm{e}^{\mathrm{i}2\pi ux}.$$

Noticing that

$$\int_{-\infty}^{+\infty}\varphi(u)\mathrm{e}^{\mathrm{i}2\nu\pi u}\mathrm{d}u = (2\pi)^{-1}\int_{-\infty}^{+\infty}\frac{\mathrm{e}^{\mathrm{i}(\nu+x)u}}{\Psi(\mathrm{i}u)}\mathrm{d}u = G(x+\nu),$$

(4) is proved.

A special case of (4). Assertion (4) is obviously true for $L_n(x)$. In this special case $P_n(\lambda)$ is an even polynomial of order $2n$, $P_n(0) = 1$, and $P_n(\lambda) = 0$ has only real zeros. We have

$$L_n(x) = \sum C_\nu^{(n)}G_n(x+\nu), \tag{4.11}$$

where $\{C_\nu^{(n)}\}$ is the sequence of Fourier coefficients of the function $(F_n(0, 2\pi u))^{-1}$, which satisfies

$$\sum|C_\nu^{(n)}| < +\infty.$$

(5) First we must point out an important fact. viz., the function $F(0, u)$ is 2π-periodic and analytic. The 2π-periodicity is obvious. The analyticity may be proved as follows. Actually, we know that for every $k \in \mathbf{Z}$,

$(\Psi(\mathrm{i}(2k\pi + u)))^{-1}$ is analytic on $[0, 2\pi]$. For a fixed R, $0 < R < |a_1|$, the series

$$\sum_{k=-\infty}^{+\infty} \frac{1}{\Psi(\sigma + \mathrm{i}(2k\pi + u))} \tag{4.12}$$

converges uniformly for σ, $|\sigma| \leqslant R$ and u, $0 \leqslant u \leqslant 2\pi$. The last assertion may be proved by using a theorem in [12], according to which for any fixed $R > 0$ and any positive integer N the relation

$$|(\Psi(\sigma + \mathrm{i}\tau))^{-1}| = O(|\tau|^{-N}) \tag{4.13}$$

holds uniformly on $|\sigma| \leqslant R$. Taking $N = 3$ in (4.13) and applying the Weierstrass theorem we readily draw the required conclusion. Because of $F(0, u) \neq 0$, $(F(0, u))^{-1}$ is also 2π-periodic and analytic.

Now we are in a position to prove (5). By a theorem of Bary (see [17], P.88, (25.1)) we claim that there is a positive λ such that $|C_\nu| = O(\mathrm{e}^{-\lambda|\nu|})$, where $\{C_\nu\}$ is the Fourier coefficients of $(F(0, 2\pi u))^{-1}$. Now we can make use of the formulae proved in (4)

$$L(x) = \sum_{\nu=-\infty}^{+\infty} C_\nu G(x + \nu).$$

Without loss of generality we may assume $x > 0$, $N \leqslant x < N + 1$, N is some positive integer. Write

$$L(x) = \sum_{i=1}^{3} L_i(x),$$

where

$$L_1(x) = \sum_{-3N/2 < \nu < -N/2} C_v G(x + \nu), \quad L_2(s) = \sum_{\nu \geqslant 0} C_v G(x + \nu),$$

$$L_3(x) = \sum_{v < 0,\, \nu \in [-3N/2, -N/2]} C_v G(x + \nu).$$

According to Karlin[18], there is a $\mu > 0$ for which

$$G(x) = O(\mathrm{e}^{-\mu|x|}), \quad |x| \to \infty. \tag{4.14}$$

Besides this, it is obvious that $\|G\|_\infty \leqslant M$ for some constant M.

(a) when $\nu \geqslant 0$, then $N < x + \nu$, and

$$|L_2(x)| \leqslant \sum_{v \geqslant 0} |C_v| \cdot |G(x + \nu)| \leqslant (A \sum |C_\nu|) \mathrm{e}^{-\mu x}.$$

(b) When $\nu < 0$,

(b.1) if $-N/2 \leqslant \nu < 0$, then $N/2 < x + \nu$, and in this case

$$\left|\sum_{-N/2\leqslant\nu<0} C_\nu G(x+\nu)\right|\leqslant\left(A\sum|C_\nu|\right)e^{-\mu x/4}.$$

(b.2) if $\nu<-3N/2$, then $|x+\nu|\geqslant N/2-1$, for $N\geqslant 4$ we have

$$\left|\sum_{\nu<-3N/2} C_\nu G(x+\nu)\right|\leqslant\left(A\sum|C_\nu|\right)e^{-\mu x/4}.$$

(c) When $-3N/2<\nu<-N/2$, from that

$$|L_1(x)|\leqslant M\sum|C_\nu|,\quad |C_\nu|=O(e^{-\lambda|\nu|})$$

we get

$$\sum_{-3N/2<\nu<-N/2}|C_\nu|=O\left(\sum_{\nu=N/2}^{3N/2}e^{-\lambda|v|}\right)=O(e^{-\lambda x/4}).$$

To sum up, we have obtained that

$$|L(x)|\leqslant|L_1(x)|+|L_2(x)|+|L_3(x)|=O(e^{-B|x|}),$$

where $B=\min\{\lambda/4,\mu/4\}$, and (5) is proved.

Remark According to Hirschman and Widder[12] the function $G(t)$ is the Green function relating to the differential operator $\Psi(D)$. (4.8) shows that $L(x)$ is a spline with knots $\{\nu\}_{\nu\in\mathbf{Z}}$, which satisfies

(1) $L(x)\in C^\infty(\mathbf{R})$,

(2) $\Psi(D)L(x)|_{(\nu,\nu+1)}=0,\ v\in\mathbf{Z}$.

For an $f\in S_G^\infty$ we set

$$S(f,x)=\sum f(\nu)L(x-\nu). \tag{4.15}$$

By (4.9) the cardinal series standing on the right hand side of (4.15) converges. Moreover, we have

$$S(f,\nu)=f(\nu)\qquad\forall v\in\mathbf{Z}.$$

Thus $S(f)$ is a cardinal interpolation operator defined on S_G^∞. In what follows we are going to prove

$$\sup_{f\in S_G^\infty}\|f-S(f)\|_\infty=\|E(\cdot)\|_\infty. \tag{4.16}$$

In order to do this we need some auxiliary results.

Lemma 4 *There exist a constant $c>0$ independent of n, and a θ with $0<\theta<1$ such that*

$$|C_\nu^{(n)}|\leqslant c\theta^{|\nu|}, \tag{4.17}$$

where

$$C_\nu^{(n)} = (2\pi)^{-1}\int_{-\pi}^{\pi}\frac{e^{i\nu u}}{F_n(0,u)}du.$$

Proof　Suppose $v \neq 0$. Integrating by parts yields

$$|C_\nu^{(n)}| = \frac{1}{|\nu|^k}\left|(2\pi)^{-1}\int_{-\pi}^{\pi}\left(\frac{1}{F_n(0,u)}\right)^{(k)}e^{-i\nu u}du\right|.$$

Denote

$$M_n^{(k)} = \left\|\left(\frac{1}{F_n(0,u)}\right)^{(k)}\right\|_\infty.$$

We take a rectifiable simple closed curve C which, on the one hand, strictly enclose the closed interval $[-\pi,\pi]$, and on the other hand, is contained in a domain D, where the functions $(F_n(0,w))^{-1}(n=1,2,3,\cdots)$ are analytic. Such a domain D exists, because there is a domain D_0 strictly containing $[-\pi,\pi]$ such that $(F(0,w))^{-1}$ is analytic in D_0, $F(0,w)\neq 0$, and the relation

$$\lim_{n\to\infty}F_n(0,w) = F(0,w)$$

holds uniformly on any closed subdomain contained within D_0. So there must be some domain $D\supset[-\pi,\pi]$, $D\subset D_0$, and $F_n(0,w)\neq 0$ in D for all n. Thus $(F_n(0,w))^{-1}$are analytic in D. Now by Cauchy's formulae (The following argument may be found in[17])

$$\left(\frac{1}{F_n(0,u)}\right)^{(k)} = \frac{k!}{2\pi i}\int_C\frac{dz}{F_n(0,z)(z-u)^{k+1}}.$$

Suppose the length of C is l. Write

$$M_n = \max|F_n(0,z)|^{-1},\quad z\in \mathbf{C},$$

and denote by δ the distance between C and $[-\pi,\pi]$. Clearly $\delta>0$. Then we have

$$\left|\left(\frac{1}{F_n(0,u)}\right)^{(k)}\right| \leqslant M_n\frac{k!\,l}{2\pi\cdot\delta^{k+1}}.$$

Notice that M_n is bounded, because

$$\lim_{n\to\infty}M_n = \max_{z\in C}|F(0,z)^{-1}| \overset{\text{df}}{=} M < +\infty.$$

So we have

$$\left\|\left(\frac{1}{F_n(0,u)}\right)^{(k)}\right\|_\infty \leqslant \frac{k!\,M'l}{2\pi\cdot\delta^{k+1}},$$

where $M' > 0$ is some constant. We choose a $B > \max\left\{\delta^{-1}, \dfrac{M'l}{2\pi\delta^2}\right\}$.
Then from

$$\frac{M'l}{2\pi\delta^2}\cdot\frac{1}{\delta^{k-1}} < B^k$$

we get

$$\left\|\left(\frac{1}{F_n(0,u)}\right)^{(k)}\right\|_\infty \leqslant B^k \cdot k!,$$

and therefore

$$|C_\nu^{(n)}| \leqslant B^k \cdot k! \cdot |\nu|^{-k} < \left(\frac{Bk}{|\nu|}\right)^k, \quad v \neq 0.$$

Take a number $p > 0$ such that $Bp^{-1} < 1$ and put $\theta_1 = Bp^{-1}$. For each ν, $|\nu| \geqslant p$ we choose a positive integer k such that

$$k \leqslant |\nu| \cdot p^{-1} < k+1.$$

Then we have

$$|C_\nu^{(n)}| \leqslant \left(\frac{Bk}{|\nu|}\right)^k = \left(\frac{B}{p}\cdot\frac{p}{|\nu|}\cdot k\right)^k \leqslant \theta_1^k$$
$$= \theta_1^{-1}\cdot\theta_1^{k+1} < \theta_1^{-1}\cdot\theta_1^{|\nu|/p}$$

By taking $\theta_1^{-1} = C$ and choosing a θ, $\theta_1^{1/p} < \theta < 1$ we get

$$|C_\nu^{(n)}| < C\cdot\theta^{|\nu|}, \quad |\nu| \geqslant p. \tag{4.18}$$

When $|\nu| < p$, on account of $|C_\nu^{(n)}| \leqslant B|\nu|^{-1}(\nu \neq 0)$, we can obtain (4.18) for such ν provided that the constant C standing on the right hand side of (4.18) is appropriately enlarged. Finally, when $\nu = 0$,

$$C_0^{(n)} = (2\pi)^{-1}\int_{-\pi}^{\pi} F_n(0,u)^{-1}\mathrm{d}u$$

is bounded, for we have

$$\lim_{n\to\infty}(2\pi)^{-1}\int_{-\pi}^{\pi} F_n(0,u)^{-1}\mathrm{d}u = (2\pi)^{-1}\int_{-\pi}^{\pi} F(0,u)^{-1}\mathrm{d}u.$$

So the constant C may be chosen so that $|C_\nu^{(n)}| \leqslant C$. Lemma 4 is proved.

Corollary *The sequence* $\left\{\sum_{\nu=-\infty}^{+\infty}|C_\nu^{(n)}|\right\}_{n=1}^{+\infty}$ *is bounded.*

We need some estimation for $L_n(x)$, which holds uniformly for all n.

Lemma 5 *There exists a constant* $A > 0$ *independent of* n *such that*

$$\left|L_n(x)\right| \leqslant A \mid x \mid^{-\nu}, \quad x \in \mathbf{R}, x \neq 0, \nu \geqslant 3,\ n = 1,2,3,\cdots \tag{4.19}$$

The proof of (4.19) is completely similar to that of (5) in Theorem 3. We omit the details.

Remark The estimation (4.19) is of interest only for sufficiently large value of $\mid x \mid$. On account of uniform boundedness of $L_n(x)$ on $\mathbf{R}$ (by (3), Theorem 3) for small value of $\mid x \mid$ instead of (4.19) we may take

$$\left|L_n(x)\right| \leqslant A(1+\mid x \mid^{\nu})^{-1}. \tag{4.20}$$

According to Micchelli[13], for each n there are constants A_n, $B_n > 0$ such that

$$\left|L_n(x)\right| \leqslant A_n \mathrm{e}^{-B_n \mid x \mid}.$$

We conjecture that there would be two constants $A, B > 0$ independent of n for which holds

$$\left|L_n(x)\right| \leqslant A\mathrm{e}^{-B\mid x \mid}.$$

But the weaker estimation (4.19) is sufficient for proving (4.16). Now we are in a position to prove

Theorem 4

$$\sup_{f \in S_G^{\infty}} \| f - S(f) \|_{\infty} = \| E(\cdot) \|_{\infty}.$$

Proof for any $f \in S_G^{\infty}$ we have $f = G * \varphi$, $\| \varphi \|_{\infty} \leqslant 1$. Denote $f_n = G_n * \varphi$. Then

$$\| f - S(f) \|_{\infty} \leqslant \| f - S_n(f_n) \|_{\infty} + \| S_n(f_n) - S(f) \|_{\infty}.$$

We have proved (refer to the proof of Theorem 2) that

$$\| f - S_n(f_n) \|_{\infty} \leqslant \| G - G_n \|_1 + \| E_n(\cdot) \|_{\infty},$$

$$\| S(f) - S_n(f_n) \|_{\infty} \leqslant \| S(f - f_n) \|_{\infty} + \| (S - S_n)(f_n) \|_{\infty}.$$

From that

$$\| f - f_n \|_{\infty} \leqslant \| G(\cdot) - G_n(\cdot) \|_1 \to 0,$$

and the assertion (5) of Theorem 3 it is easy to prove that

$$\lim_{n \to \infty} \| S(f - f_n) \|_{\infty} = 0.$$

So it remains to prove that

$$\lim_{n\to\infty} \| (S - S_n)(f_n) \|_\infty = 0.$$

First, we notice that

$$|f_n(\nu)| \leqslant \mathbf{C}, \quad \forall n \in \mathbf{Z}_+, \nu \in \mathbf{Z},$$

and $\| L_n(\cdot) - L(\cdot) \|_\infty \to 0$. Fix $x \in \mathbf{R}$. For any positive integer T we have

$$|(S - S_n)(f_n, x)| \leqslant \sum_{\nu=-\infty}^{+\infty} |f_n(\nu)| \cdot |L(x-\nu) - L_n(x-\nu)| \leqslant$$

$$C \sum_{\nu=-\infty}^{+\infty} |L(x-\nu) - L_n(x-\nu)| = C\Big\{ \sum_{|x-\nu|\leqslant T} + \sum_{|x-\nu|>T} \Big\},$$

$$\sum_{|x-\nu|>T} \leqslant \sum_{|x-\nu|>T} |L_n(x-\nu)| + \sum_{|x-\nu|>T} |L(x-\nu)| \leqslant 2A \sum_{|x-\nu|>T} |x-\nu|^{-k}$$

(because $|L(x)| \leqslant A|x|^{-k}$ also holds for any fixed $k \geqslant 3$). Thus from

$$2A \sum_{|x-\nu|>T} |x-\nu|^{-k} \leqslant 4A \sum_{\nu=1}^{\infty} \frac{1}{(T+\nu)^k} < \frac{4A}{(k-1)T^{k-1}}.$$

We see that for any $\varepsilon > 0$, there exists $T_0 \in \mathbf{Z}_+$ such that

$$4AC[(k-1)T_0^{k-1}]^{-1} < \varepsilon/2,$$

where T_0 is independent of x. Cosequently we get

$$C \sum_{|x-\nu|>T_0} < \varepsilon/2 \tag{4.21}$$

Moreover, for the same $\varepsilon > 0$, there exists $n_0 \in \mathbf{Z}_+$ such that for every $n > n_0$

$$\| L(\cdot) - L_n(\cdot) \|_\infty < \varepsilon/4CT_0,$$

and hence

$$C \sum_{|x-\nu|\leqslant T_0} |L(x-\nu) - L_n(x-\nu)| \leqslant 2T_0 C \cdot \frac{\varepsilon}{4T_0 C} = \varepsilon/2.$$

To sum up, we have proved that for any $\varepsilon > 0$, there exists $n_0 \in \mathbf{Z}_+$ such that for any $n > n_0$

$$\| (S - S_n)(f_n) \|_\infty < \varepsilon. \tag{4.22}$$

Therefore

$$\| f - S(f) \|_\infty \leqslant \lim \| E_n(\cdot) \|_\infty = \| E(\cdot) \|_\infty.$$

But $E(x) \in S_G^\infty$, so we get (4.16).

Theorem 4 is thus proved.

Set

$$S_{G,0}^{\infty} = \{f: f \in S_G^{\infty}, f(v) = 0, \forall\ v \in \mathbf{Z}\}. \quad (4.23)$$

Corollary

$$\sup_{f \in S_{G,0}^{\infty}} \| f \|_{\infty} = \| E(\cdot) \|_{\infty}. \quad (4.24)$$

$E(x)$ is an extremal function for (4.24). It can be represented by

$$E(x) = \frac{2}{\pi} \sum_{\nu=0}^{\infty} \frac{\sin(2\nu+1)\pi x}{(2\nu+1)\Psi(\mathrm{i}(2\nu+1)\pi)} \quad (E(v) = 0, v \in \mathbf{Z}). \quad (4.25)$$

§ 5. An Example

Consider the function class $A_{\infty,\delta}$ (see[19], P.151). $f(x) \in A_{\infty,\delta}$ means that

(1) $f(x)$ is real-valued and analytic on $\mathbf{R}$,

(2) $f(x+\mathrm{i}y)$ is analytic in the strip $-\delta < y < \delta, -\infty < x < +\infty$,

(3) $|\mathrm{Re} f(x+\mathrm{i}y)| \leqslant 1$ for every $z = x + \mathrm{i}y$ in the strip

$$-\delta < y < \delta, -\infty < x < +\infty.$$

It is well known that

$$f(x) = \int_{-\infty}^{+\infty} G_\delta(x-t)\varphi(t)\mathrm{d}t, \quad (5.1)$$

where $G_\delta(t) = (2\delta)^{-1}\mathrm{sech}\,\frac{\pi t}{2\delta}$, $\| \varphi \|_{\infty} \leqslant 1$, and $G_\delta(t) \in PF$, $G_\delta(-t) = G_\delta(t)$. The reciprocal of the Laplace transform of $G_\delta(t)$ is $\Psi(s) = \cos \delta s$. Thus we have

$$F(0,u) = \sum_{k=-\infty}^{+\infty} \frac{1}{\cosh(\delta(2k\pi + u))} > 0. \quad (5.2)$$

Denote by $\{C_\nu\}$ the Fourier coefficients of $(F(0,2\pi u))^{-1}$. Then we have

$$\sum_{\nu=-\infty}^{+\infty} |C_\nu| < \infty,$$

and in this case

$$L(x) = (2\delta)^{-1} \sum_{v=-\infty}^{+\infty} C_v \mathrm{sech}\,\frac{\pi}{2\delta}(x+\nu). \quad (5.3)$$

The standard function is

$$E_\delta(x) = 2(\delta)^{-1}\int_{-\infty}^{+\infty} \frac{\operatorname{sgn}\sin \pi t}{\cosh \frac{\pi}{2\delta}(x-t)}\mathrm{d}t. \tag{5.4}$$

We have

Theorem 5

$$E(A_{\infty,\delta}) = \| E_\delta(\cdot)\|_\infty = \frac{4}{\pi}\sum_{k=0}^{\infty}\frac{(-1)^k}{(2k+1)\cosh(2k+1)\delta\pi}.$$

The set $\{\nu\}_{\nu\in\mathbf{Z}}$ *is an optimal set of sampling points,* $I^*(f) = \{f(v)\}_{v\in\mathbf{Z}}$ *is an optimal information operator, and* $S(f)$ *an optimal linear algorithm.*

Remark 1 We have merely treated the special case $G(-t) = G(t)$. It is expected to remove the evenness condition.

Remark 2 Similar results may be obtained for some classes of sets of sampling points with more general character. Actually, denote by Θ the class of denumerable sets of real numbers of the following type. We say $\xi = \{\xi_j\} \in \Theta$, if

(1) $\xi_j < \xi_{j+1}$, $\forall j \in \mathbf{Z}$,

(2) $\varliminf_{a\to\infty}\frac{\operatorname{card}\{\xi \cap [-a,a]\}}{2a} \leqslant 1$,

where card $\{\xi \cap [-a,a]\}$ means the number of points of ξ contained in $[-a,a]$. All results obtained in this paper are still valid.

References

[1] Micchelli, C. A., Rivlin, T. J. A Survey of Optimal Recovery. In C. Micchelli edited: *Optimal Estimation in Approximation Theory*. New York: Plenum Press, 1977. 1 ~ 68

[2] Micchelli, C. A., Rivlin, T. J., Winograd, S. The Optimal Recovery of smooth functions. Numer, Math., 1976, 26: 191 ~ 200

[3] Великин, В. Л. Оптимальная интерполяция периодических дифференцируемых функций с ограниченной старшей производной. Мат. Заметки, 1977, 22(5): 633 ~ 670

[4] Осипенко, К. Ю. Наилучшее приближение аналитических функций по информации об их значениях в конечном числе точек. Мат. Заметкн, 1976,

19(1):29 ~ 40

[5] Sun Yongsheng. On Optimal interpolation for a differentiable function class (Ⅰ). ATA, 1986, 2(4):49 ~ 54

[6] Sun Yongsheng. Optimal recovery on a differentiable function class(in Chinese). KEXUE TONGBAO, 1986, 31(11):809 ~ 812

[7] Sun Yongsheng, Huang Daren. Optimal Interpolation on Some Class of Differentiable Functions. In *Approximation Theory V*. New York: Academic Press, 1987. pp. 583 ~ 586

[8] Karlin, S. Oscillatory Perfect Splines and Related Extremal Problems. In S. Karlin ed.: *Studies in Spline Functions and Approximation Theory*, New York: Academic Press, 1976. pp. 371 ~ 458

[9] Чахкиев, М. А. Линейиые дифференциальные операторы с вещественным снектором и оптимальные квадратурные формулы. Изв. АНСССР сер. мат., 1984, 48(5):1078 ~ 1108

[10] Fang Gensun, One-sided *N*-widths for a class of periodic functions. Scientia Sinica, AXXXI1988, (12):1409 ~ 1423

[11] Sun Yongsheng, Huang Daren. On *N*-width of a generalized Bernoulli kernel. ATA., 1985, 1(2):83 ~ 92

[12] Hirschman, I. I., Widder, D. V. *The Convolution Transform*. New York: Princeton Press, 1955. 42 ~ 49

[13] Micchelli, C. A. Cardinal $\mathscr{L}$-Splines. In S. Karlin ed.: *Studies in Spline Functions and Approximation Theory*. New York: Academic Press, 1976. 203 ~250

[14] Новиков, С. И. Приближенне одного кдасса дифференцируемых фуикций $\mathscr{L}$-сйлайнами. Мат. Заметкн, 1983, 33(3):393 ~ 407

[15] Tang Xuhui. Approximation of some classes of differentiable functions with cardinal $\mathscr{L}$-splines on **R**. ATA, 1987, 3(2 ~ 3):1 ~ 17

[16] Ахиезер, Н. И., *Лекции по теории аппроксимации*. М.: изд НАУКА, 1965. 141

[17] Вари, И. К. *Тригонометрические ряды*. М.: изд НАУКА, 1961. 88

[18] Karlin, S. *Total positivity*. Vol. 1. Stanford, Stanford Univ. Press, 1967. 333

[19] Тиман, А. Ф. *Теория ириближения функцнй действительного неременного*. М.. Физматгиз, 1960. 151

Edited by Li Xiaoguang

全实轴上某些光滑函数类用高阶基样条的最佳逼近*

Best Approximation of Some Classes of Smooth Functions on the Whole Real Axis by Cardinal Splines of Higher Order

§ 1. Introduction

Let r be a positive integer, and $1 \leqslant p \leqslant +\infty$. Denote

$$L_p^r(\mathbf{R}) = \{f : f \in L^p(\mathbf{R}) \cap C(\mathbf{R}),\ f^{(r-1)} \text{ is abs. cont. on every finite interval},\ f^{(r)} \in L^p(\mathbf{R})\}, \tag{1.1}$$

$$W_p^r(\mathbf{R}) = \{f : f \in L_p^r(\mathbf{R}),\ \| f^{(r)} \|_p \leqslant 1\}, \tag{1.2}$$

in which $\| \cdot \|_p$ is the usual $L^p(\mathbf{R})$ - norm.

Let $T = \{t_j\}_{j \in \mathbf{Z}}$ be a real sequence which satisfies the following conditions:

(1) $t_j < t_{j+1},\ \forall j \in \mathbf{Z}$.

(2) $\lim\limits_{j \to -\infty} t_j = -\infty,\ \lim\limits_{j \to +\infty} t_j = +\infty$.

Given

$$S_{m,T} = \{s(t) : s(t) \in C^{m-2}(\mathbf{R}),\ D^m s(t) |_{(t_j, t_{j+1})} = 0,\ \forall j \in \mathbf{Z}\}, \tag{1.3}$$

where m is a positive integer. It is the class of cardinal splines of order m (i.e., degree m − 1) with simple knots $t_j, j \in \mathbf{Z}$. Given a function $g \in L^p(\mathbf{R})$.

---

* 本文的俄文译稿发表于《Математические Заметки》1990, No10.

Supported by the National Fund of Natural Sciences.

* 本文与李淳合作.

We set

$$e(g, S_{m,T})_p = \inf \| g - f \|_p, \tag{1.4}$$

where f is taken over the set $S_{m,T} \cap L^p(\mathbf{R})$. The quantity $e(g, S_{m,T})_p$ is known as the best approximation of g in metric $L^p(\mathbf{R})$ by $S_{m,T} \cap L^p(\mathbf{R})$. The quantity

$$E(W_p^r(\mathbf{R}); S_{m,T})_p = \sup_{g \in W_p^r} e(g, S_{m,T})_p \tag{1.5}$$

is said to be the best approximation of $W_p^r(\mathbf{R})$ in $L^p(\mathbf{R})$ by $S_{m,T} \cap L^p(\mathbf{R})$. The objective of this paper is to obtain the exact expressions for $E(W_p^r(\mathbf{R}); S_{m,T})_p$ under the condition $T = \{j\}_{j \in \mathbf{Z}}$, $p = 1, 2, +\infty$ and $m \geqslant r$. It is worthy to point out two crucial elements in our mrthods of approach:

(1) *To establish a duality theorem* for the $L^p(\mathbf{R})$-best approximation by the cardinal spline space. As far as we know, N.P. Korneichuk and others [1] established some duality formulas for the L^p-best approximation of a periodic function which is approximated by periodic polynomial splines. Such duality relations play very important role in obtaining sharp estimates for many optimal problems on the class of periodic differentiable functions. For the case under our consideration in this paper, the analogous duality relations are remained to be established. We solve this problem by applying the B-spline basis of $S_{m,T} \cap L^p(\mathbf{R})$.

(2) *To obtain the exact estimation of the minimal information diameter for the differentiation operator* $D^k (1 \leqslant k \leqslant r-1, r \leqslant 2)$ on $W_p^r(\mathbf{R})$ over the set of in formation operators $\{I_\xi : \xi \in \Theta\}$, where $p = 1, 2, +\infty$, and $\Theta = \{\xi : \xi = \{\xi_j\}_{j \in \mathbf{Z}}\}$ is a set of real sequences $\xi = \{\xi_j\}_{j \in \mathbf{Z}}$ satisfying the following conditions:

(1) $\xi_j < \xi_{j+1}$, $\forall j \in \mathbf{Z}$.

$$(2)\ \varlimsup_{a \to +\infty} \frac{\operatorname{card}\{\xi \cap [-a, a]\}}{2a} \leqslant 1, \tag{1.6}$$

Where $\operatorname{card}\{\xi \cap [-a, a]\}$ is the number of $\xi_j \in \xi$ contained in $[-a, a]$. A periodic version of this problem has been solved by Sun [2] recently. It has very close connection with the problem of exact constants in

Landau-Kolmogorov inequality. (see[3])

§ 2. A duality theorem

Lemma 2.1 *Let p be such that $1 \leqslant p < +\infty$, and $f \in L_p^r(\mathbf{R})$. Then*

$$\lim_{|x| \to \infty} f^{(k)}(x) = 0, \ k = 0, \cdots, r-1. \tag{2.1}$$

Proof $\forall k \in \{0, \cdots, r-1\}$, and $x_1 < x_2$. From $f, f^{(r)} \in L^p(\mathbf{R})$ by stein's inequality [4] we have $f^{(k)} \in L^p(\mathbf{R})$. Then for $1 < p < +\infty$

$$\begin{aligned} & \left| \, |f^{(k)}(x_2)|^p - |f^{(k)}(x_1)|^p \right| \\ = & \left| \int_{x_1}^{x_2} p \, |f^{(k)}(x)|^{p-1} \operatorname{sgn}[f^{(k)}(x)] f^{(k+1)}(x) \mathrm{d}x \right| \\ \leqslant & \ p \int_{x_1}^{x_2} |f^{(k+1)}(x)| \cdot |f^{(k)}(x)|^{p-1} \mathrm{d}x \\ \leqslant & \left(\int_{x_1}^{x_2} |f^{(k+1)}(x)|^p \mathrm{d}x \right)^{\frac{1}{p}} \cdot p \cdot \left(\int_{x_1}^{x_2} |f^{(k)}(x)|^p \mathrm{d}x \right)^{\frac{1}{p'}} \\ = & \ p \left(\int_{x_1}^{x_2} |f^{(k+1)}(x)|^p \mathrm{d}x \right)^{\frac{1}{p}} \cdot \left(\int_{x_1}^{x_2} |f^{(k)}(x)|^p \mathrm{d}x \right)^{1-\frac{1}{p}}. \end{aligned} \tag{2.2}$$

$(1/p + 1/p' = 1)$. For $p = 1$

$$\begin{aligned} & \left| \, |f^{(k)}(x_1)| - |f^{(k)}(x_2)| \, \right| \leqslant |f^{(k)}(x_1) - f^{(k)}(x_2)| \\ = & \left| \int_{x_1}^{x_2} f^{(k+1)}(x) \mathrm{d}x \right| \leqslant \int_{x_1}^{x_2} |f^{(k+1)}(x)| \mathrm{d}x. \end{aligned} \tag{2.3}$$

So that for $1 \leqslant p < +\infty$

$$\lim_{x_2 > x_1 \to \infty} \left| \, |f^{(k)}(x_2)|^p - |f^{(k)}(x_1)|^p \right| = 0,$$

and therefore the limit $\lim_{x \to +\infty} |f^{(k)}(x)|^p$ exists. Now from $|f^{(k)}|^p \in L(\mathbf{R})$ we derive $\lim_{x \to +\infty} f^{(k)}(x) = 0$. Similarly, $\lim_{x \to -\infty} f^{(k)}(x) = 0$. Lemma 2.1 is proven.

Let $T = \{t_j\}$ be given as in § 1. Following [5], we introduce the B-spline function as follows:

$$M_{j,r}(x) = r[t_j, \cdots, t_{j+r}](\cdot - x)_+^{r-1}, \ \forall j \in \mathbf{Z} \tag{2.4}$$

It is well-known that $M_{j,r}(x)$ has compact support $\overline{\text{supp } M_{j,r}} = [t_j, t_{j+r}]$.

Lemma 2.2 *Suppose that* $1 \leqslant p \leqslant +\infty$, $r \in \mathbf{Z}_+$, *and* $f \in L_p^r(\mathbf{R})$. *Then*

$$\int_{\mathbf{R}} M_{j,r}(x) f^{(r)}(x) \mathrm{d}x = 0, \ \forall j \in \mathbf{Z} \tag{2.5}$$

iff

$$f(t_j) = \begin{cases} 0, \text{for } 1 \leqslant p < +\infty, \\ c, \text{for } p = +\infty, \end{cases} \quad \forall j \in \mathbf{Z}, \tag{2.6}$$

where c is a number independent of j.

Proof From the integral formulae

$$\int_R M_{j,r}(x) f^{(r)}(x) \mathrm{d}x = r! [t_j, \cdots, t_{j+r}] f \tag{2.7}$$

(see [5])(2.5) is trivally derived.

Now we assume the validity of (2, 5), we prove (2.6) by mathematical induction. When $r = 1$, from (2.5) and (2.7) we derive $f(t_{j+1}) - f(t_j) = 0$, $\forall j \in \mathbf{Z}$. So there must be a number c such that $f(t_j) = c$ for all $j \in \mathbf{Z}$. Therefore, when $1 \leqslant p < +\infty$ by lemma 2.1

$$c = \lim_{j \to \pm\infty} f(t_j) = \lim_{x \to \infty} f(x) = 0.$$

Thus the proposition (2.6) holds true for $r = 1$.

Suppose that the proposition holds for $r - 1$. By (2.5), (2.7), and the definition of devided differences we have

$$[t_j, \cdots, t_{j+r-1}] f - [t_{j+1}, \cdots, t_{j+r}] f = (t_j - t_{j+r}) [t_j, \cdots, t_{j+r}] f = 0,$$

$\forall j \in \mathbf{Z}$.

So that there is a number c_1 such that

$$[t_j, \cdots, t_{j+r-1}] f = c_1, \ \forall j \in \mathbf{Z}. \tag{2.8}$$

If $c_1 \neq 0$, without loss of generality we may assume $c_1 > 0$. Then

$$[t_{j+1}, \cdots, t_{j+r-1}] f - [t_j, \cdots, t_{j+r-2}] f = (t_{j+r-1} - t_j) [t_j, \cdots, t_{j+r-1}] f$$
$$= c_1 (t_{j+r-1} - t_j) \geqslant c_1 (t_{j+r-1} - t_{j+r-2}), \ \forall j \in \mathbf{Z}.$$

To sum from $j = 1$ to k on each side of the above inequality we get

$$[t_{k+1}, \cdots, t_{k+r-1}] f \geqslant [t_1, \cdots, t_{r-1}] f + c_1 (t_{k+r-1} - t_{r-1}). \ (\forall k \in \mathbf{Z}_+) \tag{2.9}$$

On the other hand, in view of

$$\operatorname{supp} M_{k+1,r-2} \subset (t_{k+1}, t_{k+r-1}),$$

we have

$$[t_{k+1}, \cdots, t_{k+r-1}]f = \frac{1}{(r-2)!}\int_{t_{k+1}}^{t_{k+r-1}} M_{k+1,r-2}(x) f^{(r-2)}(x)\mathrm{d}x. \quad (2.10)$$

(1) When $p = +\infty$, from $f, f^{(r)} \in L^{\infty}(\mathbf{R})$ we get $f^{(r-2)} \in L^{\infty}(\mathbf{R})$. we notice that

(a) $M_{k+1,r-2}(x) > 0, x \in (t_{k+1}, t_{k+r-1}), k \in \mathbf{Z}$.

(b) $\int_{t_{k+1}}^{t_{k+r-1}} M_{k+1,r-2}(x)\mathrm{d}x = 1, \forall k \in \mathbf{Z}$ (2.11)

(see [5]). Consequently

$$|[t_{k+1}, \cdots, t_{k+r-1}]f| \leqslant \| f^{(r-2)} \|_{\infty}/(r-2)! \quad \forall k \in \mathbf{Z}$$

which contradicts with (2.9) (because $t_k \to +\infty$ when $k \to +\infty$).

(2) When $1 \leqslant p < +\infty$, integration by parts yields

$$[t_{k+1}, \cdots, t_{k+r-1}]f = \frac{1}{(r-2)!}\int_{t_{k+1}}^{t_{k+r-1}} f^{(r-2)}(x)\mathrm{d}\left(\int_{t_{k+1}}^{x} M_{k+1,r-2}(t)\mathrm{d}t\right)$$

$$= \frac{f^{(r-2)}(t_{k+r-1})}{(r-2)!} - \frac{1}{(r-2)!}\int_{t_{k+1}}^{t_{k+r-1}}\left(\int_{t_{k+1}}^{x} M_{k+1,r-2}(t)\mathrm{d}t\right) f^{(r-1)}(x)\mathrm{d}x. \quad (2.12)$$

From $f, f^{(r)} \in L^p(\mathbf{R})$, by Stein's inequality we get $f^{(r-1)} \in L^p(\mathbf{R})$. So that from (2.11), (2.12) we obtain

$$|[t_{k+1}, \cdots, t_{k+r-1}]f|$$

$$\leqslant \frac{|f^{(r-2)}(t_{k+r-1})|}{(r-2)!} + \frac{1}{(r-2)!}\int_{t_{k+1}}^{t_{k+r-1}} |f^{(r-1)}(x)|\mathrm{d}x$$

$$\leqslant \frac{|f^{(r-2)}(t_{k+r-1})|}{(r-2)!} + \frac{1}{(r-2)!}\left(\int_{t_{k+1}}^{t_{k+r-1}} |f^{(r-1)}(x)|^p \mathrm{d}x\right)^{\frac{1}{p}} \cdot (t_{k+r-1} - t_{k+1})^{\frac{1}{p'}}$$

$$\leqslant \frac{|f^{(r-2)}(t_{k+r-1})|}{(r-2)!} + \frac{\| f^{(r-1)} \|_p}{(r-2)!}(t_{k+r-1} - t_{r-1})^{\frac{1}{p'}},$$

where $k \geqslant r-3$, $\frac{1}{p} + \frac{1}{p'} = 1$. In view of $0 \leqslant \frac{1}{p'} < 1$, $\lim\limits_{k\to+\infty} t_{k+r-1} = +\infty$, and $\lim\limits_{k\to+\infty} f^{(r-2)}(t_{k+r-1}) = 0$, we see that when $k \to +\infty$ the above inequality leads to contradiction with (2.9).

To sum up, in both cases we must have $c_1 = 0$, *i.e.*,

$$[t_j, \cdots, t_{j+r-1}]f = 0, \forall j \in \mathbf{Z}.$$

Thus by the induction hypothesis we conclude the validity of (2.6) for r. Lemma 2.2 is proven.

In what follows we write

$$N_{j,r}(x) = (t_{j+r} - t_j)M_{j,r}(x)/r, \forall j \in \mathbf{Z} \tag{2.13}$$

This is the so-called normalized B-spline. Then Lemma 2.2 may be reformulated in the following version.

Lemma 2.2′ *Suppose* $1 \leqslant p \leqslant +\infty$, $r \in \mathbf{Z}_+$, $f \in L_p^r(\mathbf{R})$. *Then*

$$\int_{\mathbf{R}} N_{j,r}(x) f^{(r)}(x)\mathrm{d}x = 0, \ \forall j \in \mathbf{Z}$$

iff

$$f(t_j) = \begin{cases} 0, & \text{when } 1 \leqslant p < +\infty, \\ c & \text{when } p = +\infty. \end{cases}$$

Now we are in a position to prove the duality theorem concerning the best approximation of a function $g \in L^p(\mathbf{R})$ by $S_{r,T} \cap L^p(\mathbf{R})$. By[5], $\{N_{j,r}(x)\}_{j\in\mathbf{Z}}$ is the Schauder basis of $S_{r,T} \cap L^p(\mathbf{R})$.

Theorem 1 *Suppose that* $1 \leqslant p \leqslant +\infty$, $\frac{1}{p} + \frac{1}{p'} = 1$, $g \in L^p(\mathbf{R})$ $(1 \leqslant p < +\infty)$, *and* $g \in L^\infty(\mathbf{R}) \cap C(\mathbf{R})$, $(p = +\infty)$. *Then we have*

$$\inf_{\alpha\in\mathbf{R}^{\mathbf{Z}}} \| g - \sum \alpha_j N_{j,r} \|_p$$
$$= \sup\left\{\int_{\mathbf{R}} g(x) f^{(r)}(x)\mathrm{d}x : f \in W_{p'}^r(\mathbf{R}),\ f(t_j) = 0, \forall j \in \mathbf{Z}\right\}, \tag{2.14}$$

where $\alpha = (\alpha_j) \in \mathbf{R}^Z$ *such that* $\sum \alpha_j N_{j,r} \in S_{r,T} \cap L^p(\mathbf{R})$.

Proof In what follows $\int_{\mathbf{R}} g(x)h(x)\mathrm{d}x$, $\int_a^b g(x)h(x)\mathrm{d}x$ will be writen simply as $\int gh$, $\int_a^b gh$ respectively.

First, we notice that

$$\sup\left\{\int g f^{(r)} : f \in W_{p'}^r(\mathbf{R}), f(t_j) = 0, \forall j \in \mathbf{Z}\right\}$$

$$= \sup\left\{\int gf^{(r)} : f \in W_{p'}^r(\mathbf{R}), \int N_{j,r}f^{(r)} = 0, \forall j \in \mathbf{Z}\right\} \quad (2.15)$$

Actually, when $1 < p \leqslant +\infty$ (2.15) is immediately derived from Lemma 2.2, because then $1 \leqslant p' < +\infty$. When $p = 1, p' = +\infty$, then from Lemma 2.2 it is derived that

$$\text{the left side of (2.15)} \leqslant \text{the right side.} \quad (2.16)$$

On the other hand, for every $f \in W_{p'}^r(\mathbf{R})$ such that $\int N_{j,r}f^{(r)} = 0, \forall j \in \mathbf{Z}$ there is a number c such that $f(t_j) = c$ for all $j \in \mathbf{Z}$. Put $f_1 = f - c$. Then $f_1 \in W_{p'}^r(\mathbf{R})$ and $f_1(t_j) = 0$ for all $j \in \mathbf{Z}$. So that we have

$$\int gf^{(r)} = \int gf_1^{(r)} \leqslant \text{the left side of (2.15)},$$

and therefore

$$\sup\left\{\int gf^{(r)} : f \in W_{p'}^r(\mathbf{R}), \int N_{j,r}f^{(r)} = 0, \forall j \in \mathbf{Z}\right\}$$
$$\leqslant \sup\left\{\int gf^{(r)} : f \in W_{p'}^r(\mathbf{R}), f(t_j) = 0, \forall j \in \mathbf{Z}\right\}. \quad (2.17)$$

the validity of (2.15) for $p = 1$ is proved.

In the next step we prove

$$\inf_{\alpha \in \mathbf{R}^{\mathbf{Z}}} \| g - \sum \alpha_j N_{j,r} \|_p = \sup\left\{\int gf^{(r)} : f \in W_{p'}^r(\mathbf{R}), \int N_{j,r}f^{(r)} = 0, \forall j \in \mathbf{Z}\right\}. \quad (2.18)$$

In view of $\int N_{j,r}f^{(r)} = 0$ and $\| f^{(r)} \|_{p'} \leqslant 1$, then for any $\alpha = (\alpha_j)$ such that $\sum \alpha_j N_{j,r} \in L^p(\mathbf{R})$ we have

$$\int gf^{(r)} = \int (g - \sum \alpha_j N_{j,r}) f^{(r)} \leqslant \| g - \sum \alpha_j N_{j,r} \|_p \cdot \| f^{(r)} \|_{p'}$$
$$\leqslant \| g - \sum \alpha_j N_{j,r} \|_p.$$

So that

$$\int gf^{(r)} \leqslant \inf_{\alpha \in \mathbf{R}^{\mathbf{Z}}} \| g - \sum \alpha_j N_{j,r} \|_p.$$

This means that

$$\text{the right side of (2.18)} \leqslant \text{left side.} \quad (2.19)$$

For proving the inverse inequality of (2.19) we first denote $I_N = [t_{-N},$

$t_N]$,

$$E_N = \left\{h : \text{supp } h \subset I_N, \| h \|_{p'} \leqslant 1, \int_{\mathbf{R}} N_{j,r} h = 0, \forall j \in \mathbf{Z}\right\},$$

$$(N = 1, 2, 3, \cdots) \tag{2.20}$$

and for $h \in E_N$ we put

$$f_h(x) = \frac{1}{(r-1)!} \int_{I_N} (x - t)_+^{r-1} h(t) \mathrm{d}t. \tag{2.21}$$

We claim that $f_h \in W_{p'}^r(\mathbf{R})$ and supp $f_h \subset I_N$. In fact, the validity of $f_h \in W_{p'}^r(\mathbf{R})$ is obvious, and $f_h(x) \equiv 0$ is trivally true for $x \leqslant t_{-N}$. On the other hand, on account of that every polymonial of degree $\leqslant r - 1$ is a linear combination of some $N_{j,r}$, and that $\int N_{j,r} h = 0, \forall j \in \mathbf{Z}$ we deduce that $f_h(x) \equiv 0$ for $x \geqslant t_N$. So that supp $f_h \subset I_N$. Thus we have

$$\begin{aligned} &\sup \left\{\int g f^{(r)} : f \in W_{p'}^r(\mathbf{R}), \int N_{j,r} f^{(r)} = 0, \forall j \in \mathbf{Z}\right\} \\ &\geqslant \sup \left\{\int_{I_N} g h : \| h \|_{p'(I_N)} \leqslant 1, \int N_{j,r} h = 0, \forall j \in \mathbf{Z}\right\}. \end{aligned} \tag{2.22}$$

Now we notice that $N_{j,r}(x)$ has compact support $[t_j, t_{j+r}]$, so that for $j \leqslant -N - r$, and $j \geqslant N$ the orthogongonality condition $\int_{I_N} N_{j,r} h = 0$ is fulfilled automatically. Thus the orthogonality condition is essential only for j, $-N - r < j < N$. By the duality theorem concerning the best approximation of a funotion by a finite-dimensional subspace (see[6]) we have

$$\begin{aligned} &\sup \left\{\int_{I_N} g h : \| h \|_{p'(I_N)} \leqslant 1, \int_{I_N} N_{j,r} h = 0, \forall j \in \mathbf{Z}\right\} \\ &= \inf \left\{ \| g - \sum \alpha_j N_{j,r} \|_{p(I_N)} : \alpha_j \in \mathbf{R}, -N - r < j < N\right\} \\ &= \| g - \sum \alpha_{j,N}^* N_{j,r} \|_{p(I_N)}, \end{aligned} \tag{2.23}$$

where $\alpha_{j,N}^* = 0$, for $j \leqslant -N - r$, and $j \geqslant N$. Funotion $\sum \alpha_{j,N}^* N_{j,r}(x)$ is the best approximant of g in $L^p(I_N)$ − norm. The right side of (2.22) (N = 1, 2, ⋯) defines a monotonically increasing and bounded sequence. Now we prove

$$\lim_{N\to+\infty} \| g - \sum \alpha_{j,N}^* N_{j,r} \|_{p(I_N)}$$
$$\geqslant \inf_{\alpha} \| g - \sum \alpha_j N_{j,r} \|_p \overset{\mathrm{df}}{=} d. \qquad (2.24)$$

Assume $d > 0$ ($d = 0$ case is trival). If (2.24) is false, then there is some $\varepsilon > 0$ such that for sufficiently large N it holds

$$\| g - \sum \alpha_{j,N}^* N_{j,r} \|_{p(I_N)} < d - \varepsilon. \qquad (2.25)$$

We remark that, for every fixed $j \in \mathbf{Z}$ the sequence $\{\alpha_{j,N}^*\}_{N=1}^{+\infty}$ is bounded. In fact, from [5], there exists a constant D_r which depends only on r such that for every $j \in \mathbf{Z}$,

$$|\alpha_{j,N}^*|(t_{j+r} - t_j)^{\frac{1}{p}} \leqslant D_r \| \sum \alpha_{k,N}^* N_{k,r} \|_p (t_j, t_{j+r}).$$

Let $N \geqslant \max\{-j, j+r\}$. Then $[t_j, t_{j+r}] \subset [t_{-N}, t_N] = I_N$, and

$$|\alpha_{j,N}^*| \leqslant D_r (t_{j+r} - t_j)^{-\frac{1}{p}} \| \sum \alpha_{k,N}^* N_{k,r} \|_{p(I_N)} \leqslant$$

$$2D_r (t_{j+r} - t_j)^{-\frac{1}{p}} \| g \|_{p(I_N)} \leqslant 2D_r (t_{j+r} - t_j)^{-\frac{1}{p}} \| g \|_p < +\infty.$$

Thus by applying the diagonal process we may choose a subsequence $\{N_n\}_{n=1}^{+\infty}$ of positive integers such that for every $j \in \mathbf{Z}$,

$$\lim_{n\to+\infty} \alpha_{j,N_n}^* = \beta_j \in \mathbf{R}. \qquad (2.26)$$

In what follows we argue separately in two cases:

(1) When $1 \leqslant p < +\infty$, we put

$$f_n(x) = \begin{cases} \left| g(x) - \sum \alpha_{j,N_n}^* N_{j,r}(x) \right|^p, & x \in I_{N_n}, \\ 0, & x \in \mathbf{R} \setminus I_{N_n}. \end{cases} \qquad (2.27)$$

For any $x \in \mathbf{R}$, we have $x \in I_{N_n}$ for sufficiently great n. In view of that the support of $N_{j,r}(x)$ is compact, so we see that for every x in the series $\sum \alpha_{j,N_n}^* N_{j,r}(x)$, there is only a finite number of terms different from zero. Therefore from (2.26), (2.27) we obtain

$$\lim_{n\to+\infty} f_n(x) = \left| g(x) - \sum \beta_j N_{j,r}(x) \right|^p, \quad x \in \mathbf{R}.$$

Now by Fatou's lemma

$$\int_{\mathbf{R}} \left| g(x) - \sum \beta_j N_{j,r}(x) \right|^p \mathrm{d}x \leqslant \underline{\lim}_{n\to+\infty} \int_{\mathbf{R}} f_n(x) \mathrm{d}x$$

$$= \lim_{n\to+\infty}\int_{I_{N_n}}\left|g(x)-\sum_{j\in\mathbf{Z}}\alpha^*_{j,N_n}N_{j-r}(x)\right|^p \mathrm{d}x \leqslant (d-\varepsilon)^p.$$

This shows $\sum \beta_j N_{j,r} \in S_{r,T} \cap L^p(\mathbf{R})$.

But then

$$\int_{\mathbf{R}}\left|g(x)-\sum \beta_j N_{j,r}(x)\right|^p \mathrm{d}x \geqslant d^p,$$

which leads to a contradiction.

(2) When $p=+\infty$, then every $x \in I_{N_n}$ for sufficiently large n. By assumption (2.25) we have

$$\left|g(x)-\sum \alpha^*_{j,N_n}N_{j,r}(x)\right| < d-\varepsilon.$$

By taking limit $(n\to+\infty)$ we get

$$\left|g(x)-\sum \beta_j N_{j,r}(x)\right| \leqslant d-\varepsilon,$$

because in each sum $\sum \alpha^*_{j,N_n}N_{j,r}(x)$, the number of terms different from zero is finite. The arbitrariness of $x\in\mathbf{R}$ entails

$$\| g-\sum \beta_j N_{j,r}\|_\infty \leqslant d-\varepsilon.$$

This show that $\sum \beta_j N_{j,r} \in \delta_{r,T} \cap L^\infty \mathbf{R}$. But then the relation

$$\| g-\sum \beta_j N_{j,r}\|_\infty \leqslant d$$

leads to a contradiction.

To sum up, we have proved the validity of (2.24). Now we take into account (2.22). By taking limit on the right side of (2.22), we obtain an inequality inverse to (2.19). Theorem 1 is proven.

Theorem 2 *Suppose that* $1\leqslant p\leqslant+\infty$, $\frac{1}{p}+\frac{1}{p'}=1$, $m\geqslant r$ *are positive integers. Then*

$$E(W_p^r(\mathbf{R});S_{m,T})_p$$
$$\leqslant \sup\{\|f^{(m-r)}\|_{p'} : f\in W_{p'}^m(\mathbf{R}), f(t_j)=0, \forall j\in\mathbf{Z}\}. \quad (2.28)$$

Proof From Theorem 1 we have

$$E(W_p^r(\mathbf{R});S_{m,r})_p$$
$$= \sup_{g\in W_p^r}\inf_{\alpha}\| g-\sum \alpha_j N_{j,r}\|_p$$
$$= \sup_{g\in W_p^r}\sup\left\{\int g f^{(m)} : f\in W_{p'}^m(\mathbf{R}), f(t_j)=0, j\in\mathbf{Z}\right\}$$

$$= \sup\left\{\int g f^{(m)} : g \in W_p^r(\mathbf{R}), f \in W_{p'}^m(\mathbf{R}), f(t_j) = 0, j \in \mathbf{Z}\right\}. \tag{2.29}$$

For any $g \in W_p^r(\mathbf{R}), f \in W_{p'}^m(\mathbf{R})$ and a > 0 integration by parts yields

$$\int_{-a}^{a} g(x) f^{(m)}(x)\mathrm{d}x = \Big[\sum_{k=0}^{r-1}(-1)^k g^{(k)}(x) f^{(m-1-k)}(x)\Big]_{-a}^{a} + (-1)^r \int_{-a}^{a} g^{(r)}(x) f^{(m-r)}(x)\mathrm{d}x. \tag{2.30}$$

By Lemma 2.1

$$\lim_{a \to +\infty}\Big[\sum_{k=0}^{r-1}(-1)^k g^{(k)}(x) f^{(m-1-k)}(x)\Big]_{-a}^{a} = 0.$$

And besides this, in view of $gf^{(m)}, g^{(r)} f^{(m-r)} \in L(\mathbf{R})$, so that we have

$$\int g f^{(m)} = (-1)^r \int g^{(r)} f^{(m-r)}. \tag{2.31}$$

Thus we may rewrite (2.29) as follows:

$$\begin{aligned} & E(W_p^r(\mathbf{R}); S_{m,T})_p \\ &= \sup\left\{\int g^{(r)} f^{(m-r)} : g \in W_p^r(\mathbf{R}), f \in W_{p'}^m(\mathbf{R}), f(t_j) = 0, j \in \mathbf{Z}\right\} \\ &\leqslant \sup\{\| g^{(r)} \|_p \cdot \| f^{(m-r)} \|_{p'} : g \in W_p^r(\mathbf{R}), f \in W_{p'}^m(\mathbf{R}), f(t_j) = 0, \\ &\qquad \forall j \in \mathbf{Z}\} \\ &\leqslant \sup\{\| f^{(m-r)} \|_{p'} : f \in W_{p'}^m(\mathbf{R}), f(t_j) = 0, \forall j \in \mathbf{Z}\}. \end{aligned}$$

(2.28) is proved.

It is very interesting to consider the problem under what conditions the inequality in (2.28) turns into an equality. In the next section we shall solve some special cases of this problem.

§ 3. The minimal diameter of information for D^k on $W_p^n(\mathbf{R})$ over a class of information operators

Let us consider an optimal recovery problem for the differentiation operator D^k on $W_p^n(\mathbf{R})$. The set of information operators $\{I_\xi : \xi \in \Theta\}$ has been defined in § 1. For any fixed $\xi \in \Theta$, $\forall f \in L_p^n(\mathbf{R})$ the mapping

$$I_\xi(f) : f \longrightarrow \{f(\xi_j)\}_{j \in \mathbf{Z}} \tag{3.1}$$

is an information operator. Following [7], when I_ξ is fixed, we have the diameter of information of the operator $D^k(1 \leqslant k \leqslant n-1, n \geqslant 2)$ on $W_p^n(\mathbf{R})$ in $L^p(\mathbf{R})$-metric as

$$D_\xi(W_p^n(\mathbf{R}), D^k, L^p) = 2\sup\{\|f^{(k)}\|_p : f \in W_p^n(\mathbf{R}), I_\xi(f) = 0\}. \tag{3.2}$$

The minimal diameter of information is then defined by

$$D(W_p^n(\mathbf{R}); D^k, L^p) = \inf_{\xi \in \Theta} D_\xi(W_p^n(\mathbf{R}), D^k, L^p), (1 \leqslant k \leqslant n-1). \tag{3.3}$$

When $k = 0$, instesd of (3.3) we put

$$\begin{aligned} D(W_p^n(\mathbf{R}), L^p) &= \inf_{\xi \in \Theta} D_\xi(W_p^n(\mathbf{R}), L^p) \\ &= 2\inf_{\xi \in \Theta}\sup\{\|f\|_p : f \in W_p^n(\mathbf{R}), I_\xi(f) = 0\}. \end{aligned} \tag{3.4}$$

In this section we give the exact expressions of the quantities (3.3) for $p = 1, 2, +\infty$, and moreover, we identify some $\xi^* \in \Theta$ which realize the infimum in the right side of (3.3) (Such ξ^* is said to be an optimal set of sampling points). We denote

$$e(W_p^n(\mathbf{R}), D^k, L^p) = \frac{1}{2} D(W_p^n(\mathbf{R}), D^k, L^p).$$

$k = 0, \cdots, n-1$. (When $k = 0$, $D^0 = \mathrm{I}$ is omitted for simplicity)

Theorem 3 *Suppose* $0 \leqslant k \leqslant n-1, n \geqslant 2$. *Then*

$$e(W_2^n(\mathbf{R}), D^k, L^2) = \sup_{\substack{f \in W_2^n \\ f(j)=0}} \|f^{(k)}\|_2 = \pi^{-(n-k)}. \tag{3.5}$$

Proof The $k = 0$ case had been considered by Sun and Li in [9]. Now we consider the $1 \leqslant k \leqslant n-1$ case. $\forall\, \xi = \{\xi_j\} \in \Theta$, and $\varepsilon > 0$. By definition, there exist a positive integer N_ε and a sequence of intervals $\{I_k\}_{k=1}^{+\infty}$, $I_k = [-a_k, a_k]$, $a_k \to +\infty$ such that

$$n_k = \operatorname{card}\{\xi \cap I_k\} \leqslant 2(1+\varepsilon)a_k, \forall\, k \geqslant N_\varepsilon. \tag{3.6}$$

Put

$$W_2^n(I_s)_0 = \{f \in W_2^n(\mathbf{R}) : \operatorname{supp} f \subset I_s, f(\xi_j) = 0, \forall\, \xi_j \in I_s\}.$$

By a duality theorem due to Korneichuk ([1], p. 149, (56)) we have

$$\sup\{\|g^{(k)}\|_2 : g \in W_2^n(I_s)_0\} = E(W_2^{n-k}(I_s), S_n(I_s))_2,$$

where $S_n(I_s)$ is the space of polynomial splines of degree n with minimal defect and set of knots $\xi\cap I_s$. On account of dim $S_n(I_s)\leqslant n_s+n$, we have

$$E(W_2^{n-k}(I_s);S_n(I_s))_2\geqslant d_{n_s+n}[W_2^{n-k}(I_s),L^2],$$

Where $d_{n_s+n}[W_2^{n-k}(I_s),L^2]$ is the n_s+n dimensional Kolmogorov's width of $W_2^{n-k}(I_s)$ in $L^2(I_s)$ which, by [8], is the n_s+k+1-th eigen-value of the following differential operator:

$$(-1)^{n-k}y^{(2n-2k)}=\mu^{-1}y,\quad y^{(i)}(-a_s)=y^{(i)}(a_s)=0,$$
$$(i=0,\cdots,n-k-1).$$

On the basis of Kolmogorov's result (see G. Lorentz[8]) we have

$$d_{n_s+n}[W_2^{n-k}(I_s);L^2]=\sqrt{\mu_{n_s+k+1}},$$

and

$$\mu_{n_s+k+1}=\left(\frac{2a_s}{\pi}\right)^{2(n-k)}\frac{1}{(n_s+k+1)^{2(n-k)}}(1+O(n_s^{-1}))$$

$$=\pi^{-2(n-k)}\left(\frac{2a_s}{n_s+k+1}\right)^{2(n-k)}(1+O(n_s^{-1})).$$

From (3.6), for $s\geqslant N_\varepsilon$ we have $\dfrac{2a_s}{n_s}\geqslant\dfrac{1}{1+\varepsilon}>1-\varepsilon$. So that,

$$\frac{2a_s}{n_s+k+1}=\frac{2a_s}{n_s}\cdot\frac{n_s}{n_s+k+1}>(1-\varepsilon)\frac{n_s}{n_s+k+1},$$

and therefore

$$\frac{2a_s}{n_s+k+1}>1-2\varepsilon$$

for $s>N'_\varepsilon\geqslant N_\varepsilon$ where N'_ε is some positive integer, and $0<\varepsilon<1/2$. Then for sufficiently large s it holds

$$\mu_{n_s+n+1}\geqslant\pi^{-2(n-k)}(1-2\varepsilon)^{2(n-k)}(1+O(n_s^{-1})),$$

and consequently

$$\begin{aligned}&\sup\{\|g^{(k)}\|_2:g\in W_2^n(\mathbf{R}),I_\xi(g)=0\}\\&\geqslant\sup\{\|g^{(k)}\|_2:g\in W_2^n(I_s)_0\}\\&\geqslant d_{n_s+n}[W_2^{n-k}(I_s);L^2]=\sqrt{\mu_{n_s+k+1}}\\&\geqslant\pi^{-(n-k)}(1-K\varepsilon)(1+O(n_s^{-1}))\end{aligned}$$

for sufficiently large s, where $K>0$ is some number which depends only on

n and k. Setting $s \to +\infty$ and taking limit, we get

$$\sup\{\| g^{(k)} \|_2 : g \in W_2^n(\mathbf{R}), I_\xi(g) = 0\} \geqslant \frac{1 - K\varepsilon}{\pi^{n-k}}.$$

And therefore

$$\begin{aligned} e(W_2^n(\mathbf{R}), D^k, L^2) &= \inf_{\xi \in \Theta} \sup\{\| g^{(k)} \|_2 : g \in W_2^n(\mathbf{R}), I_\xi(g) = 0\} \\ &\geqslant (1 - K\varepsilon)/\pi^{n-k}. \end{aligned}$$

From the arbitrariness of $\varepsilon > 0$ it follows

$$e(W_2^n(\mathbf{R}), D^k, L^2) \geqslant \pi^{-(n-k)}. \tag{3.7}$$

It remains to get estimate from above. Sun and Li [9] proved

$$\sup\{\| f \|_2 : f \in W_2^n(\mathbf{R}), f(j) = 0, \forall j \in \mathbf{Z}\} = \pi^{-n}. \tag{3.8}$$

By using the Hardy-Littlewood-Polya inequality for $f \in W_2^n(\mathbf{R}), f(j) = 0$, then we get

$$\| f^{(k)} \|_2 \leqslant \| f \|_2^{1-k/n} \leqslant \pi^{-n(1-k/n)} = \pi^{-(n-k)}.$$

So from

$$e(W_2^n(\mathbf{R}), D^k, L^2) \leqslant \sup\{\| f^{(k)} \|_2 : f \in W_2^n(\mathbf{R}), f(j) = 0\} \leqslant \pi^{-(n-k)}$$

the conclusion follows. Theorem 3 is proven.

Theorem 4 *Suppose* $0 \leqslant k \leqslant n-1, n \geqslant 2$. *Then*

$$e(W_\infty^n(\mathbf{R}), D^k, L^\infty) = \sup_{\substack{f \in W_\infty^n \\ f(j)=0}} \| f^{(k)} \|_\infty = \| E_n^{(k)}(\cdot) \|_\infty. \tag{3.9}$$

$E_n(\cdot)$ is the n-degree Euler spline.

Proof The scheme of proof is similar to that of the previous theorem. The $k = 0$ case had been solved by Sun [10].

Let N_ε and $\{I_s\}_{s=1}^{+\infty}$ be given as in the previous theorem. Without loss of generality we may assume a_s to be integers. Put

$$W_\infty^n(I_s)_0 = \{f \in W_\infty^n(\mathbf{R}) : \operatorname{supp} f \subset I_s, f(\xi_j) = 0, \forall \xi_j \in I_s\}. \tag{3.10}$$

By making use of Korneichuk's duality theorem we get

$$\sup\{\| g^{(k)} \|_\infty : g \in W_\infty^n(I_s)_0\} = E(W_1^{n-k}(I_s); S_n(I_s))_1,$$

where

$$E(W_1^{n-k}(I_s); S_n(I_s))_1 = \sup_{f \in W_1^{n-k}} \inf_{g \in S_n} \| f - g \|_1.$$

Thus we have

$$E(W_1^{n-k}(I_s);S_n(I_s))_1 \geqslant d_{n_s+n}[W_1^{n-k}(I_s);L]. \quad (3.11)$$

For obtaining the asymptotically exact estimate of $d_{n_s+n}[W_1^{n-k}(I_s);L]$ $(s \to +\infty)$ we refer to Korneichuk's result (see [1], p. 267, (45)). We remark that Korneichuk considered a class of functions defined on the standard interval $I = [0,1]$. By means of a linear transformation from I onto I_s, and then by simple calculation we may obtain

$$d_{n_s+n}[W_1^{n-k}(I_s);L] \geqslant \| E_{n-k}(\cdot)\|_\infty(1-K\varepsilon), \quad (3.12)$$

for sufficiently large s, where the number $K > 0$ depends only on n and k. So that

$$\begin{aligned}&\sup\{\|g^{(k)}\|_\infty : g \in W_\infty^n(\mathbf{R}), I_\xi(g) = 0\}\\ &\geqslant \sup\{\|g^{(k)}\|_\infty : g \in W_\infty^n(I_s)_0\}\\ &\geqslant \|E_{n-k}(\cdot)\|_\infty(1-K\varepsilon) \quad (3.13)\end{aligned}$$

and

$$e(W_\infty^n(\mathbf{R}), D^k, L^\infty) \geqslant \|E_{n-k}(\cdot)\|_\infty(1-K\varepsilon).$$

In view of the arbitrariness of $\varepsilon > 0$ we get

$$e(W_\infty^n(\mathbf{R}), D^k, L^\infty) \geqslant \|E_{n-k}(\cdot)\|_\infty. \quad (3.14)$$

It remains for us to get estimate from above. Sun [10] obtained

$$\sup\{\|f\|_\infty : f \in W_\infty^n(\mathbf{R}), f(j) = 0\} = \|E_n(\cdot)\|_\infty. \quad (3.15)$$

By the Landau-Kolmogorov inequality [6], if $f \in W_\infty^n(\mathbf{R})$ such that $\|f\|_\infty \leqslant \|E_n(\cdot)\|_\infty$, then

$$\|f^{(k)}\|_\infty \leqslant \|E_n^{(k)}(\cdot)\|_\infty = \|E_{n-k}(\cdot)\|_\infty. \quad (3.16)$$

So that the conclusion of Theorem 4 completely follows.

Similarly, we may prove

Theorem 5 *Suppose* $0 \leqslant k \leqslant n-1, n \geqslant 2$. *Then*

$$e(W_1^n(\mathbf{R}), D^k, L) = \sup_{f \in W_1^n}\{\|f^{(k)}\|_1 : f(j) = 0\} = \|E_{n-k}(\cdot)\|_\infty. \quad (3.17)$$

The proof of Theorem 5 is similar to that of Theorem 4. We remark that the $k = 0$ case recently has been solved by Li [11]. In the case $1 \leqslant k$

$\leqslant n-1$ we make use of Stein's inequality [4]. We omit the details.

A periodic version of theorems 3 ~ 5 we refer to Sun [2].

§ 4. Main result and some remarks

We formulate the main result of this paper as follows:

Theorem 6 *Denote* $T^* = \{j\}_{j\in\mathbf{Z}}$ *and* S_{m,T^*} *simply by* S_{m-1}, *which is the class of* $m-1$ *degree cardinal splines. If* $m \geqslant n$, $p = 1, 2, +\infty$, *then*

(1) *for* $p = 1, +\infty$

$$E(W_p^n(\mathbf{R}), S_{m-1})_p = \| E_n(\cdot) \|_\infty. \tag{4.1}$$

(2) *for* $p = 2$

$$E(W_2^n(\mathbf{R}), S_{m-1})_2 = \pi^{-n} \tag{4.2}$$

Proof By Theorem 2 we have

$$E(W_p^n(\mathbf{R}), S_{m-1})_p \leqslant \sup\{ \| f^{(m-n)} \|_{p'} : f \in W_{p'}^m(\mathbf{R}), f(j) = 0, j \in \mathbf{Z}\}. \tag{4.3}$$

When $p = 2$, according to Theorem 3 we get

$$E(W_2^n(\mathbf{R}), S_{m-1})_2 \leqslant \pi^{-n} \tag{4.4}$$

When $p = 1, +\infty$, from Theorem 4 and Theorem 5 we get

$$E(W_p^n(\mathbf{R}), S_{m-1})_p \leqslant \| E_m^{(m-n)}(\cdot) \|_\infty = \| E_n(\cdot) \|_\infty.$$

Thus the estimates from above is obtained. For the sake of obtaining sharp estimates from below we take an arbitrarily large integer N and consider a sub-set of $W_p^n(\mathbf{R})$ which is defined as follows:

$$(W_{p,N}^n)_0 = \{f \in W_p^n(\mathbf{R}) : \text{supp } f \subset [-N, N]\}. \tag{4.5}$$

It is obvious that $f^{(j)}(-N) = f^{(j)}(N) = 0, j = 0, 1, \cdots, n-1$. Then we have

$$E(W_p^n(\mathbf{R}), S_{m-1})_p \geqslant E((W_{p,N}^n)_0, S_{m-1})_{p[-N,N]}.$$

The approximating subspace $S_{m-1}|_{[-N,N]}$ is a finite-dimensional one, and $\dim\{S_{m-1}|_{[-N,N]}\} = 2N + m - 1$. So that

$$E((W_{p,N}^n)_0, S_{m-1})_{p[-N,N]} \geqslant d_{2N+m-1}[(W_{p,N}^n)_0, L_{[-N,N]}^p].$$

Now we may consider $(W_{p,N}^n)_0$ as a sub-set of the $2N$-periodic differentiable function class $\widetilde{W}^n_{p,N}$, because we are sure that

$$(W^n_{p,N})_0 = \{f \in \widetilde{W}^n_{p,N} : f(-N) = \cdots = f^{(n-1)}(N) = 0\}, \quad (4.6)$$

which means that $\widetilde{W}^n_{p,N}$ is constrained by n linear conditions. Thus we get $d_{2N+m-1}[(W^n_{p,N})_0, L^p] \geqslant d_{2N+n+m-1}[\widetilde{W}^n_{p,N}; L^p_{[-N,N]}]$,

where $p = 1, 2, +\infty$. The further arguement may be proceeded quite similarly to that has been done in proving of Theorem 3($p=2$), Theorem 4 ($p=1$), and Theorem 5($p=+\infty$). We omit the details. We have

$$(1) \lim_{N\to+\infty} d_{2N+n+m-1}[\widetilde{W}^n_{2,N}; L^2] = \pi^{-n} \quad (4.7)$$

and

$$(2) \lim_{N\to+\infty} d_{2N+n+m-1}[\widetilde{W}^n_{p,N}; L^p] = \|E_n(\cdot)\|_\infty, (p = 1, +\infty) \quad (4.8)$$

So the estimation from below follows. The proof of Theorem 6 is competed.

Remark

(1) Theorem 6 show that every linear subspace $S_{m-1}(m \geqslant n)$ is extremal for the infinite-dimensional Kolmogorov's width number $\tilde{d}_1[W^n_p(\mathbf{R}); L^p(\mathbf{R})]$ for $p = 1, 2, +\infty$, where the idea of infinite-dimensional Kolmogorov's width number has been introduced recently by Li [12]. A periodic version of Theorem 6 is due to Korneichuk, Li gun and others. We refer the reader to [1].

(2) Instead of S_{m-1} we may take the set $\{j\pi/\sigma\}_{j\in\mathbf{Z}\,(a>0)}$ as a set of simple knots. Let the relating cardinal spline space to be denoted by $S_{m-1,\pi/\sigma}$ and we may consider the best approximation of $W^n_p(\mathbf{R})$ by $S_{m-1,\pi/\sigma} \cap L^p(\mathbf{R})$. The sharp estimates for $E(W^n_p(\mathbf{R}), S_{m-1,\pi/\sigma})_p$ may be readily obtained ($p = 1, 2, +\infty$). For example, when $p = +\infty$ we have

$$E(W^n_p(\mathbf{R}), S_{m-1,\pi/\sigma})_\infty = \mathcal{K}_n \cdot \sigma^{-n} \quad (4.9)$$

Where $\mathcal{K}_n$ is the Favard constant. Recently, professor V. M. Tikhomirov declared a similar result in a lecture given in Zhejiang University. But the detailed proof has not yet been published.

References

[1] N. P. Korneichuk. *Splines in Approximation Theory*. M.: Press《Nauka》, 1984.

132(In Russian)

[2] Sun Yongsheng. Optimal Sampling on Some Classes of Smooth Functions. to appear(In Chinese)

[3] Sun Yongsheng. Optimal Interpolation on Some Classes of Smooth Functions over the Whole Real Axis. to appear

[4] E. M. Stein. Functions of Exponential Type. Ann. Math., 1957, 65:582 ~ 592

[5] de Boor, C. Splines as Linear Combinations of B-splines. A Survey, in: G. Lorentz and others edited: *Approximation* Ⅱ. N. Y.: Acad. Press, 1976, 1 ~ 47

[6] N. P. Korneichuk. *Extremal Problems in Approximation Theory*. M.: Press 《Nauka》, 1976. 113(In Russian)

[7] J. F. Traub, H. Wozniakowski. *A general theory of optimal algorithm*. N. Y.: Acad. Press, 1980. 8

[8] G. G. Lorentz. *Approximation of Functions*. Shanghai: Sci-Tech Press, 1981. 182 ~ 187(In Chinese)

[9] Sun Yongsheng, Li Chun. Optimal recovery for $W_2^r(\mathbf{R})$in $L^2(\mathbf{R})$. To appear in 《Acta Mathematica Sinica》(English Issue)

[10] Sun Yongsheng. Optimal interpolation on some classes of differentiable functions(Ⅰ). ATA. 1986, 2(4): 49 ~ 54

[11] Li Chun. $L(\mathbf{R})$-optimal Recovery on Some Differentiable Function Classes. preprinted paper

[12] Li Chun. On Infinite-dimensional Width Number in Some Function Classes (Ⅱ). to appear

Acta Mathematica Sinica, New Series
1991, Vol. 7, No. 4, pp. 309 ~ 323

$W_2^r(\mathbf{R})$ 在 $L^2(\mathbf{R})$ 中的最优恢复

Optimal Recovery for $W_2^r(\mathbf{R})$ in $L^2(\mathbf{R})$*

Abstract

This paper is devoted to studying the problem of optimal recovery for Sobolev function classes $W_2^r(\mathbf{R})$ in $L^2(\mathbf{R})$ by using the information of function values on denumerable points. For $r \geqslant 1$, we determine that the optimal sampling points are arranged equidistantly in a suitable collection of sets of sampling points and find two kinds of cardinal spline interpolation as optimal algorithms.

§ 1. Statement of Problems

Given a positive integer r, set

$W_2^r(\mathbf{R}) := \{f : f \in L^2(\mathbf{R}), f^{(r-1)}$ is absolutely continuous on any finite interval, and $f^{(r)} \in L^2(\mathbf{R})\}$, (1.1)

$$K := \{f : f \in W_2^r(\mathbf{R}), \| f^{(r)} \|_2 \leqslant 1\}, \tag{1.2}$$

where $\| g \|_2 = \left(\int_{-\infty}^{+\infty} | g(x) |^2 \mathrm{d}x \right)^{1/2}$. By $\xi = \{\xi_j\}_{j \in \mathbf{Z}}$ we denote a biinfinite number sequence which is arranged in a strictly increasing order, i.e., $\xi_j < \xi_{j+1}$, $j \in \mathbf{Z}$. Each ξ_j is called a sampling point and ξ is also refer-

* 此文与李淳合作.

red to as the set of sampling points. Let Θ be a collection of sets of sampling points ξ, where for every $\xi = \{\xi_j\}_{j\in\mathbf{Z}}$, there exists a subsequence $\{n_k\}_{k=1}^{\infty}$ of natural numbers, such that $\xi_j \in [-n_k, n_k)$, $j = -n_k, \cdots, n_k - 1$; $k = 1, 2, \cdots$. Notice that the set Θ contains a very general kind of sets of sampling points. For example, $\{\xi = \{\xi_j\}_{j\in\mathbf{Z}}: \xi_j \in [j, j+1), j \in \mathbf{Z}\}$ is a subset of Θ. For $k = 1, 2, \cdots$,

$\Xi_{2k} := \{\xi = \{\xi_j\}_{j\in\mathbf{Z}}: -k \leqslant \xi_{-k} < \cdots < \xi_{k-1} < k, \xi_{j+2k} = 2k + \xi_j, j \in \mathbf{Z}\}$

are also subsets of Θ (cf. [1]).

For each $\xi \in \Theta$, we can determine a mapping $I_\xi: W_2^r(\mathbf{R}) \to l^\infty$, $I_\xi(f) = \{f(\xi_j)\}_{j\in\mathbf{Z}}$. I_ξ is called the information operator. By $I_\xi(K)$ we denote the image of K under mapping I_ξ. Any mapping $A: I_\xi(K) \to L^2(\mathbf{R})$ is referred to as an algorithm. Following Micchelli-Rivlin[2], the intrinsic error of the optimal recovery problem $(K, I_\xi, L^2(\mathbf{R}))$ is

$$E(K, \xi, L^2) := \inf_A \sup_{f\in K} \| f - A(I_\xi(f)) \|_2. \qquad (1.3)$$

It is known from [2] that

$$E(K, \xi, L^2) \geqslant e(K, \xi, L^2) := \sup\{ \| f \|_2 : f \in K, I_\xi(f) = 0\}. \qquad (1.4)$$

Put

$$E(K, L^2) := \inf_{\xi\in\Theta} E(K, \xi, L^2). \qquad (1.5)$$

Three problems arise: (1) It is required to obtain the exact value of $E(K, L^2)$; (2) How to determine the optimal set of sampling points $\xi^* \in \Theta$ and the optimal algorithm A^* such that the infima in (1.5) and (1.3) are attained, i.e.,

$$E(K, L^2) = E(K, \xi^*, L^2) = \sup_{f\in K} \| f - A^*(I_{\xi^*}(f)) \|_2;$$

(3) If there exists an optimal algorithm, we want to ask whether it is linear and/or unique (in some reasonable sense).

In what follows we will solve the above problems completely.

§ 2. Estimation for $E(K, L^2)$ from below

For any finite interval $I = [a, b]$, we set

$$W_2^r(I)_0 := \{f \in K : f^{(k)}(a) = f^{(k)}(b) = 0,$$
$$k = 0, \cdots, r-1, f(x) = 0 \text{ for } x \in \mathbf{R}/I\}, \tag{2.1}$$

$$W_2^r(I) := \{f : f^{(r-1)} \text{ is absolutely continuous on } I, \| f^{(r)} \|_{L^2(I)} \leqslant 1\}, \tag{2.2}$$

where $\| h \|_{L^2(I)} = \left(\int_I | h(x) |^2 \mathrm{d}x \right)^{1/2}$. Obviously $W_2^r(I)_0 \subset K$. Put

$$\mathrm{e}(K, L^2) := \inf_{\xi \in \Theta} \mathrm{e}(K, \xi, L^2). \tag{2.3}$$

Proposition 2.1 $E(K, L^2) \geqslant \mathrm{e}(K, L^2) \geqslant \pi^{-r}$.

Proof Given any $\xi \in \Theta$, there exists a subsequence $\{n_k\}_{k=1}^{\infty}$ of natural numbers, such that $\xi_j \in I_k := [-n_k, n_k)$, $j = -n_k, \cdots, n_k - 1$; $k = 1, 2, \cdots$. Let

$$\Delta_{2n_k} = \{-n_k \leqslant \xi_{-n_k} < \cdots < \xi_0 < \cdots < \xi_{n_k - 1} < n_k\}.$$

We denote the space of polynomial splines of degree $r-1$ with simple knots at Δ_{2n_k} by $S_{r-1}(\Delta_{2n_k})$. By the dual theorem of the best approximation of spline subspace (cf. [3], p. 149), we have

$$\sup\{\| g \|_2 : g \in W_2^r(I_k)_0, g(\xi_j) = 0, j = -n_k, \cdots, n_k - 1\}$$
$$= \sup_{f \in W_2^r(I_k)} \inf_{s \in S_{r-1}(\Delta_{2n_k})} \| f - s \|_{L^2(I_k)} \quad (k = 1, 2, \cdots). \tag{2.4}$$

Since $\dim S_{r-1}(\Delta_{2n_k}) \leqslant 2n_k + r$, by the definition[4] of Kolmogorov n-width,

$$\sup_{f \in W_2^r(I_k)} \inf_{s \in S_{r-1}(\Delta_{2n_k})} \| f - s \|_{L^2(I_k)} \geqslant d_{2n_k + r}(W_2^r(I_k); L^2(I_k)). \tag{2.5}$$

Thus, (1.4), (2.4) and (2.5) yield

$$E(K, \xi, L^2) \geqslant \mathrm{e}(K, \xi, L^2)$$
$$\geqslant \sup\{\| g \|_2 : g \in W_2^r(I_k)_0, g(\xi_j) = 0, j = -n_k, \cdots, n_k - 1\}$$
$$\geqslant \mathrm{d}_{2n_k + r}(W_2^r(I_k); L^2(I_k)) \quad (k = 1, 2, \cdots). \tag{2.6}$$

According to G. G. Lorentz[5],

$$d_{2n_k + r}(W_2^r(I_k); L^2(I_k)) = \sqrt{\mu_{2n_k + 1}} \tag{2.7}$$

and

$$\mu_{2n_k+1} = \left(\frac{2n_k}{\pi}\right)^{2r} \cdot \frac{1}{(2n_k+1)^{2r}}(1+O(n_k^{-1})),$$

where μ_{2n_k+1} is the $(2n_k+1)^{\text{th}}$ eigenvalue of the following eigenvalue problem of the differential equation:

$$(-1)^r y^{(2r)} = \mu^{-1} y, y^{(i)}(-n_k) = y^{(i)}(n_k) = 0, i = 0, \cdots, r-1.$$

Letting $k \to \infty$ in (2.6) and (2.7), we obtain

$$E(K, \xi, L^2) \geqslant e(K, \xi, L^2) \geqslant \pi^{-r}.$$

Since $\xi \in \Theta$ is arbitrary, then

$$\inf_{\xi \in \Theta} E(K, \xi, L^2) \geqslant \inf_{\xi \in \Theta} e(K, \xi, L^2) \geqslant \pi^{-r}.$$

It is just the conclusion of this proposition.

§ 3. Estimation for $E(K, L^2)$ from Above

In this section, we will prove that $E(K, L^2)$ is not greater than π^{-r}. Hence $E(K, L^2) = \pi^{-r}$. Moreover, we get an optimal set of sampling points and a linear optimal algorithm.

Let S_{r-1} be the space of cardinal polynomial spline functions of degree $r-1$ with integer knots, i.e.,

$$S_{r-1} := \{s : s \in C^{r-2}(\mathbf{R}), s|_{(j,j+1)} \in \pi_{r-1}, j \in \mathbf{Z}\}, \qquad (3.1)$$

where π_{r-1} denotes the class of polynomials of degree $r-1$. It is known[6] for any power growth data $y = \{y_j\}_{j \in \mathbf{Z}}$(i.e., $y_j = O(|j|^\beta)$, $|j| \to +\infty$, for some $\beta \geqslant 0$), there exists a unique power growth $s_{r-1}(y;\cdot) \in S_{r-1}$ such that

$$s_{r-1}(y; j + \alpha_r) = y_j, j \in \mathbf{Z},$$

where $\alpha_r = 0$, if r is even; $r = 1/2$, if r is odd. Furthermore, s_{r-1} can be represented by the following cardinal series:

$$s_{r-1}(y;x) = \sum_{j \in \mathbf{Z}} y_j L_{r-1}(x-j), x \in \mathbf{R}, \qquad (3.2)$$

where $L_{r-1} \in S_{r-1}$, $L_{r-1}(j+\alpha_r) = \delta_{j,0}$, and $|L_{r-1}(x)| \leqslant A e^{-B|x|}$ for $x \in \mathbf{R}$. A and B are some positive constants which depend only on r. When data $y = \{y_j\}_{j \in \mathbf{Z}}$ are values of some function f on $\{j + \alpha_r\}_{j \in \mathbf{Z}}$, we will replace $s_{r-1}(y;x)$ by $s_{r-1}(f;x)$.

Lemma 3.1. *If* $y = \{y_j\}_{j\in\mathbf{Z}} \in l^2$, *then* $s_{r-1}(y;\cdot) \in L^2(\mathbf{R})$, *and*

$$\| s_{r-1}(y;\cdot) \|_2 \leqslant C \| y \|_{l^2}, \tag{3.3}$$

where C *is a constant and* $\| y \|_{l^2} := \left(\sum_{j\in\mathbf{Z}} | y_j |^2 \right)^{1/2}$.

Proof To abbreviate, we denote $L(x) = L_{r-1}(x)$. Since $| L(x) | \leqslant A e^{-B|x|}$ for $x \in \mathbf{R}$, assuming $j \geqslant k$, we have

$$\begin{aligned}
&\int_{-\infty}^{\infty} | L(x-j) | \cdot | L(x-k) | \,\mathrm{d}x \\
&\leqslant A^2 \int_{-\infty}^{\infty} e^{-B|x-j|-B|x-k|} \mathrm{d}x \\
&= A^2 \Big[\int_{-\infty}^{k} e^{-B(j-x)-B(k-x)} \mathrm{d}x + \int_{k}^{j} e^{-B(j-x)-B(x-k)} \mathrm{d}x + \\
&\quad \int_{j}^{\infty} e^{-B(x-j)-B(x-k)} \mathrm{d}x \Big] \\
&= A^2 [B^{-1} + (j-k)] e^{-B(j-k)}.
\end{aligned}$$

By symmetry,

$$\int_{-\infty}^{\infty} | L(x-j) | \cdot | L(x-k) | \,\mathrm{d}x \leqslant A^2 [B^{-1} + (k-j)] e^{-B(k-j)},$$

for $j \leqslant k$. Thus,

$$\int_{-\infty}^{\infty} | L(x-j) | \cdot | L(x-k) | \,\mathrm{d}x \leqslant A^2 [B^{-1} + | j-k |) e^{-B|j-k|}$$

(for any j, k.) (3.4)

Therefore,

$$\begin{aligned}
\| s_{r-1}(y;\cdot) \|_2^2 &= \int_{-\infty}^{\infty} | s_{r-1}(y;x) |^2 \mathrm{d}x \\
&\leqslant \int_{-\infty}^{\infty} \sum_{j\in\mathbf{Z}} | y_j | \cdot | L(x-j) | \sum_{k\in\mathbf{Z}} | y_k | \cdot | L(x-k) | \,\mathrm{d}x \\
&= \sum_{j\in\mathbf{Z}} \sum_{k\in\mathbf{Z}} | y_j | \cdot | y_k | \int_{-\infty}^{\infty} | L(x-j) | \cdot | L(x-k) | \,\mathrm{d}x \\
&\leqslant \sum_{j\in\mathbf{Z}} \sum_{k\in\mathbf{Z}} | y_j | \cdot | y_k | A^2 (B^{-1} + | j-k |) e^{-B|j-k|} \\
&= \sum_{j\in\mathbf{Z}} \sum_{k\in\mathbf{Z}} | y_{j+k} | \cdot | y_k | A^2 (B^{-1} + | j |) e^{-B|j|} \\
&= A^2 \sum_{j\in\mathbf{Z}} (B^{-1} + | j |) e^{-B|j|} \sum_{k\in\mathbf{Z}} | y_k | \cdot | y_{k+j} | \\
&\leqslant A^2 \sum_{j\in\mathbf{Z}} (B^{-1} + | j |) e^{-B|j|} \Big(\sum_{k\in\mathbf{Z}} | y_k |^2 \Big)^{1/2} \Big(\sum_{k\in\mathbf{Z}} | y_{k+j} |^2 \Big)^{1/2}
\end{aligned}$$

$$= C^2 \| y \|_{l^2}^2,$$

where $C := A\left(\sum_{j\in\mathbf{Z}} (B^{-1} + |j|)e^{-B|j|} \right)^{1/2} < +\infty$. It proves(3.3) and this Lemma.

Lemma 3.2 *If sequence* $y^n = \{y_j^n\}_{j\in\mathbf{Z}} \in l^\infty$ *satisfies*

$$\sup_{j\in\mathbf{Z}} |y_j^n| \leqslant M < +\infty \text{ and } y_j^n = 0 \text{ for } |j| < 2n, n = 1,2,\cdots,$$

then

$$\lim_{n\to\infty} \| s_{r-1}(y^n; \cdot) \|_{L^2[-n,n]} = 0.$$

Proof As in the proof of Lemma 3.1, we have to estimate the integral $\int_{-n}^{n} |L(x-j)| \cdot |L(x-k)| dx$ for $|j| \geqslant 2n$ and $|k| \geqslant 2n$. Without loss of generality we assume n is sufficiently large such that $4Bn < e^{2Bn}$. Suppose $j \geqslant 2n$, $|k| \geqslant 2n$.

Then

$$\int_{-n}^{n} |L(x-j)| \cdot |L(x-k)| dx$$

$$\leqslant \begin{cases} \int_{-n}^{n} A^2 e^{-B(j+k-2x)} dx, & \text{if } k \geqslant 2n, \\ \int_{-n}^{n} A^2 e^{-B(j-x)} \cdot e^{-B(x-k)} dx, & \text{if } k \leqslant -2n, \end{cases}$$

$$\leqslant \frac{A^2}{2B} e^{-B(j+|k|-2n)} \text{ in both cases.}$$

In the general case, we have

$$\int_{-n}^{n} |L(x-j)| \cdot |L(x-k)| dx \leqslant \frac{A^2}{2B} e^{-B(|j|+|k|-2n)}, \quad |j| \geqslant 2n, \ |k| \geqslant 2n. \tag{3.5}$$

By assumption,

$$s_{r-1}(y^n; x) = \sum_{|j|\geqslant 2n} y_j^n L(x-j) \text{ and } |y_j^n| \leqslant M \text{ for all } j, n.$$

From this and (3.5) it follows that

$$\| s_{r-1}(y^n; \cdot) \|_{L^2[-n,n]}^2$$

$$= \int_{-n}^{n} |s_{r-1}(y^n; x)|^2 dx$$

$$\leqslant \int_{-n}^{n} \sum_{|j|\geqslant 2n} \sum_{|k|\geqslant 2n} |y_j^n| \cdot |y_k^n| \cdot |L(x-j)| \cdot |L(x-k)| \, dx$$

$$\leqslant M^2 \sum_{|j|\geqslant 2n} \sum_{|k|\geqslant 2n} \int_{-n}^{n} |L(x-j)| \cdot |L(x-k)| \, dx$$

$$\leqslant \frac{M^2 A^2}{2B} \sum_{|j|\geqslant 2n} \sum_{|k|\geqslant 2n} e^{-B(|j|+|k|-2n)}$$

$$= \frac{M^2 A^2}{2B} \Big(\sum_{|j|\geqslant 2n} e^{-B|j|} \Big) \Big(\sum_{|k|\geqslant 2n} e^{-B|k|} \Big) e^{2Bn}$$

$$= \frac{M^2 A^2}{2B} \cdot \frac{4e^{-2Bn}}{(1-e^{-B})^2},$$

which implies $\lim_{n\to\infty} \| s_{r-1}(y^n; \cdot) \|_{L^2[-n,n]} = 0$.

For any finite interval $I = [a, b]$, we set

$$\widetilde{W}_2^r(I) := \{ f : f|_I \in W_2^r(I), f^{(j)}(a) = f^{(j)}(b), j = 0, \cdots, r-1,$$
$$f(x + b - a) = f(x) \quad \text{for } x \in \mathbf{R} \}, \tag{3.6}$$

where $W_2^r(I)$ is defined as in (2.2). Clearly, for any $f \in \widetilde{W}_2^r[-2n, 2n]$, there exists a unique bounded cardinal interpolant $s_{r-1}(f; \cdot) \in S_{r-1}$ with period $4n$.

Lemma 3.3 $\sup_{f\in \widetilde{W}_2^r[-2n,2n]} \| f - s_{r-1}(f) \|_{L^2[-2n,2n]} = \pi^{-r}$.

Proof Relate the two functions f and g together by the equation

$$g(x) = \left(\frac{\pi}{2n} \right)^{r-\frac{1}{2}} f\left(\frac{2n}{\pi} x \right).$$

Then

$$g^{(r)}(x) = \left(\frac{2n}{\pi} \right)^{\frac{1}{2}} f^{(r)}\left(\frac{2n}{\pi} x \right)$$

and $$\int_{-\pi}^{\pi} |g^{(r)}(x)|^2 dx = \int_{-2n}^{2n} |f^{(r)}(y)|^2 dy.$$

Thus, $g \in \widetilde{W}_2^r[-\pi, \pi]$ if and only if $f \in \widetilde{W}_2^r[-2n, 2n]$. Let

$$s_{r-1}^*(g; x) := \left(\frac{\pi}{2n} \right)^{r-\frac{1}{2}} s_{r-1}\left(f; \frac{2n}{\pi} x \right).$$

$s_{r-1}^*(g; x)$ is a 2π-periodic polynomial spline function which interpolates g at points $\left\{ \frac{(j+\alpha_r)\pi}{2n} \right\}_{j=-2n}^{2n-1}$. It is known from [4] that

$$\sup_{g\in \widetilde{W}_2^r[-\pi,\pi]} \| g - s_{r-1}^*(g) \|_{L^2[-\pi,\pi]} = \frac{1}{(2n)^r}.$$

Hence,

$$\sup_{f\in \widetilde{W}_2^r[-2n,2n]} \| f - s_{r-1}(f) \|_{L^2[-2n,2n]}$$

$$= \left(\frac{2n}{\pi}\right)^r \sup_{g\in \widetilde{W}_2^r[-\pi,\pi]} \| g - s_{r-1}^*(g) \|_{L^2[-\pi,\pi]} = \pi^{-r}.$$

The following Lemma can be found in [7].

Lemma 3.4 *If* $g \in L^1(\mathbf{R}) \cap BV(\mathbf{R})$, *then* $\sum_{k\in\mathbf{Z}} g(x+k)$ *is absolutely uniformly convergent, where* $BV(\mathbf{R}) := \{g : g$ *is bounded variation on* $\mathbf{R}\}$.

Proposition 3.1 *If* $f \in W_2^r(\mathbf{R})$, *then* $s_{r-1}(f;\cdot) \in L^2(\mathbf{R})$. *Moreover*,

$$\sup_{f\in K} \| f - s_{r-1}(f) \|_2 \leqslant \pi^{-r}.$$

Proof If $f \in W_2^r(\mathbf{R})$, then it is easy to see that $f^2 \in L^1(\mathbf{R}) \cap BV(\mathbf{R})$. By Lemma 3.4, we have $\sum_{j\in\mathbf{Z}} f^2(j+\alpha_r) < +\infty$. And then Lemma 3.1 yields $s_{r-1}(f;\cdot) \in L^2(\mathbf{R})$.

Given an $f \in K$. For any $\varepsilon > 0$, there exists an $N(\varepsilon) > 0$, such that

$$\| f - s_{r-1}(f) \|_2^2 \leqslant \varepsilon + \int_{-n}^{n} (f(x) - s_{r-1}(f;x))^2 \mathrm{d}x, \text{ for } n > N(\varepsilon). \tag{3.7}$$

In the following we will employ Cavaretta's technique[8]. Take a function

$$g(x) := \begin{cases} 1, & -1 \leqslant x \leqslant 1, \\ (-1)^r (x-2)^r \sum_{j=0}^{r-1} \binom{r+j-1}{j} (x-1)^j, & 1 < x < 2, \\ (x+2)^r \sum_{j=0}^{r-1} (-1)^j \binom{r+j-1}{j} (x+1)^j, & -2 < x < -1, \\ 0, & |x| \geqslant 2. \end{cases} \tag{3.8}$$

$g(x)$ has the properties that $g \in C^{r-1}(\mathbf{R})$, $\| g^{(k)} \|_\infty < +\infty (k = 0, \cdots, r)$ and $g(x)$ is strictly monotone on $(1,2) \cup (-2,-1)$. Given $0 < \delta < 1$, set

$$F_n(x) := \delta f(x) g\left(\frac{x}{n}\right), \quad x \in \mathbf{R}. \tag{3.9}$$

Then $F_n(x)=0$ for $|x|\geqslant 2n$, $F_n\in C^{r-1}(\mathbf{R})$ and

$$F_n^{(r)}(x)=\delta f^{(r)}(x)g(x/n)+\delta\sum_{j=1}^{r}\binom{r}{j}f^{(r-j)}(x)g^{(j)}(x/n)n^{-j}.$$

Observe that $|g(x/n)|\leqslant 1$ and $|g^{(j)}(x/n)|\leqslant C_1$ for $x\in\mathbf{R}$, $j=1,\cdots,r$, and by Stein's inequalities[9] for the case $p=2$,

$$\|f^{(r-j)}\|_2\leqslant C_2, j=1,\cdots,r-1,$$

where C_1, C_2 are some constants which are independent of n. Thus,

$$\|F_n^{(r)}\|_{L^2[-2n,2n]}=\|F_n^{(r)}\|_2\leqslant\delta\|f^{(r)}\|_2+\delta C_1C_2\cdot 2^r/n$$
$$\leqslant\delta(1+C_1C_2\cdot 2^r/n).$$

Therefore, there exists an $N_1(\delta)>0$, such that

$$\|F_n^{(r)}\|_{L^2[-2n,2n]}\leqslant 1 \text{ for } n>N_1(\delta). \tag{3.10}$$

In addition, we consider the periodic $\widetilde{F}_n(x)$ which is an extension of $F_n(x)$ from $[-2n,2n]$ to $\mathbf{R}$ with a $4n$-period, i.e., $\widetilde{F}_n(x)=F_n(x)$ for $x\in[-2n,2n)$ and $\widetilde{F}_n(x+4n)=\widetilde{F}_n(x)$ for $x\in\mathbf{R}$. Then from $F_n^{(j)}(-2n)=F_n^{(j)}(2n)=0(j=0,\cdots,r-1)$ and (3.10) it follows that $\widetilde{F}_n\in\widetilde{W}_2^r[-2n,2n]$ for $n>N_1(\delta)$. By Lemma 3.3,

$$\|\widetilde{F}_n-s_{r-1}(\widetilde{F}_n)\|_{L^2[-n,n]}\leqslant\|\widetilde{F}_n-s_{r-1}(\widetilde{F}_n)\|_{L^2[-2n,2n]}\leqslant\pi^{-r}, n>N_1(\delta). \tag{3.11}$$

Let $n>N(\varepsilon,\delta):=\max\{N(\varepsilon),N_1(\delta)\}$. Noticing that $\widetilde{F}_n(x)=F_n(x)=\delta f(x)$ for $|x|\leqslant n$, the inequalities (3.7) and (3.11) imply

$$\begin{aligned}
&\delta^2\|f-s_{r-1}(f)\|_2^2\\
&\leqslant\varepsilon\delta^2+\int_{-n}^{n}(\delta f(x)-\delta s_{r-1}(f;x))^2\mathrm{d}x\\
&=\varepsilon\delta^2+\|\widetilde{F}_n-s_{r-1}(\delta f)\|_{L^2[-n,n]}^2\\
&\leqslant\varepsilon\delta^2+(\|\widetilde{F}_n-s_{r-1}(\widetilde{F}_n)\|_{L^2[-n,n]}+\\
&\qquad\|s_{r-1}(\widetilde{F}_n)-s_{r-1}(\delta f)\|_{L^2[-n,n]})^2\\
&\leqslant\varepsilon\delta^2+(\pi^{-r}+\|s_{r-1}(\widetilde{F}_n)-s_{r-1}(F_n)\|_{L^2[-n,n]}+\\
&\qquad\|s_{r-1}(F_n)-s_{r-1}(\delta f)\|_{L^2[-n,n]})^2.
\end{aligned}\tag{3.12}$$

On the other hand, by Lemma 3.1, we have

$$\|s_{r-1}(F_n)-s_{r-1}(\delta f)\|_{L^2[-n,n]}$$

$$= \| s_{r-1}(F_n - \delta f) \|_{L^2[-n,n]}$$
$$\leqslant C\Big(\sum_{j\in \mathbf{Z}} (F_n(j+\alpha_r) - \delta f(j+\alpha_r))^2 \Big)^{1/2}$$
$$\leqslant 2C\delta\Big(\sum_{|j|\geqslant n} f^2(j+\alpha_r) \Big)^{1/2}. \tag{3.13}$$

The last inequality follows from the fact $|F_n(x)| \leqslant \delta |f(x)|$ for $x \in \mathbf{R}$ and $F_n(x) = \delta f(x)$ for $|x| \leqslant n$. Since $\tilde{F}_n(j+\alpha_r) - F_n(j+\alpha_r) = 0$ for $|j| < 2n$, $|\tilde{F}_n(j+\alpha_r) - F_n(j+\alpha_r)| \leqslant |\tilde{F}_n(j+\alpha_r)| \leqslant \delta\Big(\sum_{k\in Z} f^2(k+\alpha_r)\Big)^{1/2} < +\infty$ for all j, n, Lemma 3.2 yields

$$\lim_{n\to\infty} \| s_{r-1}(\tilde{F}_n) - s_{r-1}(F_n) \|_{L^2[-n,n]} = \lim_{n\to\infty} \| s_{r-1}(\tilde{F}_n - F_n) \|_{L^2[-n,n]} = 0. \tag{3.14}$$

According to (3.13) and (3.14), letting $n \to \infty$ in (3.12), we obtain

$$\delta^2 \| f - s_{r-1}(f) \|_2^2 \leqslant \varepsilon\delta^2 + \pi^{-2r}.$$

Since $\varepsilon > 0$ and $0 < \delta < 1$ are arbitrary, then letting $\varepsilon \to 0^+$ and $\delta \to 1^-$ in the above inequality, we get $\| f - s_{r-1}(f) \|_2 \leqslant \pi^{-r}$. Again, since $f \in K$ is arbitrary, $\sup_{f\in K} \| f - s_{r-1}(f) \|_2 \leqslant \pi^{-r}$. It proves this proposition.

Now we are at a position to give the first main result of this paper.

Theorem 3.1 *Suppose r is a positive integer. $E(K, \xi, L^2)$, $E(K, L^2)$ and $e(K, L^2)$ are defined as in (1.3), (1.5) and (2.3), respectively. s_{r-1} is a linear cardinal spline interpolating operator which is given by (3.2). Put $\xi^* := \{j + \alpha_r\}_{j\in\mathbf{Z}}$ for $\alpha_r := (1 + (-1)^{r-1})/4$. Then we have*

$$e(K, L^2) = E(K, L^2) = E(K, \xi^*, L^2) = \sup_{f\in K} \| f - s_{r-1}(f) \|_2 = \pi^{-r}. \tag{3.15}$$

It shows that ξ^* is an optimal set of sampling points and s_{r-1} is a linear optimal algorithm for $E(K, L^2)$.

Proof According to Proposition 2.1 and Proposition 3.1, it follows from the definitions of $E(K, L^2)$ and $E(K, \xi^*, L^2)$ that

$$\pi^{-r} \leqslant e(K, L^2) \leqslant E(K, L^2) \leqslant E(K, \xi^*, L^2)$$
$$\leqslant \sup_{f\in K} \| f - s_{r-1}(f) \| \leqslant \pi^{-r}.$$

Thus (3.15) is true.

Remark We would like to point out that $\xi^\alpha := \{j + \alpha\}_{j \in \mathbf{Z}}$ is an optimal set of sampling points for every $\alpha \in \mathbf{R}$ and

$$e(K, \xi^\alpha, L^2) = E(K, \xi^\alpha, L^2) = \pi^{-r}. \tag{3.16}$$

In fact, since $f(\cdot) \in K$ if and only if $f(\cdot + \alpha) \in K$ for any $\alpha \in \mathbf{R}$, then on noticing that $s_{r-1}(f) = 0$ for $I_{\xi^*}(f) = 0$, we have

$$\begin{aligned} e(K, \xi^\alpha, L) &= \sup\{\| f \|_2 : f \in K, f(j + \alpha) = 0, j \in \mathbf{Z}\} \\ &= \sup\{\| f \|_2 : f \in K, f(j + \alpha_r) = 0, j \in \mathbf{Z}\} \\ &\leqslant \sup_{f \in K} \| f - s_{r-1}(f) \|_2 \leqslant \pi^{-r}. \end{aligned}$$

However, it is trivial that $e(K, \xi^\alpha, L^2) \geqslant e(K, L^2) \geqslant \pi^{-r}$. Therefore, $e(K, \xi^\alpha, L^2) = \pi^{-r}$. It is also easy to see that $E(K, \xi^\alpha, L^2) = \pi^{-r}$. In particular, $e(K, \xi^0, L^2) = E(K, \xi^0, L^2) = \pi^{-r}$, where $\xi^0 = \{j\}_{j \in \mathbf{Z}}$.

§ 4. Linear Central Optimal Algorithm

In the previous section, we partially answered the questions which were presented in § 1. In this section, we will give another kind of cardinal spline interpolation which is a linear central optimal algorithm for $E(K, L^2)$. The concept of central algorithm may be found in [10].

Let

$$I_0 : W_2^r(\mathbf{R}) \to l^\infty, I_0(f) := \{f(j)\}_{j \in \mathbf{Z}}$$

and

$$\mathrm{Ker}\, I_0 := \{f : f \in W_2^r(\mathbf{R}), f(j) = 0 \quad \text{for } j \in \mathbf{Z}\}. \tag{4.1}$$

Using the notation $T : W_2^r(\mathbf{R}) \to L^2(\mathbf{R}), Tf := f^{(r)}$. Obviously, $T(\mathrm{Ker}\, I_0)$ is a linear subset of $L^2(\mathbf{R})$. Moreover, we have

Lemma 4.1 *$T(\mathrm{Ker}\, I_0)$ is closed in $L^2(\mathbf{R})$.*

To prove Lemma 4.1, we need the following lemmas which are also of interest in themselves.

Lemma 4.2 *Let $f \in W_2^r(\mathbf{R})$. Then $\lim_{|x| \to x} f^{(j)}(x) = 0, j = 0, \cdots, r-1$.*

Proof Since $f \in L^2(\mathbf{R})$ and $f^{(r)} \in L^2(\mathbf{R})$, from Stein's inequalities[9] for the case $p = 2$ it follows that $f^{(j)} \in L^2(\mathbf{R})$. Given any integer $j, 0 \leqslant j \leqslant r-1$, then for any $x_1 < x_2$,

$$\begin{aligned} & \left| (f^{(j)}(x_2))^2 - (f^{(j)}(x_1))^2 \right| \\ & = 2\left| \int_{x_1}^{x_2} f^{(j)}(x) f^{(j+1)}(x) \mathrm{d}x \right| \\ & \leqslant 2\left(\int_{x_1}^{x_2} |f^{(j)}(x)|^2 \mathrm{d}x \right)^{1/2} \left(\int_{x_1}^{x_2} |f^{(j+1)}(x)|^2 \mathrm{d}x \right)^{1/2}. \end{aligned} \tag{4.2}$$

Hence,

$$\left| (f^{(j)}(x_2))^2 - (f^{(j)}(x_1))^2 \right| \to 0 \text{ as } x_2 > x_1 \to +\infty,$$

and there exists $\lim\limits_{x \to +\infty} |f^{(j)}(x)|^2$. Since $f^{(j)} \in L^2(\mathbf{R})$, this must be $\lim\limits_{x \to +\infty} f^{(j)}(x) = 0$. Similarly, $\lim\limits_{x \to -\infty} f^{(j)}(x) = 0$.

Lemma 4.3 *Suppose* $\{f_n\}_{n=1}^{\infty} \subset W_2^r(\mathbf{R})$ *and is bounded in* $L^2(\mathbf{R})$ (*i.e.*, $\| f_n \|_2 \leqslant$ *const. for all* n). *If* $\{f_n^{(r)}\}_{n=1}^{\infty}$ *converges to* $g_0 \in L^2(\mathbf{R})$ *under* $L^2(\mathbf{R})$ *norm, then there exists a function* $f_0 \in W_2^r(\mathbf{R})$ *such that*

$$f_0^{(r)}(x) = g_0(x) \qquad a.\mathrm{e}.\, x \in \mathbf{R}.$$

Proof We will argue for the cases $r = 1$ and $r > 1$.

1° Case $r = 1$.

Since $\{f_n\}_{n=1}^{\infty}$ is bounded in $L^2(\mathbf{R})$, by the weak compactness theorem for $L^2(\mathbf{R})$, we can find a subsequence $\{f_{n_k}\}_{k=1}^{\infty}$ and an $f_0 \in L^2(\mathbf{R})$ such that $f_{n_k} \xrightarrow{W} f$, i.e., $\int_{-\infty}^{\infty} f_{n_k}(x) s(x) \mathrm{d}x \to \int_{-\infty}^{\infty} f_0(x) s(x) \mathrm{d}x$ for $s \in L^2(\mathbf{R})$. In particular, taking $s(x) = \chi_{[-\rho,\rho]}(x) \mathrm{e}^{-ixy} \in L^2(\mathbf{R})$, where $i = \sqrt{-1}$, $y \in \mathbf{R}$, $\rho \in \mathbf{R}_+$, $\chi_{[-\rho,\rho]}(x)$ is the characteristic function of $[-\rho, \rho]$, we get

$$(-\mathrm{i}y) \int_{-\rho}^{\rho} f_{n_k}(x) \mathrm{e}^{-\mathrm{i}xy} \mathrm{d}x \to (-\mathrm{i}y) \int_{-\rho}^{\rho} f_0(x) \mathrm{e}^{-\mathrm{i}xy} \mathrm{d}x, \; k \to \infty. \tag{4.3}$$

On the other hand, by integration by parts,

$$(-\mathrm{i}y) \int_{-\rho}^{\rho} f_{n_k}(x) \mathrm{e}^{-\mathrm{i}xy} \mathrm{d}x = f_{n_k}(x) \mathrm{e}^{-\mathrm{i}xy} \Big|_{x=-\rho}^{\rho} - \int_{-\rho}^{\rho} f'_{n_k}(x) \mathrm{e}^{-\mathrm{i}xy} \mathrm{d}x. \tag{4.4}$$

In inequality (4.2), if we replace $f^{(j)}(x)$ by $f_{n_k}(x)$ and let $x_1 = \rho$, $x_2 \to +\infty$, then

$$\left|f_{n_k}(\rho)\right|^2 \leqslant 2\left(\int_\rho^\infty \left|f_{n_k}(x)\right|^2 dx\right)^{1/2}\left(\int_\rho^\infty \left|f'_{n_k}(x)\right|^2 dx\right)^{1/2}$$

$$\leqslant 2\| f_{n_k}\|_2\left(\left(\int_\rho^\infty \left|f'_{n_k}(x)-g_0(x)\right|^2 dx\right)^{1/2}+\left(\int_\rho^\infty |g_0(x)|^2 dx\right)^{1/2}\right)$$

$$\leqslant \text{const}\left(\| f'_{n_k}-g_0\|_2+\left(\int_\rho^\infty |g_0(x)|^2 dx\right)^{1/2}\right). \quad (4.5)$$

In the same way,

$$|f_{n_k}(-\rho)| \leqslant \text{const}\left(\| f'_{n_k}-g_0\|_2+\left(\int_{-\infty}^{-\rho} |g_0(x)|^2 dx\right)^{1/2}\right)^{1/2}. \quad (4.6)$$

Therefore, (4.4), (4.5) and (4.6) yield

$$\left|\int_{-\rho}^{\rho} f'_{n_k}(x)e^{-ixy}dx - iy\int_{-\rho}^{\rho} f_{n_k}(x)e^{-ixy}dx\right|$$

$$\leqslant \text{const}\left(\left(\| f'_{n_k}-g_0\|_2+\left(\int_\rho^\infty |g_0(x)|^2 dx\right)^{1/2}\right)^{1/2}+\left(\| f'_{n_k}-g_0\|_2+\left(\int_{-\infty}^{-\rho} |g_0(x)|^2 dx\right)^{1/2}\right)^{1/2}\right). \quad (4.7)$$

Letting $k\to\infty$ in (4.7), and noticing $f'_{n_k}\xrightarrow{L^2} g_0$ and (4.3), we get

$$\left|\int_{-\rho}^{\rho} g_0(x)e^{-ixy}dx - iy\int_{-\rho}^{\rho} f_0(x)e^{-ixy}dx\right|$$

$$\leqslant \text{const}\left(\left(\int_\rho^\infty |g_0(x)|^2 dx\right)^{1/4}+\left(\int_{-\infty}^{-\rho} |g_0(x)|^2 dx\right)^{1/4}\right). \quad (4.8)$$

Thus, for any finite interval [a, b],

$$\lim_{\rho\to+\infty}\int_a^b\left|\int_{-\rho}^{\rho} g_0(x)e^{-ixy}dx - iy\int_{-\rho}^{\rho} f_0(x)e^{-ixy}dx\right|^2 dy = 0. \quad (4.9)$$

It is known[7] that there exists $f^\wedge(y) := \frac{1}{\sqrt{2\pi}}\int_{-\infty}^{\infty} f(x)e^{-ixy}dx$ almost everywhere for each $f\in L^2(\mathbf{R})$. Moreover, $f^\wedge\in L^2(\mathbf{R})$ and $\| f^\wedge\|_2 = \| f\|_2$. $f^\wedge$ is just the so-called Fourier transform of f on $L^2(\mathbf{R})$. Using the notation $f^\wedge$, by Fatou Lemma and (4.9) we obtain

$$\int_a^b |g_0^\wedge(y) - iyf_0^\wedge(y)|^2 dy = 0.$$

Thus, $g_0^\wedge(y) = iyf_0^\wedge(y)$ a.e. $y\in[a, b]$. Since $[a, b]$ is an arbitrary finite interval, $g_0^{\wedge}(y) = \mathrm{i}y f_0^{\wedge}(y)$ a. e. $y \in \mathbf{R}$. According to [7](Chapter 5), it follows that

$$f_0 \in W_2^1(\mathbf{R}) \text{ and } f'_0(x) = g_0(x) \text{ a.e. } x \in \mathbf{R}.$$

2° Case $r > 1$.

Since $\{f_n^{(r)}\}_{n=1}^{\infty}$ is convergent in $L^2(\mathbf{R})$, of course $\{f_n^{(r)}\}_{n=1}^{\infty}$ is bouned in $L^2(\mathbf{R})$. By assumption that $\{f_n\}_{n=1}^{\infty}$ is also bounded in $L^2(\mathbf{R})$, so from Stein's inequalities[9] it follows that $\{f_n^{(j)}\}_{n=1}^{\infty}$, $j = 1, \cdots, r-1$, are all bounded in $L^2(\mathbf{R})$. As a result of 1°, we can find an element $h_{r-1} \in W_2^1(\mathbf{R})$ and a subsequence $\{f_{n_k}^{(r-1)}\}_{k=1}^{\infty}$ such that $h'_{r-1} = g_0$ and $f_{n_k}^{(r-1)} \xrightarrow{W} h'_{r-1}$. By the weak compactness theorem for $L^2(\mathbf{R})$, without loss of generality, we can find $h_{r-2} \in L^2(\mathbf{R})$ and assume $f_{n_k}^{(r-2)} \xrightarrow{W} h_{r-2}$. Replacing f_{n_k} by $f_{n_k}^{(r-2)}$ in (4.4), we obtain

$$(-\mathrm{i}y)\int_{-\rho}^{\rho} f_{n_k}^{(r-2)}(x)\mathrm{e}^{-\mathrm{i}xy}\mathrm{d}x$$

$$= f_{n_k}^{(r-2)}(x)\mathrm{e}^{-\mathrm{i}xy}\Big|_{x=-\rho}^{\rho} - \int_{-\rho}^{\rho} f_{n_k}^{(r-1)}(x)\mathrm{e}^{-\mathrm{i}xy}\mathrm{d}x. \tag{4.10}$$

According to [11], if f, f' and $f'' \in L^2(0, +\infty)$, then

$$\int_0^{\infty} |f'(x)|^2\mathrm{d}x \leqslant 2\left(\int_0^{\infty} |f(x)|^2\mathrm{d}x\right)^{1/2}\left(\int_0^{\infty} |f''(x)|^2\mathrm{d}x\right)^{1/2}. \tag{4.11}$$

Replacing f by $f_{n_k}^{(r-2)}(\cdot + \rho)$, we have

$$\int_{\rho}^{\infty} |f_{n_k}^{(r-1)}(x)|^2\mathrm{d}x$$

$$\leqslant 2\left(\int_{\rho}^{\infty} |f_{n_k}^{(r-2)}(x)|^2\mathrm{d}x\right)^{1/2}\left(\int_{\rho}^{\infty} |f_{n_k}^{(r)}(x)|^2\mathrm{d}x\right)^{1/2}. \tag{4.12}$$

Thus, by the first inequality in (4.5), if we replace f_{n_k} by $f_{n_k}^{(r-2)}$ there, then

$$|f_{n_k}^{(r-2)}(\rho)|^2$$

$$\leqslant 2\left(\int_{\rho}^{\infty} |f_{n_k}^{(r-2)}(x)|^2\mathrm{d}x\right)^{1/2}\left(\int_{\rho}^{\infty} |f_{n_k}^{(r-1)}(x)|^2\mathrm{d}x\right)^{1/2}$$

$$\leqslant 2\|f_{n_k}^{(r-2)}\|_2\sqrt{2}\left(\int_{\rho}^{\infty} |f_{n_k}^{(r-2)}(x)|^2\mathrm{d}x\right)^{1/4}\left(\int_{\rho}^{\infty} |f_{n_k}^{(r-1)}(x)|^2\mathrm{d}x\right)^{1/4}$$

$$\leqslant \text{const}\left(\int_{\rho}^{\infty} |f_{n_k}^{(r)}(x)|^2\mathrm{d}x\right)^{1/4}$$

$$\leqslant \text{const}\left(\| f_{n_k}^{(r)} - g_0 \|_2 + \left(\int_\rho^\infty | g_0(x) |^2 \mathrm{d}x \right)^{1/2} \right)^{1/2}. \quad (4.13)$$

Similarly,

$$| f_{n_k}^{(r-2)}(-\rho) |$$

$$\leqslant \text{const}\left(\| f_{n_k}^{(r)} - g_0 \|_2 + \left(\int_{-\infty}^{-\rho} | g_0(x) |^2 \mathrm{d}x \right)^{1/2} \right)^{1/4}. \quad (4.14)$$

Again, as in the proof of 1°, noticing that $f_{n_k}^{(r)} \xrightarrow{L^2} g_0$, and using (4.10), (4.13) and (4.14), we can deduce that

$$\left| \int_{-\rho}^{\rho} h_{r-1}(x) \mathrm{e}^{-\mathrm{i}xy} \mathrm{d}x - iy \int_{-\rho}^{\rho} h_{r-2}(x) \mathrm{e}^{-\mathrm{i}xy} \mathrm{d}x \right|$$

$$\leqslant \text{const}\left(\left(\int_\rho^\infty | g_0(x) |^2 \mathrm{d}x \right)^{1/8} + \left(\int_{-\infty}^{-\rho} | g_0(x) |^2 \mathrm{d}x \right)^{1/8} \right),$$

and

$$(iy) h\hat{}_{r-2}(y) = h\hat{}_{r-1}(y) \quad \text{a.e. } y \in \mathbf{R}$$

Thus $h'_{r-2} = h_{r-1}$ and $h_{r-2} \in W_2^1(\mathbf{R})$. Moreover, since $h'_{r-1} \in W_2^1(\mathbf{R})$ and $h'_{r-1} = g_0$, it follows that $h_{r-2} \in W_2^2(\mathbf{R})$ and $h''_{r-2} = g_0$.

We can continue the above process for each $\{ f_n^{(j)} \}_{n=1}^\infty$, j from $r-2$ to 2, and then find an $h_0 \in W_2^r(\mathbf{R})$ satisfying $h_0^{(r)} = g_0$. Let $f_0 = h_0$. Then this lemma is proved.

Proof of Lemma 4.1 To prove $T(\text{Ker } I_0)$ is closed in $L^2(\mathbf{R})$, we must show that for $\{ g_n \}_{n=1}^\infty \subset T(\text{Ker } I_0)$, $g_n \xrightarrow{L^2} g_0 \in L^2(\mathbf{R})$ implies that there exists an $f_0 \in \text{Ker } I_0$ such that $Tf_0 = f_0^{(r)} = g_0$. By assumption, there exists $\{ f_n \}_{n=1}^\infty \subset \text{Ker } I_0$ such that $Tf_n = g_n$. Clearly, $\{ g_n \}_{n=1}^\infty$ is bounded in $L^2(\mathbf{R})$. By the Remark of Theorem 3.1, $\| f \|_2 \leqslant \pi^{-r} \| f^{(r)} \|_2$ for $f \in \text{Ker } I_0$. It follows that $\{ f_n \}_{n=1}^\infty$ is also bounded in $L^2(\mathbf{R})$. Applying Lemma 4.3 to $\{ f_n \}_{n=1}^\infty$, we conclude that there exists an $f_0 \in W_2^r(\mathbf{R})$ such that $f_0^{(r)} = g_0$. It remains to verify that $f_0(j) = 0$ for $j \in \mathbf{Z}$.

By the first inequality in (4.5), if $f \in W_2^r(\mathbf{R})$, then

$$| f(j) |^2 \leqslant 2 \left(\int_j^\infty | f(x) |^2 \mathrm{d}x \right)^{1/2} \left(\int_j^\infty | f'(x) |^2 \mathrm{d}x \right)^{1/2}$$

$$\leqslant 2 \| f \|_2 \| f' \|_2.$$

From Stein's inequalities[9] $\| f' \|_2 \leqslant \text{const} \| f \|_2^{1-\frac{1}{r}} \cdot \| f^{(r)} \|_2^{1/r}$, it follows that

$$| f(j) |^2 \leqslant \text{const} \| f \|_2^{2-\frac{1}{r}} \cdot \| f^{(r)} \|_2^{1/r}.$$

Replacing f by $f_n - f_0$, and observing that $f_n(j) = 0$ for $j \in \mathbf{Z}$, $n = 1, 2, \cdots$, we have

$$\begin{aligned} | f_0(j) |^2 &= | f_n(j) - f_0(j) |^2 \\ &\leqslant \text{const} \| f_n - f_0 \|_2^{2-\frac{1}{r}} \cdot \| f_n^{(r)} - f_0^{(r)} \|_2^{1/r} \\ &\leqslant \text{const} \| f_n^{(r)} - g_0 \|_2^{1/r}, j \in \mathbf{Z}. \end{aligned}$$

Letting $n \to \infty$, we get $f_0(j) = 0, j \in \mathbf{Z}$. This completes the proof of Lemma 4.1.

Given any $f \in W_2^r(\mathbf{R})$, denote $y := I_0(f) := \{ f(j) \}_{j \in \mathbf{Z}}$. Set

$$A(y) := \{ g \in W_2^r(\mathbf{R}) : I_0(g) = y \}. \qquad (4.15)$$

Consider the following extremal problem:

$$\| Tg \|_2 \to \min, \text{ for } g \in A(y). \qquad (4.16)$$

On the basis of Lemma 4.1, we have

Lemma 4.4

(1) *For each* $f \in W_2^r(\mathbf{R})$, $y = I_0(f)$, *there exists a unique* $g_y \in A(y)$ *such that* $\| Tg_y \|_2 = \min\limits_{g \in A(y)} \| Tg \|_2$.

(2) $g_y \in A(y)$ *is an extremal element of problem* (4.16) *if and only if* $(Tg_y, Th) = 0$ *for any* $h \in \operatorname{Ker} I_0$, *where* $(\cdot, \cdot)$ *is the inner product of* $L^2(\mathbf{R})$.

(3) *For* $f \in K$ (*i.e.*, $\| Tf \|_2 = \| f^{(r)} \|_2 \leqslant 1$), *we have*

$$\text{rad } U(f) \leqslant \sqrt{1 - \| Tg_y \|_2^2} \sup_{f \in K_0} \| f \|_2,$$

where $U(f) := \{ g \in K : I_0(g) = I_0(f) \}$,

$$\text{rad } U(f) := \inf_{g_1 \in L^2(\mathbf{R})} \sup_{g \in U(f)} \| g_1 - g \|_2$$

is the Chebyshev radius of $U(f)$ *in* $L^2(\mathbf{R})$, $K_0 := K \cap \operatorname{Ker} I_0$ *and* g_y *is given in* (1).

(4) $\sup\{ \| f - g_y \|_2 : f \in K, y = I_0(f) \} = \mathrm{e}(K, \xi^0, L^2) = E(K, \xi^0, L^2) = E(K, L^2)$, *where* $\xi^0 = \{ j \}_{j \in \mathbf{Z}}$. *This shows that the mapping*

$\sigma: y \to g_y$ *is a optimal algorithm for* $E(K, L^2)$. *Moreover*, σ *is a central algorithm* (*i.e.*, *if* $g_y + h \in U(f)$ *for some* $h \in \operatorname{Ker} I_0$, *then* $g_y - h \in U(f)$).

This lemma is an extension of Theorem 5.1 of Traub-Wozniakowski [10]. In [10], the information operator has finite cardinality (i.e., the range field of the information operator is contained in $\mathbf{R}^m$ or $\mathbf{C}^m$). Here the information operator has denumerable cardinality (i.e., its range field is contained in l^∞). The proof in [10] is still suitable for Lemma 4.4. The crucial point is that $T(\operatorname{Ker} I_0)$ is a linear closed subset, which has been proved in Lemma 4.1. We omit the details of the proof.

It remains to construct g_y explicitly. The following lemma will answer this question.

Lemma 4.5 *Let* s_{2r-1} *be the cardinal spline interpolating operator which is defined in* (3.2) *except for that we replace* r *by* $2r$ *there*. *If* $f \in W_2^r(\mathbf{R})$, *then* $s_{2r-1}^{(r)}(f) \in L^2(\mathbf{R})$. *Furthermore*, *we have the following identical relation*:

$$\| f^{(r)} \|_2^2 = \| s_{2r-1}^{(r)}(f) \|_2^2 + \| f^{(r)} - s_{2r-1}^{(r)}(f) \|_2^2, \quad f \in W_2^r(\mathbf{R}). \tag{4.17}$$

Proof Let $f \in W_2^r(\mathbf{R})$. Firstly, to see $s_{2r-1}^{(r)}(f) \in L^2(\mathbf{R})$, we will point out that there holds ture the following more general conclusion:

$$s_{2r-1}(f) \in W_2^{2r-1}(\mathbf{R}). \tag{4.18}$$

Since $\sum_{j \in \mathbf{Z}} f^2(j) < +\infty$ (see the proof of Proposition 3.1), it follows from Lemma 3.1 that $s_{2r-1}(f) \in L^2(\mathbf{R})$.

If we analyse the structure of the fundamental cardinal spline function $L_{2r-1}(x)$ in detail (e.g., see [12]), then it is easy to see that not only $| L_{2r-1}(x) | \leqslant A \mathrm{e}^{-B|x|}$ for $x \in \mathbf{R}$, but also $| L_{2r-1}^{(k)}(x) | \leqslant A \mathrm{e}^{-B|x|}$ for $x \in \mathbf{R}$, $k = 1, \cdots, 2r-2$, and $| L_{2r-1}^{(2r-1)}(x) | \leqslant A \mathrm{e}^{-B|x|}$ a.e. $x \in \mathbf{R}$, where $A, B > 0$ are constants. Since $s_{2r-1}(f) \in C^{2r-2}(\mathbf{R})$ and

$$s_{2r-1}(f; x) = \sum_{j \in \mathbf{Z}} f(j) L_{2r-1}(x - j),$$

$$s_{2r-1}^{(k)}(f; x) = \sum_{j \in \mathbf{Z}} f(j) L_{2r-1}^{(k)}(x - j), \quad k = 1, \cdots, 2r-2, \ x \in \mathbf{R},$$

and

$$s_{2r-1}^{(2r-1)}(f;x) = \sum_{j\in\mathbf{Z}} f(j) L_{2r-1}^{(2r-1)}(x-j) \quad \text{a.e. } x \in \mathbf{R},$$

it follows, in view of the proof of Lemma 3.1, that $s_{2r-1}^{(k)}(f) \in L^2(\mathbf{R})$, $k = 1, \cdots, 2r-1$, which is just the conclusion we want.

Next, we will prove

$$\int_{-\infty}^{\infty} h^{(r)}(x) s_{2r-1}^{(r)}(f;x)\mathrm{d}x = 0 \text{ for } h \in \operatorname{Ker} I_0. \tag{4.19}$$

In fact, since $h^{(r)} \in L^2(\mathbf{R})$ and $s_{2r-1}^{(r)}(f) \in L^2(\mathbf{R})$, the integration in (4.19) makes sense. For an arbitrary positive integer n, by integration by parts, we have

$$\int_{-n}^{n} h^{(r)}(x) s_{2r-1}^{(r)}(f;x)\mathrm{d}x$$
$$= \sum_{k=0}^{r-2} (-1)^k s_{2r-1}^{(r+k)}(f;x) h^{(r-k-1)}(x) \Big|_{x=-n}^{n} +$$
$$(-1)^{r-1} \int_{-n}^{n} s_{2r-1}^{(2r-1)}(f;x) h'(x)\mathrm{d}x. \tag{4.20}$$

However, since $h(j) = 0$ and $s_{2r-1}^{(2r)}(f;x) = 0$, for $j < x < j+1$, $j \in \mathbf{Z}$, so

$$\int_{-n}^{n} s_{2r-1}^{(2r-1)}(f;x) h'(x)\mathrm{d}x$$
$$= \sum_{j=-n}^{n-1} \int_{j}^{j+1} s_{2r-1}^{(2r-1)}(f;x) h'(x)\mathrm{d}x$$
$$= \sum_{j=-n}^{n-1} \left(h(x) s_{2r-1}^{(2r-1)}(f;x) \Big|_{x=j^+}^{(j+1)^-} - \int_{j}^{j+1} h(x) s_{2r-1}^{(2r)}(f;x)\mathrm{d}x \right) = 0. \tag{4.21}$$

By Lemma 4.2 and (4.18), we have

$$\lim_{|x|\to\infty} h^{(r-k-1)}(x) = 0, \quad \lim_{|x|\to\infty} s_{2r-1}^{(r+k)}(f;x) = 0, k = 0, \cdots, r-2. \tag{4.22}$$

It follows from (4.20), (4.21) and (4.22) that

$$\lim_{n\to\infty} \int_{-n}^{n} h^{(r)}(x) s_{2r-1}^{(r)}(f;x)\mathrm{d}x =$$
$$\lim_{n\to\infty} \sum_{k=0}^{r-2} (-1)^k s_{2r-1}^{(r+k)}(f;x) h^{(r-k-1)}(x) \Big|_{x=-n}^{n} = 0.$$

This proves (4.19).

Finally, observing that $f - s_{2r-1}(f) \in \operatorname{Ker} I_0$, by (4.19) we obtain $\int_{-\infty}^{\infty} (f^{(r)}(x) - s_{2r-1}^{(r)}(f;x)) s_{2r-1}^{(r)}(f;x)\mathrm{d}x = 0$, i.e.,

$$\int_{-\infty}^{\infty} f^{(r)}(x) s_{2r-1}^{(r)}(f;x)\mathrm{d}x = \int_{-\infty}^{\infty} (s_{2r-1}^{(r)}(f;x))^2 \mathrm{d}x.$$

Thus

$$\begin{aligned}
&\int_{-\infty}^{\infty} (f^{(r)}(x) - s_{2r-1}^{(r)}(f;x))^2 \mathrm{d}x \\
&= \int_{-\infty}^{\infty} |f^{(r)}(x)|^2 \mathrm{d}x - 2\int_{-\infty}^{\infty} f^{(r)}(x) s_{2r-1}^{(r)}(f;x)\mathrm{d}x + \\
&\quad \int_{-\infty}^{\infty} |s_{2r-1}^{(r)}(f;x)|^2 \mathrm{d}x \\
&= \int_{-\infty}^{\infty} |f^{(r)}(x)|^2 \mathrm{d}x - \int_{-\infty}^{\infty} |s_{2r-1}^{(r)}(f;x)|^2 \mathrm{d}x
\end{aligned}$$

This completes the proof of Lemma 4.5.

It is worth remarking that the identical relation (4.17) is very interesting and powerful. For example, by (4.17) and Lemma 4.4 we conclude that the unique solution of the minimum problem (4.16) is $g_y = s_{2r-1}(y;\ \cdot)$. Although the theory of cardinal splines was long set up, nevertheless, so far as we know, the identical relation (4.17) has not yet been established. This is rather curious as similar identical relations for the odd degree splines on a finite interval had been found at an earlier time.

Combining Lemma 4.4, Lemma 4.5 with Theorem 3.1, we obtain the second main result of this paper.

Theorem 4.1 *Suppose r is a positive integer. Let s_{2r-1} be given as in Lemma 4.5. $e(K,\xi,L^2)$, $E(K,\xi,L^2)$ and $E(K,L^2)$ are defined as in (1.4), (1.3) and (1.5), respectively. $\xi^0 = \{j\}_{j\in\mathbf{Z}}$. Then*

(1) *$s_{2r-1}(y;\ \cdot)$ is the unique solution of the minimum problem (4.16).*

(2) $\sup\limits_{f\in K} \| f - s_{2r-1}(f) \|_2 = e(K,\xi^0,L^2) = E(K,\xi^0,L^2) = E(K, L^2) = \pi^{-r}$.

s_{2r-1} is a linear central optimal algorithm for $E(K, L^2)$.

Note added in proof After we submitted the present paper, we found that the Lemma 4.5 was contained in monograph "Cardinal Spline Interpolation", written by I.J. Schoenberg, SIAM publications, Philadelphia, Pennsylvania, 1973. Nevertheless, the method we used in proof of Lemma 4.5 is different from Schoenberg's and has the advantage to extend to the cases of cardinal $\mathscr{L}$-spline and cardinal Hermite interpolation.

References

[1] Sun Yongsheng. On optimal interpolation for a differentiable function class (I). ATA, 1986, 2: 49 ~ 54

[2] Micchelli, C.A., & Rivlin, T.J. A survey of optimal recovery. In: C. Micchelli and others edited: *Optimal Estimation in Approximation Theory*. N.Y.: Plenum Press, 1977. 1 ~ 68

[3] Корнейчук, Н.Л. *Сплайны в теории приближения* . М.: 《НАУКА》, 1984. 149

[4] Pinkus, A. *N-Widths in Approximation Theory*. New York: Springer-Verlag, 1985, 9

[5] Lorentz, G.G. *Approximation of Functions*. New York: Holt, Rinehart & Winston. Inc., 1966

[6] de Boor, C. & Schoenberg, I.J. Cardinal interpolation and spline functions Ⅷ, in LNM 501, New York: Springer-Verlag, 1976

[7] Butzer, P.L. & Nessel, R.J. *Fourier Analysis and Approximation* Vol.1. Basel: Birkhauser-Verlag, 1971

[8] Cavaretta, A.S., An elementary proof of Kolmogorov's theorem. AMM., 1974, 81: 480 ~ 486

[9] Stein, E.M., Functions of exponential type. Annals of Math., 1957, 65: 582 ~ 592

[10] Traub, J.F. & Wozniakowski, H. *A General Theory of Optimal Algorithms*. New York: Academic Press, 1980. 75 ~ 77

[11] Hardy, G.H., Littlewood, J.E. & Polya, G. *Inequalities*. London: Cambridge Press, 1952

[12] Chen Hanlin, Hu Yingsheng & Micchelli, C.A. On the fundamental function for cardinal $\mathscr{L}$-spline interpolation. ATA, 1986, 2: 1 ~ 17

Approx Theary & its Appl.
Mar., 1991, 7:1

全实轴上某些光滑函数类用高阶基样条的最佳单边逼近*

Best One-sided Approximation of Some Classes of Smooth Functions on the Whole Real Axis by Cardinal Splines of Higher Order

Abstract

The objective of this paper is to study the quantity $E^+(W_p^r(\mathbf{R}); \mathscr{S}_{m,T})_p$. In this paper we establish a duality theorem for the quantity $E^+(W_p^r(\mathbf{R}); \mathscr{S}_{m,T})_p$ for $m \geqslant r$. By virtue of the duality theroem we obtain the exact estimation for $E^+(W_1^r(\mathbf{R}); \mathscr{S}_{m-1})_1$ *for all* $m \geqslant r$.

§1. Introduction

Let r be a positive integer, and $1 \leqslant p \leqslant +\infty$. Denote

$L_p^r(\mathbf{R}) = \{f : f \in L^p(\mathbf{R}) \cap C(\mathbf{R}), f^{(r-1)}$ is abs. cont. on every finite interval, $f^{(r)} \in L^p(\mathbf{R})\}$,

$$W_p^{r,-}(\mathbf{R}) = \{f : f \in L_p^r(\mathbf{R}), \| f_-^{(r)} \|_p \leqslant 1\},$$

$$W_p^r(\mathbf{R}) = \{f : f \in L_p^r(\mathbf{R}), \| f^{(r)} \|_p \leqslant 1\},$$

in which $\| \cdot \|_p$ is the common $L_p(\mathbf{R})$-norm and $f_-^{(r)}$ is the negative part of $f^{(r)}$.

---

* Supported by the NSFC.

Received Jan. 30, 1990, Revised Aug. 13, 1990.

本文与刘永平合作.

Let $T = \{t_j\}_{j \in \mathbf{Z}}, \mathbf{Z} = \{0, \pm 1, \pm 2, \cdots\}$, be a real sequence which satisfies the following conditions:

(1) $t_j < t_{j+1}, \forall j \in \mathbf{Z}$,

(2) $\lim\limits_{j \to -\infty} t_j = -\infty$ and $\lim\limits_{j \to +\infty} t_j = +\infty$.

Let m be a positive integer. Given

$$\mathscr{L}_{1,T} = \{S(t) : S'(t)|_{(t_j, t_{j+1})} = 0 \ \forall j \in \mathbf{Z}\},$$

$$\mathscr{L}_{m,T} = \{S(t) : S(t) \in C^{m-2}(\mathbf{R}), S^{(m-1)} \in \mathscr{L}_{1,T}\}, (m \geqslant 2).$$

$\mathscr{L}_{m,T} (m \geqslant 1)$ is the class of cardinal splines of order m (i, e., degree $m-1$) with simple knots $t_j, j \in \mathbf{Z}$. Specially, if $T = \{j\}, j \in \mathbf{Z}$, we denote $\mathscr{L}_{m-1}$ instead of $\mathscr{L}_{m,T}$. Given a function $g \in L^p(\mathbf{R}) (1 \leqslant p \leqslant +\infty)$, we set

$$E^+(g; \mathscr{L}_{m,T})_p = \inf\{\|g - f\|_p : f \in \mathscr{L}_{m,T} \cap L^p(\mathbf{R}), f \geqslant g \text{ a.e. on } \mathbf{R}\}. \tag{1.1}$$

The quantity $E^+(g; \mathscr{L}_{m,T})_p$ is known as the best one-sided approximation of g from above in metric $L^p(\mathbf{R})$ by $\mathscr{L}_{m,T} \cap L^p(\mathbf{R})$, and the quantity

$$E^+(W_p^r(\mathbf{R}); \mathscr{L}_{m,T})_p = \sup\{E^+(g, \mathscr{L}_{m,T})_p : g \in W_p^r(\mathbf{R})\}, \tag{1.2}$$

the best one-sided approximation of the class $W_p^r(\mathbf{R})$ from above by $\mathscr{L}_{m,T} \cap L^p(\mathbf{R})$ in metric $L^p(\mathbf{R})$.

§ 2. A duality theorem

Let $T = \{t_j\}_{j \in Z}$ be given as in § 1. Given a function f, $[t_j, \cdots, t_{j+r}]f$ denotes its r-th order divided difference over the points $t_j, \cdots, t_{j+r}$, $\forall j \in \mathbf{Z}, \forall r \in \mathbf{Z}_+$. As in[1] we define the B-spline functions as follows:

$$M_{j,r}(x) = r[t_j, \cdots, t_{j+r}](\cdot - x)_+^{r-1}, \ \forall j \in \mathbf{Z}. \tag{2.1}$$

It is well-known that $M_{j,r}(x)$ has compact support $\operatorname{supp} M_{j,r} = [t_j, t_{j+r}]$.

The following lemma is cited from [2]. For completeness we give its proof in details.

Lemma 2.1 *Suppose that* $1 \leqslant p \leqslant +\infty$, $r \in \mathbf{Z}_+$, *and* $f \in L_p^r(\mathbf{R})$. *Then*

$$\int_{\mathbf{R}} M_{j,r}(x) f^{(r)}(x) \mathrm{d}x = 0, \forall j \in \mathbf{Z}, \tag{2.2}$$

iff

$$f(t_j) = \begin{cases} 0, & \text{for} \quad 1 \leqslant p < +\infty, \\ c, & \text{for} \quad p = +\infty, \end{cases} \tag{2.3}$$

where c is a number independent of j.

Proof From the integral formulae

$$\int_{\mathbf{R}} M_{j,r}(x) f^{(r)}(x) \mathrm{d}x = r! [t_j, \cdots, t_{j+r}] f, \quad j \in \mathbf{Z}, \tag{2.4}$$

(see[1])(2.2) is trivally derived.

Now we assume the validity of (2.2). We prove (2.3) by mathematical induction. When $r = 1$, from (2.2) and (2.4) we derive $f(t_{j+1}) - f(t_j) = 0$, $\forall\ j \in \mathbf{Z}$. So there must be a number c such that $f(t_j) = c$ for all $j \in \mathbf{Z}$. It is easy to verify that

$$\lim_{x\to\infty} f^{(j)}(x) = 0 \tag{2.5}$$

holds for any $f \in L_p^r(\mathbf{R})(1 \leqslant p < +\infty)$, $j = 0, 1, \cdots, r-1$. (see also [2]) Therefore, when $1 \leqslant p < +\infty$, by (2.5) we have

$$c = \lim_{j\to\pm\infty} f(t_j) = \lim_{x\to\infty} f(x) = 0.$$

Thus the proposition (2.2) is true for $r = 1$.

Suppose that the proposition holds for $r - 1$. By (2.2), (2.4) and definition of divided differences we have

$$[t_j, \cdots, t_{j+r-1}] f - [t_{j+1}, \cdots, t_{j+r}] f = (t_j - t_{j+r}) [t_j, \cdots, t_{j+r}] f = 0, j \in \mathbf{Z}.$$

So that there is a number c_1 such that

$$[t_j, \cdots, t_{j+r-1}] f = c_1, \forall\, j \in \mathbf{Z}. \tag{2.6}$$

From ([6], formulae(2.93)) we see

$$c_1 = [t_j, \cdots, t_{j+r-1}] f = \frac{1}{(r-1)!} f^{(r-1)}(\theta_j)$$

for some $\theta_j \in [t_j, t_{j+r-1}]$. In view of $\lim\limits_{j\to+\infty} t_j = +\infty$ we have $\lim\limits_{j\to+\infty} \theta_j = +\infty$. Then , when $1 \leqslant p < +\infty$, using (2.5) for $j = r - 1$, we get

$$c_1 = \lim_{j\to+\infty} \frac{1}{(r-1)!} f^{(r-1)}(\theta_j) = 0.$$

When $p = +\infty$, making use of (2.6) and the formulae

$$[t_{j+1}, \cdots, t_{j+r-1}] f - [t_j, \cdots, t_{j+r-2}] f$$

$$= (t_{j+r-1} - t_j)[t_j, \cdots, t_{j+r-1}]f = (t_{j+r-1} - t_j)c_1, \tag{2.7}$$

to sum for $j = 1$ to k on each side of (2.7) we get

$$c_1(t_{k+r-1} - t_{r-1}) = [t_{k+1}, \cdots, t_{k+r-1}]f - [t_1, \cdots, t_{r-1}]f$$

$$= \frac{1}{(r-1)!}\{f^{(r-1)}(\theta_k) - f^{(r-1)}(\theta_1)\}$$

for some $\theta_k \in (t_{k+1}, t_{k+r-1})$, $k = 0, 1, 2, \cdots$. In view of

$\lim\limits_{k\to\infty} t_{k+r-1} = +\infty$, we immediately obtain that $c_1 = 0$.

To sum up, in both cases we have

$$[t_j, \cdots, t_{j+r-1}]f = c_1 = 0, \ \forall j \in \mathbf{Z}.$$

Now by the induction hypothesis we conclude the validity of (2.6) for r. Lemma 2.1 is proved.

In what follows we write

$$N_{j,r}(x) = \frac{1}{r}(t_{j+r} - t_j)M_{j,r}(x), \ j \in \mathbf{Z}. \tag{2.8}$$

This is the so-called normalized B-spline. Lemma 2.1 may be reformulated in the following version.

Lemma 2.1′ *Suppose* $1 \leqslant p \leqslant +\infty$, $r \in \mathbf{Z}_+$ *and* $f \in L_p^r(\mathbf{R})$. *Then*

$$\int_{\mathbf{R}} N_{j,r}(x) f^{(r)}(x)\mathrm{d}x = 0, \ \forall j \in \mathbf{Z}$$

iff

$$f(t_j) = \begin{cases} 0, & \text{when } 1 \leqslant p < +\infty, \\ c, & \text{when } p = +\infty. \end{cases}$$

Lemma 2.2 *Suppose* $g \in L^p(\mathbf{R})$, $1 \leqslant p < +\infty$. *then*

$$\sup\left\{\int_{\mathbf{R}} g(t) f^{(r)}(t)\mathrm{d}t : f \in W_{p'}^{r,-}(\mathbf{R}), \int_{\mathbf{R}} N_{j,r} f^{(r)}\mathrm{d}t = 0, \forall j \in \mathbf{Z}\right\}$$

$$= \sup\left\{\int_{\mathbf{R}} g(t) f^{(r)}(t)\mathrm{d}t : f \in W_{p'}^{r,-}(\mathbf{R}), f(t_j) = 0, \forall j \in \mathbf{Z}\right\}, \tag{2.9}$$

where $\frac{1}{p} + \frac{1}{p'} = 1$.

Proof Actually, when $1 < p < +\infty$, (2.9) is immediately derived from Lemma 2.1, because then $1 < p' < +\infty$. When $p = 1$, $p' = +\infty$, then from Lemma 2.1 it is derived that

$$\text{the right side of (2.9)} \leqslant \text{the left side.} \tag{2.10}$$

On the other hand, for every $f \in W_{p'}^{r,-}(\mathbf{R})$ such that $\int_{\mathbf{R}} N_{j,r}(x) f^{(r)}(x) = 0$, $\forall j \in \mathbf{Z}$ by Lemma 2.1 there is a number c such that $f(t_j) = c$ for all $j \in \mathbf{Z}$. Put $f_1 = f - c$, then $f_1 \in W_{p'}^{r,-}(\mathbf{R})$ and $f_1(t_j) = 0$ for all $j \in \mathbf{Z}$. So that we have

$$\int_{\mathbf{R}} g(t) f^{(r)}(t) \mathrm{d}t = \int_{\mathbf{R}} g(t) f_1^{(r)}(t) \mathrm{d}t \leqslant \text{the right side of } (2.9).$$

And therefore

$$\sup\left\{\int_{\mathbf{R}} g(t) f^{(r)}(t) \mathrm{d}t : f \in W_{p'}^{r,-}(\mathbf{R}), \int_{\mathbf{R}} N_{j,r} f^{(r)} \mathrm{d}t = 0, \forall j \in \mathbf{Z}\right\}$$

$$\leqslant \sup\left\{\int_{\mathbf{R}} g(t) f^{(r)}(t) \mathrm{d}t : f \in W_{p'}^{r,-}(\mathbf{R}), f(t_j) = 0, \forall j \in \mathbf{Z}\right\}. \quad (2.11)$$

Thus by (2.10) and (2.11) the validity of (2.9) for $p = 1$ is proved.

Lemma 2.3 *If $g \in L^p(\mathbf{R}) \cap L^\infty(\mathbf{R})(1 \leqslant p < +\infty)$, and there exists $u \in \mathscr{S}_{r,T} \cap L^p(\mathbf{R})$ such that $u(t) \geqslant g(t)$, $\forall t \in (\mathbf{R})$, then*

$$E^+(g, \mathscr{S}_{r,T})_p = \sup\left\{\int_{\mathbf{R}} g(t) f^{(r)}(t) \mathrm{d}t : f \in W_{p'}^{r,-}(\mathbf{R}),\right.$$

$$\left.\int_{\mathbf{R}} N_{j,r}(t) f^{(r)}(t) \mathrm{d}t = 0, j \in \mathbf{Z}\right\}. \quad (2.12)$$

Proof (1) We first write $I_N = [t_{-N}, t_N]$, and

$$E_N = \left\{ h(t) : g \in L^{p'}(I_N), \operatorname{supp} h \subset I_N, \| h_- \|_{p'} \leqslant 1,\right.$$

$$\left.\int_{\mathbf{R}} N_{j,r}(t) h(t) \mathrm{d}t = 0, \forall j \in \mathbf{Z}\right\}.$$

For $h \in E_N$ we put

$$f_h(x) = \frac{1}{(r-1)!} \int_{I_N} (x - t)_+^{r-1} h(t) \mathrm{d}t.$$

We claim that $f_h \in W_{p'}^{r,-}(\mathbf{R})$ and $\operatorname{supp} f_h \subset I_N$. In fact, the validity of $f_h \in W_{p'}^{r,-}(\mathbf{R})$ is obvious, and $f_h(x) = 0$ is trivally true for $x \leqslant t_{-N}$. On the other hand, on account of that every polynomial of degree $\leqslant r - 1$ is a linear combination of some $N_{j,r}$, and $\int_{\mathbf{R}} N_{j,r}(t) h(t) \mathrm{d}t = 0$, $\forall j \in \mathbf{Z}$, we deduce that $f_h(x) = 0$ for $x \geqslant t_N$. So that Supp $f_h \subset I_N$. Thus we have

$$\sup\left\{\int_{\mathbf{R}} g(t) f^{(r)}(t) \mathrm{d}t : f \in W_{p'}^{r,-}(\mathbf{R}),\right.$$

$$\int_{\mathbf{R}} N_{j,r}(t) f^{(r)}(t)\mathrm{d}t = 0, \forall j \in \mathbf{Z}\Big\}$$

$$\geqslant \sup\left\{\int_{I_N} g(t)h(t)\mathrm{d}t : h \in L^{p'}(I_N),\ \| h_- \|_{p'(I_N)} \leqslant 1, \text{and}\right.$$

$$\left.\int_{I_N} N_{j,r}(t)h(t)\mathrm{d}t = 0, \quad \forall j \in \mathbf{Z}\right\} \tag{2.13}$$

for every positive integer N.

Now we notice that $N_{j,r}(t)$ has compact support $[t_j, t_{j+r}]$, so that the orthogonality condition is essential only for j, $-N-r < j < N$. By the duality theorem concerning the best one-sided approximation of a function by a finite-dimensional subspace (see [3]) we have

$$\sup\left\{\int_{I_N} g(t)h(t)\mathrm{d}t : h \in L^{p'}(I_N),\ \| h_- \|_{p'(I_N)} \leqslant 1, \int_{I_N} N_{j,r} h\mathrm{d}t = 0, j \in \mathbf{Z}\right\}$$

$$= E^+(g; \mathscr{S}_{r,N})_{p(I_N)} = \| g - \sum \alpha^*_{j,N} N_{j,r} \|_{p(I_N)}, \tag{2.14}$$

where $\mathscr{S}_{r,N} \stackrel{\mathrm{df}}{=} \mathrm{span}\{N_{j,r}\}_{j=-N-r-1}^{N-1}$, and $u^*_N \stackrel{\mathrm{df}}{=} \sum \alpha^*_{j,N} N_{j,r}$ is the best approximant of g such that $u^*_N \geqslant g$.

(2) It is easy to verify that the sequence $\{E^+(g, \mathscr{S}_{r,N})_{p(I_N)}\}_{N=1}^{+\infty}$ is montonically increasing. In fact, for any $N \geqslant 1$

$$E^+(g, \mathscr{S}_{r,N})_{p(I_N)} = E^+(g, \mathscr{S}_{r,N+1})_{p(I_N)} \leqslant E^+(g, \mathscr{S}_{r,N+1})_{p(I_{N+1})}.$$

Now, we set that

$$L = \lim_{N\to\infty} E^+(g, \mathscr{S}_{r,N})_{p(I_N)},$$

and prove that

$$D \geqslant d \stackrel{\mathrm{df}}{=} E^+(g, \mathscr{S}_{r,T})_p. \tag{2.15}$$

Assume d > 0(d = 0 case is trival). If (2.15) is false, then there is some ε such that $0 < \varepsilon <$ d and for sufficiently large N it holds

$$\| g - \sum \alpha^*_{j,N} N_{j,r} \|_{p(I_N)} < d - \varepsilon \tag{2.16}$$

We remark that for every fixed $j \in \mathbf{Z}$ the sequence $\{\alpha^*_{j,N}\}_{N=1}^{+\infty}$ is bounded. In fact, by [1] (see [1], Theorem 5.2) there exists a constant D, depending only on r such that for every $j \in \mathbf{Z}$

$$|\alpha^*_{j,N}| (t_{j+r} - t_j)^{\frac{1}{p}} \leqslant D_r \| \sum \alpha^*_{k,N} N_{k,r} \|_{p(t_j, t_{j+r})}.$$

Let $N \geqslant \max\{-j, j+r\}$, then we have $[t_j, t_{j+r}] \subset [t_{-N}, t_N] = I_N$, and

$$|\alpha_{j,N}^*| \leqslant D_r(t_{j+r}-t_j)^{-\frac{1}{p}}\left\{\|g-\sum\alpha_{j,N}^*N_{j,r}\|_{p(I_N)}+\|g\|_p\right\}$$

$$\leqslant D_r(t_{j+r}-t_j)^{-\frac{1}{p}}(\|g\|_p+L).$$

Thus by applying the diagonal process we may choose a subsequence $\{N_n\}_{n=1}^{+\infty}$ of positive integers such that for every $j \in \mathbf{Z}$

$$\lim_{n\to\infty}\alpha_{j,N_n}^* = \beta_j \in \mathbf{R}. \tag{2.17}$$

We put

$$f_n(x)=\begin{cases}\left|g(x)-\sum\alpha_{j,N_n}^*N_{j,r}(x)\right|^p, & x\in I_{N_n},\\ 0, & x\in\mathbf{R}\setminus I_{N_n}.\end{cases} \tag{2.18}$$

For every $x \in \mathbf{R}$ we have $x \in I_{N_n}$ for sufficiently large n. In view of the compactness of the support of $N_{j,r}(x)$ we see that for each x there is only a finite number of terms in the series $\sum\limits_j\alpha_{j,N_n}^*N_{j,r}(x)$ different from zero. Therefore, from (2.17), (2.18) we obtain

$$\lim_{n\to+\infty}f_n(x)=\left|g(x)-\sum\beta_jN_{j,r}(x)\right|^p,\quad x\in\mathbf{R}.$$

Meanwhile, $\sum\beta_jN_{j,r}(x) \geqslant g(x)$ follows from $\sum\alpha_{j,N_n}^*N_{j,r}(x) \geqslant g(x)(n=1,2,3,\cdots)$. Now by Fatou's lemma, we get

$$\int_{\mathbf{R}}\left|g(x)-\sum\beta_jN_{j,r}(x)\right|^p\mathrm{d}x$$

$$\leqslant \varliminf_{n\to+\infty}\int_{\mathbf{R}}f_n(x)\mathrm{d}x$$

$$=\lim_{n\to+\infty}\|g-\sum\alpha_{j,N_n}^*N_{j,r}\|_{p(I_N)}^p$$

$$\leqslant(\mathrm{d}-\varepsilon)^p.$$

This shows that $\sum\beta_jN_{j,r}\in\mathscr{S}_{r,T}\cap L^p(\mathbf{R})$. But then

$$\int_{\mathbf{R}}\left|g(x)-\sum\beta_jN_{j,r}(x)\right|^p\mathrm{d}x\geqslant d^p,$$

which leads to a contradiction. So we have proved $L \geqslant d$. Thus from (2.13) ~ (2.15) we have obtained

$$E^+(g,\mathscr{S}_{r,T})_p=d\leqslant L\leqslant\sup\left\{\int_{\mathbf{R}}g(t)f^{(r)}(t)\mathrm{d}t:f\in W_{p'}^{r,-}(\mathbf{R}),\right.$$

$$\left.\int_{\mathbf{R}}N_{j,r}f^{(t)}\mathrm{d}t=0,\ \forall j\in\mathbf{Z}\right\}. \tag{2.19}$$

On the other hand, for any $u(t) \in \mathscr{S}_{r,T} \cap L^p$ such that $u(t) \geqslant g(t)$ we have

$$\begin{aligned}
&\int_{\mathbf{R}} g(t) f^{(r)}(t)\mathrm{d}t \\
&= \int_{\mathbf{R}} [g(t) - u(t)] f_+^{(r)}(t)\mathrm{d}t - \int_{\mathbf{R}} [g(t) - u(t)] f_-^{(r)}(t)\mathrm{d}t \\
&\leqslant - \int_{\mathbf{R}} [g(t) - u(t)] f_-^{(r)}(t)\mathrm{d}t,
\end{aligned}$$

where $f_+^{(r)} = \max\{f^{(r)}, 0\}$. Therefore

$$\sup \int_{\mathbf{R}} g(t) f^{(r)}(t)\mathrm{d}t \leqslant \sup \left| \int_{\mathbf{R}} [g - u] f_-^{(r)}\mathrm{d}t \right| \leqslant \| g - u \|_p, \tag{2.20}$$

for every $u(x) \in \mathscr{S}_{r,T} \cap L^p(\mathbf{R})$ such that $u(t) \geqslant g(t)$, where the supremum is taken over all $f \in W_{p'}^{r,-}(\mathbf{R})$ constrained by the condition $\int_{\mathbf{R}} N_{j,r} f^{(r)}\mathrm{d}t = 0,\ \forall j \in \mathbf{Z}$. Taking infimum on the right side of (2.20) over all $u(t) \in \mathscr{S}_{r,T} \cap L^p(\mathbf{R})$ such that $u(t) \geqslant g(t)$ we have

$$\begin{aligned}
&E^+(g, \mathscr{S}_{r,T})_p \\
&\geqslant \sup\left\{ \int_{\mathbf{R}} g(t) f^{(r)}\mathrm{d}t : f \in W_{p'}^{r,-}(\mathbf{R}), \int_{\mathbf{R}} N_{j,r}(t) f^{(r)}(t)\mathrm{d}t = 0,\ \forall j \in \mathbf{Z} \right\}.
\end{aligned} \tag{2.21}$$

Thus from (2.19), (2.21) and Lemma 2.1, Lemma 2.3 follows.

Theorem 1 *If r is a positive integer and $m \geqslant r$, then*

$$\begin{aligned}
&E^+(W_p^r(\mathbf{R}); \mathscr{S}_{m,T})_p \\
&= \sup\left\{ \int_{\mathbf{R}} g^{(r)}(t) f^{(m-r)}(t)\mathrm{d}t : g \in W_p^r(\mathbf{R}), f \in W_{p'}^{m,-}(\mathbf{R}), f(t_j) = 0,\ \forall j \in \mathbf{Z} \right\}
\end{aligned}$$

holds for $1 \leqslant p < +\infty$. (2.22)

Proof By Lemma 2.2 and Lemma 2.3 we have

$$\begin{aligned}
&E^+(W_p^r(\mathbf{R}); \mathscr{S}_{m,T})_p \\
&= \sup\{E^+(g, \mathscr{S}_{m,T})_p : g \in W_p^r(\mathbf{R})\} \\
&= \sup_{g \in W_p^r(\mathbf{R})} \sup\left\{ \int_{\mathbf{R}} g f^{(m)}\mathrm{d}t : f \in W_{p'}^{m,-}(\mathbf{R}), f(t_j) = 0,\ \forall j \in \mathbf{Z} \right\} \\
&= \sup\left\{ \int_{\mathbf{R}} f^{(m)} g\mathrm{d}t : g \in W_p^r(\mathbf{R}), f \in W_{p'}^{m,-}(\mathbf{R}), f(t_j) = 0,\ \forall j \in \mathbf{Z} \right\}.
\end{aligned} \tag{2.23}$$

For any $g \in W_p^r(\mathbf{R})$, $f \in W_{p'}^{m,-}(\mathbf{R})$ and $a > 0$, integreation by parts yields

$$\int_{-a}^{a} g(x) f^{(m)}(x) \mathrm{d}x = \Big[\sum_{k=0}^{r-1} (-1)^k g^{(k)}(x) f^{(m-k-1)}(x)\Big]_{-a}^{a} + (-1)^r \int_{-a}^{a} g^{(r)}(x) f^{(m-r)}(x) \mathrm{d}x. \tag{2.24}$$

By (2.5) we see that $\lim_{x\to\infty} g^{(k)}(x) = 0$ and $f^{(m-k-1)} \in L^\infty(\mathbf{R}) \cap C(\mathbf{R})$ for $k = 0, \cdots, r-1$. So we get

$$\lim_{a\to+\infty} \Big[\sum_{k=0}^{r-1} (-1)^k g^{(k)}(x) f^{(m-k-1)}(x)\Big]_{-a}^{a} = 0.$$

In addition, in view of $g(t) f^{(m)}(t)$, $g^{(r)}(t) \cdot f^{(m-r)}(t) \in L(\mathbf{R})$, so that we have

$$\int_{\mathbf{R}} g(t) f^{(m)}(t) \mathrm{d}t = (-1)^r \int_{\mathbf{R}} g^{(r)}(t) f^{(m-r)}(t) \mathrm{d}t. \tag{2.25}$$

Thus we may rewrite (2.23) as follows:

$$E^+ (W_p^r(\mathbf{R}); \mathscr{S}_{m,T})_p$$

$$= \sup\Big\{\int_{\mathbf{R}} g^{(r)} f^{(m-r)} \mathrm{d}t : g \in W_p^r(\mathbf{R}), f \in W_{p'}^{m,-}(\mathbf{R}), f(t_j) = 0, \forall j \in \mathbf{Z}\Big\},$$

which is (2.22). We complete the proof of Theorem 1.

Corollary *If* $p = 1 (p' = +\infty)$, *then*

$$E^+ (W_1^r(\mathbf{R}); \mathscr{S}_{m,T})_1 \leqslant \sup\{E_1(f^{(m-r)})_\infty : f \in W_\infty^{m,-}(\mathbf{R}), f(t_j) = 0, \forall j \in \mathbf{Z}\},$$

where

$$E_1(\varphi)_\infty \stackrel{\mathrm{df}}{=} \inf_{c\in\mathbf{R}} \| \varphi - c \|_\infty. \tag{2.26}$$

§ 3 The exact estimation for $E^+ (W_1^r(\mathbf{R}); \mathscr{S}_{m-1})_1 (m \geqslant r)$

As in [3], denote the Bernoulli function as follows:

$$B_r(t) = \sum_{\nu=1}^{\infty} \nu^{-r} \cos\Big(\nu t - \frac{\pi r}{2}\Big) \text{ and } D_r(t) = 2(-1)^{r-1} B_r(t),$$

$$H_r \stackrel{\mathrm{df}}{=} E_1(D_r)_\infty = \frac{1}{2}\{\sup_t D_r(t) - \inf_t D_r(t)\}.$$

The following lemma gives the upper estimation for $E^+ (W_1^r(\mathbf{R}); \mathscr{S}_{m-1})_1$.

Lemma 3.1 *Let r be a positive integer. Then*

$$\sup\{E_1(g)_\infty : g \in W_\infty^{r;-}(\mathbf{R}), g(j) = 0, \forall j \in \mathbf{Z}\} \leqslant \frac{H_r}{\pi^r}. \quad (3.1)$$

Proof For a fixed $r \in \mathbf{Z}_+$ we first define the function $\chi_r(t)$ as follows (see [4])

$$\chi_r(t) = \begin{cases} 1, & -1 \leqslant t \leqslant 1, \\ (-1)^r (t-2)^r \sum_{j=0}^{r} \binom{r}{j} (t-1)^j, & 1 \leqslant t \leqslant 2, \\ (t+2)^r \sum_{j=0}^{r} \binom{r}{j} (t+1)^j, & -2 \leqslant t \leqslant -1, \\ 0, & |t| > 2, \end{cases} \quad (3.2)$$

where $\binom{r}{j} = \dfrac{r!}{(r-j)!j!}, j = 0, 1, \cdots, r$.

For every $g \in W_\infty^{r;-}(\mathbf{R})$ such that $g(j) = 0 (j \in \mathbf{Z})$, taking $I_N = [-N, N] (N \in \mathbf{Z}_+)$ and $\alpha \in (0,1)$, we define $G_N(t) = \alpha g(t) \chi_r\left(\dfrac{t}{N}\right)$, $t \in [-2N, 2N,]$, and $G_N(t + 4N) = G_N(t) (t \in \mathbf{R})$. $G_N(t)$ is a 4N-periodic function, which possesses the following properties.

(1) $G_N(j) = 0, \forall j \in \mathbf{Z}$. (3.3)

(2) $G_N(t) \in W_\infty^{r;-}(\mathbf{R})$ for sufficiently large N. (3.4)

(3.3) is obvious. We prove (3.4). By Leibnitz rule

$$G_N^{(r)}(t) = \alpha \left\{ g^{(r)}(t) \chi_r\left(\frac{t}{N}\right) + \sum_{j=1}^{r} \binom{r}{j} g^{(r-j)}(t) N^{-j} \chi_r^{(j)}\left(\frac{t}{N}\right) \right\} \quad (3.5)$$

for any $t \in [-2N, 2N]$, and denoting

$$C_1 \overset{\text{df}}{=} \max_{1 \leqslant j \leqslant r} \| \chi_r^{(f)} \|_\infty, \quad C_2 \overset{\text{df}}{=} \max_{0 \leqslant j \leqslant r-1} \| g_r^{(j)} \|_\infty,$$

notice that $C_1, C_2 < +\infty$, then we get

$$\left\| \sum_{j-1}^{r} \binom{r}{j} g^{r-j}(\cdot) N^{-j} \chi_r^{(j)}\left(\frac{\cdot}{N}\right) \right\|_\infty \leqslant \frac{C_1 C_2}{N} \sum_{j=1}^{r} \binom{r}{j} \leqslant C_1 C_2 2^r \cdot N^{-1}. \quad (3.6)$$

On account of $g^{(r)}(t) \geqslant 1, 0 \leqslant \mathscr{X}_r(t) \leqslant 1$, and $\alpha(1 + C_1 C_2 2^r N^{-1}) < 1$ for sufficiently large N, we see that

$$G_N^{(r)}(x) = \alpha g^{(r)}(x) \geqslant \alpha > -1 \text{ when } -N \leqslant x \leqslant N,$$

$$G_N^{(r)}(x) \geqslant \alpha\{g^{(r)}(x) - C_1C_2 2^r N^{-1}\} \geqslant -\alpha(1 + C_1C_2 2^r N^{-1}) > -1$$

when $N \leqslant |x| \leqslant 2N$ for sufficiently large N. So that $G_N^{(r)}(x) > -1$ ($\forall x \in \mathbf{R}, N \geqslant N_0, N_0$ is some sufficiently large number). This is (3.4).

Therefore, we get

$$E_1(\alpha g)_{\infty(I_N)} \leqslant \frac{1}{2}\{\max_t G_N(t) - \min_t G_N(t)\}. \tag{3.7}$$

Notice that the 2π-periodic function $\widetilde{G}_n(u) \overset{\text{df}}{=} G_N\left(\frac{2Nu}{\pi}\right)$ satisfies $\widetilde{G}_N\left(\frac{j\pi}{2N}\right) = 0(-2N+1 \leqslant j \leqslant 2N,)$, and $\widetilde{G}_N^{(r)}(u) = \left(\frac{2N}{\pi}\right)^r G_N^{(r)}\left(\frac{2Nu}{\pi}\right) > -\left(\frac{2N}{\pi}\right)^r$, then applying the same arguement as that of in [3], Theorem 9.5.1, we have

$$\begin{aligned}
&\frac{1}{2}\{\max_t G_N(t) - \min_t G_N(t)\}\\
&= \frac{1}{2}\left\{\max_u \widetilde{G}_N(u) - \min_u \widetilde{G}_N(u)\right\} \leqslant\\
&\left(\frac{2N}{\pi}\right)^r \frac{1}{2}\frac{1}{(2N)^r}\left\{\max_u D_r(u) - \min_u D_r(u)\right\}\\
&= \frac{H_r}{\pi^r}.
\end{aligned} \tag{3.8}$$

From (3.7) and (3.8) we derive

$$E_1(\alpha g)_{\infty(I_N)} \leqslant \frac{H_r}{\pi^r}.$$

Further, $E_1(g)_\infty = \lim\limits_{\substack{\alpha \to 1- \\ N \to +\infty}} E_1(\alpha g)_{\infty(I_N)} \leqslant \frac{H_r}{\pi^r}$.

So

$$\sup\{E_1(g)_\infty : g \in W_\infty^{r;-}(\mathbf{R}), g(j) = 0, \forall j \in \mathbf{Z}\} \leqslant \frac{H_r}{\pi^r}$$

follows.

Now we are going to prove the sharpness of (3.1), which will be derived from the following series of lemmas.

Lemma 3.2 *Let $m \in \mathbf{Z}_+$. Then for every $s(t) \in \mathscr{S}_{m-1} \cap L(\mathbf{R})$*

$$\int_{\mathbf{R}} s(t)\mathrm{d}t = 2\sum_{j \in \mathbf{Z}} s(2j + \alpha_m) \tag{3.9}$$

holds, where $\alpha_m = \frac{1}{4}(1+(-1)^m)$.

Proof $\forall s(t) \in \mathscr{S}_{m-1} \cap L(\mathbf{R})$ we put $s_\pi(t) = \frac{1}{\pi} s\left(\frac{t}{\pi}\right)$ and

$$\varphi(s_\pi, t) = \sum_{j\in\mathbf{Z}} s_\pi(t+2j\pi). \tag{3.10}$$

Then, when $m = 1$, it is obvious that $\varphi(s_\pi, t) = \alpha \operatorname{sgnsin} t + \gamma$ for some $\alpha, \gamma \in \mathbf{R}$, When $m > 1$, $\varphi(s_\pi, t)$ is a 2π-periodic spline function with simple knots $\{j\pi\}$, $j \in \mathbf{Z}$, and

$$\int_{-\pi}^{\pi} \varphi^{(m-1)}(s_\pi, t)\mathrm{d}t = \varphi^{(m-2)}(s_\pi, -\pi) - \varphi^{(m-2)}(s_\pi, -\pi) = 0.$$

Therefore, $\varphi^{(m)}(s_\pi, t) = \alpha \operatorname{sgnsin} t$, and $\varphi(s_\pi, t) = \alpha\Phi_{1,m-1}(t) + \gamma$ for some $\alpha, \gamma \in \mathbf{R}$, where $\Phi_{1,m}(t)$ denotes the m-th periodic integral of sgnsin t with periodic mean value zero. In view of $\Phi_{1,m-1}(\alpha_m\pi) = 0$ we have $\varphi(s_\pi, \alpha_m\pi) = \gamma$. Thus we obtain

$$\begin{aligned}\int_{\mathbf{R}} s(t)\mathrm{d}t &= \frac{1}{\pi}\int_{\mathbf{R}} s\left(\frac{t}{\pi}\right)\mathrm{d}t = \sum_{j\in\mathbf{Z}}\int_{2j\pi}^{2(j+1)\pi} s_\pi(t)\mathrm{d}t \\ &= \sum_{j\in\mathbf{Z}}\int_0^{2\pi} s_\pi(t+2j\pi)\mathrm{d}t = \int_0^{2\pi}\varphi(s_\pi, t)\mathrm{d}t \\ &= \int_0^{2\pi}(\alpha\Phi_{1,m-1}(t)+\gamma)\mathrm{d}t \\ &= 2\pi\gamma = 2\pi\varphi(s_\pi, \alpha_m\pi) = 2\sum_{j\in\mathbf{Z}} s(2j+\alpha_m).\end{aligned}$$

Lemma 3.2 is proved.

Lemma 3.3 *Let* $m, r \in \mathbf{Z}_+$. *Then*

$$E^+(W_1^r(\mathbf{R}); \mathscr{S}_{m-1})_1 \geqslant \sup_{f\in W_1^r(R)}\left|2\sum_{j\in\mathbf{Z}} f(2j) - \int_{\mathbf{R}} f(t)\mathrm{d}t\right|. \tag{3.11}$$

Proof For every $f \in W_1^r(\mathbf{R})$ and $\varepsilon > 0$, there exists an $S_\varepsilon \in \mathscr{S}_{m-1} \cap L(\mathbf{R})$ such that $S_\varepsilon(t) \geqslant f(t)$ and $\|f - S_\varepsilon\|_1 < E^+(f, \mathscr{S}_{m-1})_1 + \varepsilon$. Therefore, by Lemma 3.2 we get

$$\begin{aligned}\varepsilon + E^+(f, \mathscr{S}_{m-1})_1 > \|f - S_\varepsilon\|_1 &= \int_{\mathbf{R}} S_\varepsilon(t)\mathrm{d}t - \int_{\mathbf{R}} f(t)\mathrm{d}t \\ &= 2\sum_{j\in\mathbf{Z}} S_\varepsilon(2j+\alpha_m) - \int_{\mathbf{R}} f(t)\mathrm{d}t.\end{aligned}$$

When ε tends to zero, we get

$$E^+(f, \mathscr{S}_{m-1})_1 \geqslant 2\sum_{f\in\mathbf{Z}} f(2j+\alpha_m) - \int_{\mathbf{R}} f(t)\mathrm{d}t.$$

Taking into account that the set $W_1^r(\mathbf{R})$ is symmetric with respect to 0, and moreover, it is invariant with respect to a shift of arguement, we conclude that

$$E^+(W_1^r(\mathbf{R});\mathscr{S}_{m-1})_1 \geqslant \sup_{f\in W_1^r(\mathbf{R})}\left|2\sum_{j\in\mathbf{Z}} f(2j) - \int_{\mathbf{R}} f(t)\mathrm{d}t\right|,$$

which is (3.11).

Lemma 3.4 *If* $f \in W_1^r(\mathbf{R})$, *then*

$$\left|2\sum_{j\in\mathbf{Z}} f(2j) - \int_{\mathbf{R}} f(t)\mathrm{d}t\right| = \pi^{-r}\left|\int_{\mathbf{R}} D_r(\pi u) f^{(r)}(u)\mathrm{d}u\right|. \quad (3.12)$$

Proof $\forall f \in W_1^r(\mathbf{R})$ putting

$$f_N(t) = \left(\frac{\pi}{2N}\right)^{r-1} f\left(\frac{2N}{\pi}t\right)\chi_r\left(\frac{2t}{\pi}\right), \ -\pi \leqslant t \leqslant \pi, \quad (3.13)$$

we regard $f_N(t)$ as a 2π-periodic function, where $\chi_r(t)$ is defined in (3.2). Consider that

$$f_N^{(r)}(t) = \frac{2N}{\pi} f^{(r)}\left(\frac{2Nt}{\pi}\right)\chi_r\left(\frac{2t}{\pi}\right) + g_N(t), \ t \in [-\pi, \pi], \quad (3.14)$$

where the function $g_N(t)$ is given by

$$g_N(t) = \sum_{j=1}^{r}\binom{r}{j}\left(\frac{\pi}{2N}\right)^{j-1}\left(\frac{2}{\pi}\right)^{j}\chi_r^{(j)}\left(\frac{2t}{\pi}\right) f^{(r-j)}\left(\frac{2Nt}{\pi}\right), \quad (3.15)$$

then, from the fact that

$$C_1 \overset{\mathrm{df}}{=} \max_{1\leqslant j\leqslant r}\|\chi_r^{(j)}\|_\infty < +\infty, \ C_2 \overset{\mathrm{df}}{=} \max_{0\leqslant j\leqslant r-1}\|f^{(j)}\|_1 < +\infty,$$

$$\int_{-\pi}^{\pi}\left|\frac{2N}{\pi} f^{(r-j)}\left(\frac{2Nt}{\pi}\right)\right|\mathrm{d}t = \int_{-2N}^{2N}|f^{(r-j)}(t)|\mathrm{d}t \leqslant \|f^{(r-j)}\|_1,$$

we can derive

$$\int_{-\pi}^{\pi}|g_N(t)|\,\mathrm{d}t \leqslant \sum_{j=1}^{r}\binom{r}{j}\left(\frac{\pi}{2N}\right)^{j}\left(\frac{2}{\pi}\right)^{j}\|\chi_r^{(j)}\|_\infty\|f^{(r-j)}\|_1 \leqslant$$

$$\frac{\pi}{2}C_1C_2\cdot 2^r\cdot N^{-1}, \quad (3.16)$$

here $2N \geqslant \pi$. Therefore, by the equality (7.11) of Chapter 7 in [3] we have

$$\left|\int_{-\pi}^{\pi} f_N(t)\mathrm{d}t - \frac{2\pi}{2N}\sum_{k=-N+1}^{N} f_N\left(\frac{2k\pi}{2N}\right)\right|$$

$$= \frac{2}{(2N)^r}\left|\int_{-\pi}^{\pi} B_r(2Nu) f_N^{(r)}(u)\mathrm{d}u\right|$$

$$= \frac{1}{(2N)^r}\left|\int_{-\pi}^{\pi} D_r(2Nt) f_N^{(r)}(t)\,dt\right|. \tag{3.17}$$

A change of variable yields

$$\int_{-\pi}^{\pi} D_r(2Nt) f_N^{(r)}(t)\,dt$$

$$= \int_{-\pi}^{\pi} D_r(2Nt)\,\frac{2N}{\pi} f^{(r)}\left(\frac{2Nt}{\pi}\right)\chi_r\left(\frac{2t}{\pi}\right)dt + \int_{-\pi}^{\pi} D_r(2Nt) g_N(t)\,dt$$

$$= \int_{-2N}^{2N} D_r(\pi u) f^{(r)}(u)\chi_r\left(\frac{u}{N}\right)du + \int_{-\pi}^{\pi} D_r(2Nt) g_N(t)\,dt. \tag{3.18}$$

It is easy to see that

(1) $\left|D_r(\pi u) f^{(r)}(u)\chi_r\left(\frac{u}{N}\right)\right| \leqslant \| D_r \|_\infty \cdot | f^{(r)}(u) |$

for any $u \in \mathbf{R}$ and $N \in \mathbf{Z}_+$.

(2) $\lim\limits_{N\to+\infty} D_r(\pi u) f^{(r)}(u)\chi_r\left(\frac{u}{N}\right) = D_r(\pi u) f^{(r)}(u)$

for every $u \in \mathbf{R}$.

Then, by Lebesgue's dominated convergence theorem we get

$$\lim_{N\to+\infty}\int_{-2N}^{2N} D_r(\pi u) f^{(r)}(u)\chi_r\left(\frac{u}{N}\right)du = \int_{\mathbf{R}} D_r(\pi u) f^{(r)}(u)\,du. \tag{3.19}$$

In addition, from (3.16) we see

$$\left|\int_{-\pi}^{\pi} D_r(2Nt) g_N(t)\,dt\right| \leqslant \| D_r \|_\infty \int_{-\pi}^{\pi} | g_N(t) |\,dt$$

$$\leqslant C_1 C_2{}^r \pi \| D_r \|_\infty \cdot N^{-1}. \tag{3.20}$$

Thus, by (3.17) ~ (3.20) it follows that

$$\left|2\sum_{j\in\mathbf{Z}} f(2j) - \int_{\mathbf{R}} f(t)\,dt\right|$$

$$= \lim_{N\to+\infty}\left(\frac{2N}{\pi}\right)^r \left|\frac{2\pi}{2N}\sum_{j=-N+1}^{N} f_N\left(\frac{2j\pi}{2N}\right) - \int_{-\pi}^{\pi} f_N(t)\,dt\right|$$

$$= \frac{1}{\pi^r}\lim_{N\to+\infty}\left|\int_{-\pi}^{\pi} D_r(2Nt) f_N^{(r)}(t)\,dt\right|$$

$$= \frac{1}{\pi^r}\lim_{N\to+\infty}\left|\int_{-2N}^{2N} D_r(\pi u) f^{(r)}(u)\chi_r\left(\frac{u}{N}\right)du + \int_{-\pi}^{\pi} D_r(2Nt) g_N(t)\,dt\right| = \pi^{-r}\left|\int_{\mathbf{R}} D_r(\pi u) f^{(r)}(u)\,du\right|.$$

(3.12) is proved.

Now we are in a position to formulate and prove the main theorem.

Theorem 2 *If $m, r \in \mathbf{Z}_+$ and $m \geqslant r$, then*

$$E^+(W_1^r(\mathbf{R});\mathscr{S}_{m-1})_1 = \frac{H_r}{\pi^r}. \tag{3.21}$$

Proof We first show that

$$E^+(W_1^r(\mathbf{R});\mathscr{S}_{m-1})_1 \leqslant \frac{H_r}{\pi^r}. \tag{3.22}$$

(1) Case $m = r$ By (2.26) and (3.1) we have

$$E^+(W_1^r(\mathbf{R});\mathscr{S}_{r-1})_1 \leqslant \sup\{E_1(g)_\infty : g \in W_\infty^{r,-}(\mathbf{R}), g(j) = 0, \forall j \in \mathbf{Z}\} \leqslant \frac{H_r}{\pi^r}. \tag{3.23}$$

(2) Case $m > r$. By making use of the one-sided Kolmogorov's inequality (see [5], Theorem 2.5.3)

$$E_1(f^{(m-r)})_\infty \leqslant \mathrm{d}_{m,m-r}(E_1(f)_\infty)^{\frac{r}{m}} \| f_-^{(m)} \|_\infty^{1-\frac{r}{m}}, \ (1 \leqslant r \leqslant m-1), \tag{3.24}$$

where $\mathrm{d}_{m,m-r} = H_r \cdot H_m^{-\frac{r}{m}}$, and by (2.26) we have

$$E^+(W_1^r(\mathbf{R});\mathscr{S}_{m-1})_1$$
$$\leqslant \mathrm{d}_{m,m-r} \sup \{(E_1(f)_\infty)^{\frac{r}{m}} : f \in W_\infty^{m,-}(\mathbf{R}), f(j) = 0, \forall j \in \mathbf{R}\}.$$

Then by Lemma 3.1 we get

$$E^+(W_1^r(\mathbf{R});\mathscr{S}_{m-1})_1 \leqslant \mathrm{d}_{m,m-r}\left(\frac{H_m}{\pi^m}\right)^{\frac{r}{m}} = \frac{H_r}{\pi^r}. \tag{3.25}$$

Thus from (3.23), (3.24), the inequality (3.22) follows.

On the other hand, by Lemma 3.3, 3.4 we have

$$\begin{aligned} E^+(W_1^r(\mathbf{R});\mathscr{S}_{m-1})_1 &\geqslant \sup_{f\in W_1^r}\left|2\sum_{j\in\mathbf{Z}} f(2j) - \int_{\mathbf{R}} f(t)\mathrm{d}t\right| \\ &= \frac{1}{\pi^r}\sup_{f\in W_1^r}\left|\int_{\mathbf{R}} D_r(\pi u) f^{(r)}(u)\mathrm{d}u\right| \\ &= \frac{1}{\pi^r}E_1(D_r)_\infty = H_r \cdot \pi^{-r}, \end{aligned} \tag{3.26}$$

where we have used the fact that $\int_{\mathbf{R}} f^{(r)}(u)\mathrm{d}u = 0$, for any $f \in W_1^r(\mathbf{R})$.

Thus, by (3.22) and (3.26), we complete the proof of Theorem 2.

Remark A periodic version of Theorem 1 and Theorem 2 was given by N.P.Korneichuk, A.A.Ligun and V.G.Doronin in monograph [3].

Referenece

[1] de Boor, C. Splines as Linear Combinations of *B*-splines. A Survey, in: G. Lorentz and others edited: *Approximation* Ⅱ. N. Y: Acad. Press, 1976. 1 ~ 47

[2] Sun Yongsheng & Li Chun. Best Approximation of Some Classes of Smoogh Functions on the Whole Real Axis by Cardinal Splines of Higher Order. Mat. Zametki, 1990, 48(4): 100 ~ 109(In Russian)

[3] Корнейчук, Н. П., Лигун, А. А. Доронин. В. Г. *Аппроксимация с ограниченями*. Киев: ДУМКА, 1982. 199

[4] Cavaretta, A. S. An Elementary Proof of Kolmogorov's Theorem. AMM, 1974, 5: 480 ~ 486

[5] Корнейчук, Н. П. *Сплайны в Теории Приближения*. М.: НАУКА, 1984. 86

[6] Schumaker, L. *Spline Functions: Basic Theory*. New York: Wiley-interscience publication, 1981. 47

CONSTRUCTIVE APPROXIMATION

1993, Constr. Approx. 9:391 ~ 405

Sobolev-Wiener 光滑函数类用二重取样的最优恢复*

Optimal Recovery of the Sobolev-Wiener Class of Smooth Functions by Double Sampling

Abstract

An optimal interpolation problem is considered on the Sobolev-Wiener class of smooth functions defined on the real line by double samples. We calculate the exact value of the minimal intrinsic error, identify an optimal set of sampling points and constructing an optimal linear estimator (algorithm).

1. Introduction

Given $r \geqslant 2, 1 \leqslant p, q \leqslant +\infty$. Denote

$W^r_{p,q}(\mathbf{R}) = \{f \in L_q(\mathbf{R}): f^{(r-1)}$ absolutely continuous on any finite interval, and $\| f^{(r)} \|_{p,q} \leqslant 1\}$,

where

$$\| f^{(r)} \|_{p,q} = \begin{cases} \left\{ \sum_{j \in \mathbf{Z}} \| f^{(r)}(\cdot + j) \|^q_{L_p[0,1]} \right\}^{1/q}, & 1 \leqslant q < +\infty, \\ \sup_{j \in \mathbf{Z}} \| f^{(r)}(\cdot + j) \|_{L_p[0,1]}, & q = +\infty, \end{cases}$$

---

* Date received: May 23, 1991. Date revised: September 15, 1992.
Communicated by Charles A. Micchelli.
AMS classification: 41A05, 41A35.
本文与刘永平合作.

for any $p, 1 \leqslant p \leqslant +\infty$. $W^r_{p,q}(\mathbf{R})$ is said to be a Sobolev-Wiener class (see [2]). Some extremal problems of approximation theory have been considered on $W^r_{p,q}(\mathbf{R})$ (see [5] and [6]). In this paper, we consider an optimal recovery problem on $W^r_{\infty,p}(\mathbf{R})$ in $L_p(\mathbf{R})$ by using information operators defined as follows.

Following [8], we write $\xi = \{\xi_j\}_{j\in\mathbf{Z}} \in \Theta_\sigma$, if:

(1) $\xi_j < \xi_{j+1}, \forall j \in \mathbf{Z}$;

(2) $\lim\limits_{j\to-\infty} \xi_j = -\infty$, $\lim\limits_{j\to+\infty} \xi_j = +\infty$; and

(3) $\varliminf\limits_{N\to+\infty} \dfrac{\operatorname{card}\{\xi \cap [-N, N]\}}{2N} \leqslant \sigma$;

where $\sigma \in \mathbf{R}_+$ fixed. When $\sigma = 1$ instead of Θ_1 we write Θ for simplicity. For $k, 1 \leqslant k \leqslant r$, fixed and $\xi \in \Theta_\sigma$ the k-fold sampling of $f \in W^r_{p,q}(\mathbf{R})$ over $\xi = \{\xi_j\}_{j\in\mathbf{Z}}$ is defined by

$$I^{(k)}_\xi(f) =: \{f(\xi_j), f'(\xi_j), \cdots, f^{(k-1)}(\xi_j)\}_{j\in\mathbf{Z}}. \tag{1}$$

According to [8], the minimal information diameter of the information operator, $I^{(k)}_\xi$ for $W^r_{p,q}(\mathbf{R})$ in $L_q(\mathbf{R})$ when ξ runs over the set Θ_σ, is given by

$$D^{(k)}_\sigma(W^r_{p,q}, L_q) = 2 \inf_{\xi\in\Theta_\sigma} \sup\{\|f\|_q : f \in W^r_{p,q}, I^{(k)}_\xi(f) = 0\}, \tag{2}$$

where $I^{(k)}_\xi(f) = 0$ means $f(\xi_j) = f'(\xi_j) = \cdots = f^{(k-1)}(\xi_j) = 0$ for all $j \in \mathbf{Z}$. Denote by $E^{(k)}_\sigma(W^r_{p,q}, L_q, \xi)$ the intrinsic error for the optimal recovery problem of $W^r_{p,q}(\mathbf{R})$ in the space $L_q(\mathbf{R})$ by using an information operator $I^{(k)}_\xi$ where $\xi \in \Theta_\sigma$ is fixed, and denote by $E^{(k)*}_\sigma(W^r_{p,q}, L_q, \xi)$ the linear intrinsic error for the same problem. When ξ runs over Θ_σ the minimal intrinsic error and the minimal linear intrinsic error is defined by

$$E^{(k)}_\sigma(W^r_{p,q}, L_q) =: \inf_{\xi\in\Theta_\sigma} E^{(k)}_\sigma(W^r_{p,q}, L_q, \xi),$$

and

$$E^{(k)*}_\sigma(W^r_{p,q}, L_q) =: \inf_{\xi\in\Theta_\sigma} E^{(k)*}_\sigma(W^r_{p,q}, L_q, \xi),$$

respectively.

By [8] we have

Proposition 1 *For all* $\sigma \in \mathbf{R}_+$, $k \in \mathbf{Z}_+$, $1 \leqslant k \leqslant r$,

$$\frac{1}{2}D_{\sigma}^{(k)}(W_{p,q}^{r},L_{q}) \leqslant E_{\sigma}^{(k)}(W_{p,q}^{r},L_{q}) \leqslant E_{\sigma}^{(k)*}(W_{p,q}^{r},L_{q}). \quad (3)$$

An extremely interesting problem is to clarify for which values of p, q, r, σ, k in (3) hold equalities. It has been proved that equalities hold in (3) if $p = q = +\infty$, $r \geqslant 2$, $\sigma = n \in \mathbf{Z}_{+}$, $1 \leqslant k \leqslant r$ (see [3], [9], [10], and [12]). For other values of p, q this problem remains left open. In this paper, we prove that equalities hold in (3) when $p = +\infty$, any $q \in [1, +\infty)$, $r \geqslant 2$, $\sigma = n \in \mathbf{Z}_{+}$, $k = 2$. Moreover, we identify an optimal set of sampling, construct an optimal linear algorithm realizing the optimal intrinsic error, and obtain the exact expression for the optimal intrinsic error $E_{n}^{(2)}(W_{\infty,p}^{r}, L_{p})$. The main result may be formulated as follows:

Theorem 1 *Suppose* $r \geqslant 2$, $1 \leqslant p < +\infty$. *Then we have*

$$\frac{1}{2}D_{1}^{(2)}(W_{\infty,p}^{r},L_{p}) = E_{1}^{(2)}(W_{\infty,p}^{r},L_{p}) = E_{1}^{(2)*}(W_{\infty,p}^{r},L_{p}).$$

$$= \| E_{r}^{*} + \| E_{r}^{*} \|_{\infty} \|_{L_{p}[0,1]},$$

where $E_{r}^{*}(x) = 2^{-r}E_{r}(2x)$, *and* $E_{r}(t)$ *is the usual Euler spline of degree* r *(which satisfies* $E_{r}(t+1) = -E_{r}(t)$*).* $\xi^{0} = \{\nu\}_{\nu\in\mathbf{Z}}$ *is an optimal set of sampling points, and* $T_{r2}(f)$ *is a linear optimal algorithm which will be defined in the following section.*

The paper is organized into four sections. Section 2 is a preliminary section containing some useful results. Section 3 contains an upper bound for

$$E_{n}^{(k)*}(W_{\infty,p}^{r},L_{p}).$$

It is sharp when $k = 2$. In Section 4 we get the sharp lower bound for

$$D_{1}^{(2)}(W_{\infty,p}^{r},L_{p}).$$

2. Preliminary Results

First we point out that the main tool we use to get the sharp upper bound for $E_{n}^{(2)}(W_{\infty,p}^{r},L_{p})$ is the k-fold cardinal interpolation technique developed in [12]. In what follows we recall some basic results about it.

Proposition 2 (see [12]) *Let* $r \geqslant 2$. *Then for any* k, $1 \leqslant k \leqslant r$, *there exists uniquely a* 2*-periodic perfect spline* $E_{rk}(x)$ *such that:*

(1) $E_{rk}^{(i)}(j) = 0, i = 0, \cdots, k-1, \forall j \in \mathbf{Z}$,

(2) $S^-(E_{rk}^{(r)}(x), [-1,1)) = 2k$;

(3) $E_{rk}(-x) = (-1)^k E_{rk}(x) = E_{rk}(x+1)$; *and*

(4) *when* $k = 2$ *we have*

$$E_{r2}(x) = E_r^*(x) + (-1)^{[(r-1)/2]} \| E_r^* \|_\infty,$$

where $E_r^*(x) = 2^{-r}E_r(2x)$, *and* $E_r(t)$ *is the usual Euler spline of degree* r (*which satisfies* $E_r(t+1) = -E_r(t)$).

Let the set of all knots of $E_{rk}(x)$ be $\{\eta_j : j \in \mathbf{Z}\}$, where we assume

$$\cdots < \eta_{-1} < \eta_0 < 0 \leqslant \eta_1 < \eta_2 < \cdots. \tag{4}$$

Some information about $\{\eta_j : j \in \mathbf{Z}\}$ is given in [3]. Denote

$$S_{r,k} =: \{s(x) \in C^{r-2} : s^{(r)}(x)|_{(\eta_j, \eta_{j+1})} = 0, j \in \mathbf{Z}\}. \tag{5}$$

$S_{r,k}$ is the set of cardinal splines we will use to construct the k-fold cardinal interpolation operator. Set

$$F(\mathbf{R}) =: \{f \in C(\mathbf{R}) : \text{there exists some } \beta > 0 \text{ such that } |f(x)| = O(|x|^\beta), |x| \to \infty\}.$$

$AC^{(r-1)}(\mathbf{R})_0 =: \{f : f^{(r-1)}$ absolutely continuous on every finite interval, $f^{(r)}$ is piecewisely continuous on every finite interval$\}$.

The following lemma summarizes fundamental results about the k-fold cardinal interpolation technique which we will use in what follows. For more detailed information we refer to [3].

Proposition 3 *Let* $r \geqslant 2, 1 \leqslant k \leqslant r$, *For every* $f \in F(\mathbf{R}) \cap AC^{(r-1)}(\mathbf{R})$, *there exists uniquely a cardinal spline* $T_{rk}(f, \cdot) \in S_{r,k}$ *such that*

$$T_{rk}^{(j)}(f, \nu) = f^{(j)}(\nu), \quad \forall \nu \in \mathbf{Z}, j = 0, \cdots, k-1. \tag{6}$$

The operator $T_{rk}(f)$ *may be represented by*

$$T_{rk}(f, x) = \sum_{j=0}^{k-1} \sum_{\nu \in Z} f^{(j)}(\nu) L_{rkj}(x - \nu), \tag{7}$$

where the fundamental functions $\{L_{rkj}(x)\}_{j=0}^{k-1}$ *satisfy*:

(1) $L_{rki}^{(j)}(0) = \delta_{ij}, i, j = 0, \cdots, k-1$;

(2) $L_{rki}^{(j)}(\nu) = 0, i, j = 0, \cdots, k-1, \nu \in \mathbf{Z} \setminus \{0\}$; *and*

(3) $|L_{rkj}(x)| \leqslant d\, exp(-\beta_0 |x|)$ *for all* j *and* $x \in \mathbf{R}$($\beta_0 > 0$ *is some fixed number*).

There is a remainder formula

$$f(x) - T_{r,k}(f,x) = \int_{\mathbf{R}} K_{rk}(x,t) f^{(r)}(t) \mathrm{d}t, \tag{8}$$

where

$$K_{rk}(x,t) = O(e^{-\beta_0|t|}), \forall x, t \in \mathbf{R} \tag{9}$$

$$K_{rk}(x+1,t) = K_{rk}(x,t-1), \forall x, t \in \mathbf{R}, \tag{10}$$

and

$$\int_{\mathbf{R}} |K_{rk}(x,t)| \mathrm{d}t = |E_{rk}(x)|, \forall x \in \mathbf{R}. \tag{11}$$

Denote $\xi^0 = \{\nu\}_{\nu \in \mathbf{Z}}$. Then the sample of f over ξ^0 is

$$I_{\xi^0}^{(k)}(f) = \{f(\nu), f'(\nu), \cdots, f^{(k-1)}(\nu)\}_{\nu \in \mathbf{Z}}. \tag{12}$$

To get a sharp lower bound for $D_1^{(2)}(W_{\infty,p}^r, L_p)$ we will make use of a comparison theorem due to A. A. Ligun [4].

Proposition 4 (Comparison Theorem). *Let* $n, r \in \mathbf{Z}_+, 1 \leqslant p \leqslant \infty$, *and* $\widetilde{M}_{r,2k}$ *be the set of* $2\pi-$ *periodic perfect splines of order* r *with* $2k$ ($k \leqslant n$) *free nodes in a periodic interval. Then for any nonnegative* $\varphi(t) \in \widetilde{M}_{r,2k}$ *such that* $\varphi(t) \neq c + \varphi_{n,r}(t+\gamma)$ *it holds that* $\|\varphi\|_{p[0,2\pi]} > \|\varphi_{n,r} + \|\varphi_{n,r}\|_\infty\|_{p[0,2\pi]}$, *where* $\varphi_{n,r}(t)$ *is the* rth *order periodic intergral of* $\varphi_{n,0}(t) = \operatorname{sgn}\sin nt$ *satisfying* $\int_0^{2\pi} \varphi_{n,r}(t)\,dt = 0$, $c, \gamma \in \mathbf{R}$, *chosen arbitrarily*.

3. An Upper Bound for $E_n^{(k)*}(W_{\infty,p}^r, L_p)$

We will obtain an upper bound for $E_n^{(k)*}(W_{\infty,p}^r, L_p)$ when $n = 1$. When $n > 1$ is an integer, a corresponding result may be obtained by changing the scale of the independent variable x.

Theorem 2 *For all* $r \geqslant 2, 1 \leqslant k \leqslant r, 1 \leqslant p < +\infty$, *then*

$$\sup_{f \in W_{\infty,p}^r} \|f - T_{rk}(f)\|_p = \sup\{\|f\|_p : f \in W_{\infty,p}^r, I_{\xi^0}^{(k)}(f) = 0\}$$
$$= \|E_{rk}\|_{L_p[0,1]}. \tag{13}$$

Proof For ease of reading we divide the proof into two steps.

(I) Estimation from above.

First we notice that $W^r_{\infty,p}(\mathbf{R}) \subset W^r_{p,p}(\mathbf{R}) = W^r_p(\mathbf{R})$, which is a Sobolev class. Then for every $f \in W^r_{\infty,p}(\mathbf{R})$ it holds than $\lim_{x\to\infty} f(x) = 0$. This means $f \in F(\mathbf{R})$. We have

$$\| f - T_{rk}(f) \|^p_{L_p(\mathbf{R})} \leqslant \int_{\mathbf{R}} \left(\int_{\mathbf{R}} | K_{rk}(x,t) | \cdot | f^{(r)}(t) | \, \mathrm{d}t \right)^p \mathrm{d}x. \tag{14}$$

$$\begin{aligned} &\int_{\mathbf{R}} | K_{rk}(x,t) | \cdot | f^{(r)}(t) | \, \mathrm{d}t \\ &\leqslant \left(\int_{\mathbf{R}} | K_{rk}(x,t) | \, \mathrm{d}t \right)^{1/p'} \left(\int_{\mathbf{R}} | K_{rk}(x,t) | \cdot | f^{(r)}(t) |^p \mathrm{d}t \right)^{1/p}, \end{aligned} \tag{15}$$

where $1/p + 1/p' = 1$. From (14) and (15) it follows that

$$\begin{aligned} &\| f - T_{rk}(f) \|^p_{L_p}(\mathbf{R}) \\ &\leqslant \int_{\mathbf{R}} \left\{ \left(\int_{\mathbf{R}} | K_{rk}(x,t) | \, \mathrm{d}t \right)^{p/p'} \cdot \int_{\mathbf{R}} | K_{rk}(x,t) | \cdot | f^{(r)}(t) |^p \mathrm{d}t \right\} \mathrm{d}x \\ &\leqslant \int_{\mathbf{R}} | E_{rk}(x) |^{p-1} \cdot \sum_{j\in\mathbf{Z}} \left(\int_j^{j+1} | K_{rk}(x,t) | \cdot c_j(f)^p \mathrm{d}t \right) \mathrm{d}x \\ &= \sum_{l\in\mathbf{Z}} \int_l^{l+1} | E_{rk}(x) |^{p-1} \left(\sum_{j\in\mathbf{Z}} \int_j^{j+1} | K_{rk}(x,t) | \, c_j(f)^p \mathrm{d}t \right) \mathrm{d}x \\ &= \int_0^1 | E_{rk}(x) |^{p-1} \sum_{l\in\mathbf{Z}} \sum_{j\in\mathbf{Z}} \left(\int_j^{j+1} | K_{rk}(x+l,t) | \, c_j(f)^p \mathrm{d}t \right) \mathrm{d}x, \end{aligned} \tag{16}$$

where $c_j(f) =: \operatorname{ess\,sup}_{j\leqslant t\leqslant j+1} | f^{(r)}(t) |$. On account of

$$K_{rk}(x+1,t) = K_{rk}(x,t-1) \Rightarrow K_{rk}(x+l,t) = K_{rk}(x,t-l),$$

so we have

$$\begin{aligned} \sum_{l\in\mathbf{Z}} \int_j^{j+1} | K_{rk}(x+l,t) | \, \mathrm{d}t &= \sum_{l\in\mathbf{Z}} \int_j^{j+1} | K_{rk}(x,t-l) | \, \mathrm{d}t \\ &= \sum_{l\in\mathbf{Z}} \int_{j-l}^{j+1-l} | K_{rk}(x,t) | \, \mathrm{d}t \\ &= \int_{\mathbf{R}} | K_{rk}(x,t) | \, \mathrm{d}t \\ &= | E_{rk}(x) |. \end{aligned} \tag{17}$$

Interchanging the order of summation $\sum_l \sum_j$ in (16) and taking into account(17) we get

$$\| f - T_{rk}(f) \|_{L_p(\mathbf{R})}^p \leqslant \int_0^1 | E_{rk}(x) |^{p-1} \cdot | E_{rk}(x) | \cdot \sum_{j \in \mathbf{Z}} c_j(f)^p \mathrm{d}x$$

$$\leqslant \int_0^1 | E_{rk}(x) |^p \mathrm{d}x = \| E_{rk} \|_{L_p[0,1]}^p. \tag{18}$$

Here we have used the fact $\sum_{j \in \mathbf{Z}} c_j(f)^p = \| f^{(r)} \|_{\infty, p}^p \leqslant 1.$

(Ⅱ) Estimation from below.

We only need to prove

$$\sup\{ \| f \|_{L_p(\mathbf{R})} : f \in W_{\infty, p}^r, I_{\xi^0}^{(k)}(f) = 0 \} \geqslant \| E_{rk} \|_{L_p[0,1]}. \tag{19}$$

First we notice that $I_{\xi^0}^{(k)}(E_{rk}) = 0$. Following [1], put

$$g(x) = \begin{cases} 1, & -1 \leqslant x \leqslant 1, \\ (-1)^r (x-2)^r \sum_{j=0}^{r-1} \binom{r+j-1}{j} (x-1)^j, & 1 < x < 2, \\ (x+2)^r \sum_{j=0}^{r-1} \binom{r+j-1}{j} (x+1)^j, & -2 < x < -1, \\ 0, & | x | \geqslant 2, \end{cases} \tag{20}$$

and then define

$$\chi_{r,N}(x) = \begin{cases} 1, & -2N+1 \leqslant x \leqslant 2N-1, \\ g(x-2N+2), & 2N-1 < x < 2N, \\ g(x+2N-2), & -2N < x < -2N+1, \\ 0, & | x | \geqslant 2N. \end{cases} \tag{21}$$

We see that $\chi_{r,N}(x) \in C^{r-1}(\mathbf{R})$, and

$$| \chi_{r,N}^{(r)}(x) | \leqslant M, \quad \forall x \in \mathbf{R}, \tag{22}$$

for some constant $M > 0$ independent of N. Take a fixed $\alpha \in (0,1)$ and put

$$\varphi_{r,N}(t) = \alpha E_{rk}(t) \chi_{r,N}(t), \quad N = 1, 2, 3, \cdots. \tag{23}$$

By the Leibnitz rule

$$\varphi_{r,N}^{(r)}(t) = \alpha \{ E_{rk}^{(r)}(t) \chi_{r,N}(t) + G_N(t) \}, \tag{24}$$

where $G_N(t) = 0$, $| t | \leqslant 2N-1$, and $| G_N(t) | \leqslant c$, $2N-1 \leqslant | t | \leqslant N$, $c > 0$, is a constant independent of N. This yields

$$\| \varphi_{r,N}^{(r)} \|_{\infty, p} \leqslant \alpha (4N + 2c^p)^{1/p}. \tag{25}$$

From (25) it follows

$$\varphi_{r,N}(t) =: \alpha^{-1}(4N + 2c^p)^{-1/p}\varphi_{r,N}(t) \in W^r_{\infty,p}(\mathbf{R}).$$

It is easy to verify $I^{(k)}_{\xi^0}(\varphi_{r,N}) = 0$. Thus we may conclude that the left side of (19),

$$\begin{aligned}
&\sup\{\|f\|_{L_p(\mathbf{R})} : f \in W^r_{\infty,p}, I^{(k)}_{\xi^0}(f) = 0\} \\
&\geqslant \|\varphi_{r,N}\|_{L_p(\mathbf{R})} \\
&= \alpha^{-1}(4N+2c^p)^{-1/p}\left(\int_{-2N}^{2N} |\varphi_{r,N}(t)|^p \mathrm{d}t\right)^{1/p} \\
&\geqslant (4N+2c^p)^{-1/p}\left(\int_{-2N+1}^{2N-1} |E_{rk}(t)|^p \mathrm{d}t\right)^{1/p} \\
&\geqslant (4N-2)^{1/p}\cdot(4N+2c^p)^{-1/p}\|E_{rk}\|_{L_p[0,1]}. \qquad (26)
\end{aligned}$$

When $N \to +\infty$ taking the limit in (26) we obtain(19). The proof of Theorem 1 is completed.

Corollary 1 *Suppose $r \geqslant 2, 1 \leqslant k \leqslant r$, fixed, then*

$$E^{(k)*}_1(W_{\infty,p}, L_p) \leqslant \|E_{rk}\|_{L_p[0,1]}. \qquad (27)$$

Corollary 2 *Suppose $r \geqslant 2, k = 2$, then*

$$\begin{aligned}
\frac{1}{2}D^{(2)}_1(W^r_{\infty,p}, L_p) &\leqslant E^{(2)}_1(W^r_{\infty,p}, L_p) \\
&\leqslant E^{(2)*}_1(W^r_{\infty,p}, L_p) \\
&\leqslant \|E^*_r + \|E^*_r\|_\infty\|_{L_p[0,1]}. \qquad (28)
\end{aligned}$$

In the next section we are going to prove that the inequalities in (28) turn into equalities. To this end, we only need to prove

$$\frac{1}{2}D^{(2)}_1(W^r_{\infty,p}, L_p) \geqslant \|E^*_r + \|E^*_r\|_\infty\|_{L_p[0,1]}. \qquad (29)$$

4. Lower Bound for $D^{(2)}_1(W^r_{\infty,p}, L_p)$

When $r \geqslant 2, k = 2, n = 1$, every $\xi \in \Theta$satisfies:

(1) $\xi_j < \xi_{j+1}, \xi_j \to -\infty, (j \to -\infty), \xi_j \to +\infty (j \to +\infty)$; and

(2) $\varliminf_{N\to+\infty} \dfrac{\operatorname{card}\{\xi \cap [-N,N]\}}{2N} \leqslant 1.$

Notice that $I^{(2)}_\xi(f) = \{f(\xi_j), f'(\xi_j)\}_{j\in\mathbf{Z}}$. Denote

$$\mathrm{e}(W^r_{\infty,p}, L_p) =: \inf_{\xi\in\Theta}\sup\{\|f\|_{L_p(\mathbf{R})} : f \in W^r_{\infty,p}, I^{(2)}_\xi(f) = 0\}. \qquad (30)$$

From [8], $\mathrm{e}(W^r_{\infty,p}, L_p) = \frac{1}{2}D^{(2)}_1(W^r_{\infty,p}, L_p)$. Denote Θ^* the subset of

Θ such that every $\xi \in \Theta^*$ satisfies the condition(*):

$$(*) \qquad \sup_{j\in\mathbf{Z}}(\xi_{j+1} - \xi_j) < +\infty \tag{31}$$

and

$$e^*(W^r_{\infty,p}, L_p) =: \inf_{\xi\in\Theta^*} \sup\{\|f\|_{L_p(\mathbf{R})} : f \in W^r_{\infty,p}, I^{(2)}_\xi(f) = 0\}. \tag{32}$$

First we prove

Lemma 1

$$e(W^r_{\infty,p}, L_p) = e^*(W^r_{\infty,p}, L_p). \tag{33}$$

Proof It is obvious that we have only to prove the following assertion. For any $\xi \in \Theta$, such that

$$\sup_{j\in\mathbf{Z}}(\xi_{j+1} - \xi_j) = +\infty,$$

we have

$$\sup\{\|f\|_{L_p(\mathbf{R})} : f \in W^r_{\infty,p}, I^{(k)}_\xi(f) = 0\} = +\infty, k = 2, \cdots, r.$$

We should make use of $g(t)$ once more. For any $c \geqslant 1$, put $G_c(t) =: g(c^{-1}t)$, $d_c = c^{-r}\{2c + 4\}^{1/p}\cdot\|g^{(r)}\|_\infty$, and $F_c(t) = d_c^{-1}g(c^{-1}t)$. Then from

$$\begin{aligned}\|G_c^{(r)}\|_{\infty,p} &\leqslant \Big\{\sum_{[c]\leqslant l\leqslant[2c]}(\operatorname{ess}\sup_{l\leqslant x\leqslant l+1}|G_c^{(r)}(x)|)^p + \\ &\qquad \sum_{[-2c]\leqslant k\leqslant[-c]}(\operatorname{ess}\sup_{k\leqslant x\leqslant k+1}|G_c^{(r)}(x)|)^p\Big\}^{1/p} \\ &\leqslant c^{-r}\|g^{(r)}\|_\infty(2c+4)^{1/p} = d_c.\end{aligned}$$

It follows that $F_c(t) \in W^r_{\infty,p}(\mathbf{R})$, and moreover,

$$F_c^{(j)}(t) \equiv 0, \ |t| \geqslant 2c, j = 0, \cdots, r-1,$$

$$\begin{aligned}\|F_c\|_{L_p(\mathbf{R})} &\geqslant \Big\{\int_{-c}^{c}|F_c|^p \mathrm{d}t\Big\}^{1/p} = d_c^{-1}(2c)^{1/p} \\ &= \frac{c^r(2c)^{1/p}}{\|g^{(r)}\|_\infty\cdot(2c+4)^{1/p}} \to +\infty, c \to +\infty.\end{aligned}$$

By supposition there exists a subsequence of positive integers $\{j_n\}$ such that $\lim_{n\to\infty}(\xi_{j_n+1} - \xi_{j_n}) = +\infty$. We define

$$f_n(x) = F_{c_n}\left(x + \frac{\xi_{j_n} + \xi_{j_n+1}}{2}\right),$$

$$c_n = \frac{\xi_{j_n+1} - \xi_{j_n}}{4}.$$

Then we have:

(1) $f_n \in W^r_{\infty,p}(\mathbf{R}),\ I^{(2)}_\xi(f_n) = 0$; and

(2) $\| f_n \|_{L_p(\mathbf{R})} \geqslant \dfrac{c_n^r (2c_n)^{1/p}}{\| g^{(r)} \|_\infty (2c+4)^{1/p}}$.

So that

$$\sup_{\substack{f \in W^r_{\infty,p} \\ I^{(2)}_\xi(f)=0}} \| f \|_{L_p(\mathbf{R})} \geqslant \sup_{n \in \mathbf{Z}_+} \| f_n \|_{L_p(\mathbf{R})} = +\infty.$$

Lemma 1 is proved.

Now we are going to prove the main result of this section.

Theorem 3 *Suppose* $r \geqslant 2, 1 \leqslant p < +\infty$. *Then*

$$e^*(W^r_{\infty,p}, L_p) \geqslant \| E^*_r + \| E^*_r \|_\infty \|_{L_p[0,1]}.$$

After this, combining Theorems 2 and 3 we readily complete the proof of Theorem 1. Because the proving process is somewhat involved, in what follows we divide the proof into several subsections.

(Ⅰ) For all $N > 0$, set

$$W^{r,0}_{\infty,p}(N) =: \{ f : f \in W^r_{\infty,p}, \operatorname{supp}(f) \subset [-N, N) \},$$

$$\widetilde{\prod}{}^r_{4N}(2N) =: \{ f : f \text{ is a } 2N-\text{periodic perfect spline of degree } r,\ S^-(f^{(r)}, [-N, N)) \leqslant 4N \}.$$

Notice that for all $h \in \widetilde{\prod}{}^r_{4N}(2N) \Rightarrow \| h^{(r)} \|_{\infty, p[-N,N]} \leqslant (2N)^{1/p} < (2N)^{1/p}(1 + c_p)$ for any $c_p > 0$, so that $(1 + c_p)^{-1}(2N)^{-1/p} \| h^{(r)} \|_{\infty, p[-N,N]} \leqslant 1$. But, generally speaking, $(1 + c_p)^{-1}(2N)^{-1/p} h$ cannot be in $W^{r,0}_{\infty,p}(N)$. The main idea of our approach is to construct some function belonging to $W^{r,0}_{\infty,p}(N)$, which coincides with

$$(1 + c_p)^{-1}(2N)^{-1/p} h$$

on a subinterval of $[-N, N]$ such that the ratio of that measure with respect to $2N$ tends to 1 as $N \to +\infty$. First we notice that for all $N > 0$, $\xi \in \Theta$,

$$\sup \{ \| f \|_{L_p(\mathbf{R})} : f \in W^r_{\infty,p}, I^{(2)}_\xi(f) = 0 \}$$

$$\geqslant \sup\{ \| f \|_{L_p(\mathbf{R})} : f \in W^{r;0}_{\infty,p}(N), I^{(2)}_{\xi}(f) = 0\}. \quad (34)$$

Given $\varepsilon > 0$ and $\xi \in \Theta^*$, by definition, there is a sequence of real numbers N_j such that $N_j \geqslant 2, N_j \nearrow +\infty$, and

$$\operatorname{card}\{\xi \cap [-N_j, N_j]\} < 2N_j(1+\varepsilon),\ j \in \mathbf{Z}_+. \quad (35)$$

Thus we can choose in $[-N_j, N_j)$ a set which contains

$$\xi \cap [-N_j, N_j) \text{ of } 2N(j,\varepsilon) := \left[N_j(1+\varepsilon) + \frac{1}{2}\right]$$

points, where $[x]$ denotes the largest integer not exceeding x. Let these points be denoted orderly by

$$-N_j \leqslant \xi'_1 < \xi'_2 < \cdots < \xi'_{2N(j,\varepsilon)} < N_j.$$

By [7] there exists a perfect spline $f_j(x)$ for every $j \in \mathbf{Z}_+$ such that:

(1) $f_j(x + 2N_j) = f_j(x), |f_j^{(r)}(x)| = 1, x \in [-N_j, N_j)$;

(2) $f_j(x) \geqslant 0, f_j(\xi'_i) = f'_j(\xi'_i) = 0, i = 1, \cdots, 2N(j,\varepsilon)$; and

(3) $S^-(f_j^{(r)}, [-N_j, N_j)) \leqslant 4N(j,\varepsilon)$.

(Ⅱ) Now we go back to $g(x)$ defined in (20). For every $j \in \mathbf{Z}_+$, we choose a number N'_j for which $N_j = N'_j + \sqrt{N'_j}$ and then set

$$g_j(x) = \begin{cases} 1, & |x| \leqslant N'_j, \\ g\left(\dfrac{x - N'_j}{\sqrt{N'_j}} + 1\right), & N'_j < x < N_j, \\ g\left(\dfrac{x + N'_j}{\sqrt{N'_j}} - 1\right), & -N_j < x < -N'_j, \\ 0, & |x| \geqslant N_j. \end{cases}$$

Take $\lambda \in (0,1)$, and put $F_j(x) = \lambda f_j(x) g_j(x), j \in \mathbf{Z}_+$.

Lemma 2 *$F_j(x)$ possesses the following properties:*

(1) $F_j(\xi_k) = F'_j(\xi_k) = 0, \forall \xi_k \in \xi$; and

(2) Suppose $\xi \in \Theta^*$, then there is some j_0 such that for every $j > j_0$ we have

$$\begin{cases} F_j^{(r)}(x) = \lambda f_j^{(r)}(x), & |x| \leqslant N'_j, \\ |F_j^{(r)}(x)| < 1, & N'_j < |x| < N_j, \\ F_j^{(r)}(x) = 0, & |x| \geqslant N_j. \end{cases} \quad (36)$$

Proof Property (1) is obviously true. In order to prove (2) we only have to prove $|F_j^{(r)}(x)| < 1$ when $N'_j < |x| < N_j$. For definiteness, let $x \in (N'_j, N_j)$. We have

$$F_j^{(r)}(x) = \lambda g_j(x) f_j^{(r)}(x) + \lambda \sum_{i=1}^{r} \binom{r}{i} g_j^{(i)}(x) f_j^{(r-i)}(x), \tag{37}$$

and

$$g'_j(x) = \frac{1}{\sqrt{N'_j}} g'\left(\frac{x - N'_j}{\sqrt{N'_j}} + 1\right), \cdots,$$

$$g_j^{(r)}(x) = \frac{1}{(N'_j)^{r/2}} g^{(r)}\left(\frac{x - N'_j}{\sqrt{N'_j}} + 1\right).$$

Notice that

$$\max\{\| g' \|_\infty, \cdots, \| g^{(r)} \|_\infty\} = M_1 < +\infty, \tag{38}$$

which may be verified directly , and that under the condition(*)

$$\max\{\| f_j \|_\infty, \cdots, \| f_j^{(r-1)} \|_\infty\} = M_2 < +\infty, \tag{39}$$

where $M_2 > 0$ is a constant independent of j, The last assertion will be proved in the last subsection. Then we have

$$|F_j^{(r)}(x)| \leqslant \lambda g_j(x) |f_j^{(r)}(x)| + \frac{\lambda M_1 M_2 \cdot 2^r}{\sqrt{N'_j}} = \lambda g_j(x) + \frac{\lambda M}{\sqrt{N'_j}},$$

$M = 2^r M_1 M_2$. Taking into account that $0 \leqslant g_j(x) \leqslant 1$where $N'_j \leqslant x \leqslant N_j$, we have

$$|F_j^{(r)}(x)| \leqslant \lambda\left(1 + \frac{M}{\sqrt{N'_j}}\right). \tag{40}$$

Thus for every $j > j_0$, $\lambda\left(1 + M/\sqrt{N'_j}\right) < 1$, then

$$|F_j^{(r)}(x)| < 1, N'_j < x < N_j. \tag{41}$$

The case $-N_j < x < -N'_j$ may be treated similarly, Lemma 2 is proved.

(Ⅲ)Auxiliary lemmas.

Lemma 3 *For $j \geqslant j_0$ we have*

$$(2N_j + 2)^{-1/p} F_j \in W^r_{\infty, p}(\mathbf{R}). \tag{42}$$

Proof According to Lemma 2, we have for $j \geqslant j_0$, $|F_j^{(r)}(x)| < 1$ if $|x| < N_j$, and $F_j^{(r)}(x) = 0$ if $|x| \geqslant N_j$. Thus

$$\| F_j^{(r)} \|_{\infty,p} = \Big(\sum_{k=[-N_j]}^{[N_j]} \operatorname{ess}\sup_{k \leqslant x \leqslant k+1} | F_j^{(r)}(x) |^p \Big)^{1/p}$$

$$< \Big(\sum_{k=[-N_j]}^{[N_j]} 1 \Big)^{1/p} = ([N_j] - [-N_j] + 1)^{1/p}$$

$$\leqslant (N_j - (-N_j - 1) + 1)^{1/p}$$

$$= (2N_j + 2)^{1/p}.$$

Therefore, we get

$(2N_j + 2)^{-1/p} F_j \in W_{\infty,p}^{r,0}(N_j) \subset W_{\infty,p}^r(\mathbf{R})$.

This proves (42).

Lemma 4 *For any $\xi \in \Theta_*$ and $j \geqslant j_0$ as given previously, we have*

$$\sup\{ \| f \|_{L_p(\mathbf{R})} : f \in W_{\infty,p}^r, I_\xi^{(2)}(f) = 0 \}$$

$$\geqslant \lambda (2N_j + 2)^{-1/p} \| f_j \|_{L_p[-N_j, N_j]} - e_p(j), \tag{43}$$

where $e_p(j) := \lambda M_2 (N'_j)^{1/2p} (N_j + 1)^{-1/p} \to 0$ *as* $j \to \infty$.

Proof

$$(2N_j + 2)^{-1/p} \| \lambda f_j \|_{L_p[-N_j, N_j]} - (2N_j + 2)^{-1/p} \| F_j \|_{L_p[-N_j, N_j]}$$

$$\leqslant (2N_j + 2)^{-1/p} \| F_j - \lambda f_j \|_{L_p[-N_j, N_j]}$$

$$= (2N_j + 2)^{-1/p} \Big(\int_{[-N_j, -N'_j] \cup [N'_j, N_j]} | \lambda (1 - g_j(x)) f_j(x) |^p \mathrm{d}x \Big)^{1/p}$$

$$\leqslant \lambda (2N_j + 2)^{-1/p} \Big(\int_{-N_j}^{-N'_j} | f_j(x) |^p \mathrm{d}x + \int_{N'_j}^{N_j} | f_j(x) |^p \mathrm{d}x \Big)^{1/p}$$

$$\leqslant \lambda (2N_j + 2)^{-1/p} M_2 (2\sqrt{N'_j})^{1/p} = \lambda M_2 (N'_j)^{1/2p} (N_j + 1)^{-1/p}$$

$$= e_p(j).$$

Thus,

$$(2N_j + 2)^{-1/p} \| F_j \|_{L_p[-N_j, N_j]} \geqslant \lambda (2N_j + 2)^{-1/p} \| f_j \|_{L_p[-N_j, N_j]} - e_p(j),$$

and, since $(2N_j + 2)^{-1/p} F_j \in W_{\infty,p}^r(\mathbf{R})$ (Lemma 5) and $I_\xi^{(2)}(F_j) = 0$ (definition), we obtain (43). Notice that $\lim_{j \to \infty} (N'_j)(N_j)^{-1} = 1$, we have

$$e_p(j) = \lambda M_2 (N'_j)^{1/2p} (N_j + 1)^{-1/p} \to 0 \quad \text{as} \quad j \to \infty.$$

Lemma 4 is proved.

(IV) Now we are going to get a sharp lower bound. To begin with $\| f_j \|_{L_p[-N_j, N_j]}$, if we take into account

$$\left\{\int_{-N_j}^{N_j} | f_j(x) |^p \mathrm{d}x\right\}^{1/p} = \left\{N_j\int_{-1}^{1} | f_j(N_jx) |^p \mathrm{d}x\right\}^{1/p}$$

$$= N_j^{r+1/p}\left\{\int_{-1}^{1} | N_j^{-r}f_j(N_jx) |^p \mathrm{d}x\right\}^{1/p}, \quad (44)$$

and denote $\Phi(x) = N_j^{-r}f_j(N_jx)$, then we see that $\Phi(x)$ satisfies:

(1) $\Phi(x+2) = \Phi(x)$, $| \Phi^{(r)}(x) | = 1$;

(2) $S^-(\Phi^{(r)}, [-1,1)) \leqslant 4N(j,\varepsilon)$;

(3) $\Phi(\xi''_i) = \Phi'(\xi''_i) = 0, i = 1,2,\cdots,2N(j,\varepsilon)$;

where ξ''_i is contained in $[-1,1)$ such that

$$-1 \leqslant \xi''_1 < \xi''_2 < \cdots < \xi''_{2N(j,\varepsilon)} < 1; \text{and}$$

(4) $\Phi(x) \geqslant 0$.

By a comparison theorem due to A.A.Ligun (see [4] or Prop.4) we have

$$\left\{\int_{-1}^{1} | \Phi(x) |^p \mathrm{d}x\right\}^{1/p}$$
$$\geqslant \left\{\int_{-1}^{1} | E_{r,2\overline{N}_j}(x) + \| E_{r,2\overline{N}_j}(\cdot) \|_\infty |^p \mathrm{d}x\right\}^{1/p}, \quad (45)$$

where $\overline{N}_j =: N(j,\varepsilon)$, $E_{r,2\overline{N}_j}(x)$ is the Euler spline of degree r with step-size $1/2\overline{N}_j$. Changing the scale of variable by $x = (\overline{N}_j)^{-1}x'$ and denoting

$$\Psi(x) = E_{r,2\overline{N}_j}(x) + \| E_{r,2\overline{N}_j}(\cdot) \|_\infty,$$

we get

$$\left\{\int_{-1}^{1} | \Psi(x) |^p \mathrm{d}x\right\}^{1/p} = \left\{\frac{1}{\overline{N}_j}\int_{-\overline{N}_j}^{\overline{N}_j} \left| \Psi\left(\frac{x}{\overline{N}_j}\right) \right|^p \mathrm{d}x\right\}^{1/p}$$

$$= (\overline{N}_j)^{-r}\left\{\frac{1}{\overline{N}_j}\int_{-\overline{N}_j}^{\overline{N}_j} \left| (\overline{N}_j)^r\Psi\left(\frac{x}{\overline{N}_j}\right) \right|^p \mathrm{d}x\right\}^{1/p}$$

$$= (\overline{N}_j)^{-r}\left\{\int_{-1}^{1} \left| (\overline{N}_j)^r\Psi\left(\frac{x}{\overline{N}_j}\right) \right|^p \mathrm{d}x\right\}^{1/p}$$

$$= 2^{1/p}(\overline{N}_j)^{-r}\left\{\int_{0}^{1} \left| \overline{N}_j^r\Psi\left(\frac{x}{\overline{N}_j}\right) \right|^p \mathrm{d}x\right\}^{1/p}$$

$$= 2^{1/p}(\overline{N}_j)^{-r} \| E_r^* + \| E_r^* \|_\infty \|_{L_{p[0,1]}}. \quad (46)$$

From (43) ~ (46) it follows that

$$\sup\{ \| f \|_{L_p}(\mathbf{R}) : f \in W^r_{\infty,p}, I_\xi^{(2)}(f) = 0\}$$

$$\geqslant \lambda \cdot 2^{1/p} \cdot N_j^{r+1/p}(1 + c_p(j))^{-1}(2N_j)^{-1/p} \cdot (\overline{N}_j)^{-r} \| E_r^* +$$

$$\| E_r^* \|_\infty \|_{L_{p[0,1]}} - e_p(j) \to \frac{\lambda}{1+\varepsilon} \| E_r^* + \| E_r^* \|_\infty \|_{L_{p[0,1]}}, \text{ as } j \to +\infty.$$

Letting $\lambda \to 1-$, and $\varepsilon \to 0+$, we get the sharp lower bound. The proof of Theorem 3 is completed.

It remains for us to give a proof of (39). This will be done in the next subsection.

(V) A proof of (39).

For convenience, we formulate (39) as Lemma 5.

Lemma 5 *Assume $\xi \in \Theta^*$. Then the function f_j satisfies* (39).

Proof For ease of exposition we first reduce the problem under considereation to the 2π-periodic case. To this end, we set

$$x = \frac{N_j}{\pi} x', \quad \left(\frac{\pi}{N_j}\right)^r f_j\left(\frac{N_j}{\pi} x'\right) =: L_j(x').$$

$L_j(x')$ is 2π-periodic perfect spline of degree r such that:

(1) $S^-(L_j^{(r)}, [0, 2\pi)) \leqslant 4N(j, \varepsilon)$; and

(2) $L_j(x') \geqslant 0$, $L_j(\xi_i^*) = L'_j(\xi_i^*) = 0$, where $\xi_i^* = (\pi/N_j)\xi'_i$. So that

$$0 < \xi_{i+1}^* - \xi_i^* = \frac{\pi}{N_j}(\xi'_{i+1} - \xi'_i) \leqslant \frac{M\pi}{N_j} \leqslant \pi[N_j M^{-1}].$$

Put $N = [N_j M^{-1}]$, which may be arbitrarily large. We assume $N > 4$. Then we have

$$0 < \xi_{i+1}^* - \xi_i^* \leqslant \frac{1}{4} \cdot \frac{\pi}{[4^{-1}N]} = \frac{1}{4} \frac{\pi}{N^*}, N^* = [4^{-1}N].$$

Construct a 2π-periodic perfect spline of degree r $S_{N^*,r}(x)$ such that:

(1) $S_{N^*,r}(x + \pi/N^*) = S_{N^*,r}(x)$;

(2) $S^-(S_{N^*,r}^{(r)}, [0, 2\pi)) = 4N^*$; and

(3) $S_{N^*,r}(k\pi/N^*) = S'_{N^*,r}(k\pi/N^*) = 0$, $k = 0, \cdots, 2N^* - 1$, and $S_{N^*,r}(x) \geqslant 0$.

We claim that $\| L_j \|_\infty \leqslant \| S_{N^*,r} \|_\infty$. Suppose $\| L_j \|_\infty > \| S_{N^*,r} \|_\infty$. Then there is a point $\xi \in [0, 2\pi]$ and a number $\lambda \in (0,1)$ such that $\lambda L_j(\xi) > \| S_{N^*,r} \|_\infty$. By an appropriate translation of the independent variable in $L_j(x)$, if necessary, we can make $S_{N^*,r}(x) - \lambda L_j(x) =:$

$R_j(x)$ satisfy the following conditions:

(1) There exists a set of intervals $\{(z_k, z'_k)\}(k = 0,1,\cdots,2N^* - 1)$ such that

$$z_0 < 0 < z'_0 < z_1 < \frac{\pi}{N^*} < z'_1 < \cdots < z_{2N^*-1} < \frac{2N^* - 1}{N^*}\pi < z'_{2N^*-1} < 2\pi,$$

and $R_j(z_k) = R_j(z'_k) = 0, k = 0,1,\cdots,2N^* - 1$.

(2) In the interval $(k_0\pi/N^*, (k_0+1)/N^*\pi)$ containing ξ, in addition to z'_{k_0}, z_{k_0+1}, there are two points $z^*, z^{**}, z'_{k_0} < z^* < z^{**} < z_{k_0+1}$, $R_j(z^*) = R_j(z^{**}) = 0$.

(3) At each point of $\{z_k, z'_k, z^*, z^{**}\}$ $R_j(x)$ changes sign.

Applying Rolle's theorem successively to $R_j(x), R'_j(x), \cdots, R_j^{(r-1)}(x)$, we conclude that

$$S^-(R_j^{(r)}, [0,2\pi)) \geqslant 4N^* + 2.$$

On the other hand, we have

$$R_j^{(r)}(x) = S_{N^*,r}^{(r)}(x) - \lambda L_j^{(r)}(x),$$

where by [4]

$$S_{N^*,r}^{(r)}(x) = (-1)^k, x \in \left(\frac{k\pi}{2N^*} + \alpha, \frac{k+1}{2N^*}\pi + \alpha\right),$$

$k = 0,\cdots,2N^* - 1$, α is a constant, and $|\lambda L_j^{(r)}(x)| < 1$, so that

$$S^-(R_j^{(r)}, [0,2\pi)) = 4N^*.$$

We arrive at a contradiction. Thus we have proved $\| L_j \|_\infty \leqslant \| S_{N^*,r} \|_\infty$, and therefore

$$\| f_j \|_\infty \leqslant \left(\frac{N_j}{\pi}\right)^r \| L_j \|_\infty \leqslant \left(\frac{N_j}{\pi}\right)^r \| S_{N^*,r} \|_\infty \leqslant \left(\frac{4M}{\pi}\right)^r (N^*)^r \| S_{N^*,r} \|_\infty.$$

On account of $\| S_{N^*,r} \|_\infty = O((N^*)^{-r})$ we get $\| f_j \|_\infty \leqslant c_0, c_0 > 0$, is independent of j. Then by using the Landau-Kolmogorov inequality (see [11]) we get

$$\| f_j^{(l)} \|_\infty \leqslant c_0, l = 1,\cdots,r-1.$$

Thus (39) follows.

To sum up, we have completely proved Theorem 1.

A similar result may be obtained for $n \in \mathbf{Z}_+, n > 1$, if we change the scale of the independent variable.

Acknowledgments Thanks are due to the referees for critical comments on an earlier version of this paper which enabled us to improve the presentation. Yongsheng Sun was supported by NSFC, and Yongping Liu was supported by YNSFC.

References

[1] A. S. Cavaretta Jr.. An elementary proof of Kolmogorov's inequality. AMM, 1974, 5: 480 ~ 486

[2] J. J. F. Fournier, J. Stewart. Amalgams of L^p and l^q. Bull. Amer. Math. Soc., 1985, 13(1): 1 ~ 20

[3] Daren Huang, Yongsheng Sun. Optimal interpolation on some classes of differentiable functions. ATA. 1988, 4(3): 13 ~ 22

[4] A.A. Ligun. Exact inequalities for perfect splines and its applications (in Russian). Izv. Vyssh. Uchben. Zaved. Mat., 1985, 5: 32 ~ 38

[5] Yongping Liu. Infinite-dimensional widths and optimal recovery of some smooth function classes of $L_p(\mathbf{R})$ in metric $L(\mathbf{R})$, (*to appear*):

[6] Yongping Liu, Yongsheng Sun. Infinte-dimensional width and optimal interpolation problems on Sobolev - Wiener space. Sci. Sinica Ser. A, 1992, 6: 582 ~ 591

[7] V.P. Motornyi. On the best quadrature formula of the form $\sum p_k f(x_k)$ for some classes of differentiable periodic functions (in Russian). Izv. Akad. Nauk SSSR Ser. Mat., 1974, 38(3): 583 ~ 614

[8] Yongsheng Sun. Optimal interpolation on some classes of smooth functions defined on the whole real axis. Adv. in Math., 1991, 20(1): 1 ~ 14(in Chinese)

[9] Yongsheng Sun. Optimal recovery on some classes of differentiable functions. ATA, 1986, 2(4): 49 ~ 54

[10] Yongsheng Sun. Optimal recovery on some classes of differentiable functions. Kexue Tongbao (Bull. of Sciences, in Chinese), 1986, 11: 809 ~ 812

[11] Yongsheng Sun. *Theory of Approximation*, vol. 1. Beijing: 1989, Beijing Normal University Press, 454(in Chinese)

[12] Yongsheng Sun, Daren Huang: Optimal interpolation on some classes of differentiable functions. In: (C.K. Chui et al. eds.), *Approximation Theory*, V. New York: Academic Press, 1986, 583 ~ 586

Journal of Beijing Normal University (Natural Science)
Sep. 1999 Vol. 35 Sup.

定义在 $\mathbf{R}^d$ 上的某些光滑函数类在逼近论中的极值问题*

Extremal Problems in Approximation Theory for Some Classes of Smooth Functions Defined on $\mathbf{R}^d$

Abstract

Most work to date on optimal problems in approximation theory has been for function classes defined on a compact interval which are compact subsets in normed linear spaces. Over the last years, however, extremal problems in approximation theory for some classes of functions defined on the real line are extensively investigated. This survey is based on recent work of different authors and on several previous papers listed at the end of the paper. Some unpublished papers are also used.

Part One: The Case $d = 1$

0 Introduction

In the main stream of the development of the approximation theory

* Supported by NSF China (19671012) and the Doctoral Programme Foundation of Education Ministry.

Part One of this paper was completed in 1992 (unpublished). Its first four sections are written by Sun Yongsheng, the fourth and fivth sections in cooperation with Liu Yongping and Chen Dirong.

seeking for exact solutions of extremal problems has always occupied an outstanding position. Its history goes back to Tchebyshev, Markov, Zolotaliev, etc., in the last century. But now it is of one opinion that the fundamental work in modern development of the theory was done by J. Favard. In 1936 Favard[1], Achiezer and Krein[2] proved

Theorem 1 *Let $r \in \mathbf{Z}_+$ be a fixed positive integer and $\widetilde{W}^r_\infty$ be a class of 2π-periodic functions such that, $f \in \widetilde{W}^r_\infty$ if and only if, f is 2π-periodic, $f^{(r-1)}$ is absolutely continuous and $\| f^{(r)} \|_\infty \leqslant 1$. Then*

$$E(\widetilde{W}^r_\infty, T_{2n-1}, C) \triangleq \sup_{f \in \widetilde{W}^r_\infty} E_n(f)_c = \sup_{\substack{f \in \widetilde{W}^r_\infty \\ f \perp T_{2n-1}}} \| f \|_c = \mathcal{K}_r \cdot n^{-r},$$

where $T_{2n-1} \triangleq \mathrm{span}\{1, \cos x, \cdots, \cos(n-1)x, \sin x, \cdots, \sin(n-1)x\}$, $E_n(f)_c \triangleq \min_{g \in T_{2n-1}} \| f - g \|_c$, $\| f \|_c \triangleq \max_x | f(x) |$, *and $\mathcal{K}_r$ is the Favard's constant. $f \perp T_{2n-1}$ means* $\int_0^{2\pi} f(t) \begin{matrix} \sin kt \\ \cos kt \end{matrix} \mathrm{d}t = 0, k = 0, 1, \cdots, n-1.$

Besides this, Favard proved

$$E^L(\widetilde{W}^r_\infty, T_{2n-1}, C) = E(\widetilde{W}^r_\infty, T_{2n-1}, C),$$

where $E^L(\widetilde{W}^r_\infty, T_{2n-1}, C)$ denotes the best linear approximation of $\widetilde{W}^r_\infty$ by means of T_{2n-1} in C norm, i.e.,

$$E^L(\widetilde{W}^r_\infty, T_{2n-1}, C) \triangleq \inf_A \sup_{f \in \widetilde{W}^r_\infty} \| f - A(f) \|_c,$$

where A runs over the set of all linear continuous operators from $\widetilde{W}^r_\infty$ to T_{2n-1}.

J. Favard's work marked a starting point of a series of modern works in seeking for exact solutions of extremal problems in approximation theory. Among others, the works of S. M. Nikolskii[3] in 40 decades, and the works of N. P. Korneichuk[4] in 60 ~ 70 decades are outstanding, which deeply influenced the further development of this research direction.

The flourishing of spline functions in 50 ~ 60 decades strongly vitalized the approximation theory. Especially, it yields many useful tools to solve extremal problems of approximation theory, such as the n-width problem, the optimal recovery problem, etc. A basic result about the exact estimation of best approximation of periodic differentiable functions by polynomial splines is due to V. M. Tikhomirov[5] and A. A. Ligun[6].

Theorem 2 *Denote*

$$\tilde{S}_{m,n} \triangleq \{ s(\cdot) : s \in \tilde{C}_{2\pi}^{m-1}, y^{(m+1)} \Big|_{\left(\frac{j\pi}{n}, \frac{j+1}{n}\pi\right)} = 0, \forall j \in Z \}.$$

Then for every $m \geqslant r-1$ we have

$$\begin{aligned} E(\widetilde{W}^r_\infty, \tilde{S}_{m,n}, C) &\triangleq \sup_{f \in \widetilde{W}^r_\infty} \min_{s \in \tilde{S}_{m,n}} \| f - s \|_c \\ &= \sup_{f \in \widetilde{W}^r_\infty} \| f - s_{r-1,n}(f) \|_c \\ &= \mathscr{K}_r \cdot n^{-r}, \end{aligned}$$

where $n, r \in \mathbf{Z}_+$, and $s_{r-1,n}(f)$ is the interpolating operator from $\widetilde{W}^r_\infty$ to $\tilde{S}_{r-1,n}$.

In approximation theory to put forward the n- width problem marked a new turning point in its development. The pioneer work in this direction is due to A. N. Kolmogorov[7]. From the end of 50 decades V. M. Tikhomirov published a series of papers about the n- width problem which raised the resarch of optimal problems of approximation theory to a higher level. A fundamental result of V. M. Tikhomirov[8] about the exact estimation of the n- width of $\widetilde{W}^r_\infty$ in C reads as follows.

Theorem 3 *Let $n, r \in \mathbf{Z}_+$. Then*

$$\begin{aligned} d_{2n-1}(\widetilde{W}^r_\infty, C) &= d_{2n}(\widetilde{W}^r_\infty, C) = \\ d^{2n-1}(\widetilde{W}^r_\infty, C) &= d^{2n}(\widetilde{W}^r_\infty, C) = \\ \delta_{2n-1}(\widetilde{W}^r_\infty, C) &= \delta_{2n}(\widetilde{W}^r_\infty, C) = \mathscr{K}_r \cdot n^{-r}, \end{aligned}$$

where d_N, d^N, δ_N denotes the N-Kolmogorov width, the N-Gelfand width, and the N-linear width respectively.

V. M. Tikhomirov identified optimal subspaces for these N- widths, and moreover, he found some subspaces of spline functions as optimal subspaces of d_N and δ_N.

From 70 to 80 decades a further progression had been made in the research of extremal problems of approximation especially for periodic function classes. Russian school made great contribution. The works of N. P.

Korneichuk, A. A. Ligun, V. P. Motornii, etc., played a leading role[4,9~11]. Meanwhile, motivated by the $\mathscr{L}$-spline theory of I. J. Schoenberg and C. A. Micchelli, Y. S. Sun, D. R. Huang and G. S. Fang[12,13] systematically extended some fundamental results of Russian school to the more general case, where the periodic function class is a convolution class with the generalized Bernoulli function

$$G_r(x) \triangleq \frac{1}{2\pi} \sum_{\nu} \frac{e^{i\nu x}}{P_r(i\nu)}, \tag{1}$$

where $P_r(\lambda) = \lambda^r + \sum_{j=1}^{r} a_j \lambda^{r-j}$ is an arbitrary real polynomial of degree r. The summation in (1) is extended over all $\nu \in \mathbf{Z}$ such that $P_r(i\nu) \neq 0$. It is known that when $P_r(\lambda)$ has only real zeros the function $G_r(x)$ is a special example of the so-called B - kernel (due to A. Pinkus[14]). But little is known about $G_r(x)$ when $P_r(\lambda)$ has imaginary conjugate roots. Among other things Fang[15] and M. A. Chakhiev[16] clarified the relation of $G_r(x)$ with the NCVD function and the B - kernel respectively.

Now we turn to the next case, when the approximation problem of functions are defined on the real line $\mathbf{R}$. S . Bernstein[17] first posed and studied this problem in 40 decades. The approximating function class used by Bernstein is that of the transcendental integral functions of exponential type. We first recall some definitions and notations about it.

An integral function

$$f(z) = \sum_{k=0}^{\infty} \frac{a_k}{k!} z^k$$

is said to be a transcendental integral function of exponential type with type $\sigma (\geqslant 0)$, provided that

$$\overline{\lim_{n \to +\infty}} \mid a_n \mid^{1/n} = \sigma .$$

From the integral function theory it is known that

$$\sigma = \overline{\lim_{|z| \to \infty}} \frac{\ln | f(z) |}{| z |}.$$

Denote by E_σ the set of all transcendental integral functions with type $\leqslant \sigma$, and

$B_\sigma \triangleq \{f \in E_\sigma: |f(x)| \leqslant C$ on $\mathbf{R}$, where C is depending on $f\}$.

S. Bernstein[17] discovered that, to consider the best uniform approximation of a continuous function defined and bounded on $\mathbf{R}$, B_σ may be used as a nice approximating set, which in many respects is analogous to that of trigonometric polynomials T_{2n-1} in the 2π-periodic case. M. G. Krein[18] established the fundamental result which is the analogy of Theorem 1 for the real line case.

Theorem 4 $\forall\, r \in \mathbf{Z}_+, \sigma > 0,$ *we have*

$$E(\widetilde{W}^r_\infty(\mathbf{R}), B_\sigma, L_\infty) = \mathcal{K}_r \cdot \sigma^{-r},$$

where $\widetilde{W}^r_\infty(\mathbf{R})$ *is the set of all functions* f *defined on* $\mathbf{R}$ *such that*

(1) $f \in L_\infty(\mathbf{R})$.

(2) $f^{(r-1)}$ *is locally abs. cont. on* $\mathbf{R}$ *and* $\|f^{(r)}\|_\infty \leqslant 1$,

and

$$E(\widetilde{W}^r_\infty(\mathbf{R}), B_\sigma, L_\infty) \triangleq \sup_{f \in \widetilde{W}^r_\infty} \inf_{g \in B_\sigma} \|f - g\|_\infty.$$

Since 40 decades of this century the best and linear approximation of functions defined on $\mathbf{R}$ by B_σ were intensitively studied by many mathematicians. Among the contributors we may mention S. Bernstein, H. I. Achiezer[19], M. G. Krein[18], S. M. Nikolskii, I. I. Ibragimov[20], etc. But their works are mainly limited in the circle of constructive theory of functions. For a long time the studying of extremal problems in approximation on $\mathbf{R}$ progressed slowly. This is mainly because of the following situation. First, the notion of Kolmogorov n-width $d_n(A, X)$ makes sense essentially for compact sets in X. In such case $d_n(A, X) \downarrow 0$. For Sobolev classes $W^r_\infty(\mathbf{R})$ or $W^r_p(\mathbf{R})$ which are defined on the real line $\mathbf{R}$, and are non-compact in $C(\mathbf{R})(L_p(\mathbf{R}))$, it is meaningless to consider the finte-dimensional n- widths. It is natural to introduce the notion of infinite -dimensional widths. To this end, we must first decompose the set of infinite-dimensional linear subspace with dimension number $\aleph_0$ of a fixed linear normed space into a set of subclasses by means of some ordered index set. V. M. Tikhomirov[21], and independently Li Chun[22] proposed the notion of average dimension of a linear subspace (dimension $\aleph_0$) and that of locally finite-dimensional linear subspace. These notions lead to two versions of infinite-dimensional width①.

Second, the essence of n-width problem consists in comparing the approximating properties of different subspaces of the same dimension. For a long time B_σ is almost the unique one to be used until I. J. Schoenberg, C. A. Micchelli and others established the cardinal $\mathscr{L}$-spline function space on $\mathbf{R}$ with its interpolation theory in 60 ~ 70 decades.

Last, some basic classes of differentiable functions, such as the Sobolev class $W_p^r(\mathbf{R})$ have close connection with the subclasses of functions of compact support, and in turn, the latter essentially coincides with the periodic function class. So we must remark that works having been done earlier for periodic case are necessary preparation for the exploration of the $\mathbf{R}$ case.

In recent years, we have begun to study systematically the optimal problems of approximation theory for smooth function classes defined on $\mathbf{R}$. We closely connected the infinite - dimensional widths problem (in the sense of Li) with the optimal recovery theory, and in this respect we obtained a series of fundamental results. Meanwhile, we learned that V. M. Tikhomirov, G. G. Magaril-I'llyaev, Din-Zong and others, from the end of 70 decades began to published a series of interesting papers about the average dimension of linear space. Professor V. M. Tikhomirov read a lecture *Extremal Problems and Approximation Theory* during his visit to Zhejiang Universtiy in the fall in 1989 which played an important role in pushing forward the investigation of this direction in our country.

1. Some basic classes of smooth functions defined on the real line

1.1 $L_p^r(\mathbf{R})$ and its generalization

Let $r \in \mathbf{Z}_+$ be fixed, and $p \in [1, +\infty]$. We say that $f \in L_p^r(\mathbf{R})$,

① Professor V. M. Tikhomirov told us that his idea goes back to 50 decades.

provided that

(1) $f \in L_p(\mathbf{R}) \cap C(\mathbf{R})$.

(2) $f^{(r-1)}$ is locally abs. cont. on $\mathbf{R}$, and $f^{(r)} \in L_p(\mathbf{R})$.

Denote

$$W_p^r(\mathbf{R}) \triangleq \{f : f \in L_p^r(\mathbf{R}),\ \| f^{(r)} \|_p \leqslant 1\},$$

where $\|\cdot\|_p$ is the usural $L_p(\mathbf{R})$-norm. The following facts are well-known.

Lemma 1 *Let p be a real number such that $1 \leqslant p < \infty$. Then for $f \in L_p^r(\mathbf{R})$ it holds*

$$\lim_{x \to \infty} f^{(k)}(x) = 0,\ k = 0, \cdots, r-1.$$

Lemma 2 *If $f \in L_p^r(\mathbf{R})$, $1 \leqslant p \leqslant \infty$, $r \geqslant 2$, then $f^{(k)} \in L_p(\mathbf{R})$, $k = 1, \cdots, r-1$.*

We are need of the following generalization of $L_p^r(\mathbf{R})$. Let $P_r(\lambda) = \prod_{j=1}^{r} (\lambda - \lambda_j)$ be a real polynomial of degree r with r real zeros $\lambda_1, \cdots, \lambda_r$. Write $D = \mathrm{d}/\mathrm{d}x$. The differential operator induced by $P_r(\lambda)$ will be denoted by $P_r(D)$.

We say that $f \in L_p(P_r)$, if

(1) $f \in L_p(\mathbf{R}) \cap C(\mathbf{R})$.

(2) $f^{(r-1)}$ is locally abs. cont. on $\mathbf{R}$, and $\| P_r(D) f \|_p < +\infty$.

When $\lambda_1 = \cdots = \lambda_r = 0$, $P_r(D) = D^r$ and $L_p(P_r) = L_p^r$. Lemma 1, 2 hold true for $L_p(P_r)$. It is worthy to point out that, if $\lambda_1 \cdots \lambda_r \neq 0$, then $L_p(P_r)$ is a convolution class.

Lemma 3[23] *If $\lambda_1 \cdots \lambda_r \neq 0$, then $f \in L_p(P_r)$ if and only if there exists a function $\varphi \in L_p(\mathbf{R})$ such that*

$$f(x) = \int_{\mathbf{R}} G_r(x-t) \varphi(t) \mathrm{d}t.$$

where

$$G_r(x) \triangleq \frac{1}{2\pi} \int_{\mathbf{R}} \frac{\mathrm{e}^{iux}}{P_r(\mathrm{i}u)} \mathrm{d}u,$$

$$\varphi(x) \xlongequal{\text{a.e.}} P_r(D) f(x) \quad \text{on } \mathbf{R}.$$

$G_r(x)$ is said to be the *Bernoulli function* determined by $P_r(D)$. It is the analogy of the 2π-periodic generalized Bernoulli function

$$\widetilde{G}_r(x) \triangleq \frac{1}{2\pi}\sum_{\nu}\frac{e^{i\nu x}}{P_r(i\nu)}, \quad (P_r(i\nu) \neq 0), \tag{2}$$

which also makes sense when $\lambda_1 \cdots \lambda_r = 0$. In what follows we will denote

$$W_p(P_r) \triangleq \{f: f \in L_p(P_r), \| P_r(D)f \|_p \leqslant 1\}.$$

It often happens to consider the class of periodic functions analogous to $L_p(P_r)$. The 2π-periodic class is given as follows.

When $P_r(0) \neq 0$,

$$\widetilde{L}_p(P_r) \triangleq \left\{f: f(x) = \int_0^{2\pi}\widetilde{G}_r(x-t)\varphi(t)\mathrm{d}t, \varphi \in \widetilde{L}_p\right\}.$$

When $P_r(0) = 0$,

$$\widetilde{L}_p(P_r) \triangleq \left\{f: f(x) = c + \int_0^{2\pi}\widetilde{G}_r(x-t)\varphi(t)\mathrm{d}t,\right.$$

$$\left.\varphi \in \widetilde{L}_p, \int_0^{2\pi}\varphi(t)\mathrm{d}t = 0\right\}.$$

Analogously we write

$$\widetilde{W}_p(P_r) \triangleq \{f \in \widetilde{L}_p(P_r): \| \varphi \|_p \leqslant 1\}.$$

To consider extremal problems for functions defined on **R**, in some cases it is more reasonable to use mixed norms than the usual L_p-norm. So we introduce some mixed normed linear space as follows.

1.2 The amalgam $(L_p, l^q), 1 \leqslant p, q \leqslant +\infty$

We say that $f \in (L_p, l^q)$, if

(1) f is locally L_p summable on **R**.

(2) If we denote

$$c_n(f) \triangleq \left[\int_n^{n+1}|f(x)|^p\mathrm{d}x\right]^{1/p} \quad (1 \leqslant p < \infty)$$

$$\triangleq \operatorname{ess}\sup_{n \leqslant x \leqslant n+1}|f(x)| \quad (p = +\infty)$$

for any $n \in \mathbf{Z}$, then the sequence $\{c_n(f)\}_{n\in\mathbf{Z}} \in l^q$. (L_p, l^q) is a linear vector space, which may be endowed with a mixed norm as follows:

When $1 \leqslant p, q < +\infty$ we set

$$\| f \|_{pq} \triangleq \left\{\sum_{k\in\mathbf{Z}} c_k(f)^q\right\}^{1/q}.$$

When p or q is $+\infty$, the corresponding norm to be used is ess-sup norm or sup norm. Then it is easily verified that (L_p, l^p) becomes a B space. When $p = q$, it turns into the usual $L_p(\mathbf{R})$. The idea of considering the amalgam (L_p, l^q) as opposed to the usual Lebesgue space $L_p = (L_p, l^p)$ is a natural one because it allows us to separate the global behavior from the local behavior of a function. This idea goes back to 1926 and N. Wiener who considered the special cases (L_1, l^2) and (L_2, l^∞) in [24] and (L_∞^*, l^1) and (L_1, l^∞) in [25]. The first systematic study of these spaces was undertaken in 1975 by F. Holland[26]. A survey about the amalgams and their various applications is due to Fournier and Stewart in 1985[27].

Lemma 4 (Fournier, Stewart[27])

(1) $q_1 \leqslant q_2 \Rightarrow (L_p, l^{q_1}) \subset (L_p, l^{q_2})$,

(2) $p_1 \leqslant p_2 \Rightarrow (L_{p_2}, l^q) \subset (L_{p_1}, l^q)$,

(3) $q \leqslant p \Rightarrow (L_p, l^q) \subset L_p \cap L_q$,

(4) $p \leqslant q \Rightarrow L_p \cup L_q \subset (L_p, l^q)$,

(5) $1 \leqslant p, q < +\infty \Rightarrow (L_p, l^q)^* = (L_{p'}, l^{q'})$,

where $1/p + 1/p' = 1/q + 1/q' = 1$, *and* X^* *denotes the dual space of* X.

For convenience, it is reasonable to consider some linear space of differentiable functions defined on $\mathbf{R}$ with mixed norm.

Let $r \in \mathbf{Z}_+$ be a fixed number, $1 \leqslant p, q, s \leqslant +\infty$, We say that $f \in L_{pqs}^r(\mathbf{R})$, if

(1) $f \in L_s(\mathbf{R}) \cap C(\mathbf{R})$.

(2) $f^{(r-1)}$ is locally abs. cont. on $\mathbf{R}$, and $f^{(r)} \in (L_p, l^q)$.

In what follows $L_{pqs}^r(\mathbf{R})$ will be termed as a S - W class (Sobolev-Wiener class). When $s = q$, we write $L_{pq}^r(\mathbf{R})$ instead of $L_{pqs}^r(\mathbf{R})$. When $p = q = s$, it turns into the usual class $L_p^r(\mathbf{R})$. As in the previous case we will also denote

$$W_{pqs}^r(\mathbf{R}) \triangleq \{f : f \in L_{pqs}^r(\mathbf{R}), \| f^{(r)} \|_{pq} \leqslant 1\},$$

and $W_{pq}^r(\mathbf{R}) = W_{pqq}^r(\mathbf{R})$. Analogously, instead of D^r we take $P_r(D)$ as in the previous case and obtain $L_{pqs}(P_r)$, $W_{pqs}(P_r)$, $L_{pq}(P_r)$, $W_{pq}(P_r)$, etc.

I. J. Schoenberg[28], A. Jakimovski and D. C. Russell[29] considered some cardinal interpolation problem on the S-W class.

2. Best approximation and best linear approximation of functions defined on R by means of a cardinal $\mathscr{L}$-spline linear space

The cardinal $\mathscr{L}$-spline theory which was founded and extensively studied by I.J.Schoenberg, C.A.Micchelli, etc., in 60 ~ 70 decades, yields a fundamental tool in approximating functions on the real line. In recent years, we have succeeded in discovering the optimal property of cardinal $\mathscr{L}$-splines spaces in solving many extremal problems concerning function classes on **R**. To begin with, we first recall some fundamental facts about the Schoenberg-Micchelli theory.

2.1 Cardinal $\mathscr{L}$-splines defined by an ordinary differential operator with constant real coefficients

Given a real polynomial

$$P_{l+1}(x) = a_0 + \cdots + a_l x^l + x^{l+1} = \prod_{j=0}^{l}(x - t_j), t_j \in \mathbf{R}.$$

Denote $D = \mathrm{d}/\mathrm{d}x$ and put

$$P_{l+1}(D) = \prod_{j=0}^{l}(D - t_j).$$

Let

$$\mathrm{e}^T \triangleq \{\mathrm{e}^{t_j} : j = 0, \cdots, l\}.$$

Definition 1 *The class of functions*

$$\mathscr{S}_l(T) \triangleq \{s : s \in C^{l-1}(\mathbf{R}), s\mid_{(\nu,\nu+1)} \in \Pi_l(T), \nu \in \mathbf{Z}\}, \qquad (3)$$

where $\Pi_l(T) \triangleq \{y : P_{l+1}(D)y = 0\}$ *is called the class of cardinal $\mathscr{L}$-splines defined by the differential operator* $P_{l+1}(D)$.

$\mathscr{S}_l(T)$ is an infinite-dimensional linear space. When we take it as an approximating set, we must properly narrow its scope, for example, may take $\mathscr{S}_l(T) \cap L_p(\mathbf{R})$.

Among many basic problems concerning $\mathscr{S}_l(T)$ perhaps the cardinal $\mathscr{L}$-spline interpolation problem is the most important one.

Let the set of knots of interpolation be given as $\{\alpha+\nu\}_{\nu\in\mathbf{Z}}$, where $\alpha\in[0,1)$ is fixed. I. J. Schoenberg[28] introduced the linear manifold of null-splines as follows.

Definition 2 *Let* $\alpha\in[0,1)$ *be fixed. Denote*

$$\mathscr{S}_l^0(T)\triangleq\{s:s\in\mathscr{S}_l(T),\ s(\alpha+\nu)=0,\ \nu\in\mathbf{Z}\}.$$

Every $s\in\mathscr{S}_l^0(T)$ *is called a null-spline.*

$\mathscr{S}_l^0(T)$ is a linear manifold of the linear space $\mathscr{S}_l(T)$. It plays an important role in solving interpolating problems on $\mathscr{S}_l(T)$. I. J. Schoenberg[28,30], C. A. Micchelli[31] proved the following.

Theorem 5

(1) $\dim\mathscr{S}_l^0(T)=N$, *where* $N=l$ *if* $0<\alpha<1$, *and* $N=l-1$, *if* $\alpha=0$.

(2) *There exists a system of basic functions* $s_1,\cdots,s_N$ *such that for some* N *negative numbers* $\lambda_1(\alpha)<\lambda_2(\alpha)<\cdots<\lambda_N(\alpha)<0$ *it holds*

$$S_j(x)=A_l(x_j,\lambda_j),\ x\in[0,1],\ j=1,\cdots,N,$$

$$S_j(x+1)=\lambda_jS_j(x),\quad x\in\mathbf{R},\ j=1,\cdots,N.$$

where the functions $A_l(x,\lambda_1),\cdots,A_l(x,\lambda_N)$ are the so-called *eigen-nullsplines*, and $A_l(x,\lambda)$ is the l-th *exponential Euler polynomial* relating to $P_{l+1}(D)$ (See [31]). It may be represented by

$$A_l(x,\lambda)\triangleq\frac{1}{2\pi\mathrm{i}}\int_C\frac{\mathrm{e}^{xz}}{(\mathrm{e}^z-\lambda)P_{l+1}(z)}\mathrm{d}z,$$

C is a contour on the complex plane which includes inside itself all zeros of $P_{l+1}(z)=0$, but excludes any pole of $\mathrm{e}^{xz}/(\mathrm{e}^z-\lambda)$.

The following theorem contains a series of further informations about $A_l(x,\lambda)$ which may be of use in the sequel.

Theorem 6 *Denote*

$$G(z)=G(z,x,\lambda)\triangleq\frac{\mathrm{e}^{xz}}{\mathrm{e}^z-\lambda},$$

$$D_jf\triangleq(D-t_j)f,\ f^{[j]}\triangleq D_0\cdots D_{j-1}f,\ f^{[0]}=f,\ j=0,\cdots,l.$$

Then we have

(1) $A_l^{[i]}(1)=\lambda A_l^{[i]}(0)+\delta_{il},\ i=0,\cdots,l.$

(2) $D_l A_l = A_{l-1}$.

(3) $A_l(x, \lambda) = G(t_0, t_1, \cdots, t_l; x, \lambda)$ *which denotes the devided difference of* $G(\cdot)$ *with respect to* $\{t_0, t_1, \cdots, t_l\}$.

(4) $A_l(x+1) - \lambda A_l(x) \triangleq \varphi_l(x)$, *which is the divided difference of* $e^{\cdot x}$ *with respect to* $\{t_0, t_1, \cdots, t_l\}$.

(5) If $\lambda \geqslant 0$ *and* $\lambda \notin e^T$, *then* $A_l(x, \lambda)$ *has not zero in* $(0,1)$. *If* $\lambda < 0$, *then* $A_l(x, \lambda)$ *has just one zero in* $[0,1)$ *which is a simple one*.

(6) *Let* $\alpha \in (0,1)$. *Then* $A_l(\alpha, \lambda)$ *has just* l *zeros such that* $\lambda_1(\alpha) < \cdots < \lambda_l(\alpha) < 0$. *When* $\alpha = 0$, $A_l(0, \lambda)$ *has just* l-1 *zeros such that* $\lambda_1(\alpha) < \cdots < \lambda_{l-1} < 0$.

Now we are going to formulate the fundamental theorem of cardinal interpolation on $\mathscr{S}_l(T)$.

Definition 3 *The sequence* $Y = \{y_j\}_{j \in \mathbf{Z}}$ *is said to be of power-growth, if there exists some* $\gamma \geqslant 0$ *such that* $|y_\nu| = 0(|\nu|^\gamma)$, $(\nu \to \infty)$. *Analogously, a function* f *defined on* $\mathbf{R}$ *is said to be power-growth, if there exists some* $\gamma \geqslant 0$ *such that* $f(x) = 0(|x|^\gamma)$, $(x \to \infty)$.

Theorem 7 *Let* $\xi \in [0,1)$ *be the zero of* $A_l(x, -1)$. *Then for any* $\alpha \in [0,1) \setminus \{\xi\}$, *and any sequence* Y *of power-growth, there exists exactly one* $s \in S_l(T)$ *such that* s *is of power-growth with the same* γ *as* Y, *and*

$$s(\alpha + \nu) = y_\nu, \quad \nu \in \mathbf{Z}.$$

Furthermore, there is a $L(x) \in S_l(T)$ *which satisfies*

(1) $L(\alpha + \nu) = \delta_{o\nu}, \quad \nu \in \mathbf{Z}$.

(2) $|L(x)| \leqslant Ae^{-B|x|}$, $x \in \mathbf{R}$, *where* A, B *are some positive constants*.

(3) $s(x) = \sum_{\nu \in \mathbf{Z}} y_\nu L(x - \nu)$.

To insure $\alpha \neq \xi$, we may take α as the zero of $A_{l+1}(x, -1)$ in $[0,1)$ where A_{l+1} is the exponential Euler polynomial which corresponds to $DP_{l+1}(D)$. It is worthy to note that $A_{l+1}(x, -1)$ is the Euler spline relating to $P_{l+1}(D)$.

We must also make use of the fundamental cardinal $\mathscr{L}$-spline interpolating formulate with remainder term. An integral form of the remainder term was first found by C. de Boor, I. J. Schoenberg[32] (for $P_{l+1}(D) = D^{l+1}$), and by S. I. Novikov[33] (for self-adjoint $P_{l+1}(D)$). The following general form is due to C. Micchelli[31], Tang[34] and Li[35].

As before, we assume $\alpha \in [0,1)$, and $A_l(\alpha, -1) \neq 0$. Let the N zeros of $A_l(\alpha, \lambda) = 0$ be given in the manner $\lambda_1 < \cdots < \lambda_m < -1 < \lambda_{m+1} < \cdots < \lambda_N < 0$. $(0 \leqslant m \leqslant N)$. Denote

$$\beta_0 \triangleq \min\{\ln|\lambda_m|, \ln|\lambda_{m+1}|^{-1}\},$$

$$\varepsilon \triangleq \{f: \text{for some } \beta, 0 < \beta \leqslant \beta_0, f(x) = O(e^{\beta|x|})\},$$

$$\varepsilon^{l+1} \triangleq \{f \in C(\mathbf{R}): f^{(l)} \text{loc. abs. cont. on } \mathbf{R}, \text{ and } f^{(l+1)} \text{loc. essentially bounded on } \mathbf{R},\ f^{(l+1)} \in \varepsilon\}.$$

Theorem 8

(1) $\forall f \in \varepsilon^{l+1}$ *there exists* $s_l = s_l(f, x) \in \mathscr{S}_l(T)$ *such that*

$$s_l(f, x) = \sum_{\nu \in \mathbf{Z}} f(\alpha + \nu) L(x - \nu),$$

$$f(x) - s_l(f, x) = \int_{\mathbf{R}} G(x, t) P_{l+1}(D) f(t) \mathrm{d}t, \tag{4}$$

where

$$G(x, t) \triangleq g(x - t) - \sum_{\nu \in \mathbf{Z}} g(\alpha + \nu - t) L(x - \nu),$$

$g \in S_l(T)$ *is some $\mathscr{L}$-spline of power-growth.*

(2) $G(x, t)$ *satisfies*

a) $G(x + 1, t) = G(x, t - 1), x, t \in \mathbf{R}$.

b) $\operatorname{sgn} G(x, t) = \sigma \operatorname{sgn} \sin \pi(x - \alpha) \operatorname{sgn} \sin \pi t$, *where* $\sigma \in \{1, -1\}$ *is fixed*, $A_{l+1}(\alpha, -1) = 0$.

c) $|G(x, t)| \leqslant C(x) e^{-\beta_0 |t|}$, *where $C(x)$ is some function bounded on* $\mathbf{R}$.

d)
$$\int_{\mathbf{R}} |G(x, t)| \mathrm{d}t = |E_{l+1}(x)|,$$

$$\int_{\mathbf{R}} |G(x, t)| \mathrm{d}x = |E^*_{l+1}(t + 1 - \alpha)|,$$

where E_{l+1}, E^*_{l+1} are the Euler splines relating to $P_{l+1}(D)$ and $P^*_{l+1}(D)$

$\triangleq (-1)^{l+1}P_{l+1}(-D)$ *respectively*.

Remark 1 Tang[34] considered the kernel function $G(x,t)$. Li[35] considered $G(x,t)$ together with the kernel function $G^*(x,t)$ corresponding to $P_{l+1}^*(D)$, and obtained

$$G(x-1+\alpha, t-1+\alpha) = (-1)^{l+1}G^*(t,x).$$

When $P_{l+1}(D) = D^{l+1}$, or $P_{l+1}(D) = P_{l+1}^*(D)$, most part of Theorem 8 is knowm (de Boor, Schoenberg[32], C. Micchelli[31], S. I. Novikov[33]).

Remark 2 Changing scale. Let $\sigma > 0$, and $P_{l+1}(\lambda)$ be given as before. We introduce the cardinal $\mathscr{L}$-spline class with a set of simple knots $\{\sigma^{-1}j\}_{j\in\mathbf{Z}}$

$$\mathscr{S}_{l,\sigma} \triangleq \{s \in C^{l-1}(\mathbf{R}) : s|_{(\sigma^{-1}j,\sigma^{-1}(j+1))} \in \Pi_l, j \in \mathbf{Z}\}.$$

When $\sigma = 1$, we denote $\mathscr{S}_{l,1} = \mathscr{S}_l$. It is easy to reduce the interpolation on $\mathscr{S}_{1,\sigma}$ to the $\mathscr{S}_l(T)$ case via scale transformation. To this end, we put $P_{l+1,\sigma^{-1}}(x) \triangleq P_{l+1}(\sigma x)$. On the cardinal $\mathscr{L}$-spline space relating to $P_{l+1,\sigma^{-1}}(D)$ with the set of simple knots $\{j\}_{j\in\mathbf{Z}}$ from Theorem 3 there exists a fundamental function $L(x)$ satisfying the following conditions:

(1) $(P_{l+1,\sigma^{-1}}(D)L)(x) = 0, \forall x \in (j, j+1), j \in \mathbf{Z}$.

(2) $L(\alpha+\nu) = \delta_{o\nu}$.

(3) $|L(x)| \leqslant A\mathrm{e}^{-B|x|}, x \in \mathbf{R}$.

We put $L_\sigma(x) \triangleq L(\sigma x)$. Then we see

a) $P_{l+1}(D)L_\sigma(x) = (P_{l+1,\sigma^{-1}}(D)L)(\sigma x) = 0, \forall x \in (\sigma^{-1}j, \sigma^{-1}(j+1))$,

b) $L_\sigma(\sigma^{-1}\nu + \sigma^{-1}\alpha) = \delta_{o\nu}$,

c) $|L_\sigma(x)| \leqslant A\mathrm{e}^{-B'|x|}, B' = \sigma B$.

$L_\sigma(x)$ is the fundamental function relating to $\mathscr{S}_{l,\sigma}$.

We may also give the remainder formulae relating to the interpolating process with respect to $\mathscr{S}_{l,\sigma}$

2.2 Best approximation and best linear approximation of some classes of smooth functions on R by $\mathscr{S}_{l,\sigma}$

Let $X(\mathbf{R})$ be a normed linear space, which consists of some measurable functions on $\mathbf{R}$. The endowed norm is denoted by $\|\cdot\|_X$. Let a class of functions be $\mathfrak{M} \subset X(\mathbf{R})$. Denote $\mathfrak{R}_\sigma \triangleq X(\mathbf{R}) \cap \mathcal{S}_{l,\sigma}$, $\sigma > 0$. Consider the approximation of $\mathfrak{M}$ by $\mathfrak{R}_\sigma$ in the metric $X(\mathbf{R})$-norm. We are interested in the following two versions of approximation.

Definition 4 For any $f \in \mathfrak{M}$, let

$$e(f, \mathfrak{R}_\sigma, X) \triangleq \inf_{g \in \mathfrak{R}_\sigma} \| f - g \|_X$$

be the best approximation of f by means of $\mathfrak{R}_\sigma$ in $X(\mathbf{R})$ metric. And

$$E(\mathfrak{M}, \mathfrak{R}_\sigma, X) \triangleq \sup_{f \in \mathfrak{M}} e(f, \mathfrak{R}_\sigma, X)$$

be the best approximation of the class $\mathfrak{M}$ by means of $\mathfrak{R}_\sigma$ in $X(\mathbf{R})$ metric.

The quantity $E(\mathfrak{M}, \mathfrak{R}_\sigma, X)$ is a function depending on the variable $\sigma > 0$.

Definition 5 *Denote by sp* $\{\mathfrak{M}\}$ *the linear manifold spanned by* $\mathfrak{M}$ *in* $X(\mathbf{R})$, *and* $A : \mathrm{sp}\,\{\mathfrak{M}\} \to \mathfrak{R}_\sigma$ *a linear bounded operator. Then*

$$E\{\mathfrak{M}, A, X\} \triangleq \sup_{f \in \mathfrak{M}} \| f - Af \|_X$$

is the linear deviation of $\mathfrak{M}$ *from* $\mathfrak{R}_\sigma$ *by means of a linear approximating method A. When A runs over the set of all such linear operators, we call*

$$E^L(\mathfrak{M}, \mathfrak{R}_\sigma, X) \triangleq \inf_A E(\mathfrak{M}, A, X)$$

the best linear approximation of $\mathfrak{M}$ *by means of* $\mathfrak{R}_\sigma$.

Of course, the quantity $E^L(\mathfrak{M}, \mathfrak{R}_\sigma, X)$ also depends on the variable $\sigma > 0$, and $\forall_\sigma > 0$ it holds

$$E(\mathfrak{M}, \mathfrak{R}_\sigma, X) \leqslant E^L(\mathfrak{M}, \mathfrak{R}_\sigma, X).$$

We are interested in the following problems.

1) For some basic classes of functions, for instance, for Sobolev class or S-W class to calculate accurately $E(\mathfrak{M}, \mathfrak{R}_\sigma, X)$, $E^L(\mathfrak{M}, \mathfrak{R}_\sigma, X)$ in $X = L_p(\mathbf{R})$ norm, or to determine asymptotically.

2) To clarify the situation, under which $E(\sigma) = E^L(\sigma)$ (for all $\sigma > 0$, or for all $\sigma = n \in \mathbf{Z}_+$ in some cases), or $E(\sigma) \approx E^L(\sigma)$ ($\sigma \to +\infty$) holds.

3) To construct linear bounded operator $A^\circ : \mathrm{sp}\,\{\mathfrak{M}\} \to \mathfrak{R}_\sigma$ satisfying

$$E^L(\mathfrak{M}, \mathfrak{R}_\sigma, X) = E(\mathfrak{M}, A^\circ, X).$$

Such A° (if it exists) will be called an optimal linear operator.

Recently, these problems are solved for $X = L_p(\mathbf{R})$, $\mathfrak{M} = W_p(P_r)$ or

$W_{pqs}(P_r)$. In what follows without loss of generality we formulate results only for $\sigma = 1$.

Theorem 9 *Let $p \in [1, +\infty]$, Then*

$$E(W_p(P_r), \mathscr{S}_{r-1}, L_p) = E^L(W_p(P_r), \mathscr{S}_{r-1}, L_p) = \sup_{f \in W_p(P_r)} \| f - s_{r-1}(f) \|_p = \lambda_1(p, p, P_r),$$

where

$$\lambda_n(p, q, P_r) \triangleq \sup\{ \| h \|_{q[-1.1]} : h \in \widetilde{W}_p(P_r, [-1,1]), (P_r(D)h(x))\sin \pi x \geqslant 0\}, (n \in \mathbf{Z}_+), \quad (5)$$

while $\widetilde{W}_p(P_r, [-1,1])$ is the 2-periodic differentiable function class defined by the differential operator $P_r(D)$.

It may be compared with $\widetilde{W}_p(P_r)$ which is a class of 2π-periodic differentiable functions defined by $P_r(D)$. Li[36] and Chen[37] proved $\lambda_n(p, q, P_r)$ is attainable and characterized the optimal functions. When $p = 1, 2, +\infty$ it may be calculated exactly.

Special Cases

(1) $p = 1$.

$$E(W_1(P_r), \mathscr{S}_{r-1}, L_1) = E^L(W_1(P_r), \mathscr{S}_{r-1}, L_1) = \sup_{f \in W_1(P_r)} \| f - s_{r-1}(f) \|_1 = \lambda_1(1, 1, P_r) = \| E_r^*(\cdot) \|_\infty.$$

This result is due to Li[38].

(2) $p = 2$.

$$E(W_2(P_r), \mathscr{S}_{r-1}, L_2) = E^L(W_2(P_r), \mathscr{S}_{r-1}, L_2) = \sup_{f \in W_2(P_r)} \| f - s_{r-1}(f) \|_2 = | P_r(\mathrm{i}\pi) |^{-1}.$$

This result is due to Sun and Li[39] for $P_r(D) = D^r$, and Chen[40] for general case. See also G. G. Magaril-I'llyaev[41].

(3) $p = \infty$.

$$E(W_\infty(P_r), \mathscr{S}_{r-1}, L_\infty) = E^L(W_\infty(P_r), \mathscr{S}_{r-1}, L_\infty) = \sup_{f \in W_\infty(P_r)} \| f - s_{r-1}(f) \|_\infty$$

$$= \| E_r(\cdot) \|_\infty .$$

This result is due to C. Micchelli[31] and S. I. Novikov[33] for a self-adjoint operator $P_r(D)$, and Tang[34] for general case.

Remark 3 A more general version of Theorem 5 may be formulated as follows.

Theorem 9′ *For any* $p, 1 \leqslant p \leqslant \infty, \sigma > 0$, *we have*

$$\begin{aligned} E(W_p(P_r), \mathscr{S}_{r-1,\sigma}, L_p) &= E^L(W_p(P_r), \mathscr{S}_{r-1,\sigma}, L_p) \\ &= \sup_{f \in W_p(P_r)} \| f - s_{r-1,\sigma}(f) \|_p \\ &= \lambda_1(p, p, P_{r,\sigma^{-1}}), \end{aligned} \tag{6}$$

where $P_{r,\sigma^{-1}}(D) = P_r(\sigma D)$.

The demonstraton of Theorem 5 is rather involved. It is crucial to use the analytic technique, by which A. Pinkus[42] to attack the n-width $d_n(W_p^r, L_p)$. Besides these, we must take account of some special properties of the interpolating $\mathscr{L}$-spline function $s_{r-1}(f)$ when $f \in W_p(P_r)$.

When p and q take different values, the problem becomes more difficult. Li Chun[36] first obtained a fundamental result for the 2π-periodic case.

Let $P_r(x) = x^\mu \prod_{j=1}^{l} (x^2 - t_j^2), t_j \geqslant 0, \mu = 0$ or 1, and $r = \mu + 2l$.

To consider the 2π-periodic class of differentiable functions $\widetilde{W}_p(P_r)$ we introduced a 2π-periodic $\mathscr{L}$-spline function class as follows:

$$\widetilde{\mathscr{S}}_{r-1,n} \triangleq \{ f : f \in \widetilde{C}^{r-2}, P_r(D) f(x) = 0,$$

$$x \in \left(\frac{j\pi}{n}, \frac{j+1}{n}\pi \right), j = 0, \cdots, 2n-1 \right\}.$$

There is an interpolating operator $R_{r-1,n}(f)$ from $\widetilde{C}_{2\pi}$ to $\widetilde{\mathscr{S}}_{r-1,\sigma}$ such that

$$R_{r-1,n}\left(f, \alpha_r + \frac{j\pi}{n} \right) = f\left(\alpha_r + \frac{j\pi}{n} \right), j = 0, \cdots, 2n-1,$$

$$\alpha_r \triangleq \frac{1}{4}(1 + (-1)^{r-1}) \frac{\pi}{4}.$$

Li Chun[36] proved.

Theorem 10 *Suppose* $1 < q \leqslant p < +\infty$. *Then*

$$E(\widetilde{W}_p(P_r),\mathscr{S}_{r-1,n},\widetilde{L}_q) = E^L(\widetilde{W}_p(P_r),\mathscr{S}_{r-1,n},\widetilde{L}_q)$$
$$= \sup_{f\in \widetilde{W}_p(P_r)} \| f - R_{r-1,n}(f)\|_q$$
$$= \tilde{\lambda}_n(p,q,P_r), \qquad (7)$$

where we use the symbol $\tilde{\lambda}_n(p,q,P_r) = \lambda_n(p,q,P_{r,\pi})$.

When $P_r(D) = D^r$, two special cases, i.e. p arbitrary, $q = 1$, and $p = \infty$, q arbitrary had been considered by N. P. Korneichuk[9].

After Li's[36], Chen Dirong obtains more general result[43] which contains Theorem 6 as a special example. We shall formulate Chen's result in the next section.

Now we can raise a question: If we can establish some result analogous to Theorem 6 for the Sobolev class of differentiable functions $W_p(P_r)$ which are defined on **R** under the condition $q > p$? Liu[44] pointed out that, if $1 \leqslant q < +\infty$, then

$$E(W_\infty(P_r)\cap L_q,\mathscr{S}_{r-1},L_q)$$
$$= E^L(W_\infty(P_r)\cap L_q,\mathscr{S}_{r-1},L_q)$$
$$= \sup_{f\in W_p(P_r)} \| f - s_{r-1}(f)\|_q = \infty.$$

In order to obtain meaningful result, Liu[44] proposed to consider the *S-W* class $W_{pq}(P_r)$ in $L_q(\mathbf{R})$ norm. This had been done first in the two cases $p = \infty, 1\leqslant q<+\infty$, and $1 < p \leqslant +\infty, q = 1$ by Liu[44,45], and a complete analogy of the 2π-periodic case considered by N. P. Korneichuk is established. Recently, Fang and Liu[46] solved completely the general case $p \geqslant q$. Their final result may be formulated as follows.

Theorem 11 *Suppose* $p \geqslant q$. *Then*

$$E(W_{pq}(P_r),\mathscr{S}_{r-1}\cap L_q,L_q)$$
$$= E^L(W_{pq}(P_r),\mathscr{S}_{r-1}\cap L_q,L_q)$$
$$= \sup_{f\in W_{pq}(P_r)} \| f - s_{r-1}(f)\|_q = \lambda_1(p,q,P_r).$$

Special cases,

$$\lambda(p,1,P_r) = \| E_r(\cdot)\|_{p'[0,1]},$$
$$\lambda_1(\infty,p,P_r) = \| E_r(\cdot)\|_{p[0,1]}.$$

Theorem 11 contains Theorem 5 as its special case. In demonstrating the above theorem, the formulae (4) plays very important role. Especially, to get the estimation from above, we rely heavily upon the sign variation properties of the kernel $G(x, t)$. $p = q = 2$ is a single exception. For then the Fourier transform technique is fully effective. It is more difficult to get the sharp estimation from below. Here we need to use a localization technique, and to rely upon some results concerrning the sharp estimation of n-width $d_n(\widetilde{W}_p^r, L_q)(p \geqslant q)$, which was obtained by Buslave and Tikhomirov[47].

2.3 Best approximation of classes of functions on R by the higher-degree cardinal $\mathscr{L}$-spline subspace

For simplicity, in this subsection we are limited to the case $P_r(D) = D^r$. Suppose $m \geqslant r$. Denote

$$\mathscr{S}_{m-1} \triangleq \{s \in C^{m-2}(\mathbf{R}) : D^m s(t)\mid_{(\nu, \nu+1)} = 0, \nu \in \mathbf{Z}\}.$$

We take $\mathscr{S}_{m-1} \cap L_p(\mathbf{R})$ as the approximating set. The function class to be approximated is $W_p^r(\mathbf{R})$. We consider the case $m > r$. There are two main points in considering the best approximation $E(W_p^r(\mathbf{R}), \mathscr{S}_{m-1}, L_p)$: 1) to establish the dual formulae of the best approximation $e(f, \mathscr{S}_{m-1}, L_p)$. 2) to make use of the exact constants in Landau-Kolmogorov's inequalities.

Concerning 1), Sun and Li[48] proved the following dual formulae.

Let $T \triangleq \{t_j\}_{j\in\mathbf{Z}}$ be a real two-sided sequence which satisfies

1) $t_j < t_{j+1}$, $\forall j \in \mathbf{Z}$.

2) $\lim\limits_{j\to+\infty} t_j = +\infty$, $\lim\limits_{j\to-\infty} t_j = -\infty$.

For any fixed $m \in \mathbf{Z}_+$ we set

$$\mathscr{S}_{m-1,T} \triangleq \{s \in C^{m-2}(\mathbf{R}) : D^m s(t)\mid_{(t_j, t_{j+1})} = 0, j \in \mathbf{Z}\}.$$

Write down the $\mathscr{B}$-splines

$$M_{j,r}(x) \triangleq r[t_j, \cdots, t_{j+r}](\cdot - x)_+^{r-1}, j \in \mathbf{Z},$$

$$N_{j,r}(x) \triangleq (t_{j+r} - t_j)M_{j,r}(x)/r.$$

From C. de Boor[49], $\{N_{n,r}\}_{j\in\mathbf{Z}}$ is a Schauder basis in $\mathscr{S}_{r-1,T} \cap L_p(\mathbf{R})$. We have

Theorem 12 *Suppose* $1 \leqslant p \leqslant \infty, 1/p + 1/p' = 1, g \in L_p(\mathbf{R})$, $(1 \leqslant p < \infty)$, *and* $g \in C(\mathbf{R}) \cap L_\infty(\mathbf{R}), (p = \infty)$. *Then*

$$\begin{aligned} & e(g, \mathscr{S}_{r-1,\,T} \cap L_p, L_p) \\ & \triangleq \inf_{\alpha_j} \Big\| g - \sum_{j \in \mathbf{Z}} \alpha_j N_{jr} \Big\|_p \\ & = \sup\Big\{\Big|\int_{\mathbf{R}} g(x) f^{(r)}(x)\mathrm{d}x : f \in W_{p'}^r(\mathbf{R}), f(t_j) = 0, j \in \mathbf{Z}\Big\}, \end{aligned} \tag{8}$$

where $\alpha \triangleq \{\alpha_j\}_{j \in \mathbf{Z}}$ *is any sequence such that* $\sum \alpha_j N_{jr} \in \mathscr{S}_{r-1,\,T} \cap L_p(\mathbf{R})$.

From (8) it is derived that

Corollary 1 *Suppose* $1 \leqslant p \leqslant +\infty, 1/p + 1/p' = 1, m \geqslant r$. *Then*

$$\begin{aligned} & E(W_p^r(\mathbf{R}), \mathscr{S}_{m-1,\,T} \cap L_p, L_p) \\ & \leqslant \sup\{\| f^{(m-r)} \|_{p'} : f \in W_{p'}^m(\mathbf{R}), f(t_j) = 0, j \in \mathbf{Z}\}. \end{aligned} \tag{9}$$

If we take $t_j = j$, *then we get from* (9)

$$\begin{aligned} & E(W_p^r(\mathbf{R}), \mathscr{S}_{m-1}, L_p) \\ & \leqslant \sup\{\| f^{(m-r)} \|_{p'} : f \in W_{p'}^m(\mathbf{R}), f(j) = 0, j \in \mathbf{Z}\}. \end{aligned} \tag{10}$$

Sun and Li[48] proved, when $p = 1, 2, +\infty$, the inequalities in (10) turn into equalities. Then, by making use of the famous Landau-Kolmogorov inequality ($p' = \infty$), the Hardylittlewood-Polya inequality ($p' = 2$), and the Stein inequality ($p' = 1$), we obtain exact estimations which indentify with the $m = r$ case.

The case $p = 2, m = 2r$ may be treated in another way. It is noted that in this case $\mathscr{S}_{m-1}$ is the so-called natural spline function class. When we use $\mathscr{S}_{m-1} \cap L_2(\mathbf{R})$ to approximate $W_2^r(\mathbf{R})$ in $L_2(\mathbf{R})$ metric, we have the following result completely analogous to the periodic case.

Theorem 13[39] *For any* $f \in W_2^r(\mathbf{R})$, *denote*

$$Y = I_0(f) \triangleq \{f(j)\}_{j \in \mathbf{Z}}$$

and $$A(Y) \triangleq \{g \in W_2^r(\mathbf{R}) : I_0(g) = Y\}.$$

Then

1) *there exists uniquely a* $g_Y \in A(Y)$ *such that*

$$\| g_Y^{(r)} \|_2 = \min \| g^{(r)} \|_2.$$

2) *denote* $U(f) \triangleq \{g \in W_2^r(\mathbf{R}) : I_0(g) = I_0(f)\}, f \in W_2^r(\mathbf{R})$ *is fixed. Then the radius of* $U(f)$ *is*

$$\operatorname{rad} U(f) \leqslant (1 - \| g_Y^{(r)} \|_2^2)^{1/2} \cdot \sup_{\substack{f \in W_2^r(\mathbf{R}) \\ f(j)=0}} \| f \|_2.$$

3)

$$\sup_{f \in W_2^r} \| f - g_Y \|_2 = E(W_2^r(\mathbf{R}), \mathscr{S}_{2r-1}, L_2)$$

$$= \sup\{ \| f \|_2 : f \in W_2^r(\mathbf{R}), f(j) = 0\}$$

$$= \pi^{-r}.$$

4) *there exists an interpolating operator* $s_{2r-1}(f)$ *from* $W_2^r(\mathbf{R})$ *into* $\mathscr{S}_{2r-1}$ *such that*

a) $s_{2r-1}(f;,j) = f(j), \forall j \in \mathbf{Z}$.

b) $s_{2r-1}^{(r)}(f) \in L_2(\mathbf{R})$ *and*

$$\| f^{(r)} \|_2^2 = \| s_{2r-1}^{(r)} \|_2^2 + \| f^{(r)} - s_{2r-1}^{(r)} \|_2^2.$$

c)

$$E(W_2^r(\mathbf{R}), \mathscr{S}_{2r-1}(f), L_2) = \sup_{f \in W_2^r} \| f - s_{2r-1}(f) \|_2 = \pi^{-r},$$

$$s_{2r-1}(f) = g_Y(f).$$

Now suppose $p > q$. When we consider the approximation of $W_2^r(\mathbf{R})$ in $L_q(\mathbf{R})$ by means of $\mathscr{S}_{m-1} \cap L_q(\mathbf{R})$ it is natural to take $W_{pq}^r(\mathbf{R})$ instead of $W_p^r(\mathbf{R})$. Liu and Sun[44] extended the dual formula (8) to the *S*-*W* class.

Theorem 14 *Suppose* $1 \leqslant p \leqslant q \leqslant +\infty$. *Then for any* $f \in L_q(\mathbf{R})$ *it holds that*

$$E(f, \mathscr{S}_{m-1,T}(\mathbf{R}) \cap L_{pq}(\mathbf{R}), L_{pq}) \leqslant$$
$$\sup\left\{\int_{\mathbf{R}} f(t) g^{(m)}(t) \mathrm{d}t : g \in W_{p'q'}^m(\mathbf{R}), g(t_j) = 0, j \in \mathbf{Z}\right\}.$$

Corollary 2 *If* $1 \leqslant p \leqslant q \leqslant \infty$, *then for any* $m \geqslant r$ *we have*

$$E(W_q^r(\mathbf{R}) \mathscr{S}_{m-1,T} \cap L_{pq}, L_{pq}) \leqslant$$
$$\sup\{ \| g^{(m-r)} \|_{q'} : g \in W_{p'q'}^m(\mathbf{R}), g(t_j) = 0, j \in \mathbf{Z}\}. \qquad (11)$$

There is a theorem "dual" to Theorem 14, we don't mention it in details. But the inequality "dual" to (11) is of use. It reads as follows:

If $1 < q \leqslant p \leqslant \infty, q < +\infty$, then

$$E(W_{pq}^r(\mathbf{R}), \mathscr{S}_{m-1,T} \cap L_q, L_q(\mathbf{R}))$$
$$\leqslant \sup\{\| g^{(m-r)} \|_{p'q'} : g \in W_{q'}^m(\mathbf{R}), g(t_j) = 0, j \in \mathbf{Z}\}. \quad (12)$$

Liu[44,45] got some exact estimations by making use of (12).

Theorem 15 *Suppose* $1 \leqslant p \leqslant +\infty$. *Then for every* $m \geqslant r$ *it holds*

1) $E(W_{p1}^r(\mathbf{R}), \mathscr{S}_{m-1}, L_1(\mathbf{R}))$
$$= E(W_{p1}^r(\mathbf{R}), \mathscr{S}_{r-1}, L_1(\mathbf{R}))$$
$$= E^L(W_{p1}^r(\mathbf{R}), \mathscr{S}_{r-1}, L_1(\mathbf{R}))$$
$$= \sup_{f \in W_{p1}^r(\mathbf{R})} \| f - s_{r-1}(f) \|_1 = \| E_r(\cdot) \|_{p'[0,1]}. \quad (13)$$

2) $E(W_{\infty p}^r(\mathbf{R}), \mathscr{S}_{r-1}, L_p(\mathbf{R}))$
$$= E^L(W_{\infty p}^r(\mathbf{R}), \mathscr{S}_{r-1}, L_p(\mathbf{R}))$$
$$= \sup_{f \in W_{\infty p}^r} \| f - s_{r-1}(f) \|_p = \| E_r(\cdot) \|_{p[0,1]}. \quad (14)$$

We don't know if
$$E(W_{\infty p}^r(\mathbf{R}), \mathscr{S}_{m-1}, L_p(\mathbf{R})) = E(W_{\infty p}^r(\mathbf{R}), \mathscr{S}_{r-1}, L_p(\mathbf{R}))$$
for $m > r$?

The periodic analogy of (13), (14) is due to A. A. Ligun[50]. See also N. P. Korneichuk[9].

Here we may raise the following problem.

Problem 1 Suppose $1 \leqslant q \leqslant p \leqslant \infty$. We would like to know if
$$E(W_{pq}^r(\mathbf{R}), \mathscr{S}_{r-1}, L_q(\mathbf{R})) = E(W_{pq}^r(\mathbf{R}), \mathscr{S}_{m-1}, L_q(\mathbf{R}))$$
holds for every $m > r$. Analogousely, we don't know whether
$$E(\widetilde{W}_p^r, \widetilde{\mathscr{S}}_{r-1,n}, \widetilde{L}_q) = E(\widetilde{W}_p^r, \widetilde{\mathscr{S}}_{m-1,n}, \widetilde{L}_q)$$
holds for $m > r$? (See (7)).

Remark 4 We have paid more attention to the approximation by higher-degree cardinal $\mathscr{L}$-splines. This is because of the following situation. From [51], we have
$$\lim_{m \to \infty} \mathrm{e}(f, \mathscr{S}_{m-1} \cap L_p, L_P(\mathbf{R})) = \mathrm{e}(f, B_{\pi,p}, L_p(\mathbf{R})), (\text{on } \mathbf{R}),$$
and
$$\lim_{m \to \infty} \mathrm{e}(f, \widetilde{\mathscr{S}}_{m-1,n}, \widetilde{L}_p) = \mathrm{e}(f, T_{2n-1}, \widetilde{L}_p)$$
(2π-periodic case), where

$$B_{\pi,p} \triangleq \{f \in B_\pi : \| f \|_{p(\mathbf{R})} < \infty\},$$

when $p = \infty, B_{\pi,p} = B_\pi$. These relations reveal the close connection between approximations by splines and by the classical subspace $B_{\pi,p}$ (or T_{2n-1}).

2.4 One-sided approximation

One-sided approximation is one version of the so-called approximation with constraints. The problem of one-sided approximation was first raise by Genelius[52] for periodic functions and was extensively studied by N. P. Korneichuk, A. A. Ligun, V. D. Doronin, etc. (see [11]). In this subsection we are going to formulate one result on the real line **R**, which is due to Sun and Liu[53].

Definition 6 *Let $f \in L_p(\mathbf{R}), 1 \leqslant p \leqslant \infty$. The quantity*

$$\begin{aligned}&e^+(f, \mathscr{S}_{m-1,T} \cap L_p(\mathbf{R}), L_p(\mathbf{R}))\\ &= \inf\{\| f - g \|_p : g \in \mathscr{S}_{m-1,T} \cap L_p(\mathbf{R}), g \geqslant f \text{ a.e.}\}\end{aligned}$$

is said to be the upper-sided best approximation of f in $L_p(\mathbf{R})$-metric by means of $\mathscr{S}_{m-1,T} \cap L_p(\mathbf{R})$.

The lower-sided best approximation may be defined analogously.

Let's denote

$$E^+(W_p^r(\mathbf{R}), \mathscr{S}_{m-1,T} \cap L_p, L_p(\mathbf{R})) \triangleq \sup_{f \in W_p^r} e^+(f, \mathscr{S}_{m-1,T} \cap L_p, L_p(\mathbf{R})).$$

A dual formula was established by Sun and Liu[53] which is of use in obtaining exact estimation for $E^+(W_p^r(\mathbf{R}), \mathscr{S}_{m-1,T} \cap L_p, L_p(\mathbf{R}))$.

Theorem 16 *Suppose $m \geqslant r \geqslant 1$. Then*

$$\begin{aligned}&E^+(W_p^r(\mathbf{R}), \mathscr{S}_{m-1,T} \cap L_p, L_p(\mathbf{R}))\\ &= \sup\Big\{\int_{\mathbf{R}} g^{(r)}(t) f^{(m-r)}(t)\,dt : g \in W_p^r(\mathbf{R}),\\ &\qquad f \in W_{p'}^{m,-}(\mathbf{R}), f(t_j) = 0, j \in \mathbf{Z}\Big\},\end{aligned}$$

where $1 \leqslant p < \infty$, and

$$W_q^{r,-}(\mathbf{R}) \triangleq \{f : f \in L_q^r(\mathbf{R}), \| f_-^{(r)} \|_q \leqslant 1\},$$

$f_-^{(r)}$ is the negative part of $f^{(r)}$.

Substituting $t_j = j$ we get

Corollary 3 *When $p = 1$, it holds*

$$E^{+}(W_1^r(\mathbf{R}), \mathscr{S}_{m-1}, L_1(\mathbf{R}))$$
$$\leqslant \sup\{e_1(f^{(m-r)})_\infty : f \in W_\infty^{m,-}(\mathbf{R}), f(j) = 0, j \in \mathbf{Z}\},$$

where $e_1(\phi)_\infty \triangleq \inf\limits_{C\in\mathbf{R}} \|\phi - C\|_\infty$.

Sun and Liu[53] obtained.

Theorem 17 *For* $m \geqslant r \geqslant 1$ *we have*

$$E^{+}(W_1^r(\mathbf{R}), \mathscr{S}_{m-1} \cap L_1(\mathbf{R}), L_p(\mathbf{R})) = \frac{H_r}{\pi^r},$$

where

$$H_r \triangleq \frac{1}{2}\{\sup_t D_r(t) - \inf_t D_r(t)\},$$
$$D_r(t) \triangleq 2\sum_{\nu=1}^{\infty} \nu^{-r}\cos\left(\nu t - \frac{\pi r}{2}\right).$$

Some extension of Theorem 17 was obtained by Zhang Jianhua[54].

2.5 Cardinal $\mathscr{L}$-spline interpolation at k-fold knots of interpolation

In this subsection we mention some new results about the cardinal $\mathscr{L}$-interpolation at k-fold knots ($k \geqslant 2$), i.e., the Hermite type interpolation. It is known that the basic result in this direction of research is due to Lipow and Schoenberg[30]. Huang and Sun[55] revealed an optimal property of the Hermite type cardinal interpolating operator when they considered the optimal recovery problem for $W_\infty^r(\mathbf{R})$ in $L_\infty(\mathbf{R})$ by means of set of multi-fold sampling point sets.

The following lemma is due to Bojanov and Huang[56]. But its special case $k = 2$ was known by V. L. Velikin[57].

Lemma 5 *Let* $r \geqslant 2$. *Then for every integer* $1 \leqslant k \leqslant r$ *there exists uniquely a* 2-*periodic perfect spline functon* $E_{rk}(x)$ *such that*

1) $E_{rk}^{(r)}(0+) = 1$.

2) $E_{rk}^{(i)}(j) = 0, i = 0, 1, \cdots, k-1, j \in \mathbf{Z}$.

3) $E_{rk}(-x) = (-1)^k E_{rk}(x) = E_{rk}(x+1)$.

4) $S^{-}(E_{rk}^{(r)}(x), [-1, 1)) = 2k$.

5) *All zeros of* $E_{rk}(x)$ *are contained in* $\mathbf{Z}$.

6) *If* $r + k$ *is odd, then* $\mathbf{Z}$ *is a subset of the set of knots of* $E_{rk}(x)$.

7) *If* $r\cdot k$ *is odd, then* $\{j+1/2\}_{j\in\mathbf{Z}}$ *is a subset of the set of knots of* $E_{rk}(x)$.

8) *When* $r+k$ *is odd, then*

$$\operatorname{sgn} E_{rk}(x)=(-1)^{(r+1-k)/2}, x\in(0,1).$$

When $r+k$ *is even, then*

$$\operatorname{sgn} E_{rk}(x)=(-1)^{(r-k)/2}, x\in(0,1).$$

Some examples of $E_{rk}(x)$.

Example 1 $k=1$. $E_{r,1}(x)$ is nothing but the r-degree Euler spline function $E_r(x)$. It is well-known that when r is odd, $\mathbf{Z}$ is its set of knots, and when r is even, the set of knots is $\{j+1/2\}_{j\in\mathbf{Z}}$.

Example 2 $k=2$. Denote $E_r\triangleq\|E_r(\cdot)\|_\infty$. Then

$$E_{r,2}(x)=E_r(x)+(-1)^{[(r-2)/2]}E_r,$$

$[\alpha]$ is the largest integer not exceeding α.

Now we denote the set of all knots of $E_{rk}(x)$ by $\{\eta_j\}_{j\in\mathbf{Z}}$, where we may assume $\cdots<\eta_{-1}<\eta_0<0\leqslant\eta_1<\eta_2<\cdots$. There are just $2k$ η_j contained in $[-1,1)$. Besides these, we have $\eta_{j+2k}=\eta_j+2$. This means that $\{\eta_j\}_{j\in\mathbf{Z}}$ is a periodic set of points defined on the real line.

Set

$$\mathscr{S}_{r-1}(k)\triangleq\{s\in C^{r-2}: D^r s(t)\mid_{(\eta_j,\eta_{j+1})}=0, j\in\mathbf{Z}\}.$$

When $k=1$, this is simply the set $\mathscr{S}_{r-1,T}$, where $T=\{\eta_j\}_{j\in\mathbf{Z}}$. It is noted that $\mathscr{S}_{r-1}(k)$ is a set of cardinal $\mathscr{L}$-splines with periodic simple knots. When $k=1,2$, these knots are equidistant. But for $k\geqslant3$ they are no longer equidistant. $\mathscr{S}_{r-1}(k)$ heavily depends on k, because $\{\eta_j\}_{j\in\mathbf{Z}}$ is nothing but the set of knots of $E_{nk}(x)$.

Lemma 6 *There exists a set of functions* $\{L_{rkj}(x)\}_{j=0}^{k-1}\subset\mathscr{S}_{r-1}(k)$ *such that*

1) $L_{rkj}^{(i)}(0)=\delta_{ij}, i,j=0,\cdots,k-1$.

2) $L_{rkj}^{(i)}(\nu)=0, \nu\neq0, i,j=0,\cdots,k-1$.

3) $|L_{rkj}(x)|\leqslant de^{-\beta_0|x|}, j=0,\cdots,k-1, x\in\mathbf{R}$, and $d,\beta_0>0$ *are some positive constants which are independent of* j *and* x.

4) *For any* $f \in L^r_\infty(\mathbf{R})$, *there is just one spline function* $T_{rk}(\cdot) = T_{rk}(f, \cdot) \in \mathscr{S}_{r-1}(k)$ *for which it holds*

$$T^{(i)}_{rk}(f, \nu) = f^{(i)}(\nu), i = 0, \cdots, k-1, \nu \in \mathbf{Z}.$$

5) *We have the remainder formula*

$$f(x) - T_{rk}(f, x) = \int_{\mathbf{R}} K_{rk}(x, t) f^{(r)}(t) \mathrm{d}t,$$

where the kernel $K_{rk}(x, t)$ *is analogous to* $G(x, t)$. *Especially*.

$$\int_{\mathbf{R}} | K_{rk}(x, t) | \mathrm{d}t = | E_{rk}(x) |.$$

This result is due to Huang and Sun. It extends Theorem 7 and 8.

The following theorem is proved by Huang and Sun[56].

Theorem 18 *For any* $k, 1 \leqslant k \leqslant r (r \geqslant 2)$, *we have*

$$\sup_{f \in W^r_\infty} \| f - T_{rk}(f) \|_\infty = \| E_{rk}(\cdot) \|_\infty.$$

It is meaningless to consider the above interpolation problem in $L_p(\mathbf{R})$ metric, because then it is easily verified that

$$\sup_{f \in W^r_\infty} \| f - T_{rk}(f) \|_p = \infty, (1 \leqslant p < +\infty).$$

Liu[58] pointed out that in this case it is favorable to take the $S-W$ class $W^r_{\infty p}(\mathbf{R})$ instead of $W^r_\infty(\mathbf{R})$.

Theorem 19 *For any* $k, 1 \leqslant k \leqslant r, (r \geqslant 2)$, *we have*

$$\sup_{f \in W^r_{\infty p}} \| f - T_{rk}(f) \|_p = \| E_{rk}(\cdot) \|_{p[0,1]}.$$

Recently, Liu[59] extended the above result to a more general case $W^r_{pq}(\mathbf{R})$ in $L_q(\mathbf{R})$. His result is very interesting, but it is too long to be cited here.

3 N-width problem by means of a set of approximating linear subspaces with dimension $\aleph_0$

The proper Kolmogorov, Gelfand and linear n-width are finite-dimensional, i.e., the approximating subspaces are finite-dimensional, and their dimension number $\leqslant n$, n is some positive integer (or the sections are finite-codimensional, and their codimension number $\leqslant n$, if the Gelfand width number is considered). The proper n-widths concepts essentially characterized the compact sets in a B space. More precisely, let X be a B space, $A \subset X$ is a set to be approximated by means of the set $\mathscr{L}_n$ of linear subspaces of X with dimension $\leqslant n$. Then the Kolmogorov n-width is

$$d_n(A, X) \triangleq \inf_{L \in \mathscr{L}_n} E(A, L, X).$$

If A is bounded, then

$$\lim_{n \to +\infty} d_n(A, X) = 0 \text{ if } \overline{A} \text{ is compact in } X,$$

where $\overline{A}$ is the closure of A in X. If $\overline{A}$ is non-compact, then

$$\lim_{n \to +\infty} d_n(A, X) > 0.$$

In the later case, the Kolmogorov number $\{d_n(A, X)\}$ is meaningless in characterizing the approximating property of A. In order to treat the later case, we need to extend reasonably the notion of Kolmogorov's width. In recent years. some extended notions are introduced by V.M. Tikhomirov[21], and independently by Li Chun[22]. The concepts of Gelfand width, linear width, etc., have been also extended by these authors.

3.1 Fundamental notions

Let X be a B space, and $\{P_a\}_{a \geqslant 0}$ is a set of projections from X into X such that

1) P_0 is the null-projection.

2) Denote $P_a(X) \triangleq X_a$. For every $s, t \in \mathbf{R}, 0 \leqslant s \leqslant t$, we have $X_s \subset X_t$.

3) $P_t \xrightarrow{s} I, (t \to +\infty)$, i.e., $\forall x \in \mathrm{X}$,

$$\lim_{t \to +\infty} \| P_t(x) - x \| = 0.$$

Denote $\mathfrak{F} \triangleq \{P_a\}_{a \geqslant 0}$, ц $\triangleq \{X_a\}_{a \geqslant 0}$ and $\Phi \triangleq \{\mathfrak{F}, ц\}$. In the following we shall make use of the symbol $P_a(x) = x_a, A_a = P_a(A)$ for $A \subset X$.

Some examples:

Example 3 $P_t = I$(Identity), $t > 0$. Then for every $a > 0, X_a = X$.

Example 4 Let $X = L_p(\mathbf{R}), 1 \leqslant p \leqslant \infty, \chi_a(\cdot)$ is the characteristic function of $[-a, a]$, i.e., $\chi_a(t) = 1$, for $t \in [-a, a]$, $\chi_a(\mathrm{t}) = 0$, $\forall t \in \mathbf{R} \setminus [-a, a]$. We define $P_a: X \to X$ by $f \mapsto \chi_a f$. Then $\{\mathfrak{F}, ц\}$ satisfies 1) ~ 3).

Definition 7 *Suppose* $F \subset X$ *is a linear subspace, and* $\Phi = \{\mathfrak{F}, \text{ц}\}$. *Denote* $F^0 \triangleq \{x \in F: \| x \| \leqslant 1\}$. *For any* $\varepsilon > 0$ *we define*

$$k = k(\varepsilon, a) \triangleq \min\{n \in \mathbf{Z}_+ : \exists \text{ linear subset } M \subset X_a \text{ such that } \dim M = n, \text{ and } E((F^0)_a, M, X_a) < \varepsilon\}. \quad (15)$$

If the set of integers in (15) *is empty, we set* $k(\varepsilon, a) = +\infty$.

If there exist some $\varepsilon_0 > 0$ *and* $\sigma \geqslant 0$ *such that for every* $\varepsilon \in (0, \varepsilon_0)$, $k(\varepsilon, a)$ *for any* a *is finite and moreover*

$$\lim_{a \to +\infty} \frac{k(\varepsilon, a)}{a} = \sigma, \quad (16)$$

then we denote $\overline{\dim}\, F = \sigma$, *and the number* σ *is termed as the* Φ*-average dimension of* F. *It's noted that* σ *may be any non-negative real number. In the sequel, when* σ *is finite,* F *is said to be finite* Φ*-average dimensional (or finite average dimensional).*

In what follows we denote $\text{ц}_\sigma = \{F \subset X : \overline{\dim}\, F \leqslant \sigma\}$.

Definition 8 *Let* $F \subset X$ *be a linear subspace. If for every* $a > 0$, $\dim F_a < +\infty$, *then* F *is said to be locally finite-dimensional (LFD). In addition to this, if*

$$\overline{\lim_{a \to +\infty}} \frac{\dim F_a}{a} = \sigma, \quad (17)$$

then F *is said to be locally finite-dimensional with degree* σ(σ-LFD). *We denote* $\mathscr{S}_\sigma \triangleq \{F \subset X : F$ is LFD, with degree $\leqslant \sigma\}$.

The notion of average dimension is due to Tikhomirov[21], and that of σ-LFD is due to Li Chun[22].

Remark 5 The notion of Tikhomirov has close connection with the idea of average entropy of Shannon and Kolmogorov (See [60]). It is incomparable with that of Li. But there is another version of definition about the average dimension which is slightly different from the above mentioned and it may be comparable with Li's. We formulate it as follows.

Definition 7′ *Let* $F \subset X$ *be a linear subspace, and* $\Phi = \{\mathfrak{F}, \text{ц}\}$. *The number*

$$\lim_{\varepsilon \to +\infty} \overline{\lim_{a \to +\infty}} \frac{k(\varepsilon, a)}{a} = \sigma \quad (16')$$

is said to be the Φ-average dimension of F(in X), and is denoted by $\sigma = \overline{dim}\, F$.

We see that (16)⇒(16′). i. e., condition (16′) is slightly weaker than (16). Moreover, if for $F \subset X$ (17) holds then

$$\lim_{\varepsilon\to 0+} \varliminf_{a\to+\infty} \frac{k(\varepsilon, a)}{a} = \varliminf_{a\to\infty} \frac{\dim F_a}{a} = \sigma.$$

This means that every locally finite-dimensional subspace F with degree σ (σ- LFD) has average dimension σ in the sense of Definition 7′.

Defintion 9 *Given a subset* $K \subset X$ *and* $\Phi = \{\mathfrak{F}, \text{ц}\}$. *The quantity*

$$\bar{d}_\sigma(K, X) \triangleq \inf_{F\in \text{ц}_a} E(K, F, X)$$

is said to be the Φ-average σ − *width (in the sense of Kolmogorov).*

Similarly, we may define the Φ-average linear σ-width $\bar{\delta}_\sigma(K, X)$. Any linear subspace $F^* \in \text{ц}_\sigma$ such that $E(K, F^*, X) = \bar{d}_\sigma(K, X)$($E^L(K, F^*, X)) = \bar{\delta}_\sigma(K, X)$, the average linear σ-width) is said to be optimal.

Remark 6 If $P_a = I$ for all a, then $k(\varepsilon, a)$ is independent of a. Definition 9 then is reduced to the proper case. The notion given by Definition 9 makes sense essentially for the case $k(\varepsilon, a) \to +\infty$ ($\varepsilon > 0$, fixed, $a \to +\infty$).

If in Definition 9 we take $\mathscr{L}_\sigma$ instead of ц_σ, then we come to the notion of infinite-dimensional width due to Li Chun[22].

Defintion 9′ *Let* $K \subset X$ *be a subset and* $\Phi = \{\mathfrak{F}, \text{ц}\}$. *The quanlity*

$$\tilde{d}_\sigma(K, X) \triangleq \inf_{F\in \mathscr{L}_\sigma} E(K, F, X)$$

is said to be the infinite-dimensional σ-width (in the sense of Kolmogorov). Any $F^* \in \mathscr{L}_\sigma$ *for which* $\tilde{d}_\sigma(K, X) = E(K, F^*, X)$ *is said to be optimal*. The infinite-dimensional linear σ-width is similarly defined.

$\bar{d}_\sigma$ and $\tilde{d}_\sigma$($\bar{\delta}_\sigma$ and $\tilde{\delta}_\sigma$)may be comparable with each other if we use the Definition 7′. Under such situation we have

$$\bar{d}_\sigma(F, X) \leqslant \tilde{d}_\sigma(F, X),\ \bar{\delta}_\sigma(F, X),\ \leqslant \tilde{\delta}_\sigma(F, X).$$

Now we turn to the Gelfand width. There are two versions of its definition. The one given by Magril-I'llyaev[41] seems to be more convenient.

Defintion 10 *Let X be a B space, X^* is its conjugate space, and $L \subset X$ be a linear subspace, $\Phi = \{\mathfrak{F}, Ц\}$. Suppose there is a subset $A \subset X^*$ such that*

1) *the elements of A are linearly independent, and* $\operatorname{card} A \leqslant \aleph_0$,

2) $L = L(A) \triangleq \{x \in X : f(x) = 0, \forall f \in A\}$,

3) $\forall X_a \in Ц(a > 0)$, *denote*

$$X_a^{\perp} \triangleq \{f \in X^* : f(X_a) = \{0\}\},$$

and

$$l(a, L, X) \triangleq \operatorname{card}\{f : f \in A, f \notin X_a^{\perp}\}.$$

We assume that $l(a, L, X)$ is finite. Then $L = L(A)$ is said to be Φ-locally finite-codimensional (Φ-LFC). (In fact, for any $a > 0$, $L \cap X_a$ in X_a is finite-codimensional) codim $(L \cap X_a, X_a) \leqslant l(a, L, X)$. If, in addition to this, we have

$$\varliminf_{a \to \infty} \frac{l(a, L, X)}{a} = \sigma,$$

then σ is said to be the Φ- LFD condim-number of L. We write then $\overline{\operatorname{codim}}\, L = \sigma$, or more precisely, Φ-codim $(L, X) = \sigma$.

Denote $Ц_\sigma \triangleq \{L \subset X : \overline{\operatorname{codim}} L \leqslant \sigma\}$.

Definition 11 *Let $K \subset X$ be a given subset. The quantity*

$$\bar{d}^a(K, X) \triangleq \inf_{L \in Ц_a} \sup_{x \in L \cap K} \| x \|$$

is said to be the average σ-G width. Any $L^* \in Ц_a$ such that $\bar{d}^a(K, X) = \sup_{x \in L^* \cap K} \| x \|$ is said to be optimal.

In what follows we take $X = L_p(\mathbf{R})$, $P_a : f \mapsto x_a \cdot f$ to present some examples. These examples are taken from [22, 40, 41].

Example 5 Let $X(\mathbf{R})$ be a linear set of measurable real-valued functions defined on $\mathbf{R}$ which includes all continuous functions with compact supports. $\forall \sigma > 0, \varphi \in C_0(\mathbf{R})$ (i.e., φ has compact support). Then $\operatorname{span}\{\varphi(\cdot - k/\sigma) : k \in \mathbf{Z}\} \in \mathscr{S}_\sigma$. For if we suppose $\operatorname{supp}(\varphi) \subset [-b, b]$, $b > 0$, then for any $a > 0$ we have

$$P_a(\mathscr{F}_\sigma) \subset \operatorname{span}\{\varphi(\cdot - k/\sigma): k = -[(a+b)\sigma], \cdots, [(a+b)\sigma]\},$$

where we denote

$$\mathscr{F}_\sigma \triangleq \operatorname{span}\{\varphi(\cdot - k/\sigma)\}_{k\in\mathbf{Z}}.$$

Thus, we have

$$\underline{\lim}_{a\to+\infty} \frac{1}{2a}\dim(P_a(\mathscr{F}_\sigma)) \leqslant \overline{\lim}_{a\to+\infty} \frac{1}{2a}\dim(P_a(\mathscr{F}_\sigma)) \leqslant \overline{\lim}_{a\to+\infty} \frac{2[(a+b)\sigma]+1}{2a} = \sigma.$$

Here we have used $2a$ instead of a as the index of X_a, which is the measure of $[-a, a]$. This is more adequate to the specific case $X = X(\mathbf{R})$.

Special case. Suppose $r \geqslant 1$, and $h > 0$.

$$\mathscr{S}_{r-1,h^{-1}} = \{s \in C^{r-2}(\mathbf{R}): s^{(r)}(x)\mid_{(jh,jh+h)} = 0, j \in \mathbf{Z}\}$$

is LFD. It may be easily proved that

$$\underline{\lim}_{a\to+\infty} \frac{\dim(\mathscr{S}_{r-1,h^{-1}})_a}{2a} = \lim_{a\to+\infty} \frac{\dim(\mathscr{S}_{r-1,h^{-1}})_a}{2a} = h^{-1},$$

i.e., $\mathscr{S}_{r-1,h^{-1}}$ is h^{-1}-LFD.

Example 6 Let $p \in [1, +\infty]$, $X = L_p(\mathbf{R})$. Then for any $r \geqslant 1$,

$$\overline{\dim}(\mathscr{S}_{r-1,h^{-1}} \cap L_p(\mathbf{R})) = h^{-1}.$$

See Magail-I'llyaev[41], Chen[37].

Example 7 Suppose $\sigma > 0, 1 \leqslant p \leqslant \infty$. Denote

$$B_{\sigma,p} \triangleq \{f \in E_\sigma: f \in L_p(\mathbf{R})\}$$

(when $p = \infty$, $B_{\sigma,\infty} = B_\sigma$). Then

$$\overline{\dim} B_{\sigma,p} = \sigma/\pi.$$

This example is due to Din Zung[61]. It should be noted that $B_{\sigma,p}$ is not LFD.

Example 8 Let $p \in [1, +\infty]$, $r \geqslant 1$, $X(\mathbf{R}) = L_p(\mathbf{R})$. Let $\xi = \{\xi_j\}_{j\in\mathbf{Z}}$ satisfy

1) $\xi_j < \xi_{j+1}, j \in \mathbf{Z}$,

2) $\xi_j \to -\infty (j \to -\infty)$, $\xi_j \to +\infty (j \to +\infty)$.

Corresponding to each $j \in \mathbf{Z}$, there is a non-empty subset of integers $e_j \subset \{0, 1, \cdots, r-1\}$. We consider a set of linear functionals as follows. Let the integers $k_{1,j}, \cdots, k_{|e_j|,j} \in e_j$. We denote $f_{k_{i,j}} \in A(i = 1, \cdots, |e_j|)$, if $\forall x \in L_p^r(\mathbf{R})$, $\langle f_{k_{i,j}}, x\rangle = x^{(k_i)}(\xi_j), i = 1,2,\cdots, | e_j |$. A is a denumerable set. For any $a > 0$, we have

$$\operatorname{card}\{f_{k_{i,j}} \in A : f_{k_{i,j}} \notin (L_p^r(\mathbf{R}))_a^{\perp}\} = \sum | e_j |,$$

where the summation extends over all j such that $\xi_j \in [-a, a]$, so that the cardinality is finite. This means that $L(A) = \{x \in L_p^r(\mathbf{R}) : x^{(k_i)}(\xi_i) = 0, \forall k_i \in e_j, i = 1,\cdots, | e_j |, j \in \mathbf{Z}\}$ is locally finite-codimensional.

If we take $\{\xi_j\}_{j\in\mathbf{Z}}$ satisfying

$$\lim_{a\to+\infty} \frac{\operatorname{card}\{\xi \cap [-a, a]\}}{2a} = \sigma$$

and $e_j = \{0\}$ for all $j \in \mathbf{Z}$, then $| e_j | = 1$, and

$$\overline{\operatorname{codim}}(L(A), L_p^r) = \sigma.$$

If $\xi_j = \alpha + jh, h > 0$ fixed, then we have

$$\overline{\operatorname{codim}}(L(A), L_p^r) = h^{-1}.$$

This example is taken from [41].

3.2 Theorem concerning *n*-width of a ball

Theorem concerning n-width of a ball is very important in the theory of n-widths. In this subsection we are going to discuss the possibility of extending it to the infinite-dimensional case. For definiteness, we confine ourselves to the average n-width in the sense of Tikhomirov.

Problem 2 Let $F \subset X$ be a linear subspace of Φ-$\overline{\dim}F = \sigma > 0$. For any real number $n, 0 < n < \sigma$, and linear subspace $M \subset X$, if Φ-$\overline{\dim}\, M \leqslant n$, then, is it valid that $E(F^0, M, X) = 1$? where $F^0 = \{x \in F : \| x \| \leqslant 1\}$ is the unit ball of F.

Although the counterexample of the problem has not been found, we may deem that the above conclusion is not valid under so extensive condition. We introduce a Condition S to ensure its validity. For any $\varepsilon \in (0, \varepsilon_0)$, by definition, we have

$$\lim_{a\to+\infty} \frac{k(a,\varepsilon)}{a} = \sigma > n.$$

We may choose a suffcient small $\delta > 0$ such that

$$a(n+\delta) < a(\sigma - \delta) < k(a, \varepsilon) < a(\sigma + \delta)$$

holds for any $a \geqslant a_0$, where a_0 is a sufficient large positive number.

Condition S For any $\eta \in (0,1)$, there exists a $A > 0$ such that for any $a \geqslant A$ there is a finite dimensional subspace $F(a)$ of $\dim F(a) = l \geqslant a(\sigma - \delta)$ satisfying the following conditions:

1) $\dim(F(a))_a \geqslant a(\sigma - \delta)$,

2) $\| x_a \| \geqslant \| x \| (1 - \eta)$, for any $x \in F(a)$.

The meaning of the Condition S is: for any $a \geqslant A$ we may find a linear subspace of F such that its dimension $\dim F$ approximately equals $k(a, \varepsilon)$, and moreover, $\dim F$ and the unit ball of F are "almost" unchanged under the mapping of the operator P_a ($F(a)$ contains a ball of radius $1 - \eta$, while its dimension $\geqslant a(\sigma - \delta)$).

We may prove

Proposition 1 *Let F be a linear subspace of X of Φ-$\overline{\dim}\, F = \sigma > 0$, and satisfying Condition S. Then, for any $n \in (0, \sigma)$ and linear subspace $M(\subset X)$ of Φ-$\overline{\dim}\, M \leqslant n$, it holds that $E(F^0, M, X) = 1$, consequently, $\bar{d}_n(F^0, X) = 1$.*

For applying this proposition, it is necessary to verify Condition S. One typical example is given as follows.

Let $\sigma > 0$. In example 6 we have seen that

$$\overline{\dim}(B_{\sigma\pi, p}) = \sigma.$$

Here $X = L_p(\mathbf{R})$, and $P_a: f \to \chi_a f$ is the restriction of f on the interval $[-a, a]$. Din Zung[61] proved the following facts.

Lemma 7 *Let $n \in (0, \sigma)$, $\sigma > 0$ and $\sigma - \delta > n$. For any $\beta \in (0, \sigma)$, set $t_j = j(\sigma - \beta)^{-1}$, $\forall j \in \mathbf{Z}$, and put*

$$\varphi_j(t) = \frac{\sin(\sigma - \beta)\pi(t - t_j)\sin \beta\pi(t - t_j)}{\beta(\sigma - \beta)(t - t_j)}, j \in \mathbf{Z}.$$

For any $a > 0$, set

$$S(a) = \operatorname{span}\{\varphi_j(\cdot) : |j| \leqslant [a(\sigma - \delta)]\}.$$

Then

1) $S(a) \subset B_{\sigma, p}$, $\forall a > 0$.

2) $\dim S(a) \geqslant [2a(\sigma - \delta)]$.

3) *For any* $\eta > 0$, *there exists a* $A > 0$ *such that for any* $a > A$ *and* $x(\cdot) \in S(a)$ *it holds that*

$$\| \chi_a x \| \geqslant \| x \|_p (1 - \eta).$$

This shows that $B_{\sigma\pi}$ satisfies the Condition S in $L_p(\mathbf{R})$.

Hence, we have

Corollary 4 *For any* $n \in (0, \sigma)$,

$$\bar{d}_n((B_{\sigma\pi,p})^0, L_p(\mathbf{R})) = 1.$$

If $F \subset X$ is a Φ-LFD, then the following proposition may be useful for obtaining a lower bound for the average width of F^0.

Proposition 2[41] *Let F be a LFD linear subspace of X, such that*

$$\lim_{a \to +\infty} \frac{\dim F_a}{a} = n \geqslant 0.$$

If F satisfies the following condition:

Condition M *If for any* $a > 0$, *there exists a bounded linear operator* $\Lambda_a : F_a \to F$ *such that* $P_a \cdot \Lambda_a = I_a$ (I_a *is the identity operator on* F_a).

Then

$$\bar{d}_n(F^0, X) \geqslant \varliminf_{a \to +\infty} \| \Lambda_a \|^{-1}.$$

The proposition may be used when F is the cardinal $\mathscr{L}$-spline subspace $\mathscr{S}_m \cap L_p(\mathbf{R})$ which is LFD. It is easy to verify that

$$\lim_{a \to +\infty} \frac{\dim(\mathscr{S}_m)_a}{2a} = \overline{\dim}(\mathscr{S}_m) = 1,$$

and $\mathscr{S}_m \cap L_p(\mathbf{R})$ satisfies the Condition M in $L_p(\mathbf{R})$ (See [41]). So we have

Corollary 5 *If* $0 < n < 1$, *then*

$$\bar{d}_{\mathrm{n}}((\mathscr{S}_m \cap L_p(\mathbf{R}))^0, L_p(\mathbf{R})) = 1,$$

where m is a natural number, $1 \leqslant p \leqslant +\infty$.

Concerning the theorem of average n-width of a ball, there are some other examples, but we no longer give unnecessary details here. In the following, we turn over to narratting concrete results. A complete result is the exact estimates of the average widths of Sobolev class $W_p(P_r)$ in $L_p(\mathbf{R})$ $(1 \leqslant p \leqslant \infty)$, which are obtained by Li Chun[36], Chen Dirong[37] and Magaril-I'llyaev[41] independently. Among these works, Chen's work seems to be the most extensive.

3.3 A variation problem

Let I denote a finite interval on the real line or $[0,1)$, $1 \leqslant p, q \leqslant +\infty$, $r \in \mathbf{Z}_+$. Consider the following extremal problem on the Sobolev class $W_p^r(I)$ or the periodic class $\widetilde{W}_p^r[-1,1]$:

$$\| f \|_{q(I)} \to \sup, f \in W_p^r(I) (\text{or } \widetilde{W}_p^r[-1,1]).$$

When $1 < p, q < +\infty$, this is an isoperimetric problem in the classical calculus of variations. Buslaev and Tikhomirov[47] discovered that when $p \geqslant q$ there exists an essential connection between Kolmogorov and linear widths $d_n(W_p^r(I), L_q)$, $\delta_n(W_p^r(I), L_q)$ and the eigenvalues of the Euler equation of the above variation problem. But, besides the case $p = q = 2$, for other values of p, q, the Euler equation is nonlinear, so the study of its eigenvalues and eigenfunctions is very difficult. Buslaev and Tikhomirov[47] made an outstanding contribution in its investigation. Among other things, one important result of their joint work is the complete solution of the exact estimates of $d_n(W_p^r(I), L_q)$ and $d_n(\widetilde{W}_p^r[-1,1], \widetilde{L}_q)$ for the case $p \geqslant q$. Meanwhile, in 1985, A. Pinkus[42] from a different view point completely solved the problem of exact estimation of $d_n(W_p^r(I), L_p)$ for $p \in (1, \infty)$, $p \neq 2$. In 1983, A. Pinkus[14] posed his famous conjectures about the n-widths of periodic convolution class of functions with B-kernel of NCVD kernel. Great efforts are made by Chen Hanlin, Li Chun[62], Chen Dirong[43], Fang Gensun[63] and others for confirming Pinkus' conjecture. In this respect some essential progress has been made by these authors. In this subsection, we first introduce an "isoperimetric problem" on a periodic convolution class of functions with a B-kernel or NCVD kernel. It is an extension of the same problem considered by Buslaev and Tikhomirov[47] on the Sobolev class. Its solution relies essentially on the variational arguement and has a deep connection with the n-width problem for periodic function classes as well as for function classes on $\mathbf{R}$.

Definition 12 *We say that a 2-periodic continuous function G with*

B-property, if for any $m \in \mathbf{Z}_+$, and $x_1 < x_2 < \cdots < x_m < x_1 + 2$, the linear set

$$X_m = \left\{ b + \sum_{j=1}^{m} b_j G(x - x_j) : \sum_{j=1}^{m} b_j = 0 \right\}$$

is of m-dimension, and moreover, for odd m, X_m is a weak Haar subspace.

A 2-periodic continuous function $K(x)$ is said to be (Non-degenerate) cyclic variation-diminishing (NCVD), if

1) for any $m \in \mathbf{Z}_+$ and for every choice of $x_1 < x_2 < \cdots < x_m < x_1 + 2$,

$$\dim \operatorname{span}\{K(x_1 - \cdot), K(x_2 - \cdot), \cdots, K(x_m - \cdot)\} = m,$$

2) for any 2-periodic piecewisely continuous function $h(t)$,

$$S_c(K * h) \leqslant S_c(h),$$

where

$$(K * h)(x) = \int_0^2 K(x - t) h(t) \mathrm{d}t,$$

and $S_c(f)$ denotes the cyclic sign changes of f in a periodic interval.

The CVD property of a function may be strengthened to become ECVD. Some details we refer to Pinkus[14].

An isoperimetric problem on a periodic convolution class

Let $p, q \in (1, +\infty)$, $n \in \mathbf{Z}_+$, and G, K be given as in Definition 12. Consider the extremal problems:

$$\lambda_n(p, q, K) = \sup\{ \| K * h \|_q ; h \in D_n \},$$

$$\lambda_n(p, q, G) = \sup\{ \| G * h \|_q ; h \in D_n \},$$

where

$$D_n \triangleq \{ h(t) : \| h \|_{p[-1,1]} \leqslant 1,$$

$$h\left(t + \frac{1}{n}\right) \overset{\text{a.e.}}{=} - h(t), h(t) \sin n\pi t \geqslant 0 \text{ a.e.} \}.$$

The following results are owing to Chen Dirong[43]. Similar results are also obtained by Fany Gensun[63]. We state the theorem only in B-kernel case. In NCVD kernel case the conclusion holds analogously.

Theorem 20 *For any $n \in \mathbf{Z}_+$, $\lambda_n > 0$ holds and there exists a continuous function $h_n(x) \in D_n$ such that*

1) $\| G * h_n \|_q = \lambda_n(p,q,G) = \lambda_n \| h_n \|_p$, (*that is*, $\| h_n \|_{p[-1,1]} = 1$).

2)

$$\int_0^2 G(x-y) \left| (G * h_n)(x) \right|^{q-1} \operatorname{sgn}[(G * h_n)(x)] \mathrm{d}x$$
$$= \lambda_n^q \left| h_n(y) \right|^{p-1} \operatorname{sgn}[h_n(y)], \ \forall\, y \in [0,2].$$

3) *For each* $n \in \mathbf{Z}_+$, *there is a sign* ε_n ($+1$, *or* -1 *fixed*) *and some* $\alpha_n \in [0, 1/n)$ *such that*

$$\varepsilon_n (G * h_n)(x) \operatorname{sgn} \sin n\pi(x - \alpha_n) \geqslant 0.$$

Remark 7 When G is the Bernoulli function defined by a polynomial $P_r(\lambda)$, we write $\lambda_n(p,q,P_r)$ (as in (5)) instead of $\lambda_n(p,q,G)$.

Theorem 20 includes the result of Buslaev and Tikhomirov on the Sobolev class $\widetilde{W}_p^r[-1,1]$. It is interesting to compare Theorem 20 with that of $B-T$. The kernel of the class $\widetilde{W}_p^r[-1,1]$ is

$$B_r(x) = \pi^{-r} \sum_{k=1}^{\infty} k^{-r} \cos\left(k\pi x - \frac{\pi r}{2} \right).$$

Buslaev and Tikhomirov proved: for any $p, q \in (1, +\infty)$, $n = 1$,

$$\sup\{ \| x(\cdot) \|_{q[-1,1]} : \| x^{(r)}(\cdot) \|_{p[-1,1]} \leqslant 1,$$
$$x^{(j)}(1 - (-1)^j) = 0, j = 0, 1, \cdots, r-1 \} = \lambda_1(p,q,r).$$

Theorem 21 *Let G be a B-kernel*, $1 < p < \infty$, $n \in \mathbf{Z}_+$. *Set*

$$\widetilde{B}_p = \{ a + G * h : a \in \mathbf{R}, \| h \|_{p[-1,1]} \leqslant 1, \int_0^2 h(t) \mathrm{d}t = 0 \}.$$

Then, we have

$$\begin{aligned} d_{2n}(\widetilde{B}_p, L_p) &= d^{2n}(\widetilde{B}_p, L_p) \\ &= \delta_{2n}(\widetilde{B}_p, L_p) \\ &= \lambda_n(p,p,G) \\ &\leqslant b_{2n}(\widetilde{B}_p, L_p), \end{aligned}$$

where δ_{2n}, b_{2n} *are the linear width and the Bernstein width, respectively. Moreover,*

1) $W_{2n} = \operatorname{span}\{1, G(x), G(x - 1/n), \cdots, G(x - (2n-1)/n\}$ *is an optimal subspace of* d_{2n} *and* δ_{2n}.

2) $L^{2n} = \{ f \in \widetilde{B}_p : f(i/n) = 0, i = 0, \cdots, 2n-1 \}$ *is an optimal*

subspace of d^{2n}.

3) *there exists a linear interpolation operator* (*rank* $2n$) $P_{2n}(f):\widetilde{B}_p \to W_{2n}$ *attainning* δ_{2n} (*when G is the generalized Bernoulli kernel defined by the differential operator* $P_r(D) = \prod_{j=1}^{r}(D - \lambda_j)$, $P_{2n}(f)$ *is the interpolation operator* $R_{r-1,n}(f)$ *in Theorem* 10).

The above results come from Chen[43] (When the kernel is odd, it has been seen in Chen and Li[62]). When the average width of the Sobolev class $W_p(P_r)$ is considered, its lower estimate depends essentially on the results of Theorem 21.

3.4 The exact estimate of the average width $W_p(P_r)$ in $L_p(\mathbf{R})$

This problem is completely solved independently by Chen Dirong[37], Li Chun[64] and Magaril-I'llyaev[41]. Li and Magaril-I'llyaev considered the function class $W_p^r(R)$ defined by $P_r(D) = D^r$. Li considered the infinite dimensional width by his Definition 9′. While Chen considered the more general case $P_r(D) = \prod_{j=1}^{r}(D - t_j)$, where the function class is more extensive. In addition to this, Chen's results may be further extended to some class of infinite-order differentiable functions on the real line (see 5).

Theorem 22 *Let* $\sigma > 0$, $p \in [1, +\infty]$. *Then*

$$\bar{d}_\sigma(W_p(P_r), L_p) = \bar{\delta}_\sigma(W_p(P_r), L_p) = \tilde{d}_\sigma(W_p(P_r), L_p) = \tilde{\delta}_\sigma(W_p(P_r), L_p) = \bar{d}^\sigma(W_p(P_r), L_p) = \lambda_1(p, p, P_{r,\sigma^{-1}}),$$

$S_{r-1,\sigma^{-1}} \cap L_p(\mathbf{R})$ *is an optimal subspace of* $\bar{d}_\sigma, \tilde{d}_\sigma, \bar{\delta}_\sigma, \tilde{\delta}_\sigma$, *respctively. The cardinal* $\mathscr{L}$*-spline interpolation operator* $s_{r-1,\sigma^{-1}}$ *on* $S_{r-1,\sigma^{-1}}$ *is an optimal linear operator of* $\bar{\delta}_\sigma$ *and* $\tilde{\delta}_\sigma$ *respectively, where* $P_{r,\sigma^{-1}}(D) = P_r(\sigma D)$.

Meantime, $L = \{f \in W_p(P_r): f(j/\sigma) = 0, \forall j \in \mathbf{Z}\}$ is an optimal subspace of $\bar{d}^\sigma$. When $P_r(D) = D^r$, there holds

$$\lambda_1(1, 1, P_{r,\sigma^{-1}}) = \lambda_1(\infty, \infty, P_{r,\sigma^{-1}}) = \sigma^{-r}\pi^{-r}\mathscr{K}_r,$$

$$\lambda_1(2, 2, P_{r,\sigma^{-1}}) = (\sigma\pi)^{-r}.$$

If $r = 1$, $P_1(D) = D$, then for any $p \in (1, +\infty)$,

$$\lambda_1(p, p, P_{1,\sigma^{-1}}) = p\sin\frac{\pi}{p}\Big/2\sigma\pi(p-1)^{1/p}$$

(see [41]).

When $p = 1, 2, +\infty$, more of optimal subspaces of $\bar{d}_\sigma, \tilde{d}_\sigma, \bar{\delta}_\sigma, \tilde{\delta}_\sigma$ are known. In fact, the corollary of Theorem 12 (see [48]) gives

Theorem 23 *In* Theorem 22, *if* $p = 1, 2, +\infty$, *then for each natural number* $m \geqslant r$, $S_{m-1,\sigma^{-1}} \cap L_p(\mathbf{R})$ *is an optimal subspace of* $\bar{d}_\sigma$ *and* $\tilde{d}_\sigma$ *respectively*, *where*

$$S_{m-1,\sigma^{-1}} = \{s \in C^{\deg(P)-2} : P(D)s(t) \equiv 0,$$
$$t \in (j/\sigma, (j+1)/\sigma)), \forall j \in \mathbf{Z}\},$$

while $P(t)$ *is a polynomial which has only real zeros and has* $P_r(t)$ *its factor*.

Moreover, *Theorem* 13 *shows that when* $p = 2$ the natural spline subspace is optimal for both $\bar{d}_\sigma, \tilde{d}_\sigma$ and $\bar{\delta}_\sigma, \tilde{\delta}_\sigma$ (see [40]). In addition to this, when $p = 2$, Chen and Magaril-I'llyaev[65] found the classical optimal subspace $B_{\sigma\pi,2}$ and constructed a linear optimal operator by making use of Fourier transform.

On the proof of Theorem 22

The main steps of the proof of Theorem 22 consist of:

1) The upper estimate. For any $\sigma > 0$, we have

$$\begin{aligned}\bar{d}_\sigma(W_p(P_r), L_p) &\leqslant \bar{\delta}_\sigma(W_p(P_r), L_p)\\ &\leqslant E^L(W_p(P_r), S_{r-1,\sigma^{-1}} \cap L_p, L_p)\\ &= \sup_{f\in W_p(P_r)} \| f - s_{r-1,\sigma^{-1}}(f) \|_p\\ &= \lambda_1(p, p, P_{r,\sigma^{-1}}).\end{aligned}$$

The last step may be seen in Theorem 9′.

The same conclusion as above holds for $\tilde{d}_\sigma$ and $\tilde{\delta}_\sigma$.

2) The lower estimate is more complicated. Several crucial points are:

a) Locallization. For any given interval $I_N = [-N, N] \subset R$, consider the set

$$W_{p(I_N)}(P_r) \triangleq \{f \in W_p(P_r) : \operatorname{supp} f \subset [-N, N]\}.$$

b) To substitute LFD subspace for an approximating subspace with finite average dimension.

Let A be a linear subspace of $L_p(\mathbf{R})$ of average dimension $\overline{\dim}(A) \leqslant \sigma$. According to Definition 5, there exists a finite-dimensional linear subspace L_N of $L_p(I_N)$ on I_N such that the dimension of L_N is asymptotically equal to $k(\varepsilon, N) \lessapprox 2N\sigma$, as $N \to +\infty$.

c) Consider $E(W_{p(I_N)}(P_r), L_N, L_p(I_N))$. By a change of variable, we may reduce the function class $W_{p(I_N)}(P_r)$ to the "normal" case $I_1 = [-1, 1]$. Then from Theorem 21 we obtain the asymptotic value of the lower estimates of average width.

d) Letting $N \uparrow \infty$, the sharp estimates from below are obtained.

When $\tilde{d}_\sigma$ is considered, a) is not necessary. Hence, in this case it is relatively easy to get the estimate from below.

3.5 The case $p > q$.

As it has been mentioned that when $p > q \geqslant 1$ in order to get some meaningful results it is necessary to consider the $S - W$ class $W_{pq}(P_r)$.

This moment, by Theorem 11 the upper estimate may be obtained. But the lower estimate problem is more complicated. Except the two cases $1 \leqslant q \leqslant p = \infty$ and $1 = q \leqslant p \leqslant +\infty$, in general, first it is necessary to prove the validity of equalities in Theorem 21 for $p > q$ i.e.,

$$d_{2n}(\tilde{B}_p, L_q) = d^{2n}(\tilde{B}_p, L_q) = \delta_{2n}(\tilde{B}_p, L_q) = \lambda_n(p, q, G). \quad (18)$$

Fang Gensun in a recent work[63] proved (18). On this basis, Fang and Liu[46], Jiang and Liu[66] completely solved the exact estimate problem of average n-width of $W_{pq}^r(\mathbf{R})$ in $L_q(\mathbf{R})$.

Theorem 24 *Let $1 < q \leqslant p < \infty$, $n \in \mathbf{Z}_+$. Then*

$$\begin{aligned} \bar{d}_n(W_{pq}^r, L_q) &= \bar{\delta}_n(W_{pq}^r, L_q) = \tilde{d}_n(W_{pq}^r, L_q) \\ &= \bar{d}^n(W_{pq}^r, L_q) = \tilde{\delta}_n(W_{pq}^r, L_q) \\ &= \sup_{f \in W_{pq}^r} \| f - s_{r-1, n^{-1}}(f) \|_q = n^{-r}\lambda_1(p, q, r), \end{aligned}$$

where $S_{r-1, n^{-1}} \cap L_q(R)$ is an optimal subspace, while the cardinal $\mathscr{L}$-spline interpolation operator $s_{r-1, n^{-1}}(f)$ is an optimal linear operator attainning $\bar{\delta}_n$ and $\tilde{\delta}_n$.

For any real number $\sigma \geqslant 1$, the strong asymptotic equalities are obtained:

$$\begin{aligned}\bar{d}_\sigma(W^r_{pq}, L_q) &\approx \bar{\delta}_\sigma(W^r_{pq}, L_q) \approx \bar{d}^\sigma(W^r_{pq}, L_q) \\ &\approx \tilde{d}_\sigma(W^r_{pq}, L_q) \approx \tilde{\delta}_\sigma(W^r_{pq}, L_q) \\ &\approx \sigma^{-r}\lambda_1(p, q, r), (\sigma \to +\infty). \qquad (19)\end{aligned}$$

A "dual" version of Theorem 23 may be formulated below.

Theorem 24′ *Let* $1 < q \leqslant p < \infty, n \in \mathbf{Z}_+$. *Then*

$$\bar{d}_n(W^r_p, L_{qp}) = \bar{\delta}_n(W^r_p, L_{qp}) = \tilde{d}_n(W^r_p, L_{qp}) = \bar{d}^n(W^r_p, L_{qp})$$

$$= \tilde{\delta}_n(W^r_p, L_{qp}) = \sup_{f \in W^r_p} \| f - s_{r-1, n^{-1}}(f) \|_{qp}$$

$$= n^{-r}\lambda_1(q', p', r) = n^{-r}\lambda_1(p, q, r) \quad (1/p + 1/p' = 1/q + 1/q' = 1).$$

$S_{r-1, n^{-1}} \cap L_{qp}(\mathbf{R})$ *is an optimal subspace, while the cardinal* $\mathscr{L}$*-spline interpolation operator* $s_{r-1, n^{-1}}(f)$ *is an optimal linear operator attainning* $\bar{\delta}_n$ *and* $\tilde{\delta}_n$.

For any $\sigma \geqslant 1$, there is a strong asymptotic equality similar to (19).

Remark 8 The two special cases $1 < q \leqslant p = \infty$ and $1 = q \leqslant p < \infty$, had already been discussed in [44, 45, 66, 67]. It is not difficult for us to extend Theorem 24 and Theorem 24′ to the class $W_{pq}(P_r)$, where $P_r(D) = \prod_{j=1}^{r}(D - \lambda_j), \lambda_j \in \mathbf{R}$. In this case, to get the sharp estimate from below, we must essentially rely on Fang's results[63] (see (18)) instead of the result of Buslaev and Tikhomirov[47].

Problem 3 Let $1 \leqslant q \leqslant p \leqslant \infty$. For any nature number $m > r$, is $S_{m-1, n^{-1}} \cap L_q(\mathbf{R})$ an optimal subspace of $\bar{d}_n(W^r_{pq}, L_q)$ and $\tilde{d}_n(W^r_{pq}, L_q)$ or not? When $p = q \in (1, \infty) \setminus \{2\}$, this problem has a deep relation with the Kolmogorov-Stein inequality.

3.6 The case $p < q$

This case remains to be studied. So far what we have known is very little, except the results published by G. G. Margaril-I'llyaev[41] and Liu Yongping[68]. This moment we may posed the following

Problem 4 Prove that

$$\bar{d}_\sigma(W^r_1(\mathbf{R}), L_2) \approx \tilde{d}_\sigma(W^r_1(\mathbf{R}), L_2) \approx \frac{1}{\pi^r\sqrt{2r-1}}\sigma^{-r+1/2} \quad (\sigma \to +\infty).$$

It's known that these asympotic equalities hold true for $W_1^r[0,1]$ in $L_2[0,1]$(See[69]).

4. Optimal recovery by using linear information with countable infinite cardinality

4.1 Basic concepts and the formulation of problems

Let $D \subset \mathbf{R}$ be an interval and $X(D)$ a linear subset of $C(D)$ with norm $\|\cdot\|_X$. Following Traub[70][107], let $A \subset X(D)$ be a set of problem elements, and a mapping $S: \mathscr{D} \to G$ be a solution operator, where $A \subset \mathscr{D} \subset X(D)$ and G is a linear normed space. We shall call Sx the solution element for $x \in A$. Let Y be a linear space and $I: A \to Y$ an information operator. For $x \in A$, Ix is called the information of x. When $D = [a,b]$ is finite and A is compact in $X(D)$, for recovering S it is enough to use information operator of finite rank (i.e., dim span $I(A) < \infty$). In this section we mainly concern about the case $D = \mathbf{R}$ and $A \subset X(\mathbf{R})$ being noncompact. Therefore, the information operator with infinite cardinality $\aleph_0$ should be used. Now we introduce a special family of such information operators (which are called samplings) as follows.

Let $T = \{t_j\}_{j \in \mathbf{Z}}$ with $t_j < t_{j+1}$ for $j \in \mathbf{Z}$. The mapping

$$I_T: x(t) \to \{x(t_j)\}_{j \in \mathbf{Z}}$$

is called the sampling of x at T for $x \in X(\mathbf{R})$. Let $Y = \{s = \{s_j\}_{j \in \mathbf{Z}}: s_j \in \mathbf{R}\}$. Then $I_T: X(\mathbf{R}) \to Y$ is a linear information operator.

Definition 13 *Given* $\sigma > 0$. *We say* $T \in \Theta_\sigma$ *if*

1) $t_j < t_{j+1}, j \in \mathbf{Z}$,

2) $\lim\limits_{j \to -\infty} t_j = -\infty, \lim\limits_{j \to \infty} t_j = \infty$,

3) $\lim\limits_{a \to \infty} \dfrac{\operatorname{card}\{T \cap [-a,a]\}}{2a} \leqslant \sigma$.

We denote Θ_1 by Θ for simplicity. The union of all $\Theta_\sigma, \sigma > 0$, is denoted by Ξ.

The condition 3) of Definition 13 is a density condition. Ξ contains sufficient many sets of countable points on $\mathbf{R}$.

Example 9 A subset $\tilde{\Theta}$ of Θ is defined as follows. $T = \{t_j\}_{j\in\mathbf{Z}} \in \tilde{\Theta}$ if and only if

1) there exists a $k \in \mathbf{Z}_+$ such that

$$-k \leqslant t_{-k+1} < t_{-k+2} < \cdots < t_{k-1} < t_k < k.$$

2) $t_{j+2k} = 2k + t_j$ for $j \in \mathbf{Z}$.

Similarly, we can define $\tilde{\Theta}_\sigma$.

Following Traub and Wozniakowski[70][107], we consider the problem of optimal recovery (S, A, I, G): optimal recovery of solution operator S on the set of problem elements A in the metric of G by using information I. Now we list some basic concepts as follows.

Definition 14 *Given $A \subset X(D)$, $S: A \to G$, $T \in \Theta_\sigma$, $\sigma > 0$ fixed, $I_T: C(\mathbf{R}) \to Y$ and $y \in Y$, we define*

$$U(y) = \{Sx : x \in A, I_T(x) = y\},$$

$$r(U(y)) = \inf_{y\in G}\ \sup_{x\in A, I_T(x)=y} \| Sx - y \|$$

and call $r(U(y))$ the radius of $U(y)$ in G.

The radius of information I_T for S is

$$r(I_T, S) = \sup_{y\in I_T(A)} r(U(y)).$$

The diameter of information I_T for S is

$$d(I_T, S) = \sup_{y\in I_T(A)} d(U(y)) \triangleq \sup_{y\in I_T(A)}\ \sup_{x_1, x_2\in A, I_T(x_i)=y} \| Sx_1 - Sx_2 \|.$$

Proposition 3 *If A is a balanced convex subset of $X(D)$ and S is a linear operator, then we have*

$$d(I_T, S) = 2 \sup_{h\in A, I_T(h)=0} \| Sh \|,$$

where 0 is the zero element in Y.

In this section we only consider linear problems and mainly concern about the value of $d(I_T, S)$.

Definition 15 *We call*

$$D_\sigma(A, S, G) \triangleq \inf_{T\in\Theta_\sigma} d(I_T, S)$$

the σ-minimal information diameter.

Special case Let $D = \mathbf{R}$, $G = X(\mathbf{R})$ and S be the identity operator.

The *σ-widths of mesh of A* in $X(\mathbf{R})$ is defined by

$$D_\sigma(A, X) \triangleq \inf_{T \in \Theta_\sigma} \mathrm{d}(I_T, S)$$

(The term of σ-width of mesh is refered to Babenko[71]).

Definition 16 *Any operator $\varphi: I_T(A) \to G$ is called an algorithm. The set of all algorithms is denoted by Φ_T. Φ_T^L is the subset of Φ_T composed of the algorithms which are additive and homogenous. The quantities*

$$E(A, S, I_T, G) \triangleq \inf_{\varphi \in \Phi_T} \sup_{x \in A} \| Sx - \varphi(I_T(x)) \|$$

and

$$E_L(A, S, I_T, G) \triangleq \inf_{\varphi \in \Phi_T^L} \sup_{x \in A} \| Sx - \varphi(I_T(x)) \|$$

are the intrinsic error and linear intrinsic error of the recovery problem (S, A, I, G) respectively.

Letting T runs over Θ_σ we denote

$$E_\sigma(A, S, G) \triangleq \inf_{T \in \Theta_\sigma} E(A, S, I_T, G),$$

and

$$E_\sigma^L(A, S, G) \triangleq \inf_{T \in \Theta_\sigma} E^L(A, S, L_T, G).$$

In what follows we denote $D_\sigma(A, S, G)$, $E_\sigma(A, S, G)$ and $E_\sigma^L(A, S, G)$ by $D(\sigma)$, $E(\sigma)$ and $E^L(\sigma)$ respectively.

From Traub-Wozniakowski[70][107] it follows

Proposition 4 *For $\sigma > 0$*

$$\frac{1}{2} D(\sigma) \leqslant E(\sigma) \leqslant E^L(\sigma).$$

What we consider in this section are the following:

Problems 5

1) To obtain the exact values of $D(\sigma)$, $E(\sigma)$ and $E^L(\sigma)$ for some important function classes defined on **R**, such as Sobolev classes and Sobolev-Wiener classes, or, at least, to clarify the asymptotic behaviour of those when $\sigma \to \infty$.

2) It is most interesting to clarify that under what conditions it holds

$$\frac{1}{2} D(\sigma) = E(\sigma) = E^L(\sigma).$$

For $\sigma > 0$ or $\sigma \in \mathbf{Z}_+$, or, under what condition it holds asymptotically

$$\frac{1}{2}D(\sigma) \approx E(\sigma) \approx E^L(\sigma), \sigma \to +\infty.$$

3) We call $T \in \Theta_\sigma$ an optimal sequence of sampling if T realizes $D(\sigma)$. Using an optimal sequence of sampling, the algorithm φ and linear algorithm φ^L which realize $E(\sigma)$ and $E^L(\sigma)$, respectively, are called an optimal algorithm and a linear optimal algorithm, respectively. It is most interesting to find an optimal sequence of sampling and the associated optimal and linear optimal algorithms.

The above problems were initiated to study by Sun[72] (See [73]). Many progress have been made when A are Sobolev classes or Sobolev-Wiener classes and $S = E$, the identity operator, or the differential operator D^k. The problem $(I, A, I_T, L_q(\mathbf{R}))$ is the problem of optimal interpolation, and $(D^k, A, I_T, L_q(\mathbf{R}))$ is the problem of optimal recovery of D^k. These problems are connected deeply with those discussed in section 2 and 3.

4.2 Results on Sobolev classes

Let $X(\mathbf{R}) = G = L_p(\mathbf{R})$, $S = E$(indentity), $A = W_p(P_r)$, $p \in [1, \infty]$, where $P_r(D) = \prod_{j=1}^{r}(D - \lambda_j)$, $\lambda_j \in \mathbf{R}$. This case were discussed by many authors (Sun[72] for $p = \infty$; Li[38] for $p = 1$; Sun and Li[39], and Chen[40] for $p = 2$; Li[36] and Chen[37] for $p \in (1, \infty)$).

Theorem 25 *For* $\sigma > 0$ and $p \in [1, \infty]$ *we have*

$$\begin{aligned}\frac{1}{2}D_\sigma(W_p(P_r), E, L_p) &= E_\sigma(W_p(P_r), E, L_p)\\ &= E_\sigma^L(W_p(P_r), E, L_p)\\ &= \lambda_1(p, p, P_{r,\sigma^{-1}}),\end{aligned}$$

where $P_{r,\sigma^{-1}}(D) = P_r(\sigma D)$ *and* λ_1 *is given in section* 3.

Moreover, there exists an optimal sequence of sampling $T_\sigma^* = \{\alpha + j\sigma^{-1}\}_{j\in\mathbf{Z}}$, where α depends on $P_{r,\sigma^{-1}}$ only. The interpolation operator

$$s_{r-1,\sigma^{-1}}(f, x) = \sum_{j\in\mathbf{Z}} f(\alpha + j\sigma^{-1})L_\sigma(x - j\sigma^{-1})$$

is an optimal algorithm. For $p = 2$, $\{j\sigma^{-1}\}_{j\in\mathbf{Z}}$ is an optimal sequence of sampling and the natural spline interpolation operator is also an optimal algorithm.

A special case of this problem $P_r(D) = D^r$ was also considered by G. G. Magaril-I'llyaev[41]. Now we turn to the case $S = D^k (1 \leqslant k \leqslant r-1, r \geqslant 2)$.

By making use of Landau-Kolmogorov type inequalities, Sun and Li[39] proved:

Theorem 26 *If* $r \geqslant 2$ *and* $1 \leqslant k \leqslant r-1$, *then* $(\sigma = 1)$

1) $$\frac{1}{2} D(W_p^r, D^k, L_p) = \sup_{f\in W_p^r, f(j)=0} \| f^{(k)} \|_p = \| E_r^{(k)}(\cdot) \|_\infty, p = 1, \infty,$$

2) $$\frac{1}{2} D(W_2^r, D^k, L_2) = \sup_{f\in W_2^r} \| f^{(k)} \|_2 = \pi^{-(r-k)}.$$

We know that for $p = 2$ it holds that

$$\frac{1}{2} D(W_p^r, D^k, L_p) = E(W_p^r, D^k, L_p) = E^L(W_p^r, D^k, L_p).$$

We don't know whether it holds for $p = 1, \infty$.

4.3 Results on Sobolev-Wiener classes

The following results were proved by Fang and Liu[46]

Theorem 27 *Let* $1 < q \leqslant p < \infty$,

1) $$\frac{1}{2} D_n(W_{pq}^r, E, L_q) = E_n(W_{pq}^r, E, L_q) = E_n^L(W_{pq}^r, E, L_q) = \sup_{f\in W_{pq}^r} \| f - s_{r-1, n^{-1}}(f) \|_q = n^{-r}\lambda_1(p, q, r), \text{ for } n \in \mathbf{Z}_+.$$

There exists an optimal sequence of sampling $T_n^* = \{a + jn^{-1}\}_{j\in\mathbf{Z}}$, *where* a *depends on* r *and* n. *The interpolation operator* $s_{r-1, n^{-1}}(f)$ *at* T_n^* *is an optimal algorithm.*

2) *For* $\sigma > 0$,

$$\sigma^{-r}\lambda_1(p, q, r) \leqslant \frac{1}{2} D_\sigma(W_{pq}^r, E, L_q) \leqslant E(W_{pq}^r, E, L_q) \leqslant E_\sigma^L(W_{pq}^r, E, L_q)$$

$$\leqslant \sigma^{-r}\lambda_1(p,q,r)+o(\sigma^{-r}),\ \sigma\to\infty.$$

Liu[45], Liu and Sun[44] considered the cases $1=q<p<\infty$ and $1\leqslant q<p=\infty$, respectively.

The dual case of Theorem 27 is

Theorem 27′ *Let* $1<q\leqslant p<\infty$.

1) *For* $n\in \mathbf{Z}_+$,

$$\begin{aligned}\frac{1}{2}D_n(W_p^r,E,L_{qp}) &= E_n(W_p^r,E,L_{qp})\\ &= E_n^L(W_p^r,E,L_{qp})\\ &= \sup_{f\in W_p^r}\|f-s_{r-1,n}{}^{-1}(f)\|_{qp}\\ &= n^{-r}\lambda_1(p,q,r).\end{aligned}$$

$$\begin{aligned}2)\ \sigma^{-r}\lambda_1(p,q,r) &\leqslant \frac{1}{2}D_\sigma(W_p^r,E,L_{qp})\\ &\leqslant E_\sigma(W_p^r,E,L_{qp})\\ &\leqslant E_\sigma^L(W_p^r,E,L_{qp})\\ &\leqslant \sigma^{-r}\lambda_1(p,q,r)+o(\sigma^{-r}),\ \sigma\to\infty.\end{aligned}$$

4.4 Results about multifold sampling

Let $r\geqslant 2, 1\leqslant k\leqslant r$, $T\in\Theta_\sigma$ and $X=X(\mathbf{R})\cap C^r(\mathbf{R})$. The mapping

$$I_T^k(f)=\{f(t_j),f'(t_j),\cdots,f^{(k-1)}(t_j)\}_{j\in\mathbf{Z}}$$

is a k-fold sampling of f at T. It may be viewed as an information operator. Given a set of problem elements $A\subset X$ and a solution operator S, we consider the optimal recovery problem (S,A,I_T^k,G). Assume A is balanced and convex in X and S is linear. Analogously to 4.1 we can define $r(I^k,S)$ and $d(I_T^k,S)$ respectively. Furthermore, we give

Definition 17 *For* $1\leqslant k\leqslant r$, *we set*

$$D_\sigma^k(A,S,G)=\inf_{T\in\Theta_\sigma}\mathrm{d}(I_T^k,S),$$

$$E_\sigma(A,S,G)_k=\inf_{T\in\Theta_\sigma}E(A,S,I_T^k,G),$$

$$E_\sigma^L(A,S,G)_k=\inf_{T\in\Theta_\sigma}E^L(A,S,I_T^k,G).$$

When $k=1$ we simply write $D_\sigma(A,S,G)$, $E_\sigma(A,S,G)$, $E_\sigma^L(A,S,G)$. These notations coincide with those of 4.1 when $k=1$. Similar to

Proposition 4 we have

Proposition 5 *For* $\sigma > 0$,

$$\frac{1}{2}D^k(\sigma) \leqslant E(\sigma)_k \leqslant E^L(\sigma)_k,$$

where $D^k(\sigma)$, $E(\sigma)_k$ *and* $E^L(\sigma)_k$ *denote* $D^k_\sigma(A, S, G)$, $E_\sigma(A, S, G)_k$ *and* $E^L_\sigma(A, S, G)_k$, *respectively*.

Now we may pose and study the problems as we did in 4.1. For differentiable function classes defined on a compact set, such as finite closed interval, unit circle or disc on complex plane these problems are investigated by many authors. In these cases it is enough to use finite set of sampling points. The first result is

Theorem 28 *Let* $r \geqslant 2, 1 \leqslant k \leqslant r$.

1) $\frac{1}{2}D^k(W^r_\infty, E, L_\infty) = E(W^r_\infty, E, L_\infty)_k = E^L(W^r_\infty, E, L_\infty)_k = \| E_{rk}(\cdot) \|_\infty$, *where* E_{rk} *is defined in* 2.

2) $T^* = \{j\}_{j\in \mathbf{Z}}$ *is an optimal sequence of sampling and the spline interpolation operator* $T_{rk}(f)$ *is an optimal algorithm*.

This theorem gives results for $\sigma = 1$. For any $\sigma > 0$, similar results hold by changing scale.

The upper bounds are obtained by Theorem 18. About the lower bounds there are two crucial points.

1) Extend $f\mid_{[-N,N]}$ with supp $f \subset [-N, N]$ to $\mathbf{R}$ by $2N$ period for $N \in \mathbf{Z}_+$.

2) Solve the problem of least L_∞ – norm on a class of periodic perfect splines with k-fold free knots.

In 2) we should use a result of Bojanov and Huang[56]. Given $r, k, 1 \leqslant k \leqslant r, N \in \mathbf{Z}_+$, we set

$$K_{2N} = \{f : f \in W^r_\infty, f(t + 2Nk) = f(t), \mid f^{(r)}(t) \mid = 1,$$
$$S_c(f^{(r)}, [-Nk, Nk)) \leqslant 2Nk\},$$

where $S_c(f)$ is the number of sign changes of f on a period.

Lemma 8

$$\min\{\|f\|_\infty : f\in K_{2N}, I_T^k(f)=0 \text{ for some } T\in\widetilde{\Theta}\} = \|E_{rk(\cdot)}\|_\infty,$$

where $T=\{t_j\}$ *satisfies*

1) $jk\leqslant t_j-\alpha<(j+1)k$ *for some* α *and* $j\in\mathbf{Z}$,

2) $t_{j+2kN}=t_j+2N$, $j\in\mathbf{Z}$.

By Lemma 8 and via the above two steps we can prove

$$\frac{1}{2}D^k(W_\infty^r,E,L_\infty):=\inf_{T\in\Theta}\sup\{\|f\|_\infty : f\in W_\infty^r, I_Tf=0\}$$
$$\geqslant\|E_{rk}(\cdot)\|_\infty.$$

For other of p and q, Sun and Liu[53] obtained

Theorem 29 *Let* $r\geqslant 2, 1\leqslant k\leqslant r, 1\leqslant p\leqslant\infty$.

1) $$\frac{1}{2}D^k(W_{\infty p}^r,E,L_p)\leqslant E(W_{\infty p}^r,E,L_p)_k$$
$$\leqslant E^L(W_{\infty p}^r,E,L_p)_k$$
$$\leqslant\|E_{rk}(\cdot)\|_{L_p[0,1]}.$$

2) *Equalities in* 1) *hold for* $k=2$.

The upper estimate depends on Theorem 19. The basic idea of the lower estimate is still the locallization and asymptotic approach. For any interval $[-N,N]$, we take a subset of $W_{\infty p}^r(\mathbf{R})$ in which every function f is a nonnegative perfect spline function with supp $f\subset[-N,N]$ and $[4N]$ knots, and the number of double zeros of f linearly depends on $[2N]$. According to A. A. Ligun's result[50], the minimum L^p-norm problem on this perfect spline class has solution. As soon as we get the strong asymptotic value of the minimum norm as $N\to+\infty$, the desired result will be obtained.

Developing the above results, Liu Yongping[59] obtained the exact result on the class $W_{pq}^r(\mathbf{R})$ for the case $p\geqslant q=1, 1\leqslant k\leqslant r$ and k even. He proved

Theorem 30 *Let* $1\leqslant p\leqslant+\infty, 1\leqslant k\leqslant r$ *and* k *be even. Then*

1) $$\frac{1}{2}D_n{}^k(W_{p1}^r,E,L)=E_n(W_{p1}^r,E,L)_k=E_n^L(W_{p1}^r,E,L)_k.$$

2) $$\frac{1}{2}D_n^k(W_p^r,E,L_{1p})=E_n(W_p^r,E,L_{1p})_k=E_n^L(W_p^r,E,L_{1p})_k.$$

Meantime, in the two cases he also gave an optimal set of sampling points and a linear optimal algorithm. The linear optimal algorithm is a Hermite-type cardinal spline interpolation operator, whose construction has a close connection with the eigen-functions of a nonlinear differential operator. And, the quantities of 1) and 2) are corresponding eigenvalues. The statement of the results is too complicated to be cited here.

4.5 Optimal quadrature problem

At present optimal quadrature is one of active areas in the optimal recovery theory. It is essentially related with some extremal problems in approximating of integrals. In last thirty years, very many profound results on the classes of smooth functions defined on a finite interval have been obtained, which are systemtically collected in Sun and Fang's book[13]. Now we begin to study the optimal quadrature problem on some classes of smooth functions defined on the whole real axis. So far, the known works are Fang and Liu[74] and Liu[75] in which the optimal quadrature problems with the information of Hermite-Birkhoff type on the Sobolev class $W_1^r(\mathbf{R})$ and the S-W class $W_{p1}^r(\mathbf{R})(1 < p \leqslant \infty)$ were studied. At the same time, Liu[75] considered the same problem on the function class $W^r_{\infty 1}H^\omega(\mathbf{R})$ (for the definition, see below). In these cases they have obtained the exact values of the optimal error and constructed the optimal quadrature formulas.

For a fixed natural number n, denote by $\mathscr{F}_{nk}(k = 1 \text{ or } 2)$ the sets of all sequence pairs (T, V), $T = \{t_j\}_{j\in\mathbf{Z}}$, $V = \{\nu_j\}_{j\in\mathbf{Z}}$, which satisfy the following conditions.

1) $t_j < t_{j+1}, t_j \in \mathbf{R}, \forall j \in \mathbf{Z}$,

2) $\nu_j < \mathbf{Z}_+, \nu_j \leqslant r, \forall j \in \mathbf{Z}$,

3) $\{\operatorname{card}(T \cap [j, j+1])\}_{j\in\mathbf{Z}}$ is a bounded sequence,

4) $\varlimsup\limits_{N\to\infty} \dfrac{\| (T, V) \|_{N,k}}{2N} \leqslant n$, where

$$\| (T, V) \|_{N,k} = \begin{cases} \sum\limits_{-N<t_j<N} \nu_j, & \text{if } k = 1; \\ \sum\limits_{-N<t_j<N} (\nu_j + \sigma_j), & \text{if } k = 2, \end{cases}$$

while $\sigma_j = 0$ or 1 according to ν_j even or not.

Denote by $\mathscr{A}$ the set of all matrices $A = \{a_{ij}\}_{j=0,j\in\mathbf{Z}}^{r}$ for which there exists a constant M (depending on A) such that $|a_{ij}| \leqslant M, i \in \mathbf{Z}, j = 0, 1, \cdots, r$.

We consider the following extremal problem:

$$\mathscr{R}(\mathscr{M},\mathscr{F}) =: \inf_{(T,V)\in\mathscr{F}} \inf_{A\in\mathscr{A}} \sup_{f\in\mathscr{M}} |R(f,T,V,A)|,$$

for $(\mathscr{M},\mathscr{F}) = (W_{p1}^r, \mathscr{F}_{n1})$ or $(W_{\infty 1}^r H^\omega, \mathscr{F}_{n2})$, where

$$\int_{\mathbf{R}} f(t)\mathrm{d}t = \sum_{i\in\mathbf{Z}} \sum_{j=0}^{\nu_i - 1} a_{ij} f^{(j)}(t_i) + R(f,T,V,A), \tag{20}$$

$$W_{\infty 1}^r H^\omega(\mathbf{R}) = \{f \in L^r(\mathbf{R}): \| f^{(r)}(\cdot + h) - f^{(r)}(\cdot) \|_{\infty 1} \leqslant \omega(|h|)\},$$
while $\omega(t)$ is a concave modulus of continuity.

We first show that the restriction of condition 3) is reasonable. In fact, we have

Lemma 9 *Let $T = \{t_j\}_{j\in\mathbf{Z}}$ satisfy* 1) *and* 2). *Then the following two assertions are equivalent*:

a) $\sum_{j\in\mathbf{Z}} |f(t_j)| < \infty$, for any $f \in L_1^r(\mathbf{R})$,

b) $\{\mathrm{card}\,(T \cap [j,j+1])\}_{j\in\mathbf{Z}}$ *is a bounded sequence*.

Furthermore, if b) *holds, then*

$$\sum_{i\in\mathbf{Z}} \sum_{j=0}^{\nu_i - 1} |a_{ij}| |f^{(j)}(t_j)| < \infty$$

holds for any matrix $A \in \mathscr{A}$, and any $f \in L_1^r(\mathbf{R})$.

Theorem 31 *Let $k = 1$. Among all quadrature formulas* (20) *which satisfy* 1) – 4), *the quadrature formula*

$$\int_{\mathbf{R}} f(t)\mathrm{d}t = \frac{1}{n}\sum_{i\in\mathbf{Z}} f\left(\frac{i}{n}\right) + R(f)$$

is the optimal quadrature formula on $W_{p1}^r(\mathbf{R})(1 \leqslant p \leqslant \infty, r \in \mathbf{Z}_+)$ and the optimal error

$$\mathscr{R}(W_{p1}^r, \mathscr{F}_{n1}) = \sup_{f\in W_{p1}^r} \left| \int_{\mathbf{R}} f(t)\mathrm{d}t - \frac{1}{n}\sum_{i\in\mathbf{Z}} f\left(\frac{i}{n}\right) \right|$$
$$= n^{-r} E_1(D_r)_{p'[0,1]}, 1/p + 1/p' = 1,$$

where

$$D_r(t) = -2\sum_{\nu=1}^{\infty} \frac{\cos(2\nu\pi + r\pi/2)}{(2\nu\pi)^r}.$$

Theorem 32 *Let $k = 2$. Among all quadrature formulas (20) which satisfy the condition 1) – 4) the quadrature formula*

$$\int_{\mathbf{R}} f(t)\mathrm{d}t = \frac{1}{n}\sum_{i\in\mathbf{Z}} f\left(\frac{i}{n}\right) + R(f)$$

is the optimal quadrature formula on $W^r_{\infty 1}H^{\omega}(R)(r = 1, 3, 5, \cdots)$ and the optimal error

$$R(W^r_{\infty 1}H^{\omega}, \mathcal{F}_{m2}) = \sup_{f\in W^r_{\infty 1}H^{\omega}} \left| \int_{\mathbf{R}} f(t)\mathrm{d}t - \frac{1}{n}\sum_{i\in\mathbf{Z}} f\left(\frac{i}{n}\right) \right|$$
$$= 2\pi \| f_{nr}(\omega, \cdot) \|_{\infty},$$

where

$$f_{nr}(\omega, x) = a_r + \int_0^1 D_r(x - t) f_{n0}(\omega, t)\mathrm{d}t,$$

while a_r is so chosen that $\int_0^1 f_{nr}(\omega, x)\mathrm{d}x = 0$, and $f_{n0}(\omega, x)$ as an odd function of period $1/n$ is defined as follows:

$$f_{n0}(\omega, x) = \begin{cases} \frac{1}{2}\omega(2x), & 0 \leqslant x \leqslant \frac{1}{4n}, \\ \frac{1}{2}\omega\left(\frac{1}{n} - 2x\right), & \frac{1}{4n} \leqslant x \leqslant \frac{1}{2n}. \end{cases}$$

In addition, Fang and Liu[76] considered the optimal quadrature problem on a convolution class of a PF density kernel with simple knots. We omit the details.

5. Results on a convolution class

The cardinal splines of finite degree introduced in 2 are finite order differentiable functions determined by a differential operator $P_r(D) = \prod_{i=1}^{r}(D - t_i)(t_i \in \mathbf{R}, i = 1, \cdots, r)$. This function class is very important in solving some extremal problems in approximation theory on Sobolev-Wiener class defined on **R**, which we have demonstrated in 2 – 4. In this section, the similar problems for infinite order smooth function classes will be studied.

N.I. Achiezer and M.G. Krein studied the following class in 1930's. Let $\mathscr{U}$ denote the set of all analytic functions $F(z)(z = x + iy, i = \sqrt{-1})$ which satisfy that

1) $F(x)$ is real for $x \in \mathbf{R}$,

2) $F(z)$ is analytic on the strip region:

$$x \in \mathbf{R},\ -\delta < y < \delta(\delta > 0) \text{ and } -1 \leqslant \operatorname{Re}F(z) \leqslant 1.$$

It is proved by [19] that for $F(z) \in \mathscr{U}$, $\lim\limits_{y \to \delta} h(x,y) = \lim\limits_{y \to -\delta} h(x,y)$, where $h(x,y) \triangleq \operatorname{Re} F(x + iy)$. Denote by $h(x)$ the resulting function, then for almost $x \in \mathbf{R}$.

$$F(x) = \frac{1}{2\delta}\int_{\mathbf{R}} h(t)(\operatorname{ch}((t-s)\pi/2\delta))^{-1}\mathrm{d}t.$$

For any $h \in L_\infty(\mathbf{R})$, $F(x)$ defined by the above equality has an analytic extension $F(z) \in \mathscr{U}$. Denote,

$$\mathscr{U}_0 = \{G * h : \| h \|_\infty \leqslant 1\},$$

where $G(x) = (1/2\delta)(\operatorname{ch}(x\pi/2\delta))^{-1}$ and

$$(G * h)(x) = \int_{\mathbf{R}} G(x-t)h(t)\mathrm{d}t.$$

N.I. Achiezer[19] and M.G. Krein[18] obtained the exact value of Chebyshev approximation of $\mathscr{U}_0$ by B_σ. As pointed out by Karlin[77], $G(x)$ is a totally positive kernel. A convolution class with a totally positive kernel is somewhat the analogy of a convolution class with an NCVD kernel, since both kernels have the variation diminishing properties (See Theorem 35). An important step in our study is to extend the theory of (finite order) cardinal $\mathscr{L}$-splines relating to a differential operator $P_r(D)$ to an infinite order differential operator $P(D)$, where $P(z)$ is an entire function satisfying certain conditions. This extension was started by Sun Yongsheng[78] and developed in Li Chun[79] and Chen Dirong[80]. The main point is to take $P(z)$ from the so-called Laguerre-Polya class so that some of its properties, especially that of as the limits of sequences of finite order cardinal $\mathscr{L}$-splines play essential role.

5.1 Polya frenquency

Definition 18 *An entire function $P(z)$ belongs to the Laguerre-Polya class ($P \in L - P$) if*

$$P(z) = \exp(-cz^2 + bz)\prod_{k=1}^{\infty}\left(1 - \frac{z}{a_k}\right)\exp(z/a_k),$$

where $c \geqslant 0, b, a_k \in \mathbf{R}$ and $0 < c + \sum_{k=1}^{\infty} a_k^{-2} < \infty$.

The $L - P$ class is essentially connected with totally positive functions.

Definition 19 *A function G is said to be totally positive if*

$$\det \| G(x_i - y_i) \|_{i,j=1}^{n} \geqslant 0$$

for all $x_1 < x_2 < \cdots < x_n, t_1 < t_2 < \cdots < t_n, n = 1, 2, \cdots$.

A totally positive function $G(x)$ is said to be a *Polya frequency density* (PF density), if $\int_{\mathbf{R}} G(x)\mathrm{d}x = 1$. We write $G \in$ PF for simplicity.

Schoenberg proved[30]

Theorem 33 *G is a* PF *density if and only if there exists a $P(z) \in L - P$ such that*

$$\frac{1}{P(z)} = \int_{\mathbf{R}} G(x)\exp(-xz)\mathrm{d}x \tag{21}$$

From (21) and the theory of Laplace transform, we have

Theorem 34 *G is a* PF *density if and only if for some $P(z) \in L - P$*

$$G(x) = \frac{1}{2\pi}\int_{\mathbf{R}} \frac{\exp(iux)}{P(iu)}\mathrm{d}u. \tag{22}$$

The above theorems tell us that there is a one-to-one relation between $G \in$ PF and $P \in L - P$. According to Hirschman and Widder[23], G is the Green function relating to the infinite order differential operator $P(D)$. We set deg $G = r$ if $P(z) = \mathrm{e}^{bz}\prod_{k=1}^{r}\left(1 - \frac{z}{a_k}\right)$ and ∞ otherwise. It is helpful to compare (2) with (22). We know that $\widetilde{G}_r$ is cyclic variation diminshing. The similar result holds for PF densities. To mention the theorem we introduce

Definition 20 *Let $G \in$* PF. *If the convolution transform $g = G * h$ satisfies*

$$V(g) \leqslant V(h)$$

for any $h \in C(\mathbf{R}) \cap L^{\infty}(\mathbf{R})$, *then* G *is said to be variation-diminishing, where*

$$V(g) = \sup S^{-}\{g(x_1), \cdots, g(x_n)\},$$

and $n \in \mathbf{Z}_{+}$, $x_1 < \cdots < x_n$ *are arbitrary.*

Theorem 35 $G \in \mathrm{PF}$ *if and only if* G *is variation diminishing.*

The above concepts and theorems can be found in Hirschman and Widder[23].

The following theorem reveals a relation between NCVD kernel and PF density.

Theorem 36 (Pinkus[14], p.61) *If* $G \in \mathrm{PF}$, *then its periodization* $\widetilde{G}(x) = \sum_{k \in \mathbf{Z}} G(x - 2k)$ *is an* NCVD *kernel.*

Schoenberg[30] had given a result in which he characterized PF densities with B-splines. To introduce his theorem we begin with B-splines.

For any $r + 1$ points which are dependent of r, $x_{0r} \leqslant x_{1r} \leqslant \cdots \leqslant x_{rr}$, $x_{0r} < x_{rr}$, let $M_r(x) = M_r(x; x_{0r}, \cdots, x_{rr})$ be the B-splines of degree $r - 1$ with knots $\{x_{ir}\}_{i=0}^{r}$.

Let the corresponding distribution functions be

$$F_r(x) = F_r(x; x_{0r}, \cdots, x_{rr}) = \int_{-\infty}^{x} M_r(t; x_{01}, \cdots, x_{rr}) \mathrm{d}t. \quad (23)$$

Schoenberg[30] proved

Theorem 37 *Let* $F(x) = \int_{-\infty}^{x} G(t)\mathrm{d}t$. *Then* $G \in \mathrm{PF}$ *if and only if there exists a sequence of distribution functions* (23) *such that*

$$\lim_{r \to \infty} F_r(x) = F(x)$$

holds everywhere.

Corollary 6 *If there exists a sequence of B-splines* $\{M_r\}_{r=1}^{\infty}$, *such that*

$$\lim_{r \to \infty} M_r(x) = G(x)$$

holds at all continuity points of G, then $G \in \mathrm{PF}$, where the B-spline M_r is of degree $r - 1$ for any r.

We end this section by formulating the following theorem, which is extensively used in this section.

Theorem 38 (Hirschman[23]) *For any $P \in L - P$, there is a sequence of real polynomials $\{P_r(z)\}_{r=1}^{\infty}$ satisfying the following conditions.*

1) $\deg P_r = r$,

2) $P_r(0) = 1$ *and* $P_r(z)$ *has real zeros only,*

3) $\lim_{r\to\infty} P_r(z) = P(z)$ *holds uniformly in the circle* $|z| \leqslant A$ *for every* $A > 0$.

5.2 Cardinal $\mathscr{L}$-splinies defined by the element in $L - P$ class

The first question we considered is to construct suitably the cardinal $\mathscr{L}$-splines with infinite degree. Since there is essential difference between the finite order differential operator and the infinite one, it is meaningless to replace $P_r(D)$ simply with $P(D)(P \in L - P)$ in (3).

Recall a property of the periodic splines $\widetilde{\mathscr{S}}_{r-1,n}$ as follows: $s \in \widetilde{\mathscr{S}}_{r-1,n}$ if and only if s has the form

$$s(x) = \sum_{k=1}^{2n} c_k \widetilde{G}_r(x - k/n),$$

where $\widetilde{G}_r$ is the Bernoulli function relating to $P_r(D)$. This property inspire us to define the set of cardinal $\mathscr{L}$-splines of infinite order as follows:

$$\mathscr{S}(P) = \left\{ s(x) = \sum_{k\in\mathbf{Z}} c_k G(x-k) : \sum_{k\in\mathbf{Z}} c_k G(x-k) \text{ converges on } \mathbf{R} \right\}.$$

For our purpose it is enough to consider some subsets of $\mathscr{S}(P)$, about which we shall give the detail discussion in the next subsection. In this subsection, we establish the theory of cardinal splines corresponding to $P(D), P \in L - P$. The fundamental function of cardinal interpolation is constructed and some properties are given. Other problems, such as the determination of sets of points at which splines interpolate and the remainder formula, are considered as well. Although the unicity of interpolation spline which holds for $\mathscr{S}_{r-1}$ is not established so far, we can still use interpolation operator to solve some extremal problems of approximation on the convolution class.

In what follows we assume $P \in L - P$ and $\{P_r\}_{r=1}^{\infty}$ is given as in

Theorem 38. Set

$$A(x,\lambda)=\lambda^{x+1}\sum_{k\in\mathbf{Z}}\frac{e^{2k\pi ix}}{P(\ln\tau+(u+2k\pi)i)},\tag{24}$$

where $\lambda=\tau e^{iu}$, $\tau>0$, $u\in[-\pi,\pi]$. This definition parallels to that of $A_{r-1}(x,\lambda)$ given in section 2.

Let $\lambda=e^z$. Then $A_{r-1}(x,e^z)$ is a meromophic function of z and $\lim_{r\to\infty}A_{r-1}(x,e^z)=A(x,e^z)$ holds uniformly in any compact subset of $\mathbf{C}$ excluding the poles of $A(x,e^z)$. Therefore, $A(x,e^z)$ is a meromophic function (This also can be proved by its definition), all zeros of which belong to $z:e^z<0\}$.

Lemma 10 (Li Chun[79]) *For $P\in L-P$, $A(x,-1)$ has exactly one zero $y_0\in[0,1)$. Moreover, y_0 is a simple zero of $A(x,-1)$.*

For any $x_0\in[0,1)\setminus\{y_0\}$, from above we have some vertical strip region free of zeros of $A(x_0,e^z)$. Denote by β_0 the maximum of β such that $|\operatorname{Re} z|<\beta_0$ contains no zero of $A(x_0,e^z)$. Obviously, $\beta_0>0$ depends on x_0.

Let

$$g(z)=\frac{1}{P(z)A(x_0,e^z)},$$

$$g_r(z)=\frac{1}{P_r(z)A_{r-1}(x_0,e^z)}.$$

Note that the zeros of $P(z)$ are removable singular points of $g(z)$. Therefore, $g(z)$ is analytic in $|\operatorname{Re} z|<\beta_0$. Define the fundamental function of the cardinal interpolation as follows:

$$L(x)=\frac{1}{2\pi i}\int_{-i\infty}^{i\infty}e^{xz}g(z)dz.$$

Obviously,

$$L(x)=\frac{1}{2\pi}\int_{\mathbf{R}}\frac{e^{iux}}{P(iu)A(x_0,e^{iu})}du=\frac{1}{2\pi}\int_0^{2\pi}\frac{A(x,e^{iu})}{A(x_0,e^{iu})}du.$$

The following theorem is fundamental.

Theorem 39

1) $L(x_0+j)=\delta_{j,0}$, $j\in\mathbf{Z}$,

2) $\forall \beta \in (0, \beta_0)$, *there exists a constant* $A = A(\beta)$ *such that*

$$|L(x)| \leqslant A\exp(-\beta|x|), x \in \mathbf{R},$$

3) $L \in C^\infty(\mathbf{R})$, *for* $\deg G = \infty$,

4) $\lim_{r\to\infty} \| L_{r-1} - L \|_p = 0$, *for* $p \in [1, \infty]$, *where* L_{r-1} *is the fundamental function in* $\mathscr{S}_{r-1}$,

5) $L(x) = \sum_{k\in\mathbf{Z}} c_k G(x-k)$, and $G(x) = \sum_{k\in\mathbf{Z}} d_k L(x-k)$, *where* $\{c_k\}, \{d_k\} \in l_1$.

Theorem 39 was proved by Sun Yongsheng[78] for G being even. For the general case it was proved by Li Chun[79] and Chen Dirong[80], respectively. It should be emphasised that this theorem gives $L \in \mathscr{S}(P)$ with exponential decay. 1) and 2) were proved by American scholars for cardinal $\mathscr{L}$-splinies of finite order (see [31]). But their method of proof is not suitable to the case $\deg G = \infty$.

By Theorem 39 we have

Corollary 7 *Let* $\beta \in (0, \beta_0)$. *Then for any sequence* $\{y_k\}_{k\in\mathbf{Z}}$ *satisfying* $|y_k| = 0(e^{\beta|k|})$, *the cardinal series*

$$\sum_{k\in\mathbf{Z}} y_k L(x-k) \tag{25}$$

converges uniformly on $\mathbf{R}$. *Furthermore, if we denote by* $s(x)$ *the summation function of* (25), *then* $s(x) \in \mathscr{S}(P)$.

Among all zeros of $P(z)$, without lose of generality, let a_1 be the smallest one in absolute value. Put $\alpha_0 = \min\{\beta_0, |a_1|\}$, β_0 given as in Theorem 39. Given $\alpha > 0$, we define the function class

$$\varepsilon(\alpha) = \{h: |h(x)| \leqslant M\exp(\alpha|x|), \forall x \in \mathbf{R},$$
$$M \text{ is a positive constant depending on } h\}.$$

It has been proved (Hirschman and Widder[23]) that $G \in \varepsilon(\beta)$ for any $\beta \in (0, |a_1|)$. Therefore, for any $h \in \varepsilon(\alpha)$, $\alpha < |a_1|$, we have $f = G * h \in \varepsilon(\alpha)$. From this it follows that for $h \in \varepsilon(\alpha)$, $\alpha < \alpha_0$, the cardinal splines series

$$\sum_{k\in\mathbf{Z}} f(x_0 + k) L(x-k) \tag{26}$$

converges uniformly to a function $s(f, x)$ on $\mathbf{R}$, where $f = G * h$.

Moreover, $s(f,x) \in \mathscr{S}(P)$ interpolates f at $\{x_0 + k\}_{k\in\mathbf{Z}}$. To give the remainder formula, we define

$$E(x, e^z) = \frac{A(x, e^z)}{A(x_0, e^z)},$$

$$K(x, z) = \frac{e^{xz} - \exp(x_0 z) \cdot E(x, e^z)}{P(z)}.$$

Obviously, $E(x, e^{iu}) = \sum_{k\in\mathbf{Z}} e^{iku} L(x-k)$. By the method of Micchelli[31] we have

Lemma 11

1) $K(x+1, z) = e^z K(x, z)$,

2) $K(x, z)$ *is analytic on* $\{z : |\operatorname{Re} z| \leqslant \beta_0\}$,

3) $K(\cdot, z) \in C^\infty(\mathbf{R})$ *for deg* $G = \infty$.

If we define

$$B(x, t) = \frac{1}{2\pi}\int_{\mathbf{R}} e^{-itu} K(x, iu)\, du,$$

then by the method of Lemma 11 we have $B(x, \cdot) \in \varepsilon(\beta)$, for fixed $0 < \beta < \beta_0$. This conclusion was proved by using the properties of eigensplines (c) of Theorem 8.

Theorem 40 *For* $h \in \varepsilon(\beta)$, $\alpha < \alpha_0$, *let* $f = G * h$. *If* $s(f, x)$ *is given as in* (26), *then*

$$f(x) - s(f, x) = \int_{\mathbf{R}} B(x, t) h(t)\, dt.$$

The proof of this theorem mainly relies on the equality

$$B(x, t) = G(x - t) - \sum_{k\in\mathbf{Z}} G(x_0 + k - t) L(x - k).$$

One of application of Theorem 40 is to solve in part the unicity problem of interpolation spline.

Corollary 8 *If* $s(x) = \sum_{k\in\mathbf{Z}} c_k G(x - k)$ *satisfies* $|c_k| = 0$ $(\exp(\beta|k|)$ *for some* $0 < \beta < \alpha_0)$, *then* $s(x)$ *has the form*

$$s(x) = \sum_{k\in\mathbf{Z}} s(x_0 + k) L(x - k).$$

We have discussed the existence of interpolating spline functions in $\mathscr{S}(P)$ for functions of exponential growth and constructed remainder formula. The following theorem concerns about results for function in $L_p(\mathbf{R})$, *i.e.*,

Theorem 41 *Theorem* 40 *also holds if the condition* $h \in \varepsilon(\alpha)$ *is replaced by* $h \in L_p(\mathbf{R})$.

5.3 An orthogonal schauder basis

In many approximation problems on the convolution class

$$\mathscr{B}_p(G) = \{G * h : \| h \|_p \leqslant 1\}, \quad G \in \mathrm{PF},$$

we have to make use of the subspace $\mathscr{S}_p(G)$ of $\mathscr{S}(P)$:

$$\mathscr{S}_p(G) = \overline{\operatorname{span}\{G(x-j)\}_{j\in\mathbf{Z}}},$$

where $\overline{A}$ is the closure of set A in $L_p(\mathbf{R})$. We construct an orthogonal basis for $\mathscr{S}_p(G)$ and then determine the average dimension of $\mathscr{S}_p(G)$. This section is cited from Chen Dirong[80].

For convenience, we introduce some notations. Let $c = \{c_k\}_{k\in\mathbf{Z}}$, $\mathrm{d} = \{\mathrm{d}_k\}_{k\in\mathbf{Z}}$ be two sequences, and f a function defined on $\mathbf{R}$. Put

$$a = \{a_k\} = c * d, \text{where } a_k = \sum_{j\in\mathbf{Z}} c_{k-j} d_j,$$

$$(f * c)(x) = \sum_{k\in\mathbf{Z}} c_k f(x-k),$$

$$\| f \|_* = \sup\left\{\sum_{k\in\mathbf{Z}} |f(x+k)| : x \in [0,1)\right\}.$$

If $f \in L_p(\mathbf{R})$, we denote by $\mathscr{S}_p(f)$ the closure of span $\{f(x-j)\}_{j\in\mathbf{Z}}$ in $L_p(\mathbf{R})$.

For $c \in l_p$, $d \in l_1$, the following so called Young's inequality holds

$$\| c * \mathrm{d} \|_{l_p} \leqslant \| c \|_{l_p} \| d \|_{l_1},$$

where $\| \cdot \|_{l_p}$ is the usual norm in $l_p(\mathbf{R}^{\mathbf{Z}})$.

Lemma 12 *If* $\| f \|_* < \infty$ *and* $c \in l_p$, *then*

1) $\| f * c \|_p \leqslant \| f \|_* \| c \|_{l_p}$,

2) $\mathscr{S}_p(f * c) \subseteq \mathscr{S}_p(f)$,

3) $(f * d) * c = f * (c * d)$, for $d \in l_1(\mathbf{R}^{\mathbf{Z}})$.

Given $G \in \mathrm{PF}$, Φ is defined by

$$\hat{\Phi}(u) = \hat{G}(u) M(u) = \frac{M(u)}{\sqrt{2\pi} P(\mathrm{i}u)}, \tag{27}$$

where $\hat{\Phi}(u) = \frac{1}{\sqrt{2\pi}}\int_{\mathbf{R}} e^{itu}\Phi(u)\,dt$ is the Fourier transform of Φ and

$$M(u) = \left(\sum_{k\in\mathbf{Z}} \frac{1}{|P(iu + 2k\pi i)|^2}\right)^{-1/2}. \tag{28}$$

We also consider the two function $M_r(u)$ and $\Phi_r(u)$ that are defined in (28) and (27), respectively, by replacing P with P_r. It is worthly to point out that both $\{\Phi(x-j)\}_{j\in\mathbf{Z}}$ and $\{\Phi_r(x-j)\}_{j\in\mathbf{Z}}$ are orthogonal sequences.

Lemma 13 *If $G \in PF$ and Φ, Φ_r defined as above, then*

1) $\sup \|\Phi_r\|_* < \infty$,

2) $\lim\limits_{r\to\infty} \|\Phi - \Phi_r\|_* = 0$.

The proof of this lemma is difficult. We should use some profound properties of functions in L − P class. Moreover, we have to prove these properties hold "uniformly" for P_r, $r = 1, 2, \cdots$.

By (27) there exists a $c \in l_1(\mathbf{R}^{\mathbf{Z}})$ such that

$$\Phi(x) = \sum_{k\in\mathbf{Z}} c_k G(x-k). \tag{29}$$

Similar to this, by $\widetilde{G}(u) = (M(u)^{-1})\,\hat{\Phi}(u)$, we have

$$G(x) = \sum_{k\in\mathbf{Z}} d_k \Phi(x-k), \tag{30}$$

for some $d = \{d_k\} \in l_1(\mathbf{R}^{\mathbf{Z}})$. Therefore, Lemma 12, together with (29) and (30), tells us that $\mathscr{S}_p(G) = \mathscr{S}_p(\Phi)$. The following theorem characterizes $\mathscr{S}_p(\Phi)$.

Theorem 42 *Let Φ be given as above. Then*

1) $f = \Phi * c \in L_p(\mathbf{R})$ *if and only if* $c \in l_p$. *Moreover*

$$\|\Phi\|_*^{-1} \|c\|_{l_p} \leqslant \|\Phi * c\|_p \leqslant \|\Phi\|_* \|c\|_{l_p}. \tag{31}$$

2) $\mathscr{S}_p(\Phi) = \{\Phi * c : c \in l_p\}$.

The second inequality of (31) is given by Lemma 12. To prove the first, one needs the orthogonality of $\{\Phi(x-k)\}_{k\in\mathbf{Z}}$. From (30) it follows that $\mathscr{S}_p(\Phi) \subset \{\Phi * c : c \in l_p\}$, which together with Lemma 12 proves the assertion 2).

Corollary 9 $\{\Phi(x-k)\}_{k\in\mathbf{Z}}$ *is an orthogonal Schauder basis for*

$\mathscr{S}_p(G)$.

The first inequality of (31) is essential for determining the following quantity (cf. Definition 7 and Example 5).

$$k(\mathscr{S}_p(\Phi), a, \varepsilon) = \min\{n \in \mathbf{Z}_+ : \text{there exists a linear subspace } M \text{ of dimension } n \text{ such that } E(\mathscr{S}_p(\Phi))_a^o, M, L_p[-a, a]) < \varepsilon\}.$$

Let $\varepsilon > 0$. By Lemma 13, there exists a natural number r such that $\| \Phi - \Phi_r \|_* \| \Phi \|_* < \varepsilon$. Therefore, for any $s = \Phi * c \in \mathscr{S}_p(\Phi)$ by (31), $\| c \|_{l_p} \leqslant \| \Phi \|_* \cdot \| s \|_p$.

Set $s_r = \Phi_r * c$. Then $\| s - s_r \|_p < \varepsilon$. Thus

$$E((\mathscr{S}_p(\Phi))_a^o, (\mathscr{S}_p(\Phi_r)_a, L_p[-a, a]) < \varepsilon.$$

Since $\mathscr{S}_p(\Phi_r) = \mathscr{S}_{r-1} \cap L_p(\mathbf{R})$, we obtain $\dim(\mathscr{S}_p(\Phi_r))_a \leqslant 2a + r$. This means $k(\mathscr{S}_p(\Phi), a, \varepsilon) \leqslant 2a + r$ and

$$\lim_{a\to\infty} \frac{k(\mathscr{S}_p(\Phi), a, \varepsilon)}{2a} \leqslant 1.$$

Similar but more complicated discussion gives

$$\lim_{a\to\infty} \frac{k(\mathscr{S}_p(\Phi), a, \varepsilon)}{2a} \geqslant 1.$$

Therefore, we have

Theorem 43 *For* $G \in \mathrm{PF}$, $\overline{\dim}\mathscr{S}_p(G) = \overline{\dim}\mathscr{S}_p(\Phi) = 1$.

When $\deg G < \infty$, Theorem 43 has been proved in section 3 (Example 6). Since the subspace $\mathscr{S}_p(G)$ is LFD for $\deg G < \infty$, thus the proof of Example 5 is far simpler than that of Theorem 43.

5.4 Dual theorem for cardinal spline interpolation and estimations of errors for interpolation ($p = 1, \infty$)

If $G \in \mathrm{PF}$, so is G^*, where $G^*(x) = G(-x)$. The entire function $P^* \in L - P$ relating to G^* satisfies $P^*(z) = P(-z)$ and the function $A^*(x, \mathrm{e}^z)$ defined in (24) corresponding to P^* is $A^*(1 - x_0, \mathrm{e}^z)$. Therefore, for x_0 such that with $A(x_0, -1) \neq 0$ we have $A^*(1 - x_0, \mathrm{e}^z) \neq 0$. For $h \in \varepsilon(\alpha)$, $\alpha < \alpha_0$, or $h_1 \in L_p(\mathbf{R})$, let $g = G^* * h$. The results of section 5.2 guarantee the existence of a spline

$$s^*(g, t) = \sum_{k\in\mathbf{Z}} g(k - x) L^*(t - k), \tag{32}$$

which interpolates g at $\{k - x_0\}_{k\in\mathbf{Z}}$, where $L^* \in \mathscr{S}_p(G^*)$ is the analogy of L defined in Theorem 39. The remainder formula holds

$$g(t) - \mathscr{S}^*(g,t) = \int_{\mathbf{R}} B^*(t,x)h_1(x)\mathrm{d}x.$$

Similar to Li Chun[35] (Note that the definition of P^* is somewhat different from that of Li Chun[35]) we have

$$B^*(t,x) = B(x + x_0, t + x_0). \tag{33}$$

By (33) we may prove the duality formula as follows.

Theorem 44 *Let $f = G * h$ and $g = G^* * h_1$, where $h \in L_p$, $h_1 \in L_{p'}$, $1/p + 1/p' = 1$. If $s(f,x)$ and $s^*(g,t)$ are given by* (26) *and* (32), *respectively, then*

$$\int_{\mathbf{R}}(f(x) - s(f,x))h_1(x - x_0)\mathrm{d}x = \int_{\mathbf{R}}(g(t) - s^*(g,t))h(t + x_0)\mathrm{d}t.$$

From Theorem 44 it follows that $\lambda_p = \lambda^*_{p'}$, where

$$\lambda_p = \sup\{\|f - s(f)\|_p : f \in \mathscr{B}_p(G)\},$$

$$\lambda^*_{p'} = \sup\{\|f - s(f)\|_{p'} : f \in \mathscr{B}_{p'}(G^*)\}, 1/p + 1/p' = 1. \tag{34}$$

By Theorem 44 we can reduce the estimation of λ_p to that of $\lambda^*_{p'}$. In particular, this enables us to obtain λ_1, since we can get λ_∞ and λ^*_∞ easily. To get λ_∞ we appeal to the following assertion: there exists $\sigma = 1$ or -1 such that for $x, t \in \mathbf{R}$,

$$\sigma \sin \pi(x - x_0)B(x,t)\sin \pi t \geqslant 0,$$

where x_0 is the unique zero of $\Phi(x) = (G * \operatorname{sgn} \sin \pi t)(x)$ in $(0,1)$. Therefore, similar to Theorem 11 we have $\lambda_\infty = \|\Phi\|_\infty$ and, by (34), $\lambda_1 = \lambda^*_\infty = \|\Phi^*\|_\infty = \|\Phi\|_\infty$, where $\Phi^*(x) = (G^*_* \operatorname{sgn} \sin \pi t)(x)$.

To sum up, we have

Theorem 45 *Let x_0 be the zero of Φ in* $(0,1)$ *and $s(f,x)$ the sum of the series in* (26). *Then , we have $\lambda_1 = \lambda_\infty = \|\Phi\|_\infty$.*

Sun Yongsheng[78] first proved $\lambda_\infty = \|\Phi\|_\infty$ for G being even by a different method. The material of this subsection is taken from Chen Dirong[80].

5.5 Average *n*-widths of convolution classes

For $G \in \mathrm{PF}$ we have known from Theorem 36 that $\widetilde{G}(x) = \sum_{k \in \mathbf{Z}} G(x + 2k)$ is a NCVD kernel. Thus the results of 3.3 hold for $\widetilde{G}$. In what follows we denote $\lambda_n(p, q, \widetilde{G})$ by $\lambda_n(p, q, G)$.

Theorem 46 *If $p \in (1, \infty)$, there exists a x_0 such that $s(f, x)$ defined by (26) has meaning for $f \in \mathscr{B}_p(G)$. Moreover,*

$$\lambda_p \leqslant \lambda_1(p, p, G).$$

This theorem is proved by using Theorem 12 and some properties of the limit of the cardinal splines of finite degree.

Theorem 47 *If $G \in \mathrm{PF}$, $p \in [1, \infty]$, then*

$$\overline{d}_1(\mathscr{B}_p(G), L_p(\mathbf{R})) = \overline{\delta}_1(\mathscr{B}_p(G), L_p(\mathbf{R})) = \lambda_1(p, p, G).$$

Furthermore, we have

1) $\mathscr{S}_p(G)$ is optimal for $\overline{d}_1$.

2) *The interpolation operator $s(f)$ given as in Theorem 46 is an optimal operator for $\overline{\delta}_1$.*

The proof of the theorem is divided into two steps.

1) To prove that $s(f) \in \mathscr{S}_p(G)$ for any $f \in \mathscr{B}_p(G)$, which gives

$$\overline{\delta}_1(\mathscr{B}_p(G), L_p) \leqslant \lambda_1(p, p, G).$$

2) To obtain the lower bound of $\overline{d}_1$. From Theorem 22 and the inequality $E(\mathscr{B}_p(G), \mathscr{B}_p(G_r), L_p) \leqslant \| G - G_r \|_1$, it follows that $\overline{d}_1(\mathscr{B}_p(G), L_p) \geqslant \lambda_1(p, p, G)$.

For any $\sigma > 0$, $\sigma^{-1} G(\sigma^{-1} x) \in \mathrm{PF}$. By changing scale we can reduce the problem of average σ-widths to that or average 1-widths and then solve it by Theorem 47. In the following theorems we only formulate the results about $\overline{d}_\sigma$, although similar results about $\overline{\delta}_\sigma$ also hold true.

Theorem 48 *For $\sigma > 0$, $p \in (1, \infty)$ it holds*

$$\overline{d}_\sigma(\mathscr{B}_p(G), L_p) = \overline{d}_1(\mathscr{B}_p(\sigma^{-1} G(\sigma^{-1} \cdot)), L_p).$$

Moreover,

$$\{ s(\sigma x) : s \in \mathscr{S}_p(G(\sigma^{-1} \cdot)) \}$$

is optimal for $\overline{d}_\sigma(\mathscr{B}_p(G), L_p)$.

Similar results of Theorem 24 also hold for the convolution class

$$\mathscr{B}_{pq}(G) = \{G * h : \| h \|_{pq} \leqslant 1\}, 1 \leqslant q \leqslant p \leqslant \infty.$$

Theorem 49 *For any positive integer n, $1 \leqslant q \leqslant p \leqslant \infty$, we have*

$$\overline{d}_n(\mathscr{B}_{pq}(G), L_q) = \lambda_n(p, q, G).$$

Moreover $\{s(nx) : s \in \mathscr{S}_p(G(n^{-1} \cdot))\}$ *is optimal for* $\overline{d}_n(\mathscr{B}_{pq}(G), L_q)$, $1 \leqslant q \leqslant p \leqslant \infty$.

6. Open Problems

In 2.3, 3.3 three open problems had been mentioned. Here we further list some unsolved problems and point out some ideas which may be helpful. In what follows we assume P_r is a polynomial with only real zeros and $G \in PF$.

Problem 6 For $p \geqslant q, \sigma > 0$.

1) Is $B_{\sigma\pi, q}$ optimal for $\overline{d}_\sigma(W_{pq}(P_r), L_q)$?

2) Is $B_{\sigma\pi, q}$ optimal for $\overline{d}_\sigma(\mathscr{B}_{pq}(G), L_q)$?

Chen Dirong[40] proved both 1) and 2) are true for $p = q = 2$.

It is difficult to solve problem 1 completely. One way for $q = 1$ is to prove a Taikov-type inequality

$$\sup\{\| g \|_{p\infty} : g \in W_\infty(P_r), g \perp B_{\sigma\pi, 1}\} \leqslant \| \Phi_{r,\sigma} \|_{L_{p'}[0,1]},$$

where $\Phi_{r,\sigma}(x) = \int_0^2 \widetilde{G}_r(x - t) \operatorname{sgn} \sin \sigma\pi t \, dt$, $\widetilde{G}_r$ is the periodic Bernoulli function corresponding to $P_r(D)$ (See (2)), and p' satisfies $1/p + 1/p' = 1$.

Let's consider the problems of the best and one-sided best approximation of $G \in$ PF by $B_{\sigma\pi, 1}$ in $L_1(\mathbf{R})$. Denote by x and $x_+ \in B_{\sigma\pi, 1}$ the elements of the best and one-sided best approximation of G in $L_1(\mathbf{R})$ respectively.

Problem 7

1) Is there a $t_0 \in \mathbf{R}$ such that $\sin \sigma(t - t_0)\pi[G(t) - x(t)]$ does not change sign on $\mathbf{R}$?

2) Is there a t_0 such that

$$G(t_0 + 2k/\sigma) = x_+ (t_0 + 2k/\sigma) \text{ for } k \in \mathbf{Z}?$$

The affirmative answer to this problem may give a series of results about the approximation of $\mathscr{B}_p(G)$ by $B_{\sigma,p}$. For example if 1) is true, we can obtain

$$E(\mathscr{B}_\infty(G), B_{\sigma\pi}, L_\infty) = \| G * \operatorname{sgn} \sin \sigma\pi t \|_\infty.$$

In approximation of NCVD kernel by trigonometric polynomials similar results of 1) and 2) hold (see[14], Chen Dirong and Zhang Jianhua[81], respectively). It seems to us problem 3 is more difficult than the periodic case.

Problem 8 To obtain the exact value of

$$E^+ (W_{p,1}(P_r), \mathscr{S}_{r-1}, L_1) \tag{35}$$

for the case $P_r(0) = 0$.

Sun Yongsheng and Liu Yongping[53] obtained the value of (35) for $p = 1$, and $P_r(t) = t^r$.

Problem 9 To extend the theory of cardinal $\mathscr{L}$-splines to the following entire function class

$$P(z) = \exp(-cz^2 + bz) z^s \prod_{k=1}^{\infty} (1 - z/a_k) \exp(z/a_k),$$

where c, b, a_k satisfy the condition in Definition 16 and $s \geqslant 0$ is an integer.

Problem 9 unifies the cases of 2.1 and 5.1.

Problem 10 To obtain the value of $\overline{d_a}(\mathscr{B}_p(G), L_{qp})$.

It is the dual problem of Theorem 40 and easily solved if $\deg G < \infty$. The main difficulty in solving Problem 10 is to establish

$$\overline{\dim}(\mathscr{S}_p(G), L_{qp}) = 1.$$

Problem 11 To solve the problem of optimal recovery of $\mathscr{B}_p(G)$ in the sense of section 5.

Problem 11 has been solved for $p = 1, 2$, and ∞. One crucial point to solve Problem 11 is to prove

$$\overline{\dim} \operatorname{span}\{ G(t_j - t) \}_{j \in \mathbf{Z}} \leqslant 1.$$

for any $T = \{t_j\}_{j \in \mathbf{Z}} \in \Theta$, where Θ is given by Definition 13.

Part Two: The Case $d > 1$*

0 Introduction

In part one, we mentioned that many authors studied some extremal problems in approximation theory for some classes of smooth functions defined on the real line and obtained some exact results. In the following, we will summary some results for the classes of multivariate smooth functions. To do these, as in [41], we state some needed terminology here.

Let $\alpha > 0$, and P_α be the continuous linear operator in $L_p(\mathbf{R}^d)$ defined by $P_\alpha := \chi_\alpha(\cdot) \cdot x(\cdot)$, where $\chi_\alpha(\cdot)$ is the characteristic function of the cube $[-\alpha, \alpha]^d$. Let $\varepsilon > 0$, and L be a subspace of $L_p(\mathbf{R}^d)$. We define

$K_\varepsilon(\alpha, L, L_p(\mathbf{R}^d)) := \min\{n \in \mathbf{Z}_+ : d_n(P_\alpha(L \cap BL_p(\mathbf{R}^d)), L_p(\mathbf{R}^d)) < \varepsilon\}$,

where $d_n(A, X)$ denotes the n-K width of A in X. The average dimension of L in $L_p(\mathbf{R}^d)$ is defined to be

$$\overline{\dim}(L, L_p(\mathbf{R}^d)) := \lim_{\varepsilon \to \infty} \liminf_{\alpha \to \infty} \frac{K_\varepsilon(\alpha, L, L_p(\mathbf{R}^d))}{(2\alpha)^d}.$$

Let $\sigma > 0$, and C be a centrally symmetric subset of $L_p(\mathbf{R}^d)$. The average Kolmogorov σ-width (shortly, average σ-K width) of C in $L_p(\mathbf{R}^d)$ is defined to be

$$\overline{d}_\sigma(C, L_p(\mathbf{R}^d)) := \inf_L \sup_{x(\cdot) \in C} \inf_{y(\cdot) \in L} \| x(\cdot) - y(\cdot) \|_{L_p(\mathbf{R}^d)},$$

where the infimun is taken over all subspaces $L \subset L_p(\mathbf{R}^d)$ such that $\overline{\dim}(L, L_p(\mathbf{R}^d)) \leqslant \sigma$. If there is a subspace A^* of $L_p(\mathbf{R}^d)$ with $\overline{\dim}(A, L_p(\mathbf{R}^d)) \leqslant \sigma$ such that

$$\overline{d}_\sigma(\mathfrak{M}\ L_p(\mathbf{R}^d)) = E(\mathfrak{M}, A^*, L_p(\mathbf{R}^d)),$$

then A^* is called an optimal subspace for $\overline{d}_\sigma(\mathfrak{M}, L_p(\mathbf{R}^d))$.

The average linear σ-width (shortly, average σ-L width) of C in $L_p(\mathbf{R}^d)$ is defined to be

$$\overline{d}'_\sigma(C, L_p(\mathbf{R}^d)) := \inf_{(Y, \Lambda)} \sup_{x(\cdot) \in C} \| x(\cdot) - \Lambda x(\cdot) \|_p,$$

* 这一部分作者是刘永平.

where the infimum is taken over all paris (Y, Λ) such that, for each pair(Y, Λ), Y is a normed space which is continuously imbedded in $L_p(\mathbf{R}^d)$, $C \subset Y$, Λ is a continuous linear operator from Y to $L_p(\mathbf{R}^d)$ and $\overline{\dim}(\operatorname{Im} \Lambda, L_p(\mathbf{R}^d)) \leqslant \sigma$, while Im Λ denotes the range of the operator Λ.

It follows at once from the definitions that

$$\overline{d}_\sigma(C, L_p(\mathbf{R}^d)) \leqslant \overline{d'}_\sigma(C, L_p(\mathbf{R}^d)).$$

1. Best approximation by means of polyharmonic cardinal splines

1.1 Best approximation by means of polyharmonic cardinal splines

Let Δ be the usual Laplace operator defined by

$$\Delta u = \sum_{j=1}^{d} \frac{\partial^2 u}{\partial x_j^2},$$

and Δ^k, $k > 1$, denote its kth iterate, $\Delta^k u = \Delta(\Delta^{k-1} u)$. For $k \geqslant (d+1)/2$, we denote by $SH^k(\mathbf{R}^d)$ the subspace of $\mathscr{S}'(\mathbf{R}^d)$ whose elements f enjoy the properties

1) f is in $C^{2k-d-1}(\mathbf{R}^d)$,

2) $(\Delta^k f)(x) = 0$ for all $x \in \mathbf{R}^d \setminus \mathbf{Z}^d$.

Here $\mathbf{Z}^d$ denotes the integer lattice in $\mathbf{R}^d$, and $\mathscr{S}'(\mathbf{R}^d)$ and $\mathscr{S}'(\mathbf{R}^d)$ (see [82, 83]) denote respectively the Schwartz space of rapidly decreasing functions and its dual, the space of tempered distributions. According to Madych and Nelson[84], a function or a distribution f is called a k-harmonic cardinal spline if $f \in SH^k(\mathbf{R}^d)$, and called a polyharmonic spline if $f \in SH^k(\mathbf{R}^d)$ for some k.

In [84], Madych and Nelson gave the following cardinal interpolations

$$(Sf)(x) = \sum_{\nu \in \mathbf{Z}^d} f(\nu) L_k(x - \nu), \tag{36}$$

where the cardinal function $L_k \in SH^k(\mathbf{R}^d)$ satisfies $L_k(\nu) = \delta_{0\nu}$ for all $\nu \in \mathbf{Z}^d$. This is a transparent generalization of cardinal interpolation from the univariate spline theory of odd order formed by Schoenberg[85] to the multivariate case. As a special case of multivariate cardinal interpolation with radial basis functions, M. D. Buhmann[86] considered the interpolations

$$(S_{\sigma,k}f)(x) = \sum_{\nu\in \mathbf{Z}^d} f\left(\frac{\nu}{\sigma}\right) L_k(\sigma x - v), \sigma > 0, \tag{37}$$

as an approximation to f. Exploiting the polynomial precision properties of the cardinal interpolation (36), Buhmann in [86] derived orders of convergence of (37) to a sufficiently differentiable f in the uniform norm when $\sigma \to +\infty$.

Liu in [87] consider the remainder formula of the cardinal interpolation $S_{1,k}f$ as an approximation to f, which in univariate case were founded by Schoenberg[88], and Micchelli[31]. Expoilting the remainder formula in multivariate case, we derive the orders of convergence of (37) to a sufficiently differentiable f in the metric $L_p(\mathbf{R}^d)$ when $\sigma \to +\infty$. Here $L_p(\mathbf{R}^d)$ denotes the space of all p-power integrable functions, f, on $\mathbf{R}^d$ for $1 \leqslant p < +\infty$ and $L_\infty(\mathbf{R}^d)$ denotes the space of all essentially bounded functions, f, with the norm

$$\| f \|_p = \left(\int_{\mathbf{R}^d} | f(t) |^p \mathrm{d}t \right)^{1/p}, 1 \leqslant p < \infty,$$

$$\| f \|_\infty = \operatorname*{esssup}_{t\in\mathbf{R}^d} | f(t) |$$

respectively. Further, Liu[87] also consider the average σ-K width of the Sobolev class defined by the differential operator Δ^k, and obtain some asymptotic results. In univariable case, the concept of the average σ-K width was proposed by Tikhomirov[21], and some elegant results for some classes of smooth functions defined on the real axis R had been obtained by many authors (see [37, 41, 73, 88]).

For $y = (t_1, \cdots, t_d) \in \mathbf{R}^d$ and $x = (x_1, \cdots, x_d) \in \mathbf{R}^d$, set $xt = \sum_{j=1}^{d} t_j x_j$.

The Fourer transform $\hat{\varphi}$ of $\varphi \in \mathscr{S}(\mathbf{R}^d)$ will be denoted as follows

$$(F\varphi)(x) = \hat{\varphi}(x) =: (2\pi)^{-d/2} \int_{\mathbf{R}^d} \varphi(t) \mathrm{e}^{-\mathrm{i}tx} \mathrm{d}t$$

and the transform inverse to it as follows

$$(F^{-1}\varphi)(x) = (2\pi)^{-d/2} \int_{\mathbf{R}^d} \varphi(t) \mathrm{e}^{\mathrm{i}tx} \mathrm{d}t.$$

The generalized Fourier transforms (direct and inverse) for $f \in \mathcal{S}'(\mathbf{R}^d)$ are defined respectively by the equations

$$(\hat{f}, \varphi) = (f, \hat{\varphi}), \text{ and } (F^{-1}f, \varphi) = (f, F^{-1}\varphi)$$

for all $\varphi \in \mathcal{S}(\mathbf{R}^d)$, where $(g, \varphi) = \int_{\mathbf{R}^d} g(t)\varphi(t)\mathrm{d}t$, $g \in \mathcal{S}'(\mathbf{R}^d)$, $\varphi \in \mathcal{S}'(\mathbf{R}^d)$.

For $2k \geqslant d + 1$, as in [84], set

$$E_k(x) = \begin{cases} c(d, k) \| x \|^{2k-d}, & \text{if } d \text{ is odd,} \\ c(d, k) \| x \|^{2k-d} \log \| x \|, & \text{if } d \text{ is even,} \end{cases}$$

where $\| \cdot \|$ is the Euclidean norm on $\mathbf{R}^d$, $c(n, k)$ is a constant which depends only on d and k, and is chosen such that $\Delta^k E_k(x) = \delta(x)$. Here $\delta(x)$ denotes the unit Dirac distribution at the origin. The generalized Fourier transform of E_k is

$$\hat{E}_k(\xi) = (2\pi)^{-d/2}(- \| \xi \|^2)^{-k}.$$

1.1.1 Remainder formula and order of convergence

In what follows we say that a sequence a_ν, $\nu \in \mathbf{Z}^d$, is of polynomial growth if there are positive constants c and N such that

$$| a_\nu | \leqslant c(1 + \| \nu \|)^N \text{ for all } \nu \in \mathbf{Z}^d.$$

Similarly we say that a locally bounded function f decays exponentially if

$$| f(x) | \leqslant \dot{c}\mathrm{e}^{-N\| x \|} \text{ for all } x \in \mathbf{R}^d.$$

Madych and Nelson in [84] obtained the following results.

Theorem A *If $\nu = \{\nu_j\}_{j \in \mathbf{Z}^d}$ is a sequence of polynomial growth, then there is a unique k-harmonic spline f of polynomial growth such that $f(j) = \nu_j$ for all $j \in \mathbf{Z}^d$. Every k-harmonic spline f has a unique representation in terms of translates of L_k, namely*

$$f(x) = \sum_{j \in \mathbf{Z}^d} f(j) L_k(x - j).$$

Here the function L_k is defined by its Fourier transform

$$\hat{L}_k(\xi) = (2\pi)^{-d/2} \frac{\| \xi \|^{-2k}}{\sum_{j \in \mathbf{Z}^d} \| \xi - 2\pi j \|^{-2k}}, \xi \in \mathbf{R}^d,$$

satisfies the properties $L_k(\nu) = \delta_{0,\nu}$, for all $\nu \in \mathbf{Z}^d$, and decays exponentially.

For $\alpha = (\alpha_1, \cdots, \alpha_d) \in \mathbf{Z}_+^d =: \{(\alpha_1, \cdots, \alpha_d): \alpha_j = 0, 1, 2, \cdots, j = 1, 2, \cdots, d\}$, let $\frac{\partial^\alpha}{\partial x^\alpha} = \frac{\partial^{\alpha_1+\alpha_2+\cdots+\alpha_d}}{\partial x_1^{\alpha_1}\partial x_2^{\alpha_2}\cdots\partial x_d^{\alpha_d}}$, be a differential monomial form, whose total order is $|\alpha| = \alpha_1 + \cdots + \alpha_d$. Suppose that f and g are two locally integrable function on $\mathbf{R}^d$. Then we say that the generalized derivative of f, $\frac{\partial^\alpha f}{\partial x^\alpha}$ is g, if

$$\int_{\mathbf{R}^d} f(x)\frac{\partial^\alpha \varphi(x)}{\partial x^\alpha}\mathrm{d}x = (-1)^{|\alpha|}\int_{\mathbf{R}^d} g(x)\varphi(x)\mathrm{d}x, \ \forall \varphi \in C_0^\infty(\mathbf{R}^d),$$

where $C_0^\infty(\mathbf{R}^d)$ denotes the space of infinitely differentiable functions with compact support (see [82]).

For $1 \leqslant p \leqslant \infty$, and a non-negative integer r, the Sobolev space $L_p^r(\mathbf{R}^d)$ is defined as the space of function f, with $f \in L_p(\mathbf{R}^d) \cap C(\mathbf{R}^d)$ and where all $\partial^\alpha f/\partial x^\alpha$ exist and $\partial^\alpha f/\partial x^\alpha \in L_p(\mathbf{R}^d)$ in the above sense, where $|\alpha| \leqslant r$. When $r = 2k$ is even, we set

$$W_p(\Delta^k) = \{f \in L_p^{2k}(\mathbf{R}^d): \|\Delta^k f\|_p \leqslant 1\}.$$

In [87], Liu obtain the following theorems.

Theorem 50 *If $f \in L_p^{2k}(\mathbf{R}^d)$, $2k \geqslant d + 1$, then the remainder formula*

$$f(x) - S_{1,k}f(x) = \int_{\mathbf{R}^d} G_k(x,t)\Delta^k f(t)\mathrm{d}t, \ \forall x \in \mathbf{R}^d,$$

holds. Here the function $G_k(x,t)$ is defined by the formula

$$G_k(x,t) = E_k(x-t) - \sum_{\nu\in\mathbf{Z}^d} E_k(\nu - t)L_k(x-\nu), (x,t) \in \mathbf{R}^d \times \mathbf{R}^d,$$

and satisfies the following properties:

1) $|G_k(x,t) \leqslant A(x)\mathrm{e}^{-B\|t\|}$ *for all* $(x,t) \in \mathbf{R}^d \times \mathbf{R}^d$,

where B is a positive constant.

2) $G_k(x-\nu, t) = G_k(x, t+\nu)$, *for all* $\nu \in \mathbf{R}^d$.

3) $G_k(\nu, t) = 0$, *for all* $\nu \in \mathbf{Z}^d$.

4) $G_k(x,t) = G_k(t,x)$ *for all x and t in* $\mathbf{R}^d$.

Theorem 51 *For $1 \leqslant p \leqslant \infty$ and $2k \geqslant d + 1$, let $f \in L_p^{2k}(\mathbf{R}^d)$.*

Then

$$\| f - S_{\sigma,k}f \|_p \leqslant \sigma^{-2k} C_{kp} \| \Delta^k f \|_p. \tag{38}$$

Here the constant C_{kp} depends only on k and p.

Remark When $p = \infty$, (38) was also founded by Buhmann[86] by another way.

Corollary 10 For $1 \leqslant p \leqslant \infty$, let $\varphi \in L_p(\mathbf{R}^d)$ and set

$$\Phi(x) = \int_{\mathbf{R}^d} G_k(x,t)\varphi(t)\mathrm{d}t.$$

Then, $\Phi \in L_p(\mathbf{R}^d)$ and $\Phi(\nu) = 0$ for all $\nu \in \mathbf{Z}^d$.

1.1.2 The average dimension of $SH^k(\mathbf{R}^d) \cap L_p(\mathbf{R}^d)$

In [87], Liu obtain

Theorem 52 *Let $\sigma > 0, 1 \leqslant p \leqslant \infty$ and $2k \geqslant d+1$. Then*

$$\overline{\dim}(SH_\sigma^k(\mathbf{R}^d) \cap L_p(\mathbf{R}^d), L_p(\mathbf{R}^d)) = \sigma^d.$$

1.1.3 Average σ-K width of $W_p(\Delta^k)$ in $L_p(\mathbf{R}^d)$

In [87], Liu obtain

Theorem 53 Let $2k \geqslant d+1$ and $1 \leqslant p \leqslant \infty$. Then

1) $\bar{d}_\sigma(W_p(\Delta^k), L_p(\mathbf{R}^d)) \asymp \sigma^{-2k/d}, \sigma \geqslant 1$.

2) The space $SH_\sigma^{k_{1/d}}(\mathbf{R}^d) \cap L_p(\mathbf{R}^d)$ is asymptoticly optimal for $\bar{d}_\sigma(W_p(\Delta^k), L_p(\mathbf{R}^d))$. Here "$a(\sigma) \asymp b(\sigma)$" means that $C_1 \leqslant |a(\sigma)/b(\sigma)| \leqslant C_2$ for $\sigma \geqslant 1$, while C_1 and C_2 are two positive constant depending only on k and p.

1.2 Exact results on the best approximation for some periodic class in $L_2(\mathrm{Q}^d)$ and some nonperiodic class in $L_2(\mathbf{R}^d)$

Let X, Y and W be Banach spaces, V an operator from W to Y, $\mathfrak{M}$ a subset *of* W, Φ a set of operators from Y to X. In addition, let A be a given operator with domain of definition $\mathfrak{D}(A) \subset W, \mathfrak{M} \subset \mathfrak{D}(A)$, and range $\mathscr{R}(A) \subset X$. The following problem often arises in the theory of approximation

$$E_\Phi(A, \mathfrak{M}, V, X) =: \inf_{T \in \Phi} \sup_{x \in \mathfrak{M}} \| Ax - TVx \|_X. \tag{39}$$

In particular, when $X = Y = W, A = V = \mathrm{id}$(the identity operator), and Φ is the set of all opreators from X to M for a given subspace M of X, the quantity (39) coincides with

$$E(\mathfrak{M}, M, X) =: \sup_{x\in\mathfrak{M}}\inf_{y\in M} \| x - y \|_X. \tag{40}$$

When Φ^a is the set of all operators from Y to X, we simply write $E(\cdot,\cdot,\cdot,\cdot)$ instead of $E_{\Phi_a}(\cdot,\cdot,\cdot,\cdot)$.

For $X = W = L_2[0,2\pi]$, $Y = \mathbf{R}^{2m}$ (the Euclidean space of $2m$-dimension), $A_s = \frac{\mathrm{d}^s}{\mathrm{d}t^s}$ (the s-th derivative operator, $A_0 = \mathrm{id}$) and $V_{2m} = \{x(0), x(\pi/m), \cdots, x((2m-1)\pi/m)\}$ for all $x(t) \in \widetilde{W}_2^r[0,2\pi]$, Subbotin[89] obtained the following result

$$E(A_s, \widetilde{W}_2^r[0,2\pi], V_{2m}, L_2[0,2\pi]) = m^{-r+s}, \text{ for } s = 0,1,\cdots,r.$$

Here $\widetilde{W}_2^r[0,2\pi]$, $r \in \mathbf{Z}_+ = \{1,2,\cdots\}$, denotes the class of continuous 2π-periodic function $x(t)$ for which the $(r-1)$st derivative is absolutely continuous and the r-th derivative satisfies

$$\| x^{(r)} \|_{L_2[0,2\pi]} =: \left\{ \frac{1}{\pi}\int_0^{2\pi} | x^{(r)}(t) |^2 \mathrm{d}t \right\}^{1/2} \leqslant 1.$$

For $X = W = L_2(\mathbf{R})$ and $Vf = \{f(v)\}_{v\in\mathbf{Z}}$, Sun Yongsheng and Li Chun[39] obtained

$$E(A_s, W_2^r(\mathbf{R}), V, L_2(\mathbf{R})) = \pi^{-r+s}, \text{ for } s = 0,1,\cdots,r.$$

Here $\mathbf{Z} = \{0, \pm 1, \pm 2, \cdots\}$, and $W_2^r(\mathbf{R})$ denotes the class of continuous functions $f(t) \in L_2(\mathbf{R})$ on the real axis R for which the $(r-1)$st derivative is locally absolutely continuous and the r-th derivative satisfies

$$\| f^{(r)} \|_{L_2(\mathbf{R})} =: \left\{ \int_{\mathbf{R}} | f^{(r)}(t) |^2 \mathrm{d}t \right\}^{1/2} \leqslant 1.$$

As for the best approximation problem (40), Sun and Li in [39] obtained

$$E(W_2^r(\mathbf{R}), S_m \cap L_2(\mathbf{R}), L_2(\mathbf{R})) = \sup_{f\in W_2^r(\mathbf{R})} \| f - s_{2r-1}f \|_{L_2(\mathbf{R})} = \pi^{-r}.$$

Here S_m, $m \geqslant r-1$, denotes the space of cardinal polynomial splines of degree $m-1$ with integer knots, i.e.,

$$S_m =: \{s \in C^{m-2}(\mathbf{R}) : s^{(m)}(t) = 0, \forall t \in \mathbf{R} - \mathbf{Z}\},$$

and $s_{2r-1}f \in S_{2r-1}$ is the polynomial spline which interpolates f at the points $\{v\}_{v\in\mathbf{Z}}$, i.e., $(s_{2r-1}f)(v) = f(v)$, for all $v \in \mathbf{Z}$. In [90], Liu generalized the above-mentioned results from univarite case to multivarite case respectively in the following subsections.

1.2.1 Exact results on the best approximation for some periodic class in $L_2(\mathbf{Q}^d)$

Let $k \in \mathbf{Z}_+, k \geqslant (d+1)/2$. Set

$$D_k^{\Delta}(x) = \sum_{v \in \mathbf{Z}^d - \{0\}} \frac{e^{ivx}}{(-\|v\|^2)^k}, \; vx = \sum_{j=1}^{d} v_j x_j,$$

for all $v = (v_1, v_2, \cdots, v_d)$ in $\mathbf{Z}^d$ and $x = (x_1, x_2, \cdots, x_d)$ in $\mathbf{R}^d$. For $1 \leqslant p \leqslant \infty$, let

$$\widetilde{L}_p(\Delta^k) = \{c + D_k^{\Delta} * \phi : c \in \mathbf{R}, \phi \in L_p(Q^d), \int_{\mathbf{Q}^d} \phi(t)\mathrm{d}t = 0\},$$

and set

$$\widetilde{W}(\Delta^k) = \{f = c + D_k^{\Delta} * \phi \in \widetilde{L}_p(\Delta^k) : \|\phi\|_{L_p(Q^d)} \leqslant 1\}.$$

Here $\mathbf{Q}^d =: \{x = (x_1, x_2, \cdots, x_d) \in \mathbf{R}^d : -\pi \leqslant x_j \leqslant \pi, j = 1, 2, \cdots, d\}$, and $D_k^{\Delta} * \phi$ denotes the convolution of D_k^{Δ} and ϕ defined by

$$D_k^{\Delta} * \phi(x) =: \frac{1}{(2\pi)^d} \int_{\mathbf{Q}^d} D_k^{\Delta}(x-t)\phi(t)\mathrm{d}t.$$

For $N \in \mathbf{Z}_+$, we denote by $V_N f$ the information of f defined by $V_N f = \{f(v\pi/N)\}_{v \in \square_N}$. Here $\square_N =: \{(v_1, v_2, \cdots, v_d) \in \mathbf{Z}^d : -N \leqslant v_j < N, j = 1, 2, \cdots, d\}$.

So far as we know, this class of functions, $\widetilde{W}_p(\Delta^k)$, was first introduced by Bugrov[91] for $d = 2$ and several authors considered some extremal problems on it(for instance, see [92] and [93]). Here we get the following two theorems.

Theorem 54 *Let $m \geqslant k$. Then*

$$E(\widetilde{W}_2(\Delta^k), \widetilde{SH}_N^m, L_2(\mathbf{Q}^d)) = \sup_{f \in \widetilde{W}_2(\Delta^k)} \|s_{N,2k} f - f\|_{L_2(\mathbf{Q}^d)} = \frac{1}{N^{2k}}.$$

Here $\widetilde{SH}_N^m$ denotes the periodic polyharmonic splines at knots $\{v\pi/N\}_{v \in \square_N}$, i.e.,

$$\widetilde{SH}_N^m =: \{g(t) \in C^{2m-d-1}(\mathbf{R}^d) : g(t + 2\pi v) = g(t),$$

$$\forall v \in \mathbf{Z}^d, (\Delta^m g)(t) = 0, \forall t \in \mathbf{R}^d - \frac{\pi}{N}\mathbf{Z}^d \Big\},$$

and $S_{N,2k}f$ in $\widetilde{SH}_N^{2k}$ is the interpolation spline which interpolates f at points $\{v\pi/N\}_{v\in\square_N}$.

Theorem 55 *Let $s = 0, 1, 2, \cdots, k$. Then*

$$\begin{aligned} & E(\Delta^s, \widetilde{W}_2(\Delta^k), V_N, L_2(\mathbf{Q}^d)) \\ = & \sup_{f\in \widetilde{W}_2(\Delta^k)} \| \Delta^s s_{N,2k}f - \Delta^s f \|_{L_2(\mathbf{Q}^d)} \\ = & \sup\{ \| \Delta^2 f \|_{L_2(\mathbf{Q}^d)} : V_N f = 0, f \in \widetilde{W}_2(\Delta^k) \} = \frac{1}{N^{2(k-s)}}, \end{aligned}$$

where $V_N f = 0$ means that $f(v\pi/N) = 0$ for all $v \in \square_N$.

1.2.2 Exact results on the best approximation for some nonperiodic class in $L_2(\mathbf{R}^d)$

In [90], Liu obtained

Theorem 56 *Let $k \in \mathbf{Z}_+, 2k \geqslant d+1$, and $s = 0, 1, 2, \cdots, k-1$. then*

$$E(\Delta^s, W_2(\Delta^k), V, L_2(\mathbf{R}^d)) = \sup\{ \| \Delta^s f \|_{L_2(\mathbf{R}^d)} : Vf = 0, f \in W_2(\Delta^k) \} =$$

$$\sup_{f\in W_2(\Delta^k)} \| \Delta^s f - \Delta^s s_{1,2k} f \|_{L_2(\mathbf{R}^d)} = \frac{1}{\pi^{2(k-s)}},$$

where $Vf = 0$ means that $f(v) = 0$ for all $v \in \mathbf{Z}^d$, and $s_{1,2k}f \in SH^{2k}(\mathbf{R}^d)$ is the cardinal interpolation which interpolates f at points $\{v\}_{v\in\mathbf{Z}^n}$, i.e., the liniear operator $s_{1,2k}$ satisfies the interpolation condition

$$s_{1,2k} f(v) = f(v),$$

for all $v \in \mathbf{Z}^d$.

Theorem 57 *Let m and k be two integers, and $m \geqslant k \geqslant (d+1)/2$. Then*

$$E(W_2(\Delta^k), \mathscr{F}_2^m(\mathbf{R}^d), L_2(\mathbf{R}^d)) = \sup_{f\in W_2(\Delta^k)} \| f - s_{1,2k}f \|_{L_2(\mathbf{R}^d)} = \frac{1}{\pi^{2k}}.$$

Corollary 11 *Let $2k \geqslant d+1$ and $m \geqslant k$. Then*

$$\begin{aligned} & E(W_p(\Delta^k), \mathscr{F}_p^m, L_p(\mathbf{R}^d)) \\ = & \sup\Big\{ \int_{\mathbf{R}^d} \Delta^k f(t)(\Delta^{m-k} g)(t)\mathrm{d}t : f \in W_p(\Delta^k), g \in W_p(\Delta^m), Vg = 0 \Big\}, \end{aligned}$$

for $1 < p < \infty$.

Here we have put $\mathscr{F}_2^m(\mathbf{R}^d)$ in the place of $SH^m(\mathbf{R}^d)\cap L_2(\mathbf{R}^d)$.

2. Average width problems on some classes of multivariate smooth functions on $\mathbf{R}^d$

2.1 The results in the spaces with mixed norm

Let $\boldsymbol{p}:=(p_1,p_2,\cdots,p_d), 1\leqslant p_j\leqslant\infty, i=1,2,\cdots,d, I=(a,b), -\infty\leqslant a<b\leqslant\infty$, and $L_{\boldsymbol{p}}(I^d)$ denote the Banach space of measurable functions $x(\cdot)$ on I^d with the mixed norm

$$\| x(\cdot)\|_{L_{\boldsymbol{p}}(I^d)}:=\left(\int_I \mathrm{d}t_d\left(\int_I \mathrm{d}t_{d-1}\cdots\left(\int_I |x(t)|^{p_1}dt_1\right)^{p_2/p_1}\cdots\right)^{p_2/p_1}\right)^{1/p_d}.$$

When $\boldsymbol{p}:=(p,p,\cdots,p)$, $L_{\boldsymbol{p}}$ coincides with the usual space $L_p(I^d)$. For each writing, we denote $\mathbf{1}=(1,\cdots,1)$, $\mathbf{2}=(2,2,\cdots,2)$, $\infty=(\infty,\infty,\cdots,\infty)$. If $\boldsymbol{p}=(p_1,\cdots,p_d)$, $\boldsymbol{q}=(q_1,\cdots,q_d)$, $1\leqslant p_j,q_j\leqslant\infty, j=1,\cdots,d$, then $\boldsymbol{p}\leqslant\boldsymbol{q}\,(\boldsymbol{p}<\boldsymbol{q})$ means $p_j\leqslant q_j\,(p_j<q_j)$, $j=1,\cdots,d$.

For $\boldsymbol{r}=(r_1,\cdots,r_d)\in\mathbf{N}^d$, let $W_{\boldsymbol{p}}^{\boldsymbol{r}}(\mathbf{R}^d)$ the Sobolev classes, i.e.,

$$W_{\boldsymbol{p}}^{\boldsymbol{r}}(\mathbf{R}^d)=\left\{x(\cdot):\| x(\cdot)\|_{L_{\boldsymbol{p}}(\mathbf{R}^d)}+\sum_{j=1}^d\left\|\frac{\partial^{r_j}x(\cdot)}{\partial x_j^{r_j}}\right\|_{L_{\boldsymbol{p}}(\mathbf{R}^d)}\leqslant 1\right\}.$$

In [94, 95], Magaril-I'llyaev obtained the following theorem.

Theorem 58 *Let* $\boldsymbol{r}=(r_1,\cdots,r_d)\in\mathbf{N}^d$, $\mathbf{1}<\boldsymbol{p}:=(p_1,\cdots,p_d)=\boldsymbol{q}=:(q_1,\cdots,q_d)<\infty$, *or* $\mathbf{1}\leqslant\boldsymbol{p}:=(p_1,\cdots,p_d)\leqslant\boldsymbol{q}=:(q_1,\cdots,q_d)\leqslant\mathbf{2}$ *if* $\boldsymbol{p}\neq\boldsymbol{q}$ *and* $1-\sum_{j=1}^d\frac{1}{r_j}\left(\frac{1}{p_j}-\frac{1}{q_j}\right)>0$. *Then*

$$\overline{d}_\sigma(W_{\boldsymbol{p}}^{\boldsymbol{r}}(\mathbf{R}^d)), L_{\boldsymbol{q}}(\mathbf{R}^d))\asymp\sigma^{-Y},$$

where $Y=\left(1-\sum_{j=1}^d\frac{1}{r_j}\left(\frac{1}{p_j}-\frac{1}{q_j}\right)\right)\left(\sum_{j=1}^d\frac{1}{r_j}\right)^{-1}$.

Remark 9 Tikhomirov and Margril-I'llyaev considered [95, 96] the average width problems in $L_p(G)$ for a Local Compact Abel group G and obtained some exact results some classes in $L_2(G)$.

2.2 Results on the isotropic Besov classes

Suppose that $k\in\mathbf{N}$, $r>0$ and $k-r>0$. For each $f\in L_p(\mathbf{R}^d)$,

$$\Delta_t^k f(x)=\sum_{l=0}^k(-1)^{l+k}\binom{k}{l}f(x+lt)$$

is the kth difference of the function f at the point x with step t. When $t = (0, \cdots, 0, t_j, 0\cdots, 0)$, we denote $\Delta_t^k f(x)$ by $\Delta_{t_j}^k f(x)$.

Definition 21 *Let $k \in \mathbf{N}$, $r > 0$, $k - r > 0$, $1 \leqslant \theta \leqslant \infty$, and $1 \leqslant p < \infty$. We say $f \in B_{p\theta}^r(\mathbf{R}^d)$ if f satisfies the following conditions.*

1) $f \in L_p(\mathbf{R}^d)$,

2) $\| f \|_{b_{p\theta}^r(\mathbf{R}^d)} :=$

$$\begin{cases} \left\{\int_{\mathbf{R}^d}\left(\frac{\| \Delta_t^k f(\cdot) \|_p}{| t |^r}\right)^\theta \frac{\mathrm{d}t}{| t |^d}\right\}^{1/\theta} < \infty, & 1 \leqslant \theta < \infty, \\ \sup_{| t | \neq 0} \frac{\| \Delta_t^k f(\cdot) \|_p}{| t |^r} < \infty, & \theta = \infty. \end{cases}$$

By [82], the linear space $B_{p\theta}^r(\mathbf{R}^d)$ is a Banach space with the norm

$$\| f \|_{B_{p\theta}^r(\mathbf{R}^d)} := \| f \|_p + \| f \|_{b_{p\theta}^r(\mathbf{R}^d)},$$

and is called to be isotropic Besov space.

For a real vector $\boldsymbol{M} = (\boldsymbol{M}_1, \cdots, \boldsymbol{M}_d)$, $\boldsymbol{M}_j > 0$, $j = 1, \cdots, d$, we define

$$S_{p\theta}^r b(\mathbf{R}^d) := \{f \in L_p(\mathbf{R}^d) : \| f \|_{b_{p\theta}^r(\mathbf{R}^d)} \leqslant 1\},$$

$$S_{p\theta}^r B(\mathbf{R}^d) := \{f \in L_p(\mathbf{R}^d) : \| f \|_{B_{p\theta}^r(\mathbf{R}^d)} \leqslant 1\}.$$

Let $\rho > 0$, $\nu = (\nu_1, \cdots, \nu_d)$, $\nu_i > 0$, $i = 1, \cdots, d$, and $1 \leqslant p < \infty$. Denote by $SB_\rho^p(\mathbf{R}^d)$ the restriction to $\mathbf{R}^d$ of the space of all functions of spherical exponential type ρ which belong to $L_p(\mathbf{R}^d)$ and, similarly, by $B_\nu^p(\mathbf{R}^d)$ the exponential type ν.

The approximation features of periodic Besov classes have been studied by many authors [97, 98]. In [99], Jiang and Liu study the problems of the average widths and the optimal recovery of the Besov classes $S_{p\theta}^r b(\mathbf{R}^d)$, $S_{p\theta}^r B(\mathbf{R}^d)$, and obtain the following results.

Theorem 59 *Let $k \in \mathbf{N}$, $r > 0$, $k - r > 0$, $1 \leqslant \theta \leqslant \infty$, $1 \leqslant p < \infty$, and $\sigma \geqslant 1$. Then*

$$\sigma^{-r/d} \ll \overline{d_\sigma}(\mathfrak{M}, L_p(\mathbf{R}^d)) \leqslant \overline{d'}_\sigma(\mathfrak{M}, L_p(\mathbf{R}^d)) \ll \sigma^{-r/d},$$

where $\mathfrak{M} = S_{p\theta}^r b(\mathbf{R}^d)$ or $S_{p\theta}^r B(\mathbf{R}^d)$.

Here $SB_{\rho(\sigma)}^p(\mathbf{R}^d)$ is a weakly asymptotic optimal subspace of average dimension σ for $\overline{d}_\sigma(\mathfrak{M}, L_p(\mathbf{R}^d))$, where $\rho(\sigma) \geqslant 0$ is determined by the equation $\rho(\sigma)^d$ mes $B_d(0,1) = (2\pi)^d\sigma$, while $B_d(0,1) = \{x \in \mathbf{R}^d;$ $|x| \leqslant 1\}$ *is the unit ball of the* Euclidean *space* $\mathbf{R}^d$.

In the sequel, what we see the constants $c_1, c_1', c_2, \cdots$, and so on, are dependent only on d, p, θ, and r.

2.3 Results on the anisotropic Besov classes

Definition 22 *Let* $\boldsymbol{k} = (k_1, \cdots, k_d) \in \mathbf{Z}_+^d$, $\boldsymbol{r} = (r_1, \cdots, r_d)$, $r_j > 0$, $k_j > r_j$, $j = 1, \cdots, d$, $1 \leqslant \theta \leqslant \infty$, and $1 \leqslant p < \infty$. *We say* $f \in B_{p\theta}^{\boldsymbol{r}}(\mathbf{R}^d)$ *if it satisfies the following conditions.*

1) $f \in L_p(\mathbf{R}^d)$,

2) $\| f \|_{b_{x_j p\theta}^{r_j}(\mathbf{R}^d)} :=$

$$\begin{cases} \left\{ \int_{\mathbf{R}^d} \left(\dfrac{\| \Delta_{t_j}^{k_j} f(\cdot) \|_p}{|t_j|^{r_j}} \right)^\theta \dfrac{\mathrm{d}t_j}{|t_j|} \right\}^{1/\theta} < \infty, & 1 \leqslant \theta < \infty, \\ \sup_{t_j \neq 0} \dfrac{\| \Delta_{t_j}^{k_j} f(\cdot) \|_p}{|t_j|^{r_j}} < \infty, & \theta = \infty, \end{cases}$$

for $j = 1, \cdots, d$.

By [82], the linear space $B_{p\theta}^{\boldsymbol{r}}(\mathbf{R}^d)$ is a Banach space with the norm

$$\| f \|_{B_{p\theta}^{\boldsymbol{r}}(\mathbf{R}^d)} := \| f \|_p + \sum_{j=1}^d \| f \|_{b_{x_j p\theta}^{r_j}(\mathbf{R}^d)},$$

and is called to be anisotropic Besov space. When $\theta = \infty$, $B_{p\theta}^{\boldsymbol{r}}(\mathbf{R}^d)$ and $B_{p\theta}^{r}(\mathbf{R}^d)$ coincide with the Hölder-Nikolskii spaces $H_p^{\boldsymbol{r}}(\mathbf{R}^d)$ and $H_p^{r}(\mathbf{R}^d)$, respectively.

For a real vector $\boldsymbol{M} = (M_1, \cdots, M_d)$, $M_j > 0$, $j = 1, \cdots, d$, we define

$$S_{p\theta}^{\boldsymbol{r}} b(\mathbf{R}^d) := \left\{ f \in L_p(\mathbf{R}^d) : \| f \|_{b_{x_j p\theta}^{r_j}(\mathbf{R}^d)} \leqslant \boldsymbol{M}_j, j = 1, \cdots, d \right\},$$

$$S_{p\theta}^{\boldsymbol{r}} B(\mathbf{R}^d) := \left\{ f \in L_p(\mathbf{R}^d) : \| f \|_{B_{p\theta}^{\boldsymbol{r}}(\mathbf{R}^d)} \leqslant 1 \right\}.$$

Let $\rho > 0$, $\nu = (\nu_1, \cdots, \nu_d)$, $\nu_i > 0$, $i = 1, \cdots, d$, and $1 \leqslant p < \infty$. Denote by $SB_\rho^p(\mathbf{R}^d)$ the restriction to $\mathbf{R}^d$ of the space of all functions of spherical exponential type ρ which belong to $L_p(\mathbf{R}^d)$ and, similarly, by $B_\nu^p(\mathbf{R}^d)$ the exponential type ν.

The approximation features of periodic Besov classes have been studied by many authors[97,98]. In [99], Jiang and Liu study the problems of the average widths and the optimal recovery of the Besov classes $S^r_{p\theta}b(\mathbf{R}^d)$ and $S^r_{p\theta}B(\mathbf{R}^d)$, and obtain the following results.

Theorem 60 *Let* $k = (k_1, \cdots, k_d) \in \mathbf{Z}^d_+$, $r = (r_1, \cdots, r_d)$, $k_j > r_j > 0$, $j = 1, \cdots, d$, $1 \leqslant \theta \leqslant \infty$, $1 \leqslant p < \infty$, *and* $\sigma \geqslant 1$. *Then*

$$\mu\sigma^{-a} \ll \bar{d}_\sigma(\mathfrak{U}, L_p(\mathbf{R}^d)) \leqslant \bar{d}\,'_\sigma(\mathfrak{U}, L_p(\mathbf{R}^d)) \ll \mu\sigma^{-a},$$

where $\quad \mathfrak{U} = S^r_{p\theta}b(\mathbf{R}^d)$ *or* $S^r_{p\theta}B(\mathbf{R}^d)$, $\quad a = \left(\sum_{j=1}^d \frac{1}{r_j}\right)^{-1}$,

$$\mu = \prod_{j=1}^d M_j^{a/r_j} (\text{when } \mathfrak{U} = S^r_{p\theta}B(\mathbf{R}^d), \mu = 1).$$

Here $B^p_{\rho(\sigma)}(\mathbf{R}^d)$ is a weakly asymptotic optimal subspace of average dimension $\leqslant \sigma$ for $\bar{d}_\sigma(\mathfrak{U}, L_p(\mathbf{R}^d))$, where $\rho(\sigma) = (\rho_1(\sigma), \cdots, \rho_d(\sigma))$, $\rho_i(\sigma) > 0$, is defined by $\rho_j(\sigma) = (\mu^{-1}M_j \cdot \sigma^a)^{1/r_j}$ (when $\mathfrak{U} = S^r_{p\theta}B(\mathbf{R}^d)$, $\rho_j(\sigma) = \sigma^{a/r_j}$), $j = 1, \cdots, d$.

2.4 Average widthes of $W^rH^\alpha(\mathbf{R}^d)$ in $C^k(\mathbf{R}^d)$

Let $k = (k_1, k_2, \cdots, k_d) \in \mathbf{Z}^d_+$. For $I^d = \mathbf{R}^d$ or a cube, we denote by $C^k(I^d)$ the function space

$$C^k(I^d) =: \{f \in C(I^d) : D_j^{k_j} f \in C(I^d), j = 1, 2, \cdots, d\},$$

with the norm[82]

$$\| f \|_{C^k(I^d)} =: \| f \|_{C(I^d)} + \sum_{j=1}^d \| D_j^{k_j} f \|_{C(I^d)},$$

where $D_j^{k_j} f(x) = \dfrac{\partial^{k_j} f(x)}{\partial x_j^{k_j}}$, $j = 1, 2, \cdots, d$.

Let $r = (r_1, r_2, \cdots, r_d) \in \mathbf{Z}^d_+$, $\alpha = (\alpha_1, \alpha_2, \cdots, \alpha_d)$, $0 < \alpha_j \leqslant 1$, $j = 1, 2, \cdots, d$. The anisotropic Hölder class $W^rH^\alpha(\mathbf{R}^d)$ is defined as follows

$$W^rH^\alpha(\mathbf{R}^d) =: \{f(x) \in C^r(\mathbf{R}^d) : | D_j^{r_j} f(x + he_j) - D_j^{r_j} f(x) | \leqslant | h_j |^{\alpha_j},$$
$$\forall x \in \mathbf{R}^d, \forall h \in \mathbf{R}, j = 1, 2, \cdots, d\},$$

where $e_j \in \mathbf{R}^d$ is the unit element with the j-th coordinate to be 1, $j = 1, \cdots, d$.

In [100], Luo and Sun discussed the problems of several widths of the anisotropic class $W^rH^\alpha([0,1]^d)$ of smooth functions defined on the cube $[0,1]^d$ in the metric $C^k([0,1]^d)$. After that, Luo and Liu[101] generalized the results in [100] on the function classes defined on the cube $[0,1]^d$ to those on the function classes defined on the $\mathbf{R}^d$ and obtained the average width of the function class $W^r H^\alpha(\mathbf{R}^d)$ in the space $C^k(\mathbf{R}^d)$.

Theorem 61 *Let* $0 \leqslant k_j \leqslant r_j, j = 1, 2, \cdots, d$. *Then*

$$\bar{d}_\sigma(W^r H^\alpha(\mathbf{R}^d), C^k(\mathbf{R}^d)) \asymp \sigma^{-\rho(1-\beta)},$$

where

$$\rho^{-1} = \sum_{j=1}^{d} \frac{1}{r_j + \alpha_j}, \beta = \max_{1 \leqslant j \leqslant d} \frac{k_j}{r_j + \alpha_j}.$$

$\bar{d}_\sigma(\mathfrak{M}, X)$ *denotes the* σ-*K average width of* $\mathfrak{M}$ *in* X.

2.5 Average widths of function classes with mix smoothness

Let $\mathbf{R}^d$ be the d-dimensional real Euclidean space. For $D \subset \mathbf{R}^d$, $1 \leqslant p, q < \infty$. When $x = (x_1, x_2, \cdots, x_d) \in \mathbf{R}^d$, we define that

$$x \geqslant 0 (\text{or } x > 0) \Leftrightarrow x_i \geqslant 0 (\text{or } x_i > 0), i = 1, \cdots, d.$$

Let $e \subset e_d := \{1, 2, \cdots, d\}$. We define $x^e := (x_1^e, \cdots, x_d^e)$, in which $x_i^e := x_i$ or 0 depends on $i \in e$ or not. Let $N^d := \{s \in \mathbf{Z}^d \mid s \geqslant 0\}$. For $r > 0, r \in \mathbf{Z}^d$, then Sobolev-Wiener space S_{pq}^r with a certain mixed smoothness is defined as

$$S_{pq}^r := \{f \in L_{pq}(\mathbf{R}^d) : \| f \|_{S_{pq}^r L} = \sum_{e \subset e_d} \| D^{r^e} f \|_{pq} < \infty\},$$

where $D^s f(x) := \dfrac{\partial^{|s|}}{\partial x_1^{s_1} \cdots \partial_d^{s_d}} (|s| = s_1 + s_2 + \cdots + s_d), s \in \mathbf{N}^d$. In the following, we always suppose that $s \in \mathbf{N}^d$. When $p = q$, the space $S_q^r L(\mathbf{R}^d) := S_{qq}^r L(\mathbf{R}^d)$ is the Sobolev space with mixed smoothness, it is first initiated and studied by Amanov.[82]

Denote

$$S_{pq}^r W := \{f \in L_q(\mathbf{R}^d) \mid \| f \| S_{pq}^r L \leqslant 1\}.$$

When $1 \leqslant q \leqslant p < \infty$, it is easy to verify that $S_{pq}^r W \subset S_q^r W := S_{qq}^r W$.

For $s \in \mathbf{N}^d$, let

$$Q_{2^s}^* := \{\lambda \in \mathbf{R}^d \mid \eta(s_j) 2^{s_j - 1} \leqslant |\lambda_j| < 2^{s_j}, j = 1, \cdots, d\}, \rho(s) := Q_{2^s}^* \cap \mathbf{Z}^d,$$

where $\eta(t) = \mathbf{1}$ or 0 according to $t > 0$ or not.

Let $\boldsymbol{r} = r_1 \gamma, \boldsymbol{r} \in \mathbf{N}^d, \gamma = (\gamma_1, \cdots, \gamma_d), \mathbf{1} = \gamma_1 = \cdots = \gamma_\mu <$

$\gamma_{\mu+1} \leqslant \cdots \leqslant \gamma_d (1 \leqslant \mu \leqslant d)$. Without loss of generality for any $\boldsymbol{r} > \boldsymbol{0}, \boldsymbol{r} \in \mathbf{Z}^d$, we may assume that $\boldsymbol{r} = r_1 \gamma, r_1 \in \mathbf{Z}_+$. We make use of the following notations. For every $n \in \mathbf{Z}_+$, we define

$$Q_n^{\gamma} := \bigcup_{(s,\gamma) \leqslant n} Q_2^*, (s, \gamma) = s_1 \gamma_1 + s_2 \gamma_2 + \cdots + s_d \gamma_d,$$

which is called a step hyperbolic cross. Let $f \in L_q(\mathbf{R}^d)$. Denote by $\mathfrak{F} f$ the Fourier transformation of f (generally speaking, in the sense of Schwarz distribution), by $\mathfrak{F}^{-1} f$ the inverse Fourier transformation of f. Supp $\mathfrak{F} f$ is called the spectral of f. Denote

$$E_q(Q_n^{\gamma}) = \{f \in L_q(\mathbf{R}^d) \mid \operatorname{supp} \mathfrak{F} f \subset Q_n^{\gamma}\},$$

which is a subspace of $L_q(\mathbf{R}^d)$ consisting of exponential type entire functions defined on $\mathbf{R}^d$ whose spectrals lie in step hyperbolic cross Q_n^{γ} (see [102]).

Using the Litterwood-Paley block Theorem, [102, 103] get.

Theorem 62 *Suppose* $\boldsymbol{r} = r_1 \gamma, \boldsymbol{r} \in \mathbf{N}^d, \gamma = (\gamma_1, \cdots, \gamma_d)$, $1 = \gamma_1 = \cdots = \gamma_{\mu} < \gamma_{\mu+1} \leqslant \cdots \leqslant \gamma_d (1 \leqslant \mu \leqslant d)$. *When* $1 < q \leqslant p < \infty$, $r_1 > 0$, *we have*

$$\bar{d}_{\sigma}(S_{pq}^{\boldsymbol{r}} W, L_q(\mathbf{R}^d)) \asymp \left(\frac{(\log \sigma)^{\mu-1}}{\sigma} \right)^{r}.$$

Here $E_q(Q_n^{\gamma})$ is a weakly asymptotically optimal subspace of average dimension $\leqslant \sigma$ for $\bar{d}_{\sigma}(S_{pq}^{\boldsymbol{r}} W, L_q(\mathbf{R}^d))$, where $n = n(\sigma)$ satisfies $\sigma \asymp 2^n n^{\mu-1}$.

Theorem 63 *Suppose* $\boldsymbol{r} = r_1 \gamma, \boldsymbol{r} \in \mathbf{N}^d, \gamma = (\gamma_1, \cdots, \gamma_d), 1 = \gamma_1 = \cdots = \gamma_{\mu} < \gamma_{\mu+1} \leqslant \cdots \leqslant \gamma_d (1 \leqslant \mu \leqslant d)$. *When* $1 < p \leqslant q < 2$, $r_1 > 1/p - 1/q$, *we have*

$$\bar{d}_{\sigma}(S_{pq}^{\boldsymbol{r}} W, L_q(\mathbf{R}^d)) \asymp \left(\frac{(\log \sigma)^{\mu-1}}{\sigma} \right)^{r_1 - 1/p + 1/q}.$$

Here $E_q(Q_n^{\gamma})$ is a weakly asymptotically optimal subspace of average dimension $\leqslant \sigma$ for $\bar{d}_{\sigma}(S_{pq}^{\boldsymbol{r}} W, L_q(\mathbf{R}^d))$, where $n = n(\sigma)$ satisfies $\sigma \asymp 2^n n^{\mu-1}$.

3. Optimal recovery on some classes of mutivariate smooth functions on $\mathbf{R}^d$

For $\sigma > 0$, let Θ_{σ} be the set of all sequence of points $\xi = \{\xi_{\nu}\}_{\nu \in \mathbf{Z}^d}$

satisfying the following conditions: *

1) $|\xi_\nu| \leqslant |\xi_{\nu'}|$, if and only if $|\nu| \leqslant |\nu'|$ for $\nu, \nu' \in \mathbf{Z}^d$,

2) $\xi_\nu \neq \xi'_\nu$, if and only if $\nu \neq \nu'$ for $\nu, \nu' \in \mathbf{Z}^d$,

3) $$\overline{\text{card}}\xi := \liminf_{a\to\infty} \frac{\text{card}(\xi \cap [-a,a]^d)}{(2a)^d} \leqslant \sigma.$$

Here $|\cdot|$ is the usual Euclidean norm, and card E, $E \subset \mathbf{R}^d$, denotes the number of elements in the set E.

Let $X(\mathbf{R}^d)$ be a normed space of functions on $\mathbf{R}^d$ with the norm $\|\cdot\|_X$. Put

$$E(A,B,X) := \sup_{x(\cdot)\in A} \inf_{y(\cdot)\in B} \| x(\cdot) - y(\cdot) \|_X$$

for the subsets A, B of $X(\mathbf{R}^d)$.

Let $K \subset X(\mathbf{R}^d)$. The quantity

$$d(K) := \sup_{x(\cdot),\, y(\cdot)\in K} \| x(\cdot) - y(\cdot) \|_X$$

is called the diameter of K. For $\xi \in \Theta_\sigma$, the information of $f \in K$ is defined by $I_\xi f = \{f(\xi_\nu)\}_{\nu\in\mathbf{Z}^d}$. I_ξ is called a standard sampling operator of the average cardinality $\leqslant \sigma$. The quantity

$$\Delta_\sigma(K,X) := \inf_{\xi\in\Theta_\sigma} \sup_{f\in K} d(I_\xi^{-1} I_\xi f \cap K)$$

is called the *net width* or the minimum information diameter of the set K in the space $X(\mathbf{R}^d)$. If K is a balanced and convex subset of $X(\mathbf{R}^d)$, then

$$\Delta_\sigma(K,X) = 2 \inf_{\xi\in\Theta_\sigma} \sup\{ \| f \|_X : I_\xi f = 0, f \in K \}.$$

For any $\xi \in \Theta_\sigma$, a mapping $\varphi: I_\xi(K) \to X(\mathbf{R}^d)$ is called an algorithm and $\varphi \circ I_\xi f$ is called a recoverying function of f in $X(\mathbf{R}^d)$. Denote by Φ_ξ the set of all algorithms on K. If φ can be extended into a linear operator on the linearized set of K, we call the algorithm φ to be linear. Denote by Φ_ξ^L the set of all linear algorithms on the linearized set of K. The quantity

$$E_\sigma(K,X) := \inf_{\xi\in\Theta_\sigma} \inf_{\varphi\in\Phi_\xi} \sup_{f\in K} \| f - \varphi(I_\xi f) \|_X$$

is called the *minimum intrinsic error* of the optimal recovery of the set K in the space X. Taking Φ_ξ^L in the place of Φ_ξ in the right side of the last

---

* The reader is refered to Part Ⅰ, 4.

formula, we denote by $E_\sigma^L(K,X)$ the obtained quantity and call it the *minimum linear intrinsic error*. If K is a convex and centrally symmetric subset of X, then by [70][107], there hold the following inequalities

$$\frac{1}{2}\Delta_\sigma(K,X) \leqslant E_\sigma(K,X) \leqslant E_\sigma^L(K,X).$$

In [104], Liu studied the optimal recovery of Sobolev-Wiener classes in the metric $L_q(\mathbf{R}^d)$. Continuing Liu's work in [68], Jiang and Liu[99,105] studied the optimal recovery of isotropic Besov-Wiener classes $S_{pq\theta}^r B(\mathbf{R}^d)$ in the metric of $L_q(\mathbf{R}^d)$ and isotropic Besov classes $S_{p\ \theta}^r B(\mathbf{R}^d)$ in the metric of $L_{qp}(\mathbf{R}^d)$ and obtained the following theorems.

Theorem 64 *Let $k \in \mathbf{N}, r > d/p, k - r > 0, 1 \leqslant \theta \leqslant \infty$, $1 \leqslant p < \infty$, and $\sigma \geqslant 1$. Then*

$$\sigma^{-r/d} \ll \frac{1}{2}\Delta_\sigma(S_{p\ \theta}^r B(\mathbf{R}^d), L_p(\mathbf{R}^d)) \leqslant E_\sigma(S_{p\ \theta}^r B(\mathbf{R}^d), L_p(\mathbf{R}^d)) \leqslant E_\sigma^L(S_{p\ \theta}^r B(\mathbf{R}^d), L_p(\mathbf{R}^d)) \ll \sigma^{-r/d}.$$

Theorem 65 *Let $k \in \mathbf{N}, r > d/p, k - r > 0, 1 \leqslant \theta \leqslant \infty, 1 \leqslant q \leqslant p < \infty$, and $\sigma \geqslant 1$. Then*

$$\begin{aligned}\sigma^{-r/d} &\ll \frac{1}{2}\Delta_\sigma(S_{pq\theta}^r B(\mathbf{R}^d), L_p(\mathbf{R}^d)) \\ &\leqslant E_\sigma(S_{pq\theta}^r B(\mathbf{R}^d), L_p(\mathbf{R}^d)) \\ &\leqslant E_\sigma^L(S_{pq\theta}^r B(\mathbf{R}^d), L_p(\mathbf{R}^d)) \\ &\ll \sigma^{-r/d}.\end{aligned}$$

Theorem 66 *Let $k \in N, r > d/p, k - r > 0, 1 \leqslant \theta \leqslant \infty, 1 \leqslant q \leqslant p < \infty$, and $\sigma \geqslant 1$. Then*

$$\sigma^{-r/d} \ll \frac{1}{2}\Delta_\sigma(S_{p\ \theta}^r B(\mathbf{R}^d),\quad L_{qp}(\mathbf{R}^d) \leqslant E_\sigma(S_{p\ \theta}^r B(\mathbf{R}^d),$$
$$L_{qp}(\mathbf{R}^d)) \leqslant E_\sigma^L(S_{p\ \theta}^r B(\mathbf{R}^d),\quad L_{qp}(\mathbf{R}^d)) \ll \sigma^{-r/d}.$$

In [106], Jiang Yanjie studied the optimal recovery of anisotropic Besov-Wiener classes $S_{pq\theta}^r B(\mathbf{R}^d)$ in the metric of $L_q(\mathbf{R}^d)$ and anisotropic Besov classes $S_{p\ \theta}^r B(\mathbf{R}^d)$ in the metric of $L_{qp}(\mathbf{R}^d)$ and obtained the following theorems.

Theorem 67 *Let $\boldsymbol{k} = (k_1, \cdots, k_d) \in \mathbf{N}^d, \boldsymbol{r} = (r_1, \cdots, r_d)$,*

$k_j - r_j > 0, j = 1, \cdots, d, \sum_{j=1}^{d}(1/r_j) < 1, 1 \leqslant \theta \leqslant \infty, 1 \leqslant q \leqslant p < \infty,$ *and* $\sigma \geqslant 1$. *Then*

$$\sigma^{-a} \ll \frac{1}{2}\Delta_\sigma(S^r_{pq\theta}B(\mathbf{R}^d), \quad L_q(\mathbf{R}^d)) \leqslant E_\sigma(S^r_{pq\theta}B(\mathbf{R}^d),$$

$$L_q(\mathbf{R}^d)) \leqslant E^L_\sigma(S^r_{pq\theta}B(\mathbf{R}^d), \quad L_q(\mathbf{R}^d)) \ll \sigma^{-a},$$

where

$$a = \Big(\sum_{j=1}^{d}(1/r_j)\Big)^{-1}.$$

Theorem 68 *Let* $\boldsymbol{k} = (k_1, \cdots, k_d) \in \mathbf{N}^d, \boldsymbol{r} = (r_1, \cdots, r_d),$

$k_j - r_j > 0, j = 1, \cdots, d, \sum_{j=1}^{d}(1/r_j) < 1, 1 \leqslant \theta \leqslant \infty, 1 \leqslant q \leqslant p < \infty,$ *and* $\sigma \geqslant 1$. *Then*

$$\sigma^{-a} \ll \frac{1}{2}\Delta_\sigma(S^r_{p\ \theta}B(\mathbf{R}^d), \quad L_{qp}(\mathbf{R}^d)) \leqslant E_\sigma(S^r_{pq\theta}B(\mathbf{R}^d),$$

$$L_{qp}(\mathbf{R}^d)) \leqslant E^L_\sigma(S^r_{pq\theta}B(\mathbf{R}^d), \quad L_{qp}(\mathbf{R}^d)) \ll \sigma^{-a}.$$

where

$$a = \Big(\sum_{j=1}^{d}(1/r_j)\Big)^{-1}.$$

Luo and Liu[101] obtained

Theorem 69 *Let* $0 \leqslant k_j \leqslant r_j, j = 1, \cdots, d$. *Then*

$$\sigma^{-\rho(1-\beta)} \ll \frac{1}{2}\Delta_\sigma(W^r H^\alpha(\mathbf{R}^d), \quad C^k(\mathbf{R}^d)) \leqslant E_\sigma(W^r H^\alpha(\mathbf{R}^d),$$

$$C^k(\mathbf{R}^d)) \leqslant E^L_\sigma(W^r H^\alpha(\mathbf{R}^d), \quad C^k(\mathbf{R}^d)) \ll \sigma^{-\rho(1-\beta)},$$

where

$$\rho^{-1} = \sum_{j=1}^{d}\frac{1}{r_j + \alpha_j}, \beta = \max_{1 \leqslant j \leqslant d}\frac{k_j}{r_j + \alpha_j}.$$

References

[1] Favard J. Sur l'approximation des fonctions periodiques par des polinomes trigonometriques. C R Acad Sci, 1936, 203: 1122 ~ 1124

[2] Achiezer N, Krein M. On best approximation of periodic function. Dokl AN USSR, 1937, 15

[3] Nikolskii S M. Approximation in mean. Izv Acad Nauk USSR, 1946,

10:207 ~ 256

[4] Korneichuk N P. *Extremal Problems in Approximation Theory*. [Russian]. Moscow: Acad Press, 1976.253 ~ 303

[5] Tikhomirov V M. Best methods of approximation and interpolation of differentiable functions in the space $C[-1,1]$. Mat Sbornik, 1969, 80:290 ~ 304

[6] Ligun A A. Inequalities for upper bounds of functionals. Anal Math, 1976, 2:11 ~ 40

[7] Kolmogorov A N. Über die beste Annaberung von Functionen einer gegebenen Functionenklasses. Annals of Math, 1936, 37:107 ~ 110

[8] Tikhomirov V M. Diameter of sets in function spaces and the theory of best approximation. Uspehi Mat Nauk, 1960, 15:81 ~ 120

[9] Korneichuk N P. *Splines in Approximation Theory*. [Russian]. Moscow: Acad Press, 1984.250 ~ 281

[10] Korneichuk N P. *Exact Constants in Approximation Theory*. [Russian]. Moscow: Acad Press, 1987.340 ~ 375

[11] Korneichuk N P, Ligun A A, Doronin V D. *Approximation With Constraints*. [Russian]. Kiev: Naukova Dumka, 1982.185 ~ 234

[12] Sun Yongsheng. *The Theory of Approximation* (Ⅰ). Beijing: Beijing Normal Univ Press, 1989.411 ~ 519

[13] Sun Yongsheng, Fang Gensun. *The Theory of Approximation* (Ⅱ). Beijing: Beijing Normal Univ Press, 1990.1 ~ 334

[14] Pinkus A. *N-Widths in Approximation Theory*. Berlin: Spinger- Verlag, 1985

[15] Fang Gensun. PF density, Chakhkiev condition and some extremal problems. [Chinese]. Chinese Science Bulletin, 1988, 33:404 ~ 408

[16] Chakhiev M A. Linear differential operator with real spectra and optimal quadrature formula. [Russian]. Izv Acad Nauk(Math), 1984, 48:1078 ~ 1109

[17] Bernstein S N. Best approximation of continuous functions on real line by integral functions with given order. *Collected Works* (Ⅱ). Moscow: GOSIZDAT, 1946.371 ~ 375

[18] Krein M G. On best approximation of nonperiodic differentiable functions on the whole real axis. [Russian]. Dokl A N, 1938, 18

[19] Achiezer N I. Lecture on *Approximation Theory*. Moscow: Acad Press, 1964

[20] Ibragimov I I. *Theory of Approximation by Entire Functions*. [Russian]. Baku: ELM Press, 1979

[21] Tikhomirov V M. Extremal problems and approximation theory. [Chinese]. Advances in Math. 1990, 19:449 ~ 451

[22] Li Chun. Infinite dimensional widths of function classes (Ⅰ). Chinese Science Bulletin, 1990, 36:163 ~ 166

[23] Hirschman I I, Widder D V. *The Convolution Transform*. Princeton: Princeton Press, 1955

[24] Wiener N. On the representation of functions by trigonometrical integrals. Math Z, 1926, 24:575 ~ 616

[25] Wiener N. Tauberian theorems. Ann of Math, 1932, 33:1 ~ 100

[26] Holland F. Harmonic analysis on amalgams of L_p and l^q. J London Math Soc, 1975, 10:295 ~ 305

[27] Fourier J J F, Stewart J. Amalgams of L_p and l^q. Bulletin (New Series) of the Amer Math Soc, 1985, 13:1 ~ 22

[28] Schoenberg I J. Cardinal interpolation and spline functions. JAT, 1969, 2:167 ~ 206

[29] Jakimovski A, Russell D C. Interpolation by functions with m-th derivation in pre-assigned spaces. C. K. Chui ed. *Approx Theory* Ⅲ. New York: Acad Press, 1980. 531 ~ 536

[30] Schoenberg I J. Cardinal Spline Interpolation. MRC of University of Wisconsin-Madison, 1973

[31] Micchelli C A. Cardinal $\mathscr{L}$-splines. edited by S. Karlin: *Studies in Spline Function and Approximation*. New York: Acad Press, 1976, 203 ~ 250

[32] de Boor C, Schoenberg I J. Cardinal interpolation and splines functions. Ⅷ. Lecture Notes in Math, 1976, 501:1 ~ 79

[33] Novikov S I. Approximation of one differentiable function class by $\mathscr{L}$-splines. [Russian]. Mat Zametki, 1983, 33:393 ~ 408

[34] Tang Xuhui. Approximation of some classes of differentiable functions with cardinal $\mathscr{L}$-splines on **R**. ATA, 1987, 3(2):1 ~ 17

[35] Li Chun. On some extremal problems about cardinal $\mathscr{L}$-splines. Acta Math Sinica, 1990, 33:330 ~ 343

[36] Li Chun. Doctoral thesis. Academy of Sciences of China 1990

[37] Chen Dirong. Average n-widths and optimal interpolation of Sobolev class in $L_p(\mathbf{R})$. Chinese Ann Math, 1992, 13B(4):396 ~ 405

[38] Li Chun. $L_1(\mathbf{R})$-optimal recovery on some differentiable function classes. In

Approximation Optimization and Computing Proc of ISAOC'89. Elsevier/North Holland, 1990

[39] Sun Yongsheng, Li Chun. Optimal recovery for $W_2^r(\mathbf{R})$ in $L_2(\mathbf{R})$. Acta Math Sinica (New Series), 1991, 7(4): 309 ~ 323

[40] Chen Dirong. Optimal recovery and average n-widths of some function class in $L_2(\mathbf{R})$. ATA, 1991, 7(3): 45 ~ 62

[41] Magaril-I'llyaev G G. Average dimensions, widths and optimal recovery of Sobolev classes of functions on the real axis. [Russian]. Math Sbornik, 1991, 182: 1635 ~ 1656

[42] Pinkus A. N-Widths of Sobolev class. Constructive Approximation, 1985, 1: 15 ~ 62

[43] Chen Dirong. n-widths of some periodic function classes. Science Sinica (Ser A), 1991(10): 1041 ~ 1050; Science in China (Ser A), 1992, 35: 42 ~ 54

[44] Liu Yongping, Sun Yongsheng. Infinite dimensinal Kolmogorov width and optimal interpolation on Sobolev-Wiener space. Scientia Sinica (Ser A), 1992(6): 582 ~ 591; Science in China (Ser A), 1992, 35(10): 1162 ~ 1172

[45] Liu Yongping. Infinite dimensional widths and optimal recovery of some smooth function classes of $L_p(\mathbf{R})$ in metric $L(\mathbf{R})$. Kexue Tongbao, 1993, 38(18): 1642 ~ 1682; Chinese Science Bulletin, 1993, 38(20): 1678 ~ 1682; Analysis Math, 1993, 19: 169 ~ 182

[46] Fang Gensun, Liu Yongping. Average widths and optimal interpolation of Sobolev-Wiener classes $W_{pq}^r(\mathbf{R})$ in the metric $L_q(\mathbf{R})$. JAT, 1993, 74(3): 335 ~ 352

[47] Buslaev A, Tikhomirov V M. The spectra of nonlinear differential equations and widths of Sobolev space. Math Sbornik, 1990, 181: 1587 ~ 1606

[48] Sun Yongsheng, Li Chun. Best approximation of some classes of smooth functions on the whole real axis by cardinal splines of higher order. Mat Zametki, 1990, 48: 100 ~ 109

[49] de Boor C. Splines as linear combination of B splines. In: Lorentz G G, Chui C K, Schumaker L L, ed. *Approximation Theory*. New York, Acad Press, 1976. 1 ~ 47

[50] Ligun A A. Exact inequalities for perfect splines and its applications. [Russian]. Izv VUZ Math, 1985(5): 32 ~ 38

[51] Velikin V L. Asymptotic relation between various approximation methods of periodic function by splines and trigonometric polynomials. Analysis Math, 1987, 13: 45 ~ 74

[52] Genelius T. On one sided approximation by trigonometric polynomials. Math Scan, 1956, 4: 247 ~ 258

[53] Sun Yongsheng, Liu Yongping. Best one-sided approximation of some classes of smooth function on whole real axis by cardinal splines of higher order. ATA, 1991, 7(1): 108 ~ 123

[54] Zhang Jianhua. Best one-sided approximation of generalized Sobolev class by cardinal $\mathscr{L}$-splines. preprint

[55] Huang Daren, Sun Yongsheng. Optimal interpolation on some classes of differentiable functions. ATA, 1988, 4(3): 13 ~ 22

[56] Bojanov B D, Huang Daren. Periodic monosplines and perfect splines of least norm. Constructive Approximation, 1987, 3: 363 ~ 375

[57] Velikin V L. Optimal interpolation of differentiable periodic functions with bounded highest order derivative. Mat Zametki, 1977, 22(5): 663 ~ 671

[58] Sun Yongsheng, Liu Yongping. Optimal recovery of Sobolev-Wiéner class of smooth functions by double sampling. Constructive Approximation, 1993, 9: 391 ~ 405

[59] Liu Yongping. Optimal recovery of class of smooth functions defined on the whole real axis by multifold sampling. Kexue Tongbao, 1993, 38(14): 1266 ~ 1270; Chinese Science Bulletin, 1993, 38(20): 1683 ~ 1687; Acta Math Sinica (New Series), 1995, 11: 128 ~ 141

[60] Tikhomirov V M. *Approximation Theory*. In: Current Problems in Math Fundamental Directions. [Russian]. Moscow: Acad Press, 1987. 103 ~ 270

[61] Din Z. The average ε-dimension of function class $B_{G,p}$. Mat Zametki, 1980, 28(5): 727 ~ 736

[62] Chen Hanlin, Li Chun. Exact and asymptotic estimation for n-widths of some classes of periodic functions. Constructive Approximation, 1992, 8(3): 289 ~ 307

[63] Fang Gensun. Eigenvalue of integral operator and n-widths of classes of periodic functions. preprint

[64] Li Chun. Infinite dimensional widths of function classes (Ⅱ). J Approx Theory, 1992, 69(1): 15 ~ 34

[65] Magaril-I'llyaev G G. ϕ-average widths of function classes on the real axis. Uspehi Mat Nauk, 1990, 45: 211 ~ 212

[66] Jiang Yanjie, Liu Yongping. Average B-widths and infinite-dimensional G-width of some smooth function classes on the line. Kexue Tongbao, 1998,

43(7):701 ~ 703; Chinese Science Bulletin, 1998, 43(10):810 ~ 812; Approx Theory & its Appl, 1999, 15(1):50 ~ 63

[67] Liu Yongping. Average σ-K width of some generalized Sobolev-Wiener class. ATA, 1992, 8(4):79 ~ 88

[68] Liu Yongping. Average σ-K width of convolution function class of $L(\mathbf{R})$ in $L_2(\mathbf{R})$. J Math Research & Exposition, 1995, 15(2):199 ~ 207

[69] Ismagilov R S. On n-dimensional diameters of compacts in a Hilbert space. [Russian]. Functional Analysis with its Appl, 1968, 2:32 ~ 39

[70] Traub J F, Wozniakowski H. *A General Theory of Optimal Algorithms*. New York: Acad Press, 1980, 8 ~ 41

[71] Babenko K I. *A Fundamental Course of Numerical Analysis*. [Russian]. Moscow: Nauka, 1986, 175 ~ 176

[72] Sun Yongsheng. Optimal interpolation on some classes of differentiable functions (I). ATA, 1986, 2 (4):49 ~ 54

[73] Sun Yongsheng. Optimal interpolation on some classes of smooth functions defined on the whole real axis. Adv in Math, 1990, 20(1):1 ~ 14

[74] Fang Gensun, Liu Yongping. Optimal quadrature of $W_1^r(\mathbf{R})$ on the whole real axis. Acta Math Scientia, 1996, 16(1):72 ~ 80

[75] Liu Yongping. Optimal quadrature formula on some classes of smooth functions on the whole real axis. Northeastern Math J, 1993, 9(1):113 ~ 120

[76] Fang Gensun, Liu Yongping. On the least L_∞-norm of monosplines with free knots on the real axis. J Complex, 1992, 8:467 ~ 487

[77] Karlin S. *Total Positivity* (Vol 1). Stanford: Stanford University Press, 1968

[78] Sun Yongsheng. Optimal interpolation for a convolution function class. Northeastern Math J, 1990, 6:9 ~ 25

[79] Li Chun. Cardinal spline interpolation with PF density. preprint

[80] Chen Dirong. On the cardinal spline interpolation corresponding to an infinite order differential operator. Acta Math Sinica, 1994, 10:315 ~ 324

[81] Chen Dirong, Zhang Jianhua. Best one-sided approximation of some periodic function classes. [Chinese]. Advance in Math, 1994, 23:212 ~ 221

[82] Nikolskii S M. *Approximation of Functions of Several Variables and Imbedding Theorems*. New York: Springer-Verlag, 1975, 1 ~ 255

[83] Schwartz L. *Theorie des Distributions* (Ⅰ, Ⅱ). Paris: Hermann, 1957

[84] Madych W R, Nelson S A. Polyharmonic cardinal splines. J Approx Theory,

1990, 60: 141 ~ 156

[85] Schoenberg I J. Contributions to the problem of approximation of equidistant data by analysis functions. Quart Appl Math, 1946, 4: 45 ~ 99, 112 ~ 141

[86] Buhmann M D. Multivariate cardinal interpolation with radial-basis functions. Constructive Approximation, 1990, 6: 225 ~ 255

[87] Liu Yongping. Approximation of smooth functions by polyharmonic cardinal splines in $L_p(\mathbf{R}^d)$ space. Acta Math Appli Sinica, 1998, 14(2): 157 ~ 164

[88] Schoenberg I J. On remainders and the convergence of cardinal spline interpolation for almost periodic functions. MRC TSR 1514 Dec, 1974; In Karlin, et al, ed. *Studies in Splines and Approximation Theory*. New York: Academic Press, 1975

[89] Subbotin Ju N. Extremal problems in the theory of approximation of functions on the basis of incomplete information. Trudy Mat Inst Steklov, 1980, 145: 152 ~ 168

[90] Liu Yongping. $L_2(\mathbf{R}^d)$-extremal problems on the basis of incomplete information. Analysis Math, 1995, 21: 107 ~ 124

[91] Bugrov J S. Approximation of function class defined by polyharmonic operator by trigonometric polynomials. Uspehi Math Nauk, 1958, 13(2): 149 ~ 156

[92] Luo Junbo. On the width of class of periodic functions of two variables. In: Law A G, Wang C. L, ed. *Approximation, Optimazation and Computing*: Theory and Applications. North-Holland: Elsevier Science Publishers, 1990. 147 ~ 150

[93] Trofimov V M. Approximation on some class defined by a polyharmonic operator by the truncated arithmetic means of the partial sum of Fourier series. [Russian]. In: Smirnov, Gos-Izdat, ed. *Study in Modern Problems of Function's Constructive Theory*. Moscow: Fiziko-Matematicheskoi Literaturi, 1961. 251 ~253

[94] Magaril-I'llyaev G G. Average widths of Sobolev classes on $\mathbf{R}^d$. J Approx Theory, 1994, 76(1): 65 ~ 76

[95] Tikhomorov V M. Harmonics and splines as the optimal tools of approximation and recovery. [Russian]. Uspehi Mat Nauk, 1995, 50(2): 1 ~ 50

[96] Magaril-I'llyaev G G, Tikhomirov V M. On some problems of approximation of functional classes in L_2. [Russian]. Dokl A N, 1995, 344(5): 583 ~ 585

[97] Sun Yongsheng. Estimation of width of a multivariate periodic Besov classes. Chinese Science Bulletin, 1994, 39(23): 2113 ~ 2115

[98] Romaniuk A S. On the best trigonometric and bilinear approximations of

anisotropic Besov function classes of many variables. [Russion]. Uk Math J, 1995, 47(8): 1097 ~ 1111

[99] Jiang Yanjie, Liu Yongping. Average widths and optimal recovery of multivariate Besov classes in $L^p(\mathbf{R}^d)$. Kexue Tongbao, 1998, 43: 479 ~ 482; Chinese Science Bulletin, 1998, 43(8): 638 ~ 641

[100] Luo Junbo, Sun Yongsheng. Optimal Recovery and widths of the anisotropic classes of multivariate functions. Advance in Math. (China), 1998, 27(1): 69 ~ 77

[101] Luo Junbo, Liu Yongping. Average widths and optimal recovery of the anisotropic classes $W^r H^\alpha(\mathbf{R}^d)$ in the metric $C^k(\mathbf{R}^d)$. Analysis Math, 1998, 24: 265 ~ 273

[102] Wang Heping. Average widths of Sobolev classes with mixed smoothness in $L_q(\mathbf{R}^d)$. preprint

[103] Wang Heping. Average widths of Sobolev-Wiener classes with mixed smoothness in $L_q(\mathbf{R}^d)$. preprint

[104] Liu Yongping. $L_q(\mathbf{R}^d)$-optimal recovery on the Riesz potential spaces with incomplete information. J Beijing Normal Univ (Nature Sciences), 1997, 33(2): 143 ~ 150

[105] Jiang Yanjie, Liu Yongping. Average widths and optimal recovery of multivariate Besov-Wiener classes. J Math Study, 1998, 31(4): 353 ~ 361

[106] Jiang Yanjie. Optimal recovery of anisotropic Besov-Wiener classes. preprint

[107] J. Traub, G. Wasilkowski, H. Wozniakowski, *Information-based Complexity*, New York, Acad. Press, INC, 1988, 5 ~ 113

五、带有 Gauss 测度的 B 空间内点集的平均逼近问题及多元问题

V. Average Approximation of a Subset in a B-space with Gaussian Measure and Multivariate Problems

CHINESE SCIENCE BULLETIN
July, 1992, Vol. 37 No. 14

关于 Hilbert 空间内点集的平均宽度*

Average n-Width of Point Set in Hilbert Space

Keywords: average n-width, Hilbert space, Gaussian measure.

Given a separable real Banach space F and a separable real Hilbert space X, we denote by S a bounded linear operator from F into X. Let μ be a Gaussian measure defined on the Borel field of F with mean element $m_\mu = 0$ and correlation operator C_μ. By [1], C_μ is a bounded linear operator from $F^* \to F$ such that

$$(C_\mu(L_1), L_2) = \int_F (L_1, f)(L_2, f)\mu(\mathrm{d}f), \tag{1}$$

where $L_1, L_2 \in F^*$, $(\cdot, \cdot)$ denotes a bilinear form defined on $X^* \times X$ or $X^{**} \times X^*$, for instance, $(L_1, f) = L_1(f)$, $(C_\mu(L_1), L_2) = L_2(C_\mu(L_1))$. It is obvious that $(C_\mu(L_1), L_2) = (C_\mu(L_2), L_1)$, and $(C_\mu(L), L) \geqslant 0$. The induced measure of μ by S is denoted by $\nu = \mu S^{-1}$ which is a Gaussian measure on the Borel field of X with mean element $Sm_\mu = 0$ and correlation operator $C_\nu = SC_\mu S^*$ which is a non-negative definite, symmetric and compact linear operator from X into X, and moreover, it has finite trace.

For any n-dimensional linear subspace $X_n \subset X$ and any element $f \in F$, we define the best approximation of $Sf = x$ by X_n in X by

* Project supported by the National Natural Science Foundation of China. Received January 17; revised October 7, 1991.

$$\rho(x, X_n) = \min_{y \in X_n} \| x - y \|. \tag{2}$$

$H: \mathbf{R}_+ \to \mathbf{R}_+$ is given such that $H(t)$ is strictly increasing, continuous and convex over $\mathbf{R}_+ = [0, +\infty)$. Assume $\int_F H(\| Sf \|)\mu(\mathrm{d}f) < +\infty$. Let f run over F. We define the (H, μ)-average deviation of the set $S(F) = K$ from X_n in X by

$$E(K, X_n, X) = \int_F H(\rho(Sf, X_n))\mu(\mathrm{d}f). \tag{3}$$

Definition 1 *Let X_n runs over all linear subspaces of X such that dim $X_n \leqslant n$. Then the quantity*

$$\overline{d}_n(K, X, \mu) = \inf_{X_n} \int_F H(\rho(Sf, X_n))\mu(\mathrm{d}f) \tag{4}$$

is said to be the (H, μ)-average Kolmogorov n-width of K in X. Any $X_n^* \subset X$ such that dim $X_n^* \leqslant n$ is called an optimal subspace if $E(K, X_n^*, X) = \overline{d}_n(K, X, \mu)$.

It is known that the concept of n-width has important applications in the theory of infor-mation based complexity. Some basic results have been established in [2] for the worst case setting. Recently, Traub, Wozniakowski et al. [2], in exploiting the average case setting for the theory of information-based complexity, solved the problem of minimization of the nth average radius of linear information operators in the Hilbert case. Motivated by this result, we propose in this note the notion of (H, μ)-average n-K width in Hilbert space and consider the relation between the (H, μ)-average n-K width and the nth minimal average radius of linear information operators in the Hilbert case.

From [2] an n-cardinal linear information N on F is any linear continuous mapping $F \to \mathbf{R}^n$. For any $f \in F$, $N(f)$ may be represented by $N(f) = (L_1(f), L_2(f), \cdots, L_n(f))$, where $L_1, \cdots, L_n \in F^*$. The (H, μ)-average radius of N is given by

$$r^a(N) = \int_F (H \| Sf \|)\, \mu_2(\mathrm{d}f \mid 0), \tag{5}$$

where $\mu_2(\cdot \mid y)$ is the regular conditional measure (probability) of μ with respect to $N(f) = y\ (y \in \mathbf{R}^n)$, and $\mu_2(\cdot \mid 0)$ is the conditional measure which concentrates on $N^{-1}\{0\}$. Write

$$\Lambda = \{N = [L_1, L_2, \cdots, L_n] : L_i \in F^*, L_i(C_\mu L_j) = \delta_{ij}\}.$$

Definition 2

$$r^a(n, \Lambda) = \inf\{r^a(N) : N \in \Lambda\} \tag{6}$$

is said to be the n th minimal (H, μ)-average radius of linear information operators.

Obviously, we have $r^a(n, \Lambda) \leqslant r^a(m, \Lambda)$ when $m \leqslant n$. In what follows, we consider the relation between $r^a(n, \Lambda)$ and $d_n(K, X, \mu)$.

Lemma 1 *Let $G = \overline{S(F)}$ be the closure of $S(F)$ in X. Then*

$$\overline{d}_n(K, X, \mu) = \overline{d}_n(K, G, \mu) = \inf_{X_n \subset G} \int_F H(\rho(Sf, X_n))\mu(\mathrm{d}f). \tag{7}$$

Concerning this lemma we refer to [3].

Lemma 2 $\forall\, n \geqslant 1$,

$$r^a(n, \Lambda) \leqslant \overline{d}_n(K, X, \mu).$$

Proof For any $X_n = \text{span}\,\{x_1, x_2, \cdots, x_n\} \subset \mathrm{G}$, let us construct for X_n a set of C_ν-orthonormal basis by means of the familiar Gram-Schmidt procedure. First we put $X_{n-m} = X_n \cap \ker(C_\nu), 0 \leqslant m \leqslant n$. Let the orthogonal complement of X_{n-m} in X_n be denoted by $X_m : X_n = X_m \oplus X_{n-m}$. We may regard $m > 0, X_m = \text{span}\{x_1, x_2, \cdots, x_m\}$. Set $y_i = \sum_{j=1}^{i} a_{ij} x_j (i = 1, \cdots, m)$, where a_{ij} are defined by the condition of C_ν-orthogonality: $(C_\nu y_i, y_i) = \delta_{ij}, i, j = 1, \cdots, m$. If $n - m \geqslant 1$, then we denote by $y_{m+1}, \cdots, y_n$ a set of orthonormal basis of X_{n-m}. Notice that $C_\nu y_{m+1} = \cdots = C_\nu y_n = 0$. The orthogonal projection P from X to X_n may be represented by $P_x = \sum_{j=1}^{n}(x, y_j)y'_j$, where $y'_j = \sum_{i=1}^{n} g_{ij}y_i (j = 1, \cdots, n)$, a_{ij} is some constants. When $x = S(f)$ we have

$$P(Sf) = \sum_{j=1}^{n}(Sf, y_j)y'_j, \tag{8}$$

which is an algorithm by using the information $N^*(f) = ((S^* y_1)(f), \cdots, (S^* y_n)(f)), S^* y_1, \cdots, S^* y_n \in F^*$. Write $L_i^* = S^* y_i$,

$i = 1, \cdots, m$. We see that $L_i^*(C_\mu L_j^*) = (C_\mu(S^* y_j), S^* y_i) = (SC_\mu S^* y_j, y_i) = (C_\nu y_j, y_i) = \delta_{ij}, i, j = 1, \cdots, m$, and besides this, $L_i^*(C_\mu L_j^*) = 0$ for $j = m+1, \cdots, n, i = 1, \cdots, n$, so that when we construct central algorithm (concerning this notion we refer to [2]) by using $N^* = [L_1^*, \cdots, L_n^*], L_{m+1}^*, \cdots, L_n^*$ are of no use:

$$\varphi^c(N^*(f)) = \sum_{j=1}^{n}(Sf, y_j)S(C_\mu(L_j^*)) = \sum_{j=1}^{m}(Sf, y_j)S(C_\mu(L_j^*)),$$

because $S(C_\mu(L_j^*)) = C_\nu y_j = 0, j = m+1, \cdots, n$. Thus, from [2] we have

$$\int_F H(\| Sf - P(Sf) \|)\mu(\mathrm{d}f) \geqslant \int_F H(\| Sf - \varphi^c(N^*(f)) \|)\mu(\mathrm{d}f)$$
$$= r^a(m, \Lambda) \geqslant r^a(n, \Lambda)$$

On account of $\rho(Sf, X_n) = \| Sf - P(Sf) \|$, and the arbitrariness of $X_n \subset X$, we get $r^a(n, \Lambda) \leqslant \bar{d}_n(K, X, \mu)$. Lemma 2 is proved.

Lemma 3 $\forall n \geqslant 1$,

$$r^a(n, \Lambda) = r^a(N^{**}),$$

where $N^{**}(f) = (L_1^{**}(f), \cdots, L_n^{**}(f)), L_i^{**} \in F^*$ *are defined as follows. Let* $\lambda_1^* \geqslant \lambda_2^* \geqslant \cdots \geqslant \lambda_j^* \geqslant \cdots, \lambda_j^* \geqslant 0$ *be the sequence of eigenvalues of* C_ν, *and the sequence of corresponding normalized eigenvectors be* $\eta_1^*, \eta_2^*, \cdots, \eta_j^*, \cdots$. *Set* $k = \min\{n, \sup(i, \lambda_i^* > 0)\}$. *When* $k = n, L_i^{**}(f) = (\lambda_i^*)^{-\frac{1}{2}}(S(f), \eta_i^*), i = 1, \cdots, n$. *If* $n > k$, *then when* $i = 1, \cdots, k$ *we put* $L_i^{**}(f) = (\lambda_i^*)^{-\frac{1}{2}}(S(f), \eta_i^*)$, *but* $L_i^{**}(f) \equiv 0$ for $i = k+1, \cdots, n$.

Proof See [2]. We omit the details.

Lemma 4 $X_n^* = \operatorname{span}\{\eta_1^*, \cdots, \eta_n^*\}$ *is optimal for* $\bar{d}_n(K, X, \mu)$.

Proof In this case the orthogonal projection $X \to X_n^{**}$ is a central algorithm by using the information N^{**}. This fact may be verified directly. In fact, we have

$$\varphi^c(N^{**}(f)) = \sum_{j=1}^{n} L_j^{**}(f)S(C_\mu(L_j^{**})) = \sum_{j=1}^{n}(S(f), \eta_j^*)(\eta_j^*).$$

So,

$$\int_F H(\rho(Sf, X_n^{**}))\mu(\mathrm{d}f)$$
$$= \int_F H(\| Sf - \varphi^c(N^{**}(f)) \|)\mu(\mathrm{d}f) = r^a(N^{**}).$$

To sum up, we have proved

Theorem 1. *For any given $H(t): \mathbf{R}_+ \to \mathbf{R}_+$ strictly increasing, continuous and convex over $\mathbf{R}_+ = [0, +\infty]$ and $n \geqslant 1$, it holds that*

$$\bar{d}_n(S(f), X, \mu) = r^a(n, \Lambda) = r^a(N^{**}),$$

and X_n^{**} is optimal for $\bar{d}_n$.

Theorem 1 contains $H(t) = t^p (1 \leqslant p < +\infty)$ as a special case. In this case we customarily define

$$\overline{d}_n^{\mathscr{L}_p}(S(F), X, \mu) = \inf_{X_n}\left\{\int_F (\rho(Sf, X_n))^p \mu(df)\right\}^{\frac{1}{p}}, \tag{9}$$

and

$$r^a(n, \Lambda, \mathscr{L}_p) = \inf_{N \in \Lambda}\left\{\int_F \| S(f) \|^p \mu_2(df \mid 0)\right\}^{\frac{1}{p}}. \tag{10}$$

Theorem 1 reads

$$\overline{d}_n^{\mathscr{L}_p}(S(F), X, \mu) = r^a(n, \Lambda, \mathscr{L}_p). \tag{11}$$

When $p = 2$, we have

Theorem 2 $\forall n \geqslant 1$,

$$\overline{d}_n^{\mathscr{L}_2}(S(F), X, \mu) = r^a(n, \Lambda, \mathscr{L}_2) = \left(\sum_{j=n+1}^{\infty} \lambda_j^*\right)^{\frac{1}{2}}. \tag{12}$$

Proof By Lemma 1, for any $X_n \subset G$ we have

$$\int_F \boldsymbol{\rho}^2(Sf, X_n)\mu(\mathrm{d}f) = \int_G \rho^2(x, X_n)\nu(\mathrm{d}x)$$
$$= \int_G \| x \|^2 \nu(\mathrm{d}x) - \int_G \| Px \|^2 \nu(\mathrm{d}x),$$

where P is the orthogonal projection $G \to X_n$. Noting that $\int_G \| x \|^2 \nu(\mathrm{d}x) = \mathrm{tr}(C_\nu)$, and $\| Px \|^2 = \sum_{j=1}^{n}(x, x_j)^2$ if $\{x_1, x_2, \cdots, x_n\}$ is a set of orthonormal basis of X_n, we see that

$$\int_G \| Px \|^2 \nu(\mathrm{d}x) = \sum_{j=1}^{n}\int_G (x, x_j)^2 \nu(\mathrm{d}x) = \sum_{j=1}^{n}(C_\nu x_j, x_j).$$

Consequently,

$$\int_F \rho^2(Sf, X_n)\mu(\mathrm{d}f) \geqslant \mathrm{tr}(C_\nu) - \sup_{(x_i, x_j) = \delta_{ij}} \sum_{j=1}^{n} (C_\nu x_j, x_j).$$

From [2], we have

$$\sup_{(x_i, x_j) = \delta_{ij}} \sum_{j=1}^{n} (C_\nu x_j, x_j) \leqslant \sum_{j=1}^{n} \lambda_j^*,$$

so that $\overline{d}_n^{\mathscr{L}_2}(S(F), X, \mu) \geqslant \left(\sum_{j=n+1}^{\infty} \lambda_j^* \right)^{\frac{1}{2}}$. The reverse inequality may be obtained by taking $X_n = X_n^{**}$. Theorem 2 is proved.

Remark Our results show that the average n-width has meaning for the set $K = \{Sf : f \in F\}$ where f runs over a whole linear space. Of course, we may take $K_q = \{Sf : \| f \| \leqslant q\}$ instead of K, where $q > 0$ is a fixed number. Then K_q is bounded in X. But we can prove $\overline{d}_n(K_q, X, \mu) = \{1 + O(\mathrm{e}^{-aq^2})\}\ \bar{d}_n(K, X, \mu)$, $a > 0$ is some constant. So, in average case setting, there has no essential difference between the above two cases. It seems to be interesting to consider $\bar{d}_n$ under the non-Hilbert case.

References

[1] Kuo Hui-Hsiong. *Gaussian measures in Banach spaces*. Lecture Notes in Math., 463. New York: Springer-Verlag, 1975. 15 ~ 35

[2] Traub, J. F., Wasilkowski, G. W. & Wozniakowski, H. *Information-based Complexity*. New York: Acad. Press, INC, 1988. 195 ~ 237

[3] Sun Yongsheng. *Approximation Theory* (in Chinese), Vol. 1. Beijing: Beijing Normal University Press, 1989. 297 ~ 310

[4] Micchelli, C. A. Orthogonal projections are optimal algorithms. *JAT*, 1984, 40: 101.

EAST JOURNAL ON APPROXIMATIONS
1995, Volume 1, Number 4: 431 ~ 439

带 Gauss 测度的一个 Banach 空间中最佳逼近的误差界

Average Error Bounds of Best Approximation in a Banach Space with Gaussian Measure*

This paper is concerned with some problems of best approximation in a Banach space with a probabilistic (Gaussian) measure in connection with the information-based complexity theory in the average case setting. These problems are:

(1) Average error bounds of best polynomial approximation and linear operator approximation of continuous functions on the Wiener space.

(2) μ-average n-widths on the Wiener space.

1. Average error bounds of best polynomial approximation and linear operator approximation of continuous functions on the Wiener space.

Set

$$C_0: = \{f:[0,1] \to \mathbf{R}, \text{continuous}, f(0) = 0\}.$$

As usual, $\|f\|_c: = \max_{0\leqslant t\leqslant 1}|f(t)|$. Denote by $\mathscr{B}$ the Borel class of $(C_0, \|\cdot\|)$ and by W the Wiener measure defined on $\mathscr{B}$ which, by definition [3], satisfies the following conditions:

Any element $I \in \mathscr{B}$ of the form

* Supported by the Fund of Doctoral Programs of NECC.
本文与汪成咏合作.

$$I = \{f \in C_0 : (f(t_1), f(t_2), \cdots, f(t_n)) \in E\},$$

where $0 < t_1 < t_2 < \cdots < t_n \leqslant 1$ and E is a Borel subset of R^n, is said to be a *cylinder set*, $n = 1, 2, 3\cdots$

Introduce the measure

$$W(I) = \prod_{j=1}^{n} \frac{1}{\sqrt{2\pi(t_j - t_{j-1})}} \int_E \exp\left\{-\sum_{j=1}^{n} \frac{(u_j - u_{j-1})^2}{2(t_j - t_{j-1})}\right\} du,$$

where $t_0 = u_0 = 0, du := du_1 \cdots du_n$. It is well-known [3] that this measure is a special case of the so-called Gaussian measure which may be characterized by the following property: any linear continuous functional $L(f)$ on C_0 is a random variable on the Wiener space (C_0, W) which is normally distributed, $L \sim (0, R_L)$, where

$$R_L := \int_{C_0} L^2(f) W(df)$$

is the variance. Since the appearance of the original work [4] by N. Wiener, the Wiener measure has received a great deal of attention. It is of interest for statisticians and physicians. In the recent years it also draws the attention of mathematicians working in the area of numerical analysis and theoretical computer science who are interested in the average approach to numerical and algorithmic analysis. (See Traub, Wozhniakowski, Wasilkowski [1], Wasilkowski and Lee [2].)

Given an integer $m \geqslant 0$, define the integral operator T_m on C_0 by

$$(T_0 g)(x) := g(x), \forall g \in C_0$$

(i.e., T_0 is the identity operator); For $m \geqslant 1$ and $g \in C_0$,

$$(T_m g)(x) := \frac{1}{(m-1)!} \int_0^1 (x - y)_+^{m-1} g(y) dy.$$

As an approximation tool we use the Jackson operator, which is an important linear operator in the classical approximation theory. Let us recall the definition. For given integers $r \geqslant 2$ and $n > m$, set $n' := [(n - m)/r] + 1$ and denote

$$K_{nr}^{(m)}(t) := C_{nr}^{-1} \left(\frac{\sin \frac{n't}{2}}{\sin \frac{t}{2}} \right)^{2r},$$

where

$$C_{nr} = \int_{-\pi}^{\pi} \left(\frac{\sin \frac{n't}{2}}{\sin \frac{t}{2}} \right)^{2r} \mathrm{d}t.$$

When $m = 0$ we put

$$J_n^{(0)}(g;x) = J_n(g,t) = \int_{-\pi}^{\pi} g\left(\frac{\cos u + 1}{2}\right) K_{nr}^{(0)}(t-u)\mathrm{d}u,$$

with $x = (\cos t + 1)/2$. For $m \geqslant 1$, define

$$J_n^{(m)}(g;x) = (T_m g)(x) - \left(-\frac{1}{2}\right)^m \int_{-\pi}^{\pi}\int_{u_{2m}}^{t} \cdots \int_{-\pi}^{\pi}\int_{u_2}^{u_3}\int_{-\pi}^{\pi}\left[g\left(\frac{\cos u_1 + 1}{2}\right) - g\left(\frac{\cos u_0 + 1}{2}\right)\right] \prod_{v=0}^{m-1} \sin u_{2\nu+1} \times \prod_{\nu=0}^{(m)} K_{nr}^{(m)}(u_{2\nu+1} - u_{2\nu})\mathrm{d}u_0 \mathrm{d}u_1 \cdots \mathrm{d}u_{2m}.$$

$J_n^{(m)}$ is a linear continuous operator from $C \to \mathscr{P}_n$, where $\mathscr{P}_n$ is the set of polynomials of degree $\leqslant n$.

Another tool we study here is the best approximation operator in the L_q metric. Write

$$E_n(f)_q := \min_{p \in \mathscr{P}_n} \| f - p \|_q, \quad 1 \leqslant q < \infty$$

and

$$E_n(f)_c := \min_{p \in \mathscr{P}_n} \| f - p \|_c, \quad (q = +\infty).$$

Then the p-average error of the L_p approximation by Jackson's operator over $T_m(C_0)$ is defined by

$$E_{pq}(T_m, J_n^{(m)}, W) := \left(\int_{C_0} \| T_m g - J_n^{(m)}(g) \|_q^p W(\mathrm{d}g) \right)^{1/p}.$$

Here $(1 \leqslant p < \infty, m \geqslant 0, n > m), 1 \leqslant q \leqslant \infty$ (when $q = +\infty$, $\| \cdot \|_\infty$ is taken as a sup-norm).

The p-average error of the L_q best approximation over $T_m(C_0)$ if defined in the following way:

$$E_{pq}^{(n)}(T_m; W) := \left(\int_{C_0} E_n(T_m g)_q^p W(\mathrm{d}g) \right)^{1/p}.$$

The next two statements give the asymptotic estimation of the two quantities.

Theorem 1 *Suppose that* $1 \leqslant p, q < +\infty$ *and* $n > m$. *Then*

$$E_{pq}^{(n)}(T_m, W) \leqslant E_{pq}(T_m, J_n^{(m)}, W) \leqslant M \cdot n^{-m-\frac{1}{2}},$$

where $M > 0$ *depends only on* p *and* q.

We have obtained the lower estimate

$$E_{pq}^{(n)}(T_m, W) \geqslant M^{-1} n^{-m-\frac{1}{2}}$$

in two cases:

(a) $m = 0, 1 \leqslant p, q < \infty$,

(b) $m \geqslant 1, 1 \leqslant p < \infty, 2 \leqslant q < +\infty$.

Theorem 2 *For* $q = +\infty, 1 \leqslant p < \infty$ *and* $n > m$, *we have*

$$E_{p\infty}^{(n)}(T_m, W) \asymp E_{p\infty}(T_m, J_n^{(m)}, W) \asymp n^{-m-\frac{1}{2}}(\lg n)^{\frac{1}{2}}.$$

Corollary *Suppose that* $1 \leqslant p < +\infty$. *Then*

$$\left(\int_{C_0} \omega(T_m, g, \frac{1}{n})^p W(\mathrm{d}g)\right)^{\frac{1}{p}} \asymp n^{-m-\frac{1}{2}}(\lg n)^{\frac{1}{2}},$$

where $\omega(f, \delta)$ *is the modulus of continuity of* f, *and the symbol* $a_n \asymp b_n$ *means that there exists some constant* $M > 1$ *such that* $M^{-1} \leqslant (b_n)/(a_n) \leqslant M, n = 1, 2, 3, \cdots$

2. μ-average widths on the Wiener space.

Several papers appeared recently in the mathematical literature which deal with the n-widths problem in average case setting in connection with the information-based complexity theory (see for example [5] – [7]). Let X and Y be real separable Banach spaces. Let $\mathscr{B}(X)$ be the Borel class of X. Assume that μ is a probabilistic (Borel) measure on $\mathscr{B}(X)$. Following [1], we call a bounded linear operator S from $X \to Y$ a *solution operator*. A Borel measurable mapping N from $X \to \mathscr{R}^n$ is said to be an *information operator* of cardinality n. Let ϕ be a $\mathscr{B}$-measurable mapping from $\mathscr{R}^n \to Y$. Then for a fixed N, $\phi(N(x))$ is called an algorithm based on N which may be used for approximation of Sx. The error of such an approximation is usually measured in L_p-norm, i.e., as $\| Sx - \phi(N(x)) \|_Y^p$ where $1 \leqslant p < \infty$. The μ-average error of $\| Sx - \phi(N(x)) \|_Y^p$ over X is defined by

$$E^a(S, N, \phi, \mu)_{p,Y} := \left(\int_X \| Sx - \phi(N(x)) \|_Y^p \mu(\mathrm{d}x) \right)^{1/p}.$$

By definition (see [1]), the μ-average *L_p-radius of information*, associated with N, is given by

$$r^a(S, N, \mu)_{p,Y} := \inf_\phi E^a(S, N, \phi, \mu)_{p,Y},$$

where ϕ runs over the set of all $\mathscr{B}$-measurable algorithms ϕ based on N.

Definition 1 *The n-th minimal μ-average L_p-radius of information (with respect to S) is defined as*

$$r_n^a(S, \mu)_{p,Y} := \inf_N r^a(S, N, \mu)_{p,Y},$$

where N runs over the set of all $\mathscr{B}$-measurable information operators with cardinality n.

Definition 2 *If, in the above definition, we restrict N to run over the set of all* **linear** *information operators with cardinality n, then the corresponding quantity*

$$C_n^a(S, \mu)_{p,Y} := \inf_N r^a(S, N, \mu)_{p,Y},$$

is said to be the μ-average n-Gelfand width (of S in Y under (p, Y)-error functional) and it is abbreviately denoted by μ-n G width.

Definition 3 *The μ-average n-Kolmogorov width (μ-n K width) of S in Y under (p, Y)-error functional is defined by*

$$d_n^a(S, \mu)_{p,Y} := \inf_{N,\phi} E^a(S, N, \phi, \mu)_{p,Y},$$

where the infimum is taken over the set of all couples (N, ϕ) such that N runs over all $\mathscr{B}$-measurable information operators of cardinality n and ϕ runs over all **linear** *algorithms based on one and the same N.*

Definition 4 *The μ-average n-linear width (μ-n L width) of S in Y under (p, Y)-error functional is defined by*

$$a_n^a(S, \mu)_{p,Y} := \inf_{(N,\phi)} E^a(S, N, \phi, \mu)_{p,Y},$$

where the infimum is taken over the set of all couples (N, ϕ) such that N runs over all **linear** *information operators of cardinality n and ϕ runs over all* **linear** *algorithms, based on one and the same N.*

These definitions may be seen in [8]. They signify a comprehension of the basic ideas of the classical approximation theory with that of theoretical computer science. Several results on the μ-average width have been obtained. In [7] we proved: if Y is a Hilbert space and μ is Gaussian, then

$$\mathrm{d}_n^a(S,\mu)_{p,Y} = C_n^a(S,\mu)_{p,Y} = a_n^a(S,\mu)_{p,Y} = r_n^a(S,\mu)_{p,Y} \text{ for any } p, 1 \leqslant p < \infty, n \in \mathbf{Z}_+.$$

For $X = C_0, S = T_m(m \geqslant 0)$ and $\mu = W$ Majorov [6] got some asymptotic estimates in the following two cases: (1) $m = 0, p = 1, Y = L_q[0,1], 2 \leqslant q \leqslant \infty$, for $d_n^a(T_m, W)_{p,Y}, a_n^a(T_m, W)_{p,Y}$; (2) $m \geqslant 2, p = 1, Y = L_q[0,1], 1 \leqslant q < \infty$, for the same widths as above.

Our main result is the following.

Theorem 3 *For any integer $m \geqslant 0, p, q \in [1, +\infty)$ and $Y = L_q[0,1]$, we have*

$$C_n^a(T_m, W)_{p,Y} \asymp a_n^a(T_m, W)_{p,Y} \asymp n^{-m-\frac{1}{2}}. \tag{2.1}$$

If $p \in [1, +\infty), q \in [2, +\infty), m \geqslant 0$, then

$$\mathrm{d}_n^a(T_m, W)_{p,Y} \asymp n^{-m-\frac{1}{2}}.$$

The linear subspace $\mathscr{P}_n$ is weakly asymptotically optimal for d_n^a and a_n^a, while the Jackson operator $J_n^{(m)}$ with some of its discretized versions is weakly asymptotically optimal for a_n^a

Besides these, some weakly asymptotically optimal information of standard type for C_n^a is also identified.

In what follows we outline the main step of the proof of (2.1).

(1) **The Lower Estimation of** $C_n^a(T_m, W)_{p,Y}$.

(a) The lower estimation for $C_n^a(T_m, W)_{p,Y}$, where $Y = L_q[0,1]$, $p, q \in [1, +\infty)$ is closely related to the $p = q = 2$ case, which has an essential connection with the extremal problem we are going to formulate below.

Let $Q = [0,1]$ and for $m \geqslant 0$,

$$K(x,t) = \frac{1}{m!}(x-t)_+^m, 0 \leqslant x, t \leqslant 1.$$

Define a mapping Φ in the following manner:

$$Q \to L_2[0,1]$$
$$x \to K(x, \cdot) \in L_2[0,1], x \in Q.$$

$K := \{K(x, \cdot): x \in Q\}$ is a compactum in $L_2[0,1]$. Let $X_n = \operatorname{span}\{g_1, \cdots, g_n\} \subset L_2[0,1]$ be an n-dimensional linear subspace of $L_2[0,1]$, where $g_1, \cdots, g_n$ is its orthonormal basis. Then

$$\begin{aligned} & e(K(x, \cdot), X_n, L_2): \\ &= \min_{g \in X_n} \| K(x, \cdot) - g(\cdot) \|_{L_2} \\ &= \frac{1}{m!}\left[\frac{x^{2m+1}}{2m+1} - \sum_{j=1}^{n}\left(\int_0^1 (x-t)_+^m g_j(t)\mathrm{d}t\right)^2\right]^{1/2} \\ &=: G_n(x; g_1, g_2, \cdots, g_n). \end{aligned}$$

Consider the following extremal problem:

$$\int_0^1 G_n^2(x; g_1, g_2, \cdots, g_n)\mathrm{d}x \to \inf, \tag{2.2}$$

where the infimum is taken over all sets $\{g_1, g_2, \cdots, g_n\} \subset L_2[0,1]$ of the orthonormal elements. It had been shown by R. S. Ismagilov [9] that the solution of (2.2) is closely related to the Kolmogorov n-width of the compactum K in $L_2[0,1]$. Precisely,

$$d_n(K, L_2) := \inf_{X_n} \sup_{0 \leqslant x \leqslant 1} e(K(x, \cdot), X_n, L_2).$$

Theorem 4 *For $p > 2$ and any $n \in \mathbf{Z}_+$ we have*

$$\begin{aligned} & \inf_{\{g_i\}}\left(\int_0^1 G_n^2(x; g_1, g_2, \cdots, g_n)\mathrm{d}x\right)^{1/2} \\ & \asymp \inf_{\{g_i\}}\left(\int_0^1 G_n^p(x; g_1, g_2, \cdots, g_n)\mathrm{d}x\right)^{1/p} \\ & \asymp d_n(K, L_2) = d_n(W_1^{2m+1}, L_2) \\ & \asymp \left(\sum_{j=n+1}^{\infty}\lambda_j\right)^{1/2} \\ & \asymp \pi^{-(m+1)} \cdot (2m+1)^{-1/2} \cdot n^{-m-1/2} \quad (n \to \infty), \end{aligned}$$

where

(1) *the infimum is taken over the set of all systems of northonormal elements $\{g_1, g_2, \cdots, g_n\} \subset L_2[0,1]$,*

(2) $W_1^{2m+1} := \{f : f^{(2m)}$ *is abs. cont. on* $[0,1]$, $\| f^{(2m+1)} \|_1 \leqslant 1\}$,

(3) *the symbol* $\asymp$ *means a strongly asymptotic relation,*

(4) $\{\lambda_j\}_{j\in \mathbf{Z}_+}$ *is the decreasing sequence of the eigen-values of the self-adjoint compact operator* $(T_{m+1})*(T_{m+1}) = |T_{m+1}|^2$:

$$|T_{m+1}|^2 f = \lambda f.$$

This problem is equivalent to the eigenvalue problem for the differential operator

$$\lambda D^{2m+2} f = (-1)^{(m+1)} f, f^{(k)}(0) = f^{(k+m)}(1) = 0, (0 \leqslant k \leqslant m).$$

(b) The estimation of $C_n^a(T_m, W)_{2,2}$.

Let the linear information operator $N_n(f)$ of cardinality n be given as

$$N_n(f) = [L_1(f), L_2(f), \cdots, L_n(f)],$$

where $L_i(f) = \int_0^1 f(t)\mathrm{d}g_i(t), g_i \in BV[0,1], i = 1, \cdots, n$. Without loss of generality we may assume that $g_i(1) = 0$ and

$$L_i(C_W L_j) = \int_0^1 g_i(t) g_i(t)\mathrm{d}t = \delta_{ij}, i, j = 1, \cdots, n,$$

where C_W is the correlation operator of the Wiener measure W. According to [2], we have

$$r^a(T_m, N_n, W)_{2,2}^2 = \int_0^1 G_n^2(x; g_1, g_2, \cdots, g_n)\mathrm{d}x.$$

Therefore

$$C_n^a(T_m, W)_{2,2}^2 = \inf_{\{g_i\}} \int_0^1 G_n^2(x; g_1, g_2, \cdots, g_n)\mathrm{d}x, \qquad (2.3)$$

where the infimum is taken over the set of all systems $\{g_1, g_2, \cdots, g_n\} \subset BV[0,1]$ satisfying $g_i(1) = 0$ and (2.17). Now, in view of Theorem 4, we get

Theorem 5 *For any* $m \geqslant 0$,

$$\begin{aligned} C_n^a(T_m, W)_{2,2} &= d_n^a(T_m, W) = a_n^a(T_m, W) \\ &\asymp d_n(W_1^{2m+1}; L_2) \\ &\asymp \pi^{-(m+1)} \cdot (2m+1)^{-1/2} \cdot n^{-m-1/2}, n \to \infty. \end{aligned}$$

(c) Without indulging in the details of the derivation of the lower estimate for the quantity $C_n^a(T_m, W)_{1,1}$ (i.e., $p = q = 1$) from that of $C_n^a(T_m, W)_{2,2}$ we restrict ourselves to mention two main points which seem to be crucial in our further analysis.

Lemma 1

$$C_n^a(T_m, W)_{1,1} = \inf \sqrt{\frac{1}{\pi}\int_0^1 G_n(x; g_1, g_2, \cdots, g_n)\mathrm{d}x}, \qquad (2.4)$$

where the infimum is taken over the set of all systems $\{g_1, g_2, \cdots, g_n\} \subset BV[0,1]$ *such that* $g_i(1) = 0, \int_0^1 g_i(t)g_j(t)\mathrm{d}t = \delta_{ij}, i, j = 1, \cdots, n.$

The proof of Lemma 1 may be found in [2].

The solution of the extremal problem (2.4) could be derived from that of (2.3) applying an argument based on the following

Lemma 2 Let (E, μ) *be a finite measure space. Assume that* $g \in L^1(E, \mu)$ *satisfies the conditions*:

(1) $0 \leqslant g(x) \leqslant M, \forall x \in E$, *where* $M > 0$ *is fixed*.

(2) $\int_E g(x)\mu(\mathrm{d}x) \geqslant A > 0.$

Then for any sufficiently small $\delta > 0$,

$$\mu\{x \in E : g(x) \geqslant \delta A\} \geqslant \frac{[1 - \delta\mu(E)]}{M - \delta A}.$$

(2) **The upper estimation of** $C_n^a(T_m, W)_{p,Y}$.

The upper estimate can be obtained from the following inequalities:

$$C_n^a(T_m, W)_{p,Y} \leqslant a_n^a(T_m, W)_{p,Y} \leqslant E_{pq}(T_m, J^{(m)}, W) \leqslant M \cdot n^{-m-1/2},$$

where $Y = L_q[0,1], 1 \leqslant q < +\infty$.

The details of the proof will be published elsewhere [10], [11].

References

[1] J. F. Traub, G. W. Wasilkowski, H. Wozniakowski. *Information-based – Complexity*. *New York*: Acad. Press Inc, 1988

[2] D. Lee, G. W. Wasilkowski. Approximation of linear functionals on a Banach space with a Gaussian measure. J. Complexity, 1986, 2: 12 ~ 43

[3] Hui – Hsiung Guo. Gaussian Measures in Banach Spaces. Lecture Notes in Math. Vol. 463, Springer – Verlag, 1975

[4] H. Wiener. The average value of a functional. Proc. Lond. Math. Soc., 1922, 22: 454 ~ 467

[5] V.E. Maiorov. Integral and Kolmogorov widths of classes of differentiable functions. Soviet Math. Dokl. 1991, 44, 3

[6] V.E. Maiorov. Widths on Gaussian measure spaces. Soviet Math. Dokl., 1992, 48(2): 305 ~ 309

[7] Sun Yongsheng. Average n-widths of point set in Hilbert space. Chinese Sci. Bull., 1992, 37(14): 1153 ~ 1157

[8] P. Mathe. S-numbers in information-based complexity. J. Complexity, 1990, 6: 41 ~ 66

[9] R.S. Ismagilov. n-Dimensional widths of compact in Hilbert space. Functional Analysis and its Appl., 1968, 2(2): 32 ~ 39 (in Russian)

[10] Sun Yongsheng, Wang Chengyong. Average error bounds of best approximation of continuous functions on the Wiener space. J. Complexity, 1995, 11: 74 ~ 104

[11] Sun Yongsheng, Wang Chengyong. μ-Average n-widths on the Wiener space. J. Complexity, 1994, 10: 428 ~ 436

科学通报
1994, 12, 第 39 卷, 第 23 期

一个多元周期函数的 Besov 类的宽度估计*

Estimate of Widths for a Besov Class of Multivariate Periodic Functions

$\mathbf{R}^m$ 表示 m 维欧氏空间, $X = (x_1, x_2 \cdots, x_m)$, $Y = (y_1, \cdots, y_m) \in \mathbf{R}^m$, 其数量积记作 $\langle X, Y\rangle = \sum_{j=1}^{m} x_j y_j$. 用 $L_p(T^m)$ 表示定义在 T^m 上的 m 维可测函数 $f(X) = f(x_1, \cdots, x_m)$ 的全体, 此处 $T = [0, 2\pi)$, 这就是说 $f(x_1, \cdots, x_m)$ 对每一变量 x_j 都以 2π 为周期, 并在 T^m 上 p 幂可积 $(1 \leqslant p < \infty)$ 或本性有界 $(p = \infty)$, 赋以 L_p 范数如下:

$$\| f \|_p = \left\{(2\pi)^{-m} \int_{T^m} | f(X) |^p \mathrm{d}X\right\}^{\frac{1}{p}}, 1 \leqslant p < \infty,$$

$$\| f \|_\infty = \operatorname{ess} \sup_{x \in T^m} | f(X) |, p = \infty.$$

取指标集 $e_m = \{1, \cdots, m\}$, $e \subset e_m$ 是其任一子集. 对于一给定的向量 $r = (r_1, \cdots, r_m) \in \mathbf{R}^m$, r^e 表示具有以下分量的向量:

$$r_j^e = \begin{cases} r_j, j \in e, \\ 0, j \in e_m \setminus e. \end{cases} \tag{1}$$

对于定义在 T^m 上的函数 $f(X)$, 记

$$\Delta_{h,j} f(X) := f(x_1, \cdots, x_{j-1}, x_j + h, x_{j+1}, \cdots, x_m) - f(x).$$

当 $l \geqslant 1$ 是任意正整数时, 置

$$\Delta_{h,j}^l f(X) = \Delta_{h,j}(\cdots(\Delta_{h,j})) f(X), \tag{2}$$

* 1994 - 01 - 25 收稿.
国家自然科学基金资助项目.

是 $f(X)$ 关于变元 x_j 的步长 h 的 l 阶向前差分. 今取 $k=\{k_1,\cdots,k_m\}\in\mathbf{Z}_+^m$, $h=(h_1,\cdots,h_m)$, 置

$$\Delta_h^k f(X):=\Delta_{h_1,1}^{k_1},\cdots,\Delta_{h_m,m}^{k_m}f(X) \tag{3}$$

是 f 的 $(k_1,\cdots,k_m)$ 阶, 以 $h=(h_1,\cdots,h_m)$ 为步长的混合差分.

定义1 设 $r=(r_1,\cdots,r_m)$, $r_j\geqslant 0(j=1,\cdots,m)$, $k=(k_1,\cdots,k_m)\in\mathbf{Z}_+^m$ 满足 $k_j>r_j(j=1,\cdots,m)$, $\theta\in[1,+\infty)$, 称 $f\in B_{p,\theta}^r(T^m)$, *如果* f 满足以下条件:

(1) $f\in L_p(T^m)$,

(2) 对任一非空子集 $e\subset e_m$

$$\int_{T^e}\|\Delta_{h^e}^{k^e} f(X)\|_{p(T^m)}^\theta\cdot\prod_{j\in e}\frac{\mathrm{d}h_j}{|h_j|^{1+r_j\theta}}<+\infty, \tag{4}$$

此处 $T^e=\prod\limits_{j\in e}T$. 由文献[1]知, 线性空间 $B_{p,\theta}^r(T^m)$ 可以如下地赋范:

$$\|f\|_{B_{p,\theta}^r}:=\sum_{e\subset e_m}\left\{\int_{T^e}\|\Delta_{h^e}^{k^e} f(X)\|_{p(T^m)}^\theta\prod_{j\in e}\frac{\mathrm{d}h_j}{|h_j|^{1+r_j\theta}}\right\}^{\frac{1}{\theta}}+\|f\|_{p(T^m)}. \tag{5}$$

这样一来, $B_{p,\theta}^r$ 成为 Banach 空间.

当 $\theta=\infty$ 时, $B_{p,\theta}^r(T^m)$ 定义为满足以下条件的函数全体:

(1) $f\in L_p(T^m)$,

(2) 对任一非空 $e\subset e_m$ 有

$$\|\Delta_{h^e}^{k^e} f(X)\|_{p(T^m)}\ll\prod_{j\in e}|h_j|^{r_j}. \tag{5'}$$

由文献[1]知 $B_{p\infty}^r(T^m)$ 等同于 Hölder-Никольский 类 $H_p^r(T^m)$[2], 赋以范数

$$\|f\|_{B_{p,\infty}^r}:=\sum_{e\subset e_m}\sup_{|h|>0}\left(\frac{\|\Delta_{h^e}^{k^e} f(X)\|_{p(T^m)}}{\prod\limits_{j\in e}|h_j|^{r_j}}\right)+\|f\|_{p(T^m)}, \tag{6}$$

$B_{p,\infty}^r$ 是 Banach 空间. 注意在(6)式内 $|h|:=(|h_1|,\cdots,|h_m|)$, $|h|>0$ 表示 $|h_j|>0$, $j=1,\cdots,m$.

Besov 范数(5), (6) 有几种等价形式. 为了后面的应用需要引入一种等价范数. 往下假定 $f\in L_p(T^m)$ 满足条件

$$\int_T f(X)\mathrm{d}x_j = 0, j = 1, \cdots, m, \tag{7}$$

其全体记作 $\mathring{L}_p(T^m)$.

令 $s = (s_1, \cdots, s_m), s_j \in \mathbf{Z}_+$. 记

$\square_s = \{k : k = (k_1, \cdots, k_m), 2^{s_j-1} \leqslant |k_j| < 2^{s_j}, j = 1, \cdots, m\}$,

$$\delta_s(f, X) = \sum_{k \in \square_s} C_k \mathrm{e}^{\mathrm{i}\langle k, X\rangle}, \tag{8}$$

$$C_k = C_k(f) = (2\pi)^{-m}\int_{T^m} f(X)\mathrm{e}^{-\mathrm{i}\langle k, X\rangle}\mathrm{d}X. \tag{9}$$

根据文献[1], 假定 $f \in \mathring{L}_p \cap B^r_{p,\theta} := \mathring{B}^r_{p,\theta}$, 则当 $1 \leqslant \theta < \infty$ 时有

$$\| f \|_{B^r_{p,\theta}} \asymp \left\{ \sum_{S>0} 2^{\langle r, s\rangle \theta} \| \delta_s(f, X) \|_p^\theta \right\}^{\frac{1}{\theta}}. \tag{10}$$

$\theta = \infty$ 时有

$$\| f \|_{B^r_{p,\theta}} \asymp \sup_{S>0} 2^{\langle r, s\rangle} \| \delta_s(f, X) \|_p, \tag{11}$$

其中 $1 < p < \infty$.

如上引入的 Besov 类 $\mathring{B}^r_{p,\theta}(T^m)$ 和由一个优混合偏微商确定的多元周期函数的 Sobolev 类 $W^r_{p,\alpha}(T^m)$, 以及具有某一混合光滑度的多元周期的 Hölder-Nikolskii 类 $H^r_p(T^m)$ 密切相关. 关于 $W^r_{p,\alpha}$, H^r_p 类的逼近特征的研究已有很多工作[3,4]. 这些光滑函数类的逼近特征的研究和一元情形完全不同, 突出的一点是, 对于 $W^r_{p,\alpha}(T^m)$, $H^r_p(T^m)$, $m > 1$ 时, 作为三角多项式最佳逼近子空间需选择和优混合偏导数算子相匹配的所谓双曲型三角多项式子空间(或称双曲十字型三角多项式子空间), 而对 Besov 类 $B^r_{p,\theta}(T^m)$ 的研究工作还很少[5].

设 $1 < q < \infty, 1 \leqslant \theta \leqslant \infty$. 令 $r = (r_1, \cdots, r_m) = r_1\gamma$, 此处 $\gamma = (\gamma_1, \cdots, \gamma_m), 1 = \gamma_1 = \cdots = \gamma_\nu < \gamma_{\nu+1} \leqslant \cdots \leqslant \gamma_m, r_1 > 0, 1 \leqslant \nu \leqslant m$. 记

$$\mathring{B}^r_{q,\theta} = \{f \in \mathring{L}_q(T^m) \cap B^r_{q,\theta} : \| f \|_{B^r_{q,\theta}} \leqslant 1\}, \tag{12}$$

假定 $1 < q \leqslant p < \infty$. 由文献[1] 知, 当 $r_1 > \dfrac{1}{q} - \dfrac{1}{p}$ 时, $B^r_{q,\theta}$ 连续嵌入于 $L_p(T^m)$. 在此条件下我们可以提出讨论类 $\mathring{B}^r_{q,\theta}$ 在$L_p(T^m)$空间内的逼

近特征问题. Romaniuk[5] 研究了 $\mathring{B}^r_{q,\theta}$ 在 $L_p(T^m)$ 空间内的 Kolmogorov 宽度估计, 在 $1 < q \leqslant p \leqslant 2$ 条件下证明了

$$d_M(\mathring{B}^r_{q,\theta}, L_p) \asymp [(\log M)^{\nu-1} \cdot M^{-1}]^{r_1-\frac{1}{q}+\frac{1}{p}}(\log M)^{(\nu-1)\left(\frac{1}{p}-\frac{1}{\theta}\right)_+}. \tag{13}$$

我们证得

定理 1

(1) 若 $1 < q < 2 \leqslant p < \infty, r_1 > 1/q, 1 \leqslant \theta \leqslant \infty$, 则

$$d_M(\mathring{B}^r_{q,\theta}, L_p) \asymp [(\log M)^{\nu-1} \cdot M^{-1}]^{r_1-1/q+1/2}(\log M)^{(\nu-1)(1/2-1/\theta)_+}; \tag{14}$$

(2) 若 $2 \leqslant q \leqslant p < \infty, r_1 > 1/2, 1 \leqslant \theta \leqslant \infty$, 则

$$d_M(\mathring{B}^r_{q,\theta}, L_p) \asymp [(\log M)^{\nu-1} \cdot M^{-1}]^{r_1}(\log M)^{(\nu-1)(1/2-1/\theta)_+}; \tag{15}$$

当 $\theta = \infty$ 时, $\mathring{B}^r_{q,\theta} = H^r_q(T^m)$, 此时(14),(15) 式化归为已知的结果[3].

(14),(15) 式的证明, 其上方估计的得到, 主要应用离散化技巧[3]. 而对定理 1 中的 p, q 离散化技巧的应用, 必须实质性地依赖于 Kashin[6] 中证明的关于有限维点集宽度估计的艰深结果.

考虑函数类 $\mathring{B}^r_{q,\theta}$ 在 L_p 尺度下借助于秩数固定的正交投影算子族的最佳逼近问题. 这类极值问题首先由 Temlyakov 提出[3], 它是线性宽度的一个特殊情形.

定义 2　设 $F \subset L_p(T^m), M > 0, 1 \leqslant p \leqslant \infty$. 称量

$$d^{\perp}_M(F, L_p) := \inf_{L,\, \dim L \leqslant M} \sup_{f \in F} \| f - P(f, L) \|_p, \tag{16}$$

为 F 的 M 维正交投影宽度(OP- 宽度), 此处 $P(f, L)$ 是 f 在 $L \subset L_2(T^m)$ 上的正交投影.

定理 2　设 $r_1 > \dfrac{1}{q} - \dfrac{1}{p}, 1 \leqslant \theta \leqslant \infty, 1 < q < p < \infty$, 或 $1 < q \leqslant p \leqslant 2$, 则有

$$d^{\perp}_M(\mathring{B}^r_{q,\theta}, L_p) \asymp [(\log M)^{\nu-1} \cdot M^{-1}]^{r_1-1/q+1/p}(\log M)^{(\nu-1)(1/p-1/\theta)_+}. \tag{17}$$

注　由定理2及文献[5]的定理4得到：当$1<q\leqslant p\leqslant 2, r_1>\frac{1}{q}-\frac{1}{p}, 1\leqslant\theta\leqslant\infty$时有

$$d_M(\mathring{B}^r_{q,\theta}, L_p)\asymp d'_M(\mathring{B}^r_{q,\theta}, L_p)\asymp d^{\perp}_M(\mathring{B}^r_{q,\theta}, L_p), \tag{18}$$

$d'_M(\mathring{B}^r_{q,\theta}, L_p)$是线性宽度．此时$f$的双曲型傅里叶部分和是弱渐近最优线性算子(正交投影)．

参考文献

[1] Amanov, T. I. Proc. Matem. Ins. Acad. Sci. USSR v. 77, 1965, 5 ~ 34 (In Russian)

[2] Nikolskii, S. M. Sib. M. J., 1963, 4(6): 1342 ~ 1364 (In Russian)

[3] Temlyakov, V. N. Proc. Matem. Ins. Acad. sci USSR v. 178, 1986, 3 ~ 112 (In Russian)

[4] Telyakovskii, S. A. Proc. Matem. Ins. Acad. Sci. USSR v. 182, 1988, 128 ~ 179 (In Russian)

[5] Romaniuk, A. S. Uk. M. J., 1991, 43(10): 1398 ~ 1408 (In Russian)

[6] Kashin, B. S. Izv. Akademii Nauk, USSR, Matem. seris, 1977, 41(2): 334 ~ 351 (In Russian)

科学通报
1995,3,第40卷,第6期

具有给定的混合型光滑模的多元周期函数的表现和逼近*

Representation and Approximation of Multivariate Periodic Functions with Bounded Mixed Moduli of Smoothness

1 预备事项

$\mathbf{R}^d$ 表示 d 维欧氏空间, $X=(x_1,\cdots,x_d)$, $Y=(y_1,\cdots,y_d)\in\mathbf{R}^d$, 其数量积记作 $\langle X,Y\rangle=\sum_{j=1}^{d}x_jy_j$. $f(X)=f(x_1,\cdots,x_d)$ 表示实可测函数, 对每一变量均以 2π 为周期. $\pi_d=[0,2\pi)^d$ 是 d 维周期 2π 的立方体. 对 $q,1\leqslant q\leqslant\infty$, 记 $f\in L_q(\pi_d)$, 倘若

$$\|f\|_q:=\left\{(2\pi)^{-d}\int_{\pi_d}|f(X)|^q\mathrm{d}X\right\}^{\frac{1}{q}}<\infty,1\leqslant q<\infty.$$

$$\|f\|_\infty:=\operatorname*{ess\,sup}_{x\in\pi_d}|f(X)|<\infty,q=\infty.$$

记 $f\in\mathring{L}_q(\pi_d)$, 倘若 $f\in L_q(\pi_d)$, 而且

$$\int_0^{2\pi}f(X)\mathrm{d}x_j=0,j=1,\cdots,d, \tag{1.1}$$

对几乎一切 $(x_1,\cdots,x_{j-1},x_{j+1},\cdots,x_d)\in\mathbf{R}^{d-1}$ 成立. 记 $f\in L_q(\pi_d)$, 定义

$$\Omega^l(f,t)_q:=\sup_{|h|\leqslant t}\|\Delta_h^lf(X)\|_q, \tag{1.2}$$

此处 l 是一固定正整数, $t=(t_1,\cdots,t_d)>0$(即 $t_j>0,j=1,\cdots,d$),

---

* 1994-08-25 收稿.
国家自然科学基金资助项目.
本文与汪和平合作.

$h=(h_1,\cdots,h_d)\in\mathbf{R}^d$ ，$|h|:=(|h_1|,\cdots,|h_d|)$，$|h|\leqslant t$ 含义是 $|h_j|\leqslant t_j, j=1,\cdots,d$. 最后，

$$\Delta_{h_1}^l f(X)=\sum_{j=0}^{l}(-1)^{l-j}\binom{l}{j}f(x_1+jh_1,x_2,\cdots,x_d),\Delta_h^l f(X)=\Delta_{h_{d-1}}^l\cdots(\Delta_{h_1}^l f(X),\cdots). \tag{1.3}$$

$\Omega^l(f,t)_q$ 是 f 的 l 阶 (L_q) 混合光滑模. 熟知它满足以下4个条件：

(1) $\Omega^l(f,t)_q=0$ 当 $\prod_{j=1}^{d}t_j=0$ 时.

(2) $\forall_t=(t_1,\cdots,t_d)\geqslant 0$(即 $t_j\geqslant 0, j=1,\cdots,d$)$\Omega^l(f,t)_q\geqslant 0$.

(3) 对每一变元 t_j $\Omega^l(f,t_1,\cdots,t_d)$ 是单调增的连续函数.

(4) $\forall_n=(n_1,\cdots,n_d)\in\mathbf{Z}_+^d$ 有

$$\Omega^l(f,n_1t_1,\cdots,n_dt_d)_q\leqslant\Big(\prod_{j=1}^{d}n_j\Big)^l\Omega^l(f,t)_q.$$

记 $\mathbf{R}_+^d:=\{t=(t_1,\cdots,t_d):t_j\geqslant 0, j=1,\cdots,d\}$. 定义在$\mathbf{R}_+^d$ 上的函数 $\Omega(t)$ 倘若具有性质(1) ~ (4)，在本文内称之为 l 阶混合光滑模型的函数，简记作 $\Omega\in\Phi$. 仿文献[1]，我们引入

定义1 设 $\varphi(t)=\varphi(t_1,\cdots,t_d)$ 定义域为$\mathbf{R}_+^d$，$\varphi(0)=0$，$\varphi(t)>0$ 对 $t>0$ 成立.

(1) 称 φ 满足条件 S(记作 $\varphi\in S$)，倘若存在一向量 $\alpha=(\alpha_1,\cdots,\alpha_d)$，$0<\alpha_j<1$，$(j=1,\cdots,d)$ 及一常数$c>0$，使得对任何 $\tau=(\tau_1,\cdots,\tau_d)\geqslant t=(t_1,\cdots,t_d)>0$ 均成立

$$\varphi(t)t^{-\alpha}\leqslant c\varphi(\tau)\tau^{-\alpha}. \tag{1.4}$$

此处有定义 $t^\alpha:=t_1^{\alpha_1}\cdots t_d^{\alpha_d}$ 等等. 此时称 $\varphi(t)t^{-\alpha}$ 几乎单调增.

(2) 称 φ 满足条件S_l(记作 $\varphi\in S_l$)，倘若存在向量 $\gamma=(\gamma_1,\cdots,\gamma_d)$，$0<\gamma_j<l(j=1,\cdots,d)$ 及一常数 $c>0$ 使得对任何 $\tau\geqslant t>0$ 均成立

$$\varphi(t)\cdot t^{\gamma-l\cdot 1}\geqslant c\varphi(\tau)\cdot\tau^{\gamma-l\cdot 1}, \tag{1.5}$$

此处 $\gamma-l\cdot 1=(\gamma_1-l,\gamma_2-l,\cdots,\gamma_d-l)$. 此时称函数 $\varphi(t)t^{\gamma-l\cdot 1}$ 几乎单调降.

在本文内记 $\varphi\in\Psi$，若 $\varphi\in\Phi\wedge S\wedge S_l$. 这样的 φ 的一个典型例子是 $\Omega(t)=t^r=t_1^{r_1}\cdots t_d^{r_d}$，其中 $r=(r_1,\cdots,r_d)>0$. Nikolskii 等人[2,3]利用它作为函数的 l 阶($l\cdot 1>\gamma$)混合光滑模的控制函数，以此引入了著

名的Hölder-Nikolskii类 $S_q^r H(\pi_d)$ 及Besov类 $S_{q,\theta}^r B(\pi_d)$，利用调和分析的工具证明了这些函数类中函数的表现定理. Nikolskaja[4], Galeev[5], Temlyakov[6] 研究了 $S_q^r H(\pi_d)$ 类上的逼近, Romaniuk[7,9]，孙永生[8] 研究了 $S_{q,\theta}^r B(\pi_d)$ 类上的逼近. Pustovoitov[1] 利用 Ψ 类的函数引入函数类 $H_q^\Omega(\pi_d)$，并且把 $S_q^r H(\pi_d)$ 上的一些结果扩充到该类上. 本文目的是利用 Ψ 类的函数引入Besov型类 $B_{q,\theta}^\Omega(\pi_d)$，并且在该类上建立函数的表现定理，同时得到双曲十字型的Fourier逼近和Kolmogorov宽度的渐近精确估计的一些结果.

定义2　给定 $\Omega(t)\in\Psi$, q,θ 满足 $1\leqslant q,\theta\leqslant\infty$. 称 $f\in B_{q,\theta}^\Omega(\pi_d)$，倘若

(1) $f\in \mathring{L}_q(\pi_d)$.

(2) $$\left\{\underbrace{\int_0^2\cdots\int_0^2}_{d\text{重}}\left(\frac{\Omega^l(f,t)_q}{\Omega(t)}\right)^\theta\prod_{j=1}^d\frac{\mathrm{d}t_j}{t_j}\right\}^{\frac{1}{\theta}}:=\|f\|_{B_{q,\theta}^\Omega}<\infty,1\leqslant\theta<\infty.$$

当 $\theta=\infty$ 时我们用下面的(2′)代替(2)：

(2′) $$\sup_{t>0}\frac{\Omega^l(f,t)_q}{\Omega(t)}=\|f\|_{B_{q,\infty}^\Omega}<\infty.$$

注意这里 $B_{q,\infty}^\Omega(\pi_d)=H_q^\Omega(\pi_d)$. $\|f\|_{B_{q,\infty}^\Omega}$ 是范数. 下面给出一个后面要讨论的函数类,

$$B_{q,\theta}^\Omega=\{f\in B_{q,\theta}^\Omega(\pi_d):\|f\|_{B_{q,\theta}^\Omega}\leqslant 1\}.$$

2　$1<q<\infty$ 时 $B_{q,\theta}^\Omega(\pi_d)$ 的一个表现定理

令 $s=(s_1,\cdots,s_d)\in\mathbf{Z}_+^d$，$\rho(s)=\{k\in\mathbf{Z}^d:k=(k_1,\cdots,k_d),2^{s_j-1}\leqslant|k_j|<2^{s_j},j=1,\cdots,d\}$. 设 $f\in L_q(\pi_d)$，记 $c_k=c_k(f)=(2\pi)^{-d}\int_{\pi_d}f(X)\mathrm{e}^{-i\langle k,X\rangle}\mathrm{d}X$，并且

$$\delta_s(f,X)=\sum_{k\in\rho(s)}c_k\mathrm{e}^{i\langle k,X\rangle}.\tag{2.1}$$

由Littlewood-Paley定理[3]知：若 $1<q<\infty$，则对 $f\in\mathring{L}_q(\pi_d)$ 有

$$f(X)\overset{L_q}{=}\sum_{S>0}\delta_s(f,X).\tag{2.2}$$

定理1　设 $\Omega\in\Psi$, $1<q<\infty$, $1\leqslant\theta<\infty$. 则 $f\in B_{q,\theta}^\Omega(\pi_d)$，当且

仅当

$$\left\{\sum_{s>0}\Omega^{-\theta}(2^{-s})\parallel\delta_s(f)\parallel_q^\theta\right\}^{\frac{1}{\theta}}<\infty. \tag{2.3}$$

且此时成立

$$\left\{\sum_{s>0}\Omega^{-\theta}(2^{-s})\parallel\delta_s(f)\parallel_q^\theta\right\}^{\frac{1}{\theta}}\asymp\parallel f\parallel B_{q,\theta}^{\Omega}. \tag{2.4}$$

此处$2^{-s}:=(2^{-s_1},2^{-s_2},\cdots 2^{-sd})$，两个依赖于$f$的量$A,B$，若存在与$f$无关的常数$c>0$使有$c^{-1}|B|\leqslant|A|\leqslant c|B|$，则记作$|A|\asymp|B|$. $\theta=\infty$ 时，条件(2.3) 应代之以

$$\sup_{s>0}\Omega^{-1}(2^{-s})\parallel\delta_s(f)\parallel_q<\infty. \tag{2.5}$$

见文献[1]. 定理1是Nikolskii-Lizorkin[3]一条定理的扩充. 顺便指出，定理1还可以扩充到定义在$\mathbf{R}^d$的函数上去.

3 $\mathring{B}_{q,\theta}^{\Omega}$ 在L_q内的Fourier逼近

记$\parallel s\parallel_l=\sum_{j=1}^d s_j$，$Q_n=\bigcup_{\parallel s\parallel_l\leqslant n}\rho(s)$. 令

$$T(Q_n)=\left\{t(X):t(X)=\sum_{k\in Q_n}c_k e^{i<k,X>}\right\}. \tag{3.1}$$

根据文献[3]，$T(Q_n)$内的三角多项式是双曲十字型的. 记$f\in\mathring{L}_q(\pi_d)$，置

$$S_n(f)=\sum_{\parallel s\parallel_l\leqslant n+1}\delta_s(f). \tag{3.2}$$

它是f的双曲十字型Fourier和式. 考虑

$$E_{Q_n}(f)_q:=\inf_{g\in T(Q_n)}\parallel f-g\parallel_q \tag{3.3}$$

及

$$E_{S_n}(f)_q:=\parallel f-S_n(f)\parallel_q. \tag{3.4}$$

当$1<q<\infty$时，知有(见文献[3])

$$E_{Q_n}(f)_q\asymp E_{S_{n-1}}(f)_q. \tag{3.5}$$

定理2 设单变元函数$\omega(\cdot)\in\Psi$，$\Omega(t)=\omega\left(\prod_{j=1}^d t_j\right)$，$1<q<\infty$. 则

(1) 当$1<q\leqslant 2$，$1\leqslant\theta\leqslant\infty$时，有

$$E_{Q_n}(\mathring{B}^{\Omega}_{q,\theta})_q := \sup_{f\in \mathring{B}^{\Omega}_{q,\theta}} E_{Q_n}(f)_q \asymp E_{S_{n-1}}(\mathring{B}^{\Omega}_{q,\theta})_{\mathring{q}}:$$

$$= \sup_{f\in \mathring{B}^{\Omega}_{q,\theta}} \| f - S_{n-1}(f) \|_q \asymp \omega(2^{-n}) n^{(d-1)(\frac{1}{q}-\frac{1}{\theta})_+}.$$

(2) 当 $2 < q < \infty, 1 \leqslant \theta \leqslant \infty$ 时,有

$$E_{Q_n}(\mathring{B}^{\Omega}_{q,\theta})_q \asymp E_{S_{n-1}}(\mathring{B}^{\Omega}_{q,\theta})_q \asymp \omega(2^{-n})_n^{(d-1)(\frac{1}{2}-\frac{1}{\theta})_+}.$$

定理 3 设 $\Omega(t)$ 如前, $1 < q < p < \infty, \alpha > \frac{1}{q} - \frac{1}{p}$, 此处 $0 < \alpha < 1, \omega(t)t^{-\alpha}$ 几乎单调增. 则 $E_{S_{n-1}}(\mathring{B}^{\Omega}_{q,\theta})_p \asymp \omega(2^{-n})2^{n(\frac{1}{q}-\frac{1}{p})} \cdot n^{(d-1)(\frac{1}{p}-\frac{1}{\theta})_+}, 1 \leqslant \theta \leqslant \infty$.

$\theta = \infty$ 时定理 2, 3 的结果已见于文献[1].

4 $\mathring{B}^{\Omega}_{q,\theta}(\pi_d)$ 的 Kolmogorov 宽度

假定对某个 $p, 1 < p < \infty, \mathring{B}^{\Omega}_{q,\theta} \subset L_q(\pi_d)$. 称量

$$d_M(\mathring{B}^{\Omega}_{q,\theta}, L_p) := \inf_{\mathscr{L}} \sup_{f\in \mathring{B}^{\Omega}_{q,\theta}} \inf_{g\in\mathscr{L}} \| f - g \|_p \tag{4.1}$$

为 $\mathring{B}^{\Omega}_{q,\theta}$ 在 L_p 内的 Kolmogorov M 宽度, 此处 $\mathscr{L}$ 遍取 $L_p(\pi_d)$ 内一切 $\dim\mathscr{L} \leqslant M$ 的线性子空间. Romaniuk[7,9], 孙永生[8] 对 $\Omega(t) = t_1^{r_1} \cdots t_d^{r_d}$ 的情形得到一些结果.

定理 4 设 $\Omega(t)$ 如定理 2 所给出者, $1 < q \leqslant p \leqslant 2, \alpha > \frac{1}{q} - \frac{1}{p}$, $1 \leqslant \theta \leqslant \infty$, 则

$$d_M(\mathring{B}^{\Omega}_{q,\theta}, L_p) \asymp \omega(2^{-n})2^{n\left(\frac{1}{q}-\frac{1}{p}\right)} n^{(d-1)\left(\frac{1}{p}-\frac{1}{\theta}\right)_+}.$$

此处 M 和 n 满足关系式 $M \asymp 2^n \cdot n^{d-1}$. $S_n(f)$ 是渐近最优的线性算子, $T(Q_n)$ 是(弱) 渐近极子空间.

定理 5 设 $\Omega(t)$ 如定理 2 所给出者, $1 < q \leqslant p < \infty$.

(1) 若 $1 < q < 2 \leqslant p < \infty, \alpha > \frac{1}{q}, 1 \leqslant \theta \leqslant \infty$, 则

$$d_M(\mathring{B}^{\Omega}_{q,\theta}, L_p) \asymp \omega(2^{-n})2^{n\left(\frac{1}{q}-\frac{1}{p}\right)} n^{(d-1)\left(\frac{1}{2}-\frac{1}{\theta}\right)_+}.$$

(2) 若 $2 \leqslant q \leqslant p < \infty, a > \frac{1}{2}, 1 \leqslant \theta \leqslant \infty$, 则

$$d_M(\mathring{B}^{\Omega}_{q,\theta}, L_p) \asymp \omega(2^{-n}) n^{(d-1)\left(\frac{1}{2}-\frac{1}{\theta}\right)_+}.$$

如定理 4, 此处 $M \asymp 2^n \cdot n^{d-1}$.

这些定理的证明将在另文中给出.

参考文献

[1] Pustovoitov N N. Representation and approximation of multivariate periodic functions with a given mixed modulus of continuity. Anal Math, 1994, 20 : 35 ~ 48

[2] Nikolskii S M. Functions with dominating mixed derivative which satisfies a double Hölder's condition(in Russian). Sib M J, 1963, 4(6) : 1342 ~ 1364

[3] Lizorkin P I, Nikolskii S M. Spaces of functions with mixed smoothness from view point of decomposition(in Russian). Trudy Mat Inst Steklov USSR, 1989, 187 : 143 ~ 160

[4] Nikolskaja N S. Approximation of multivariate differentiable functions by Fourier sums in metric L_p(in Russian). Sib M J, 1974, 15(2) : 395 ~ 412

[5] Galeev E M. Approximation of classes of functions with several bounded derivatives by Fourier sums (in Russian). Mat Zametki, 1978, 23(2) : 197 ~ 212

[6] Temlyakov V N. Approximation of functions with a bounded mixed derivative (in Russian). Trudy Mat Inst Steklov USSR, 1986, 178 : 1 ~ 112

[7] Romaniuk A S. Approximation of Besov classes of multivariate periodic functions in space L_q(in Russian). Uk M J, 1991, 43(10) : 1398 ~ 1408

[8] 孙永生, 一个多元周期函数的 Besov 类的宽度估计. 科学通报, 1994, 39(23) : 2 113 ~ 2 115

[9] Romaniuk A S. Kolmogorov width for classes $B^r_{p,\theta}$ of multivariate periodic functions with a lower smoothness in L_q Uk M J, 1994, 46(7) : 915 ~ 926

附录

Appendix

论文和著作目录

Bibliography of Papers and Works

论文目录

[1] Sun Yongsheng. On the best approximation of classes of functions representable in the convolution form [Russian], Dokl. Akad. Nauk USSR, 1958, 118(2): 247 ~ 250

[2] Sun Yongsheng. Best approximation of periodic differentiable functions by trigonometric polynomials [Russian], Izv. Akad. Nauk USSR Math. Series, 1959, 23 (1): 67 ~ 92

[3] 孙永生. 周期可微函数类的三角多项式最佳近迫. 北京师范大学学报(自然科学版), 1959, (1): 1 ~ 3; (3): 21 ~ 35

[4] 孙永生. 关于一类周期函数最佳逼近的注记. 北京师范大学学报(自然科学版), 1959, (3): 36 ~ 46

[5] 孙永生. 关于"周期可微函数类的三角多项式最佳逼近" 一文的补充. 北京师范大学学报(自然科学版), 1959, (4): 9 ~ 10

[6] Sun Yongsheng. On the best approximation of periodic differentiable functions by trigonometric polynomials [Russian], Izv. Akad. Nauk USSR Math. Series, 1961, 25: 143 ~ 152

[7] 孙永生. 用三角多项式近迫周期可微函数. 数学学报, 1962, 12(2): 181 ~ 202 (Sun Yongsheng. On the best approximation of periodic differentiable functions by trigonometric polynomials,

Chinese Math. Acta, 1963, 3:196 ~ 217)

[8] 孙永生. 周旋函数的最佳一致近迫和最佳线性近迫. 数学学报, 1962, 12(3):301 ~ 319 (Sun Yongsheng, On the best approximation and best linear approximation of convolutions of periodic functions, Chinese Math. Acta, 1963, 3:331 ~ 349)

[9] 孙永生. 能表示为卷积形式的周期连续函数的最佳一致逼近及最佳线性逼近. 北京师范大学学报(自然科学版), 1962, (1): 1 ~ 5

[10] 孙永生. 连续周期函数用其傅里哀级数的蔡查罗平均数的一致近迫. 数学进展, 1963, 6(4):379 ~ 387

[11] 孙永生, 张阳春. 一对共轭周期函数的最佳逼近的渐近性质. 数学学报, 1964, 14(4):578 ~ 596(Sun Yongsheng, Zhang Yangchun, Asymptotic behavior of the best approximation of two conjugate periodic functions, Chinese Math. Acta, 1964, 5:623 ~ 641)

[12] 孙永生. 关于直交函数项级数的强性(C, 1) 求和. 数学进展, 1966, 9(3):281 ~ 286

[13] 孙永生. 关于微积分的历史(1 ~ 2). 北京师范大学学报(自然科学版), 1977, (2):35 ~ 50; 1977, (3):74 ~ 86

[14] 孙永生. 关于 Oskolkov 不等式的注记. 北京师范大学学报(自然科学版), 1979, (3):71 ~ 75

[15] 孙永生. 二元周期连续函数利用其 Fourier 部分和的一致逼近. 北京师范大学学报(自然科学版), 1979, (3):16 ~ 35

[16] 孙永生. 实变函数逼近的几个问题. 见: 全国第一届函数逼近论会议论文集. 杭州大学出版社, 1979:25 ~ 34

[17] 孙永生. 周期函数用其 Fourier 部分和的平均逼近. 数学年刊, 1980, 1(2):273 ~ 282

[18] 孙永生. 关于 Jackson 不等式的注记. 北京师范大学学报(自然科学版), 1981, (1):1 ~ 6

[19] 孙永生. 关于 Cesàro 算子的逼近常数. 数学学报, 1981, 24(4): 516 ~ 537

[20] 孙永生. 关于可微函数类的 Kolmogorov 宽度. 北京师范大学学

报(自然科学版),1981,(3):25 ~ 35

[21] Sun Yongsheng. On n-width of a class of periodic analytic functions in L_1 metric, Chin. Ann. of Math.. 1981, 2(Eng. Issue): 53 ~ 58

[22] 孙永生. 关于周期函数用线性算子的平均逼近. 数学学报, 1982, 25(3):561 ~ 577

[23] 孙永生. 一个广义样条函数类上的极值问题. 科学通报, 1982, 27(11):701(Sun Yongsheng, Some extremal problems on a class of generalized spline functions, Chinese Science Bulletin, 1983, 28(3):427 ~ 428)

[24] 孙永生. 一个由线性微分算子确定的函数类的最佳逼近问题. 科学通报, 1982, 27 (20): 1279 (Sun Yongsheng, Best approximation of function classes defined by a linear differential operator, Chinese Science Bulletin, 1983, 28(8):1140~1141)

[25] 孙永生. 某些函数类上宽度的对偶定理. 中国科学(*A* 辑), 1982, 12(3):215 ~ 225(*Sun Yongsheng. Duality theorems for the width numbers of some functional classes. Scientia Sinica* (*Series A*), 1982, 25(10):1009 ~ 1020)

[26] 孙永生. 周期函数用线性算子的平均逼近. 北京师范大学数学系编: 北京师范大学 80 周年校庆论文集, 1982:1 ~ 12

[27] 孙永生. 一个广义样条函数类上的极值问题和有关的宽度问题. 中国科学(*A* 辑), 1982, 13(8):677 ~ 688 (*Sun Yongsheng. Some extremal problems on a class of periodic splines. Scientia Sinica*(*A*). 1984, 27(3):253 ~ 266)

[28] 孙永生. 光滑函数类的宽度估计问题. 见: 全国第三届函数逼近论会议论文集. 南京大学学报. 1983, (增刊):49 ~ 64

[29] 孙永生.《关于 $E_n(W_p^r)_p$ 的一些结果》一文的错误. 数学研究与评论, 1983, 3(1):160

[30] *Sun Yongsheng. Asymptotic estimation of n-widths of some classes of functions defined by an ordinary linear differential operator, Approximation theory, IV* (*College Station, Tex.*, 1983). *New York: Academic Press*, 1983. 709 ~ 713

[31] 黄达人,孙永生.关于函数类$\Omega_p^{r+1}[0,1]$的宽度估计.科学通报,1984,29(12):716 ~ 720

[32] 孙永生.周期可微函数类上的某些极值定理.北京师范大学学报(自然科学版),1984,(4):11 ~ 26

[33] *Sun Yongsheng, Huang Daren. On one-sided approximation of class of smooth functions* $\Omega_p^{2r+\sigma}$. *ATA*, 1984, 1(1):19 ~ 35

[34] *Sun Yongsheng, Huang Daren. On n-width of generalized Bernoulli kernel. ATA*, 1985, 1(2):83 ~ 92

[35] 孙永生.线性微分算子的*Landau-Kolmogorov*型不等式.科学通报,1985,30(7):481 ~ 484 (*Sun Yongsheng. Inequalities of Landau-Kolmogorov type for some linear differential operators. Chinese Science Bulletin*, 1985, 30(8):995 ~ 998)

[36] 黄达人,孙永生.光滑函数类$\Omega_p^{2r+\sigma}$的上方逼近.东北数学,1985,1(2):172 ~ 182

[37] 孙永生,黄达人.关于广义*Bernoulli*核的宽度.科学通报,1985,30(10):728 ~ 731

[38] *Sun Yongsheng. A remark on Kolmogorov's comparison theorem. Chin. Ann. of Math*, 1986, 7*B*(4):463 ~ 467

[39] *Sun Yongsheng. On optimal interpolation for a differentiable function class* (*I*). *ATA*. 1986, 2(4):49 ~ 54

[40] 孙永生.可微函数类上的最优恢复.科学通报,1986,(11):809 ~ 812 (*Sun Yongsheng. Optimal recovery on some classes of differentiable functions. Chinese Science Bulletin*, 1986, 32(16):1090 ~ 1093)

[41] *Sun Yongsheng, Huang Daren. Optimal interpolation on some class of differentiable functions. Approximation theory, V* (*Texas A&M Univ.*, 1986). *Academic Press*, 1987. 583 ~ 586

[42] 孙永生,黄达人. *Extremal properties of* $\mathscr{L}$-splines and related extremal proplems.见:全国第四届函数逼近论会议论文集.大连理工大学出版社,1987

[43] 孙永生,黄达人,房艮孙.最优恢复和宽度问题.见:全国第五届函数逼近论会议论文集.河南大学学报(增刊).1988.11 ~ 21

[44] Sun Yongsheng, Huang Daren. Fang Gensun. Extremal problems

on some classes of differentiable functions, Constructive Theory of Functions(Varna, 1987). 433 ~ 440, Publ. House Bulgar. Acad. Sci., Sofia, 1988:434 ~ 440

[45] 孙永生. 一个卷积类上的最优插值问题. 科学通报, 1988, 33(13):964 ~ 967 (Sun Yongsheng. Optimal interpolation on a convolution class of functions. Chinese Science Bulletin, 1989, 34(14):1148 ~ 1152)

[46] 孙永生. 一个可微函数类上微分算子的最优恢复问题. 北京师范大学学报(自然科学版), 1988, (3):7 ~ 10

[47] Huang Daren, Sun Yongsheng. Optimal interpolation on some class of differentiable functions. ATA, 1988, 4(3):13 ~ 22

[48] Sun Yongsheng. Optimal recovery of the derivative from a class of smooth functions. Approximation Theory VI:2 (C. K. Chui, L. L. Schumaker, J. D. Ward eds), 1989:647 ~ 650

[49] Sun Yongsheng. Optimal Recovery of differential operators on a class of smooth functions. ATA, 1990, 6(1):23 ~ 31

[50] Sun Yongsheng. Optimal interpolation on a convolution class of functions on R. 东北数学, 1990, 6(1):9 ~ 25

[51] 孙永生. 极值问题和函数逼近论(译自吉哈米诺夫:1989 年 9 月在浙江大学的报告). 数学进展, 1990, 19(4):449 ~ 451

[52] Sun Yongsheng, Li Chun. Best approximation of differentiable functions by cardinal-splines of higher order [Russian]. Matem Zametki(Math. Notices), 1990, 48(4):100 ~ 109

[53] 孙永生,《Spaces generated by Blocks》的评价, 见:北京师范大学出版社编. 北师大版图书评论(1980 ~ 1990), 北京师范大学出版社, 1990:269 ~ 272

[54] Sun Yongsheng. Optimal interpolation on some classes of smooth functions defined on the whole real axis. 数学进展, 1991, 20(1): 1 ~ 14

[55] 孙永生. 关于光滑函数类上的最优取样. 数学进展, 1991, 20(2):184 ~ 191

[56] Sun Yongsheng. Li Chun. Optimal recovery for $W_2^r(\mathbf{R})$ in $L_2(\mathbf{R})$. Acta Mathematica Sinica(New series), 1991, 7(4):309 ~

323
[57] Sun Yongsheng, Liu Yongping. Best one-sided approximation of some classes of smooth functions on the whole real axis by cardinal splines of higher order. ATA. 1991, 7(1): 108 ~ 123
[58] Chen Dirong, Sun Yongsheng. Hardy-Littlewood-Polya inequality for a linear differential operator and some related optimal problems. ATA, 1992, 8(1): 50 ~ 58
[59] 孙永生, 刘永平. 光滑函数的 Sobolev-Wiener 类利用重取样的最优恢复. 北京师范大学学报(自然科学版), 1992, 28(1): 15 ~ 18
[60] 孙永生. 线性微分算子的最优恢复. 北京师范大学学报(自然科学版), 1992, 28(4): 433 ~ 437
[61] 孙永生. 关于 Hilbert 空间内点集的平均宽度. 科学通报, 1992, (6): 488 ~ 491(Sun Yongsheng. Average n-width of a point set in Hilbert space. Chinese Sci. Bull., 1992, 37(14): 1153 ~ 1157)
[62] 刘永平, 孙永生. Sobolev-Wiener 空间上的无穷维宽度和最优插值问题, 中国科学(A 辑), 1992, 22(6): 582 ~ 591(Liu Yongping, Sun Yongsheng. Infinite-dimensional Kolmogorov width and optimal interpolation on Sobolev-Wiener space. Sci. China Ser. (A), 1992, 35(10): 1162 ~ 1172)
[63] Sun Yongsheng, Liu Yongping. N-widths for some classes of periodic functions with boundary conditions. ATA, 1992, 8(3): 21 ~ 27
[64] Sun Yongsheng, Liu Yongping. Optimal recovery of the Sobolev-Wiener class of smooth functions by double sampling. Constr. Approx. 1993, 9(4): 391 ~ 405
[65] Sun Yongsheng, Wang Chengyong. μ-Average n-widths on the Wiener space. J. Complexity, 1994, 10: 428 ~ 436
[66] 孙永生. 一个多元周期函数的 Besov 类的宽度. 科学通报, 1994, 39(23): 2113 ~ 2115
[67] Sun Yongsheng, Wang Chengyong. Average error bounds of best approximation of continuous functions on the Wiener space. J. Complexity, 1995, 11(1): 74 ~ 104
[68] 孙永生, 汪和平. 具有给定的混合型光滑模的多元周期函数的

表现和逼近.科学通报,1995,40(6):492 ~ 495

[69] 汪和平,孙永生.具有一定混合光滑性的多元函数的整函数逼近.北京师范大学学报(自然科学版),1995,31(1):29 ~ 32

[70] 汪和平,孙永生.具有一定混合光滑模的 Besov 型函数空间的函数表现定理.北京师范大学学报(自然科学版),1995,31(2):159 ~ 162

[71] 孙永生.具有一定混合光滑模的 Besov 型函数空间内的迹定理,北京师范大学学报(自然科学版),1995,31(3):290 ~ 295

[72] Wang Heping, Sun Yongsheng. Approximation of multivariate functions with a certain mixed smoothness by entire functions. Northeast. Math. J., 1995, 11(4):454 ~ 466

[73] Sun Yongsheng, Wang Chengyong. Average error bounds of best approximation in a Banach space with Gaussian measure. East J. on Approx, 1995, 1(4):431 ~ 439

[74] 孙永生, 对于我校进入 211 工程的一些想法, 师大周报, 1995-11-24

[75] 汪和平,孙永生.具有给定混合光滑模的多元周期函数空间的等价范数及求积公式.北京师范大学学报(自然科学版),1996,32(4):438 ~ 442

[76] Sun Yongsheng, Wang Heping. Representation and approximation of multivariate periodic functions with bounded mixed moduli of smoothness. Tr. Mat. Inst. Steklova, Teor. Priblizh. Garmon. Anal., 1997, 219:356 ~ 377; Translation in Proc. of Steklov Inst. Math. 1997, 219(4):350 ~ 371

[77] 罗俊波,孙永生.各项光滑型不同的多元函数类的最优恢复和宽度.数学进展,1998,27(1):69 ~ 77

[78] 蒋艳杰, 孙永生. Average widths and optimal recovery of multivariate Besov class $B^r_{p,\theta}(\mathbf{R}^d)$ in $L^r_{qp}(\mathbf{R}^d)$. 东北数学,1998,14(2):127 ~ 130

[79] Sun Yongsheng, Jiang Yanjie. Extremal problem for a convolution class of functions defined on the real line. East J. on Approx. 1998, 4(4):515 ~ 525

[80] Sun Yongsheng, Liu Yongping, Chen Dirong. Extremal problems

in approximation theory for some classes of smooth functions defined on $\mathbf{R}^d$. 北京师范大学学报(自然科学版),1999,35(增刊):79 ~ 144

[81] 孙永生.纪念傅种孙先生.数学通报,1999,(4):1,15

[82] Sun Yongsheng. Some remarks on the optimal subspaces of a convolution class with a NCVD kernel. ATA, 1999, 15(4):1 ~ 5

[83] Wang Heping, Sun Yongsheng. Approximation of functions in $S_1^r L$, and $S_1^r H$ by entire functions. ATA, 1999, 15(4):88 ~ 93

[84] Sun Yongsheng. Optimal subspaces for a periodic convolution class with a B-kernel. East J. on Approx., 2000, 6(1):53 ~ 63

[85] Sun Yongsheng, Wang Heping. A generalized version of Hardy-Littlewood-Polya inequality and some related optimal problems in approximation theory. East J. on Approx., 2001, 7(3):375 ~ 380

[86] B. Gavin, Dai Feng, Sun Yongsheng. Kolmogorov width of classes of smooth functions on the sphere S^{d-1}. 数学进展,2002,31(2):181 ~ 184(Brown G., Dai Feng, Sun Yongsheng, Kolmogorov width of classes of smooth functions on the sphere S^{d-1}, J. of Complexity, 2002, 18(4):1001 ~ 1023)

[87] 孙永生.我与师大图书馆.见:北京师范大学图书馆编:百年情结.北京师范大学出版社,2002:23 ~ 25

[88] Wang Heping, Sun Yongsheng. Representation and m-term approximation for anisotropic classes. In《Theory of Approximation of Functions and Applications》edited by S. M. Nikolskii, Moscow: Press of Inst. Math. of Russian Academy, 2003. 250 ~ 268

[89] Wang Heping, Sun Yong sheng. Kolmogorov widths between the anisotropic space and the space of functions with mixed smoothness. J. Approx. Theory, 2004, 126, 52 ~ 59

著作

1 Grebencha.算术.张禾瑞,孙永生译自俄文.商务印书馆,1953

2 Novoselov. 代数与初等函数. 张禾瑞, 赵慈庚, 孙永生, 严士健译自俄文. 高等教育出版社, 1954

3 Novoselov. 初等代数专门教程(上、下册). 赵慈庚, 董延闿, 蒋铎, 严士健, 郝鈵新, 孙永生, 袁兆鼎译. 高等教育出版社, 1956 ~ 1957

4 A. Robinson. 非标准分析. 申又枨, 王世强, 张锦文, 孙永生, 廖祖伟等译自英文. 科学出版社, 1980(第 3, 5 章为孙永生译)

5 Korneichuk. 逼近论的极值问题. 孙永生译自俄文. 上海科技出版社, 1982

6 孙永生. 泛函分析讲义. 北京师范大学出版社, 1986

7 孙永生, 中国大百科全书数学卷. 条目: 伯恩施坦, 函数逼近论, 宽度, 熵, 中国大百科全书出版社, 1988

8 孙永生. 函数逼近论(上册). 北京师范大学出版社, 1989

9 孙永生, 房艮孙. 函数逼近论(下册). 北京师范大学出版社, 1990

后　记

Postscript by the Chief Editor

在搜集编写《北京师范大学数学系史》的过程中，我深深地感到，我们数学系的5位老先生，即王世强、孙永生、严士健、王梓坤、刘绍学教授：

(1) 他们是我系的学术带头人，为我系的学科建设和人才培养花费了毕生的精力，做出了重大的贡献。5位先生均是1981年被批准为首批博士生导师，而王梓坤教授是1984年来到我校任校长。1981年我系的4位博士生导师，占全国首批数学博导数量的1/18，占我校首批博导数量的1/5(理科还有黄祖洽院士、刘若庄院士、陈光旭教授、汪堃仁院士和周廷儒院士)，一般说来，数学系在学校的多种指标中一般占1/10或更低，此次批准我系的博导数量，提高了我系在学校中的地位，且此举对我校数学系在全国数学界的地位奠定了重要基础，开创了近20多年来的良好局面。

(2) 这5位中国现代数学家，均入选《中国现代数学家传》(江苏教育出版社，第二、三、五卷，1995，1998，2002)，他们为我国现代数学的发展做出了重要的贡献。

(3) 在王梓坤教授来我校之前，另4位先生在数学系被戏称为“四大金刚”，这4位教授均毕业于北京师范大学数学系，且终身在数学系任教，这也是数学系的特殊现象。

(4) 5位先生中后4位均生于1929年，这是中国现代数学家在一所高校数学系独特的现象。

由此产生了编辑5位老先生文集的想法，并和系里的一些教师谈论过此事。《北京师范大学数学系史》出版后，与数学系主任郑学安教授和数学与数学教育研究所所长刘永平教授商量，由我主编这套文集，经与北京师范大学出版社前总编马新国教授协商同意后，将5位老先生文集列入出版社的《教授文库》的选题计划。由于出版社领导班子换届，出版计划到2003年底才落实下来。此套书列入"北京师范大学数学家文集"的选题计划。确定文选的内容包括：照片，序，论文选，全部论文和译著目录，后记。《逼近与恢复的优化——孙永生文集》是这套文集的第二部。作为孙先生的学生，我们能够协助我们敬爱的老师在他75岁生日的时候编辑出版这本文集，感到由衷的高兴。

本书的出版得到了北京师范大学出版社的大力支持和王松浦老师的热情帮助，在此表示衷心的感谢。

主编　李仲来

2004－02－08